Artemia Biology

Editors

Robert A. Browne, M.S., Ph.D.
Associate Professor, Biology
Wake Forest University
Winston-Salem, North Carolina

Patrick Sorgeloos, M.S., Ph.D.
Professor and Coordinator, Aquaculture
Artemia Reference Center
State University of Ghent
Ghent, Belgium

Clive N. A. Trotman, Ph.D., F.I.Biol.
Senior Lecturer Biochemistry
University of Otago
Dunedin, New Zealand

CRC Press
Taylor & Francis Group
Boca Raton London New York

CRC Press is an imprint of the
Taylor & Francis Group, an **informa** business

First published 1991 by CRC Press
Taylor & Francis Group
6000 Broken Sound Parkway NW, Suite 300
Boca Raton, FL 33487-2742

Reissued 2018 by CRC Press

Library of Congress Cataloging-in-Publication Data

Artemia biology / editors, Robert A Browne, Patrick Sorgeloos, Clive
N.A. Trotman.
p. cm.
Includes bibliographical references.
Includes index.
ISBN 0-8493-6729-8
III. Trotman, Clive N.A.
[DNLM: 1. Artemia--anatomy & histology. 2. Artemia--physiology.
QX 463 A786]
QL444.B815A77 1991
595.3'2--dc20
DLC
for Library of Congress 90-2023

A Library of Congress record exists under LC control number: 90002023

Publisher's Note
The publisher has gone to great lengths to ensure the quality of this reprint but points out that some imperfections in the original copies may be apparent.

Disclaimer
The publisher has made every effort to trace copyright holders and welcomes correspondence from those they have been unable to contact.

ISBN 13: 978-1-315-89079-1 (hbk)
ISBN 13: 978-1-351-06989-2 (ebk)

Visit the Taylor & Francis Web site at http://www.taylorandfrancis.com and the
CRC Press Web site at http://www.crcpress.com

INTRODUCTION

Brine shrimp are fascinating organisms. They can withstand the widest salinity range of any known metazoan. Their cysts are capable of prolonged diapause. Exquisite timing and coordination of biochemical and developmental processes are required during the post-dormancy period. The genus *Artemia* consists of a number of species, some of which reproduce sexually, others asexually, with various ploidy levels occurring among the asexual populations. Some diploid asexual populations also appear capable of at least limited recombination.

Artemia offer investigative challenges across a broad spectrum of biological areas, from molecular biology to aquacultural production. The ease with which brine shrimp can be reared and experimentally manipulated in the laboratory has made this organism a favorite for basic biological research. It also has long been a preferred food source in aquaculture.

Perhaps due to the utilization of brine shrimp in many fields of study, the *Artemia* literature, which exceeds over 4000 references, is widely scattered. It is our hope that this volume will not only be a primary reference for the large number of researchers who utilize *Artemia*, but since rapid progress is occurring in many of the fields, that it will also serve as a state-of-the-art review. In addition, almost all chapters contain considerable data that have not been previously published.

Robert A. Brown
Patrick Sorgeloos
Clive N. A. Trotman

THE EDITORS

Robert A. Browne, Ph.D., is an Associate Professor in the Department of Biology at Wake Forest University in Winston-Salem, North Carolina.

Dr. Browne obtained a B.S. and M.S. in Biology at the University of Dayton in 1972 and 1974 respectively and received the Ph.D. in 1977 from Syracuse University. He began working on *Artemia* during his three years of National Institute of Environmental Health Sciences postdoctoral fellowship in the Department of Genetics at North Carolina State University at Raleigh. Dr. Browne has been Assistant Professor (1980—1986) and Associate Professor (1986—present) at Wake Forest University and has conducted research during numerous summers at the Marine Biological Laboratory, Woods Hole, Massachusetts.

Dr. Browne received a Senior Fulbright Research Scholar Award in 1987 to conduct research on *Artemia* while in Belgium and France. He received the Award for Excellence in Research from Wake Forest University in 1988. Dr. Browne has conducted field research in *Artemia* in the American Southwest, North Africa, and South Europe. The National Geographic Society has been among the societies and foundations sponsoring his research.

Dr. Browne is the author of over 30 research papers and a book, and has presented his research at many national and international scientific meetings. He is a member of the Society of Evolution, the Ecological Society of America, the American Society of Naturalists, the Genetics Society of America, and the American Association for the Advancement of Science. His research interests are in population ecology and evolution, with research currently being conducted on the ecological genetics of *Artemia*, of endangered Hawaiian waterbirds, and mountaintop populations of small mammals.

Patrick Sorgeloos, Ph.D., is Professor of Aquaculture at the Universities of Antwerp (UIA), Brussels (VUB), and Ghent (RUG) in Belgium; he is the coordinator of the Laboratory of Aquaculture & Artemia Reference Center at the Ghent State University.

Dr. Sorgeloos obtained his training at the University of Ghent, Belgium, receiving the M.Sc. degree in 1971, the Ph.D. degree in 1975, and the postdoctoral degree of "Aggregate of Higher Education" in 1979. He attended training courses in Marine Biology at the Marine Institutes of Roscoff in France (1969), Helgoland in West Germany (1970), and Beaufort in North Carolina (1972). Since 1971 Dr. Sorgeloos has been with the Belgian National Science Foundation.

Dr. Sorgeloos' research on *Artemia* and larval nutrition in aquaculture is sponsored by the Belgian National Science Foundation, the Belgian Ministry of Science Policy, the Belgian Administration for Development Cooperation, the Flemish Association for Educational Programmes Abroad, the EEC, SINTEF-Norway, and several private companies.

Dr. Sorgeloos has organized or co-organized two international *Artemia* Symposia and several international training courses or workshops on *Artemia* biology and its use in aquaculture in Belgium, Brazil, Canada, Mexico, India, China, the Philippines, Spain, Yugoslavia, Ecuador, Panama, the Dominican Republic, and Australia. He has performed numerous consultancy missions on salt production, *Artemia*, and larviculture applications for national and international aid organizations, as well as for private companies. Dr. Sorgeloos is Vice President of the World Aquaculture Society and a Scientific Advisor for the International Foundation for Science.

Dr. Sorgeloos is the (co-)author of more than 100 papers and has been the co-author of seven books. His current major research interests relate to the culturing biology of *Artemia* and the development of larviculture diets for fish and shellfish.

Clive N. A. Trotman, Ph.D., is Senior Lecturer in the Department of Biochemistry at the University of Otago, Dunedin, in the South Island of New Zealand.

Dr. Trotman graduated with honors at Brunel University, London, in 1967, specializing in biochemistry and microbiology. His Ph.D. was awarded in 1970 for work in enzymology at the University of East Anglia, Norwich. In the same year he was appointed to a lectureship in biochemistry at the Medical School of the University of Newcastle upon Tyne. He spent a year as Visiting Lecturer in Biochemistry at the University of Otago from 1978 to 1979 before taking up his present appointment in 1980. In 1987 he was Guest Professor at the University of Antwerp (UIA), Belgium and was awarded a Corday Morgan Fellowship by the Royal Society of Chemistry, London.

Dr. Trotman spoke at the international *Artemia* conferences at Antwerp (1985) and Montreal (1988) and has published extensively with the Otago group. He is active in the New Zealand Institute of Chemistry and the Australia and New Zealand Society for Cell Biology. In 1989, he was elected to Fellowship of the Institute of Biology, London.

CONTRIBUTORS

David A. Bengtson, Ph.D.
Department of Zoology
University of Rhode Island
Kingston, Rhode Island

Sarane T. Bowen, Ph.D.
Department of Biology
San Francisco State University
San Francisco, California

Robert Browne, Ph.D.
Associate Professor
Department of Biology
Wake Forest University
Winston-Salem, North Carolina

James S. Clegg, Ph.D.
Professor and Director
Department of Zoology
Bodega Marine Laboratory
University of California, Davis
Bodega Bay, California

Godelieve R. J. Criel, Ph.D.
Senior Research Associate
Laboratory of Anatomy
State University of Ghent
Ghent, Belgium

Jesús Cruces, Ph.D.
Assistant Professor
Departamento de Bioquimica
Universidad Autónoma de Madrid
Instituto de Investigaciones Biomédicas del CSIC
Madrid, Spain

Ivo De Baere, Ph.D.
Department of Biochemistry
University of Antwerp
Antwerp, Belgium

Laurie E. Drinkwater, Ph.D.
Postdoctoral Scholar
Department of Vegetable Crops
University of California
Davis, California

Rafael Garesse, Ph.D.
Assistant Professor
Departamento de Bioquimica
Universidad Autónoma de Madrid
Instituto de Investigaciones Biomédicas del CSIC
Madrid, Spain

M. C. Geddes, Ph.D.
Senior Lecturer
Department of Zoology
University of Adelaide
Adelaide, Australia

Patrick Lavens, Ph.D.
Research Coordinator
Laboratory of Aquaculture
State University of Ghent
Ghent, Belgium

Philippe Léger, Ph.D.
Manager
Artemia Reference Center
State University of Ghent
Ghent, Belgium

Petra H. Lenz, Ph.D.
Assistant Research Professor
Pacific Biomedical Research Center
Honolulu, Hawaii

Roberto Marco, Ph.D.
Professor and Chairman
Departamento de Bioquimica
Universidad Autónoma de Madrid
Instituto de Investigaciones Biomédicas del CSIC
Madrid, Spain

Craig J. Marshall, Ph.D.
Department of Haematological Medicine
University of Cambridge
Cambridge, United Kingdom

Luc Moens, Ph.D.
Professor
Department of Biochemistry
University of Antwerp
Antwerp, Belgium

Jaime Renart, Ph.D.
Instituto de Investigaciones Biomédicas del CSIC
Departamento de Bioquimica
Universidad Autónoma de Madrid
Madrid, Spain

Herman Slegers, Ph.D.
Docent
Department of Biochemistry
University of Antwerp
Antwerp, Belgium

Patrick Sorgeloos, Ph.D.
Professor of Aquaculture
Laboratory of Aquaculture
State University of Ghent
Ghent, Belgium

Wim Tackaert, M.Sc.
Laboratory of Aquaculture and Artemia Reference Center
State University of Ghent
Ghent, Belgium

Warren P. Tate, Ph.D.
Professor
Department of Biochemistry
University of Otago
Dunedin, New Zealand

Clive N. A. Trotman, Ph.D., F.I.Biol.
Lecturer
Department of Biochemistry
University of Otago
Dunedin, New Zealand

Jozef Van Beeumen, Ph.D.
Professor
Laboratory of Microbiology
University of Ghent
Ghent, Belgium

M. L. Van Hauwaert
Chief Technician
Department of Biochemistry
University of Antwerp
Antwerp, Belgium

W. D. William, Ph.D., D.Sc.
Professor
Department of Zoology
University of Adelaide
Adelaide, Australia

Shoshana Wodak, Ph.D.
Professor
Biological Macromolecular Conformation Unit
Free University of Brussels and PGS
Ghent, Belgium

Gretel Wolf, Ph.D.
Chief Assistant
Department of Biochemistry and Zoology
University of Antwerp
Antwerp, Belgium

TABLE OF CONTENTS

Chapter 1

ARTEMIA MOLECULAR GENETICS

Roberto Marco, Rafael Garesse, Jesús Cruces, and Jaime Renart

TABLE OF CONTENTS

I. INTRODUCTION

Recombinant DNA techniques have made possible the current explosion in the field of molecular genetics.[1] The studies of the sixties and early seventies, based primarily on reassociation kinetic analyses, have been superseded by the study of individual genes that can now be isolated, amplified, and manipulated in such ways that a multitude of specific questions can be addressed. The recombinant DNA revolution has also overcome problems such as scarcity of biological materials and the difficulties of *in vivo* labeling of nucleic acids, etc.

Although the study of *Artemia* molecular genetics is in early development, several reasons justify its interest. The molecular mechanisms underlying the specific properties and adaptations of *Artemia* can now be undertaken, such as its capabilities of surviving in extreme environments (e.g., high salt or abnormal pH) or the genetic and molecular aspects of its development. The dissociation between DNA replication and cell division from the onset of cell differentiation and morphogenesis provides an unusual situation in developing organisms. It can be understood as an extreme case of the "midblastula transition"[2] that normally takes place while cleavage is going on at full pace. The embryo can be easily induced to resume development and the induction of enzymatic activities or specific mRNAs can be followed with a good reference stage, namely, the encysted early gastrula in which replication, transcription, and translation are completely turned off. (See References 3—5 for previous data on the molecular biology of *Artemia*.)

Artemia molecular genetics can fill a gap in evolutionary and comparative biology. Crustacea have not been as extensively studied as other arthropods such as the insect *Drosophila melanogaster*. In this respect it is interesting to compare the sequence data available from different groups of living organisms (Table 1). The distribution is bimodal with mammals as the best studied group (5686 sequences, 6.4×10^6 nucleotides, 31% of the total) and bacteria (especially *E. coli*) as the other peak. Among invertebrates, crustaceans represent only 0.02% of the sequences (with 0.01% for *Artemia*). Insects, and more particularly, *Drosophila* have been the more extensively investigated invertebrates. The present situation of *Artemia* is similar to that of *Arabidopsis*, a plant system that has been presented as an ideal system representative of higher plants.[6]

Many noninsect invertebrate organisms have been occasionally studied, with nematodes and sea urchins as the best examples. For other classes of arthropods, crustaceans are not well represented, but even less are *Chelicerata* (0.001% of sequences in the Genebank). The evolutionary distance among these groups is quite large and comparative studies in molecular genetics will help to establish their relationships. In fact, sequence data from a primitive crustacean, such as *Artemia*, can help to complete the phylogenetic landscape from lower eukaryotes, (e.g., protozoa and fungi), through lower invertebrates (e.g., *C. elegans*), to the higher invertebrates (e.g., arthropods).

The importance of integrating information obtained from different biological systems using different experimental approaches is increasingly recognized.[7] Studying the molecular biology and genetics, the biochemistry and cell biology, and the physiology and development of a particular type of organism in an integrated form and in comparison with other organisms is fundamental in understanding biological diversity. From a molecular point of view, *Artemia* is the best known crustacean and one of the best known aquatic organisms; thus, it can become the paradigmatic crustacean, a kind of aquatic *Drosophila*.

The widespread use of *Artemia* for aquaculture[8] further increases the need for basic research on the genus, including genetic manipulation and engineering.

In this review we present the current state of the molecular knowledge of the *Artemia* genome. We summarize what is known about *Artemia* nuclear and mitochondrial DNA sequences and the key processes of replication and transcription. This information is discussed

TABLE 1
Distribution of Sequences in Genbank[a]

Groups	Species number (109,110)	Sequences number (%)	Nucleotides number (%) (× 10^6)	Species	Sequences number (%)	Nucleotides number (%)
Mammals	4,000					
Primates		2389 (13)	3.1 (15)	Man	2338 (13)	3.0 (14)
Rodents		2770 (15)	2.7 (13)	Mouse	1794 (10)	1.5 (7)
Other		527 (3)	0.6 (3)			
Other						
Vertebrates	49,000	753 (4)	0.8 (4)	Chicken	483 (3)	0.5 (3)
Invertebrates	1,100,000	1195 (7)	1.2 (6)	Sea urchin	126 (1)	0.08 (0.4)
				Artemia	19 (0.1)	0.01 (0.05)
				Drosophila melanogaster	501 (3)	0.6 (3)
Plants	370,000	1200 (7)	1.6 (8)	*Arabidopsis*	13 (0.07)	0.1 (0.06)
				Maize	66 (0.3)	
				S. cerevisiae	664 (4)	0.8 (4)
Bacteria	3,000[b]	1573 (9)	2.3 (11)	*E. coli*	674 (4)	0.9 (4)

[a] Data taken from release 56 (June 1988);[111] 18,230 sequences and 20,795,279 nucleotides.
[b] Number of species described in the Bergey's Manual.[112]

in the light of what it is known in other animal organisms, and will serve in a final section to discuss the perspectives about the future of research in this genus.

II. ADVANTAGES AND PROBLEMS OF THE *ARTEMIA* SYSTEM

Artemia presents advantages and disadvantages as an experimental system. The advantages are commercial availability, low cost of the cysts, possibility to synchronize larvae (nauplii) in the laboratory, and a relatively extensive knowledge of the organism's biochemistry.

The major drawback for the use of *Artemia* in biochemical and molecular studies is the difficulty in culturing large quantities of particular life cycle stages, including germ cells, early and nondiapausing embryos, nauplii, and adults. Commercial suppliers have also not paid enough attention until recently to the genetic identity of their cysts, which sometimes are not even homogeneous.* Another disadvantage is the relative underdevelopment of classical genetic studies. The high chromosome number and variability of *Artemia*,[9] and the logistical problems involved in keeping separate lines alive complicate the genetic manipulation and use of classical genetic techniques, although cysts can provide a simple and inexpensive way of maintaining stocks. In conclusion, although classical genetic studies are possible, the current level is rather primitive compared to other invertebrate systems, such as *Drosophila melanogaster* or *Caenorhabditis elegans*.

Molecular work with *Artemia* encounters several methodological problems. The impermeability of cysts, the particulate-feeding of nauplii and adults, and the difficulty of rearing large amounts in axenic cultures leads to inefficient *in vivo* labeling of nucleic acids to specific activities suitable for many experiments. Proteins have been labeled *in vivo* (see, for instance, References 10—12), but specific activities are low. Furthermore, it is an organism with relatively high amounts of DNA per cell (approximately half that of humans), making it more expensive and tedious than most for screening procedures using molecular techniques. Another problem is the difficulty in obtaining purified nuclei from dormant embryos for developing embryos and nauplii. A recent report[13] of a method to obtain nuclei from nauplii based on Percoll gradient centrifugation requires further confirmation and verification with regard to the incorporation of labeled UTP (uridine triphosphate) in acid precipitable material from purified nuclei.

Finally, there is relatively little background on *Artemia* cell biology. Although interesting cellular phenomena like those involved in parthenogenesis can be studied in *Artemia*, basic information on tissue organization and physiology is still mostly missing. For example, no established cell lines of *Artemia* or any other crustacean are yet available.

III. THE GENOME OF *ARTEMIA*

The general properties of the *Artemia franciscana* genome are summarized in Table 2. *A. franciscana* (the most widely studied species) has a diploid chromosome number of 42, and a relatively large content of DNA, around 1.6×10^9 bp/haploid genome,[14,15] resulting in approximately 70 Mbp per average chromosome. Microphotographs of metaphase chromosomes[16] show that they are approximately the same size.

Interestingly, *A. franciscana* DNA has a low G+C content, namely 32.4%[15,17,18] affecting mainly the intergenic regions, as well as codon preferences (see below). As shown by reassociation kinetic analysis,[19,20] *A. franciscana* genome contains 6% of fold-back DNA, 15% of highly repeated DNA, 20% of moderately repeated DNA, and 59% of single copy DNA (Table 2).

* In the authors' laboratories, batches purchased in 1982 as being cysts from San Francisco Bay, actually were a mix of cysts from San Francisco Bay (bisexuals) and cysts from a parthenogenetic population, most probably from China.

TABLE 2
Properties of the *Artemia franciscana* Genome

		Ref.
Size	1.6×10^9 bp/haploid genome	14,15
Chromosome number	2n = 42[a]	9
Base composition	32% G+C	14,15
Reassociation kinetics data		19,20
Fold-back DNA	6%	
Highly repeated DNA	15%	
Moderately repeated DNA	20%	
Single-copy DNA	59%	

[a] In bisexual populations; different degrees of ploidy in parthenogenetic strains.

Although the analysis of defined sequences by recombinant DNA techniques is making this classification obsolete, we discuss in the following sections examples of each class of sequence, except for the fold-back DNA, which in the case of *Artemia* has not yet been studied at the molecular level. Earlier kinetic analyses indicated a relative deficiency in moderately repeated DNA in crustaceans, i.e., in sequences with repetition frequencies of a few to two hundred.[21] Subsequent data (discussed below) have shown that a normal number of intermediately repeated genes, (e.g., ribosomal, 5S RNAs, and histone genes) are present in *Artemia*. Most likely the initial underrepresentation of this class of genes was a consequence of the crudeness of the kinetic analysis.

A. HIGHLY REPEATED DNA

Highly repeated DNA is the main component of heterochromatin. In *Artemia* these sequences have been studied by Barigozzi's group[16,22] and ourselves.[14,15] One class of repeated sequences, named satellite I, has been characterized in detail. Its sequence of 113 bp is organized in tandem repeats and contain sites for Alu I, Mbo II and Taq I. Digestion of genomic DNA with other restriction endonucleases liberates the entire satellite I as large fragments. The exact size of these fragments is unknown, as no pulse-field gel electrophoresis experiment has yet been carried out. In conventional agarose gel electrophoresis, they appear close to the 23 kb Hind III fragment of λ DNA. Densitometric analysis of agarose gels show that satellite I comprises 2% of genomic DNA, representing 300,000 copies of the 113 bp-unit/haploid genome. The buoyant density of satellite I is 1.689 g/cc (deduced from its base composition). This value is very close to 1.692 g/cc, the buoyant density of the whole genome DNA (calculated by density gradient centrifugation[17]); this explains why satellite I DNA was only detected by centrifugation in gradients prepared in the presence of the dye bisbenzimidazole H33258. Barigozzi et al.[16] first demonstrated that probes derived from this satellite sequence hybridize with heterochromatic regions of the chromosomes of certain strains, but that these regions were absent in chromosomes of other strains. These results have been extended both by *in situ* hybridization and Southern blot analysis. Satellite I is present only in American bisexual populations of *Artemia*, although in varying proportions. Other strains do not have satellite I sequences (detected both by gel electrophoresis or hybridization). However, they do show different satellite bands. For example, parthenogenetic populations from Eurasia show a similar 130 bp repeat which in these populations may play the same role as satellite I in American bisexuals. These alternative satellite sequences have not yet been studied in detail, nor have they been used for *in situ* hybridizations.

B. MODERATELY REPEATED DNA

This type of sequence is primarily comprised by gene families that are repeated several times; among the best characterized are the ribosomal RNA genes and histone genes.

1. Ribosomal RNA Genes

It is well known that eukaryotic ribosomes contain four types of RNAs: the 18S in the small subunit, and the 5S, 5.8S, and 28S in the large subunit. The 5S rRNA is transcribed, by RNA polymerase III, and the rest by RNA polymerase I. The 5S and 5.8S rRNAs are approximately 120 bp and 160 bp, respectively, and this size is maintained in all eukaryotes. By contrast, 18S and 28S rRNAs vary in size, being larger in higher evolved organisms, with a value of 17S and 25S in *Artemia* (see Reference 23 for a more comprehensive review). In *Artemia*, the small rRNAs, 5S and 5.8S, have been studied by the groups of De Watcher,[24,25] by Vaughn et al.,[26] by Bagshaw et al.,[27] and by our group,[28,29] with the conclusion that 5S rRNA is 120 bp, and 5.8S is 162 bp.

The large rRNAs, 17S and 25S, have been studied in *Artemia* by Cruces et al.,[30] by Roberts and Vaughn,[31] and by Nelles et al.[32] who sequenced the 17S rRNA (1810 nucleotides in length). The 28S rRNA has been estimated by gel electrophoresis to be ≈ 3750 nucleotides in length.[30,31] *Artemia* mature 25S rRNA contains a break detectable by denaturing conditions.[30,31] This is typical of protostomes as described by Ishikawa[33] for several organisms, including *Artemia*. The 5′ coding sequence upstream of the gap has been obtained by Nelles et al.[34]

The structure of ribosomal RNA genes is a tandem array of repeating units,[35,36] each unit containing the following coding regions: 5′-17S-5.8S-25S-3′. The rDNA repeat varies according to the origin of the population studied, as previously mentioned for Satellite I DNA. We have described[35,37] a repeating unit of 16.5 kb for North American populations, both from San Francisco Bay and from Great Salt Lake (Utah). Vaughn et al.,[36] however, have reported a 13.9 kb repeat unit for a San Francisco Bay population. The repeat unit studied by these authors contains a Hind III site close to the end of the 25S coding region that could be the cause of this discrepancy. We have also described a 12.2 kb repeat unit for parthenogenetic populations from Eurasia[37,38] and in some cases, a 13.5 kb repeat.[37] This 13.5 kb is present in parthenogenetic populations in varying proportions, and is due to an extra Hind III fragment of 1.2 kb. Piot et al.[39] have also reported a 11.5 kb repeat unit, which is probably the same as the 12.2 kb repeat from parthenogenetic populations. As expected, restriction site patterns are more indicative than size; the 13.9 kb repeat described by Vaughn and co-workers[36] is quite similar to the 16.5 kb repeat studied by us (also from North American populations), much more than the 13.5 kb repeat from parthenogenetic strains. Roberts and Vaughn,[31] using a San Francisco Bay population, calculated through saturation hybridization experiments ≈320 copies per haploid genome. We have found a value of 1000 copies for animals from the same origin.[37] This difference may be due to the previously mentioned heterogeneity of commercial cysts.

The structure of the 5′ regulatory region of rRNA genes has been studied in detail.[38,40,41] The 16.5 kb unit contains at least five repeated segments of ≈600 bp each. The organization of the 12.2 kb repeat from parthenogenetic populations has also been studied by Gil and Cruces.[42] In this case, only three repetitions have been found; the first being almost identical to the first one in the 16.5 kb type of genes; the other two repetitions although homologous to the first, present important deletions affecting the 5′ region of the second and the 3′ region of the third repetition.

The initiation site of transcription in the 16.5 kb repeat has been studied by S1 protection analysis.[40-42] Both groups detected an array of initiation and/or 5′ processing sites in each of the 600 bp repetitions (the proposed initiation/processing points have some homology both to the rDNA initiation point and to the 18S rRNA processing site of *Drosophila*). The problem of how initiation at more distal repetitions is controlled has not been solved; no upstream terminators are present as in the case in the *Xenopus*[43] and mouse.[44] Further upstream initiated pre-rRNA molecules have been detected. This situation arises because transcription is initiated inside the repetitions. A similar initiation region has been found in the tse-tse fly[45] where initiation starts inside the first repetition.

2. The 5S RNA-Histone Gene Complex

Artemia presents a genomic organization not yet found in any other eukaryote: the 5S rRNA genes and the histone genes are linked together in the same repeat unit. The discovery of this linkage was made by Bagshaw and colleagues[46] and has been further studied by their group and members of the Madrid group.[27-29,47-49] These genes are repeated 100-fold,[46] organized in tandem arrays of long units.

The detailed structure of the 5S-histone repeat unit varies with the origin of the population studied, as is also the case for satellite DNA and rRNA genes. In all populations studied, two types of repeats have been found, both containing a single 5S rRNA gene and the entire complement of histone genes. In North American bisexual strains, one type of repeat unit is 9.0 kb and the second type 8.5 kb in length; the histone cluster is 3.2 kb, and the 5S rRNA gene 330 bp long (considering not only coding sequences but also 5′ and 3′ regulatory regions), plus about 4 kb of intergenic spacers of unknown function.[28]

The order of the individual histone genes in the cluster has not yet been established, since in hybridization studies[47] only long restriction fragments have been used as probes, resulting in assignment ambiguities. The most proximal histone gene to the 5S rRNA is H2A,[28] as shown by sequencing the 3′ end of the histone gene and the histone-5S intergenic spacer. The intergenic spacers are composed of several repeated elements, and their organization is different in North American bisexual vs. Eurasia parthenogenetic strains.

The 5S-histone link poses several questions about its origin. As previously mentioned, *Artemia* is the only organism in which this organization has been found to date (although other primitive crustaceans have not yet been examined). Insects do not show this organization,[50] and preliminary results indicate that decapodes (the more highly evolved crustaceans) do not have this organization. Interestingly, the 5S-histone repeat does not coexist with pure 5S rRNA and/or histone gene repeats, either in North American or Eurasiatic strains.[46,49] However, an isolated 5S rRNA gene has been found,[28,48] likely to be an orphon (as defined by Childs et al.[51]). The 5S rRNA genes have been found to be linked to other genes as well. In some fungi,[52] *Dictyostelium*,[53] and in the copepod *Calanus*,[54] 5S rRNA sequences are linked to the larger rRNA genes, as also occurs in prokaryotes.[55,56] It therefore seems that 5S rRNA genes have a special facility to transpose to other regions of the genome. The most plausible mechanism to explain the organization found in *Artemia* is that 5S rRNA invaded the histone gene repeat by a transposition event and, thereafter, molecular drive[57] would result in the elimination of the individual 5S rRNA and histone repeats, and spread the linked genes through the population.

C. SINGLE COPY DNA

Single copy DNA comprises most coding genes, and is therefore more complex than previously discussed genes. There is little information on particular genes in *Artemia*; the majority of the studies on single copy genes have been carried out for one of two reasons: (1) extension at the gene level of previous work at the enzyme and/or protein level, or (2) a search for genes whose expression is regulated during embryonic and early larval development. The approach followed in these studies has been to find first the cDNA coding for a particular protein using oligonucleotide or heterologous probes, and then look for the genes in genomic libraries. This approach has been successfully used to study the *ras* gene,[12,58] the ubiquitin genes,[59] ATPase genes,[60,61] and actin genes.[61] Table 3 shows the genes that have been studied in *Artemia* to date, in addition to the histone and ribosomal RNA genes

TABLE 3
Specific Genes Described in *Artemia*

Gene	Method of detection: cDNA	Genomic DNA	Southern blot	Ref.
Ca/Mg-ATPase	yes	yes[a]	—	60
Na/K-ATPase	yes	?	—	61
Hemoglobin	yes	—	yes	113
α-Tubulin	—	—	yes[b]	114
β-Tubulin	—	—	yes[b]	114
Actin	yes[c]	—	yes	61
ras	yes	yes	yes	12,58
Ubiquitin	yes	yes	yes	59
Homeobox domain[d]	—	—	yes	115,116
opa-Genes[d,e]	—	—	yes	116
"Zn-finger" domains[d]	—	—	yes	117
Histones	—	yes	yes	28,47,48
EF1α[f]	yes	yes	yes	62,63
EF1β	yes	—	—	118
EF1γ	yes	—	—	119
GRP33[g]	yes	—	—	120

[a] Several clones isolated, not yet studied.
[b] Hybridization analysis indicates one gene for α-tubulin and two or more genes for β-tubulin.
[c] Several isoforms have been isolated.
[d] Only the presence of homeobox, *opa*, or "Zn-finger" sequences has been detected.
[e] *Opa* sequences have been detected in several *Drosophila* genes.[121]
[f] Elongation factor 1, with three different subunits, α, β, and γ.
[g] Glycine-rich protein of 33000 Da; also known as hnRNP protein.

described in the previous section. Information about gene structure exists presently only for elongation factor(EF)1α, *ras*, and ubiquitin genes.

Using cDNA plasmid probes,[62] one of the four genes of EF1α has been cloned in a λ vector (EMBL-3). The gene spans 10 kb, containing 4 introns (from 1.35 to 2.14 kb) and 5 exons (from 40 to 234 bp).[63] Ubiquitin genes are highly conserved in all eukaryotes. We have described a polyubiquitin gene, with 8 ubiquitin elements, and an intron located in the 5′ noncoding region,[59] as has also been described for *Drosophila*.[64]

Surprisingly, the *ras* gene in *Artemia* is quite long.[58] As in other systems, the four exons are between 66 and 192 bp in length. Introns, however, are very large and variable: intron I is between 12 and 20 kb; intron II is larger than 28 kb; and intron III is 1.6 kb long. This contrasts with other invertebrate *ras* genes, such as *Drosophila*, where they are ≈ 3 kb,[65] or with mammals, where the Ha-*ras*-1 gene is 4.5 kb (although Ki-*ras*-2 is 50 kb).[66]

Actins, ATPases, EF1β and γ subunits, and the glycine-rich protein GRP33 have been studied only at the cDNA levels; tubulin, homeobox, and opa containing sequences have been detected only by Southern blot analysis (see Table 3 for references).

An interesting point obtained from these studies is the strong bias of *Artemia* on codon usage. G bases are almost absent at the third position,[67] a natural consequence of the peculiar (68% AT) composition of the *Artemia* genome.

IV. THE GENOMIC ORGANIZATION OF *ARTEMIA* MITOCHONDRIAL DNA

Mitochondrial (mt)DNA purified from a sample of *A. franciscana* has been cloned.[68,69] *Artemia* harbors one of the smallest mtDNAs (15 kb) studied to date with the exception of the nematode *Ascaris suum* which has 14,284 bp.[70] The physical map of the *A. franciscana* mtDNA was established for the following restriction enzymes: BamH I, EcoR I, Hind III, Pst I, Pvu II, and Ava I. In Figure 1, we show the physical organization of these restriction sites in *A. franciscana* mtDNA, as well as the part of the molecule that has been sequenced.

The organization of the mtDNA major genes, i.e., the 16S ribosomal RNA and 11 of the 13 protein genes, has been determined by a sequencing strategy.[69] The more rapid approach of hybridizing with *Drosophila* mtDNA probes of known genes[71,72] allowed mapping the more conserved genes 16S rRNA and cytochrome oxidase subunit I,[68] but failed in detecting the rest of the genes, most likely due to insufficient conservation of the sequence. By sequencing the ends of particular subclones (see Figure 1 for their location), and comparing the translated amino acid sequences to the *Drosophila* and vertebrate sequences, the genes shown in Figure 1 were mapped. Initially only 11 of the 13 genes were mapped.[69] Recently, by extending the same strategy, the rest of the protein coding genes, the cytochrome oxidase II, and the ATPase 8 genes were localized.[73]

The overall organization of the main genes in *Artemia* mtDNAs is exactly the same as found in insects, as exemplified by the mtDNAs of *Drosophila yakuba*[74] and *melanogaster*.[71,72] They also have identical polarity of transcription, i.e., the mtDNA strand on which the genes are codified. Starting from the putative origin of replication and transcription, located in the AT rich region of the *Drosophila* mtDNA,[75] the 16S rRNA and ND-1 are in the minus strand; apocytochrome b and ND6, in the plus strand; ND4L, ND-4, and ND-5, in the minus strand, and finally ND-3, cytochrome oxidase subunit III. ATPase subunit 6, ATPase subunit 8, cytochrome oxidase subunits II and I, and ND-2 are in the plus strand (Figure 1). This order is the same in all the arthropod mtDNAs studied to date.

The complete sequencing of the *Artemia franciscana* mtDNA is currently under-way. Presently, approximately 50% of the sequence is already available (also indicated in Figure 1). While the localization of the main protein genes is conserved among the arthropod mtDNAs, the situation is different in the case of the tRNAs. The available sequence data suggest that some are not in the same location as in *Drosophila* mtDNA. Starting from the 16S RNA, the serine [UCN]tRNA between ND-1 and apocytochrome b, and the proline and threonine tRNAs between ND-6 and ND-4L genes present in *Drosophila* have not been found in the same location in *Artemia*. Between ND-6 and ND-4L, there is a sequence with some capability of forming a tRNA structure that may correspond to an isoleucine tRNA, which is not present in *Drosophila*. Between ND-5 and ND-3, there is a cluster of five tRNAs in *Drosophila* that may be conserved in *Artemia*. To date,[73] the phenylalanine, the glutamic acid, and the serine [AGY] tRNAs have been identified, and there is room in the sequence for additional tRNAs. The glycine tRNA is also conserved in the same location between ND-3 and cytochrome oxidase III subunits, while between the ATPase 8 and the cytochrome oxidase II, there are two tRNAs, the histidine and possibly the aspartic acid tRNA, but not the lysine tRNA as in *Drosophila*. Finally, as seen in Figure 1, the methionine tRNA is conserved at the right end of ND-2 in both arthropods.

This lack of conservation of some of the tRNAs, linked to the extreme conservation of the location and orientation of the proteins genes, is intriguing in terms of both significance and mechanism. For mammals no recombination between mt DNA molecules has been detected,[76] and it is thus unclear how these molecules could maintain one level of organization while the tRNAs genes do not.

Once the mtDNA from *A. franciscana* has been cloned, the mtDNA organization in the different *Artemia* strains can be explored. As shown in Table 4, the analysis by restriction enzyme site patterns indicate that there is wide divergence between New and Old World bisexual species and among Old World bisexual and parthenogenetic ones. This is important as there is little biological information on which to base the distances among parthenogenetic animals, as they are not amenable to interbreeding crossing. It can be seen that the mtDNA

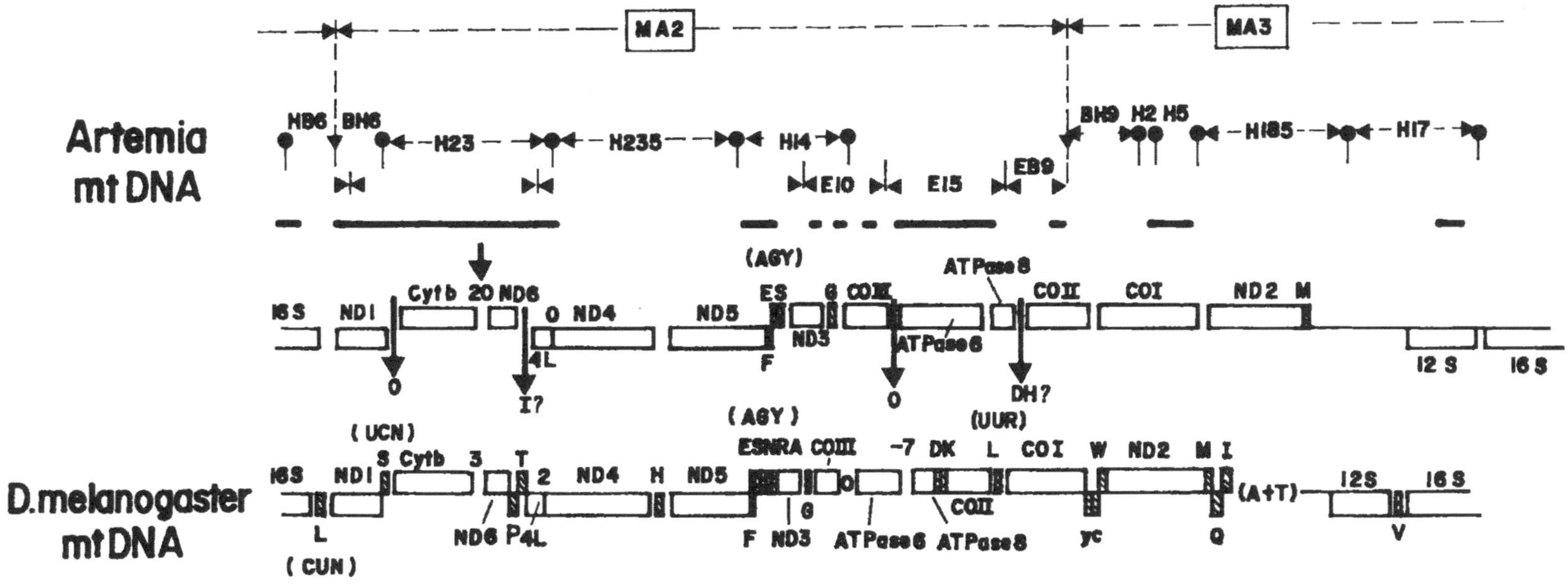

FIGURE 1. Current status of analysis of the mitochondrial DNA organization in *Artemia*. For comparison the *Drosophila melanogaster* mtDNA is also shown below. Open boxes correspond to protein and rRNA coding genes, hatched boxes to tRNA genes, labeled with the corresponding amino acid one-letter code. The interrupted thick line above the *Artemia* mtDNA map indicates the regions that are completely sequenced. The names of the different clones and subclones used in this work are shown in the upper part of the figure, between arrow heads indicating their limits in the map (B, BamH I; E, EcoR I; H, Hind III restriction sites bordering these clones).

TABLE 4
Divergence of Mitochondrial DNAs from *Artemia* of Different Geographical Origins

	San Francisco	Venezuela	Spanish bisex.	Spanish 2n	Spanish 4n
San Francisco	19	11	0	0	0
Venezuela	0.03 ± 0.01	16	0	0	0
Spanish bisex.	>0.21 ± 0.03	>0.20 ± 0.03	8	1	0
Spanish 2n	>0.31 ± 0.03	>0.30 ± 0.03	0.15 ± 0.03	13	0
Spanish 4n	>0.31 ± 0.03	>0.30 ± 0.03	>0.20 ± 0.03	>0.20 ± 0.03	12

Note: Total DNA obtained from developing cysts were digested with the following restriction enzymes as shown in Figure 1: Hind III, EcoR I, BamH I and Pst I. San Francisco correspond to the cloned *Artemia* mtDNA from San Francisco (USA), Venezuela cysts come from Bocachica, the Spanish bisexual strain from Cadiz, the Spanish diploid parthenogenetic (2n) from Cádiz, and the tetraploid (4n) both from Delta del Ebro (Spain) and Alcochete (Portugal). In the horizontal entries of the upper right half of the table, the number of restriction fragments coinciding in mobility and hybridization between different types of mtDNAs from *Artemia* strains is shown. The diagonal entries show the total number of fragments detected in each strain. In the lower left half of the table, the values of divergence of their sequences, calculated via the method of Upholt[77] are indicated.

from San Francisco is similar to the mtDNA from Venezuela; they share some restriction fragments as could be expected for geographical variation among populations of the same species. Interestingly, the pattern of fragments of the bisexual strain from Spain is completely different. Not a single fragment coincides in mobility among the 27 obtained with the 4 enzymes used. In contrast to this result, when the two American bisexual populations are compared, 22 (11 × 2) fragments of a total of 35 coincide in mobility. Applying the Upholt[77] method of calculating the divergence, $p = 3 \pm 1\%$, corresponding to ≈1.5 MY of divergence if the mutational rate of invertebrate mtDNA is similar to that of vertebrates.[78] In the case of the two bisexual species, Spanish and American, the divergence is higher than 21%, calculated by the same method. This method becomes unreliable[79] for such a level of divergence, but it indicates that they diverged a long time ago. These differences validate the classification of the bisexual strains from Europe and North America as different species.[80]

When a similar analysis was carried out with DNA isolated from Spanish parthenogenetic *Artemia*, different restriction fragment patterns are obtained compared to the two bisexual strains. The two parthenogenetic strains also differ in their restriction patterns. Again, no fragment coincides in mobility and hybridization (Table 4). The diploid parthenogenetic strain shares one fragment with the Spanish bisexual, with a divergence estimated at 14.5 ± 3%, which is too high to be accurate by these statistical methods. The divergence between the tetraploid and the diploid parthenogenetic and bisexual strains is even larger (see Table 4). It is known by classical analysis of enzyme polymorphism, as well as with respect to the meiotic mechanism of parthenogenesis, that the diploid and tetraploid strains of *Artemia* are quite different.[80]

Artemia from different parts of the world also show different organization of at least some nuclear genes, namely, satellite I sequences, the large ribosomal RNA genes and the 5S-histone genes (Table 5).

For further analyses of mtDNA divergence among *Artemia* populations see Chapter 9 (Taxonomy and Population Genetics of *Artemia* by Browne and Bowen).

V. DNA REPLICATION, MODIFICATION, AND REPAIR

DNA replication, modification, and repair processes involve a wide and complex group of reactions and events at the molecular, subcellular, and cellular level. In high eucaryotic organisms they are still poorly understood in spite of their importance and the effort inves-

TABLE 5
Genetic Differences of Different *Artemia* Populations[15,22,28,38,122]

	Bisexual strains		
	North America	Spain	Parthenogenetic strains
Satellite I-containing heterochromatin[a]	+	−	−
rDNA type of repeat[b]	16.5	13.5	12.2, 13.5
5S-histone repeat[b]	9, 8.5	Nd[c]	12, 12

[a] + and − mean presence or absence of satellite I-containing heterochromatin, respectively.
[b] Length of the repeat in kb.
[c] Not determined, although they are different from both bisexual strains from North America and parthenogenetic ones from Eurasia.

tigated in their clarification.[81,82] The available information pertaining to *Artemia* or even to crustaceans is small. At least two reasons support the interest in reviewing what is known about these processes in *Artemia*: (1) its closeness to *D. melanogaster*, which is likely to be one of the multicellular eucaryotic organisms in which we will first understand this problem, and (2) the life-history peculiarities of *Artemia*. In other biological systems (e.g., in *Drosophila*[83]) the processes of DNA replication and cell division that are very active during early embryogenesis[84] tend to overlap and continue during the remainder of embryogenesis. At these stages in *Artemia*, DNA replication and cell division is restricted to the cleavage and blastula phase, where it must be extremely active. It is repressed during gastrulation and embryogenesis[85,86] and resumes during the nauplial stages, when the larva goes through a series of growth stages to become the adult animal. Interestingly, this particular life-history strategy is associated with a particularly high resistance of the encysted embryo to desiccation, radiation injury, anoxia, etc. The overall integration of this biochemical knowledge into cell biology is still a remote goal even for the best understood eucaryotic cellular systems, such as yeast or *Drosophila*.

The presence of DNA polymerases in *Artemia* extracts and its preliminary biochemical characterization and purification has been reported.[87,88] As in other advanced eucaryotic systems, the attempt has been fraught with difficulties, especially proteolytic degradation of the enzymes. Slater and McLennan[87] demonstrated that DNA polymerase α, one of the main eucaryotic replicating enzymes,[89] is readily detectable in extracts of dormant cysts. It exists in the form of a 9 to 10S DNA polymerase α-holoenzyme complex, which increases 2- to 3-fold during pre-emergence development. Maximum activity is obtained at the onset of the phase of DNA synthesis. The activity is recovered in the soluble phase not associated with nuclei or other particulated organelles. When assayed in the presence of proteolytic inhibitors in SDS polyacrilamide gels after renaturation, the activity remains approximately constant from recently emerged to 2-day-old nauplii extracts, indicating that the enzyme is already present in the dormant cyst. Nevertheless, it is possible (as has been shown for other DNA metabolizing enzymes[90]) that they are partially active in the rehydrated gastrulae immediately prior to the resumption of development. Very little is known about the activation mechanism or about how DNA replication is actually initiated in the naupliar stem cells which will be actively dividing during the growth of the animal. A suggested regulatory mechanism is the appearance, after the resumption of development and particularly after hatching, of a particulated DNA polymerase α activity, if it can be demonstrated to be associated with the nucleus.[88] Dinucleoside compounds, of the type of Ap4A, have also been implicated in the onset of DNA replication,[91,92] Ap4A,[93] and other similar polyphosphates[94] are present in *Artemia* embryos, some of them in high concentrations. Their intriguing physiological role during *Artemia* embryogenesis remains to be clarified. They are known to participate in the control of the interconversion of purine nucleotides and to serve as a source of purines during development.[94,95]

DNA polymerase γ, the mtDNA replicating enzyme, is mainly associated with particulated organelles shortly after resumption of development.[88] Preliminary data obtained from our laboratory[73,96] indicates that mtDNA can be readily detected in dormant cysts and no large increases in mtDNA occurs during embryonic development. Although during embryogenesis there is a large increase in mitochondrial function[97] (see also Chapter 3 by Slegers), only a twofold increase in mtDNA has been detected by dot blot and Southern experiments.[96] This is consistent with the storage of mitochondria in the yolk platelets of *Artemia* dormant gastrulae previously detected using enzymatic techniques.[98] No additional DNA polymerase activity, such as polymerase δ (the other main DNA polymerase[89]) or any other enzyme or protein clearly associated with DNA replication such as primases, helicases, the DNA polymerase δ auxiliary cyclins, etc., have been detected or measured in *Artemia*. Although the whole question of the regulation of DNA replication in eucaryotic organisms is still obscure, information obtained from viral model systems suggest that initiator proteins analogous to those established for *E. coli* do exist.

In addition to the understanding of the control of DNA replication during *Artemia* development that is associated with its unique life history, it should be remembered that parthenogenetic *Artemia* strains have different degrees of ploidy (2n, 4n, 5n, and 6n). The control of chromosomal duplication has to be different in these strains as well as the mechanisms of adjustment of DNA replication to the number of replication *ori* sites in these strains.

A peculiar property of the dormant and early embryo *Artemia* is its very high radioresistance.[99] The developing nauplii are more radiosensitive and the adults are again more resistant but not at the same level as the diapause gastrulae. This would be expected since, with the exception of germinal tissues, cell division is restricted to egg cleavage and naupliar growth stages. Nevertheless, the extreme radioresistance of the gastrula and embryo needs to be explained, especially since no DNA repair polymerase activity has been detected in the cyst, although the presence of an antigenically related polypeptide[88] in SDS gels has been described. It would be interesting to assay the DNA polymerase β after irradiation of the cysts to establish whether a regulatory mechanism is responsible for this lack of activity.

DNase activities[90] are related to DNA metabolism. While for procaryotic organisms the functional role of many DNA nucleases is relatively well established,[84] their role in eucaryotes remains unclear, although they belong to the older known enzymes.[100] In *Artemia* dormant gastrulae[90] only one form of DNase I-like activity is detectable, primarily as an inactive form associated with the yolk granules, a large storage organelle specific of early developmental cellular systems. Whether the previously reported inactive DNA polymerases are also maintained by similar storage mechanisms is unknown, but it is likely at least in the case of DNA polymerase γ, since it has been reported[97,98] that significant amounts of the animal mitochondria are also stored in the yolk granules, and are released in an active form upon yolk metabolization during the resumption of development.[97] An additional DNase I-like activity, differing by several biochemical and enzymatic criteria from the stored form, appears and increases considerably during larval development. It is likely to be a digestive enzyme, associated with the gut differentiation and naupliar feeding.

Finally, as is the case for another arthropod, *Drosophila melanogaster*,[101,102] no evidence of DNA methylation has been found for *Artemia* DNA.[18] Since it has been suggested that DNA methylation is an important mechanism for controlling gene expression for vertebrate DNA (probably in a coarse way during cell differentiation and development), it remains to be established whether alternative mechanisms operate in arthropods.

VI. TRANSCRIPTION

Since the very early studies of Bellini[103] in which RNA levels were measured during development, the most significant experiments reported have been those of James et al.[104] The authors demonstrated that RNA complexity varies dramatically during development. Soon after resumption of development, there is an increase in complexity, which in newly hatched nauplii reverts to the levels found in the cysts. These type of experiments, however, could not evaluate the exact differences of RNA between cysts and nauplii. The availability of recombinant DNA probes has allowed the study of transcription of specific genes during development.

The topic of RNA polymerase levels is reviewed by Tate and Marshall in this book and will not be further discussed here.

Instead we will focus on the levels of specific mRNAs quantified in dot-blot and Southern experiments using specific probes[67] which occur during development from the encysted gastrula to the nauplius. Ca/Mg- and Na/K-ATPases, actin, *ras*, and ubiquitin mRNAs show an increase between 5 and 15 h after resumption of development (hatching being at 18—20 h in these experiments). Actin, *ras*, and ubiquitin mRNA levels decrease sharply soon after. In the case of the ATPases, there is only a slight decrease in mRNA levels until 30 h of naupliar development, at which time they again begin to increase. This last increase is also observed in the expression of actin genes, whereas *ras* and ubiquitin mRNA remain at very low levels. The low levels of some mRNAs cannot be due to experimental artifacts in the culturing of *Artemia* and/or extraction of RNAs, since all experiments were made with the same stock preparations of total RNA. In all cases there is a substantial amount of mRNA already present in the encysted gastrula (around 50% of the maximum). Levels for EF-1 mRNAs have also been measured and shown to increase sharply after hatching.[105] Tubulin and histone mRNAs follow a different pattern during development. Tubulin mRNA is already present in the encysted gastrulae and does not vary during development;[106] histone messenger, on the other hand, is only detectable after hatching, being absent in cysts and in embryos 8 h after resumption of development.[11]

Finally, a word of caution is offered, when comparing results from *Artemia* development to other model systems; *Artemia* encysted embryos are at the gastrula stage,[85,86] so they should not be directly compared to oocytes but to similar stages of development of other systems (e.g., *Xenopus, Drosophila* or sea urchins).

VII. PERSPECTIVES OF *ARTEMIA* MOLECULAR GENETICS

The exploration of the following will be of interest in a near future:

1. The study of the mechanisms controlling gene expression in *Artemia*, in particular, the regulatory sequences upstream of many genes and the specific features of genomic organization compared with those of other organisms.
2. The development of transfecting carrier systems to incorporate particular gene fragments in *Artemia* genome as currently attempted in other invertebrates, e.g., sea urchins.[107] Whether the current approach used in *Drosophila* will be useful for *Artemia* is unknown. Specifically it could be of great interest if transposable elements were found in *Artemia*.
3. In addition to housekeeping genes, two groups of genes seem to be worthwhile exploring.

a. Genes responsible for the life history of *Artemia*, in particular, its adaptation to extreme environments, high salt, and extreme pH. As described, some of the ATPases have already been cloned. By comparing their molecular organization to those of closely related species, adaptive strategies utilized by these animals might be unraveled at the genomic level. Furthermore, transformation techniques could allow tests for the controlling DNA sequences that are needed by a particular species to adapt to a different environment.

b. Genes controlling segmentation and early development. *Artemia* is a primitive crustacean with a relatively late segmentation process, occurring primarily in the postembryonic stages. The current information on homeotic and related genes in *Drosophila* and their strong conservation across the animal kingdom strongly suggests that they occur in *Artemia*. The study of *Artemia* as the organism not too distant from *Drosophila*, but with relatively different developmental strategies, would be of value. Experiments taking advantage of the ease of manipulation of the postembryonic segmentation process in *Artemia* (e.g., by nutritional manipulation of the animal) have been already performed.[108] The association of these type of experiments to molecular genetic analyses offers an interesting exchange of ideas between *Artemia* and the better-known developmental systems, such as *Caenorabdhitis elegans*, *Strongilocentrotus purpuratus*, and *D. melanogaster*.

ACKNOWLEDGMENTS

The help of Dr. Alfonso Valencia with the Genebank information in Table 1, the critical reading of the manuscript by Drs. Browne, Sillero, and Sastre, as well as the communication of unpublished results by many colleagues, is gratefully appreciated. Published and unpublished work in the laboratories of the authors were supported by grants from the Comisión Asesora para la Investigación Cientifica y Técnica, Dirección General de Investigación Cientifica y Técnica y el Fondo de Investigaciones Sanitarias de la Seguridad Social.

REFERENCES

1. **Watson, J. D., Hopkins, N. H., Roberts, J. W., Steitz, J. A., and Weiner, A. M.,** *Molecular Biology of the Gene*, Benjamin/Cummings, Menlo Park, CA, 1987.
2. **Newport, J. W. and Kirschner, M. W.,** A major developmental transition in early *Xenopus embryos*: II. Control of the onset of transcription, *Cell*, 30, 675, 1982.
3. **Sorgeloos, P., Bengtson, D. A., Decleir, W., and Jaspers, E.,** Eds., *Artemia Research and its Applications*, Vols. 1 and 2, Universa Press, Wetteren, Belgium, 1987.
4. **MacRae, T. H., Bagshaw, J. C., and Warner, A. H.,** Eds., *Biochemistry and Cell Biology of Artemia*, CRC Press, Boca Raton, FL, 1988.
5. **Warner, A. H., MacRae, T. H., and Bagshaw, J. C.,** Eds., *Cell and Molecular Biology of Artemia Development*, Plenum Press, New York, 1989.
6. **Golberg, R. B.,** Plants: novel developmental processes, *Science*, 240, 1460, 1988.
7. **Koshland, D.,** Biological systems, *Science*, 240, 1385, 1988.
8. **Léger, P. H., Bengtson, D. A., Sorgeloos, P., Simpson, K. L., and Beck, A. D.,** The nutritional value of *Artemia*, in *Artemia Research and Its Applications*, Vol. 3, Sorgeloos, P., Bengtson, D. A., Decleir, W., and Jaspers, E., Eds., Universa Press, Wetteren, Belgium, 1987, 357.
9. **Barigozzi, C. and Baratelli, L.,** New data on the chromosome number of the genus *Artemia*, *Rend. Accad. Naz. Lincei*, 73, 139, 1982.
10. **Quintanilla, M. and Renart, J.,** Post-transcriptional regulation of endoribonuclease VI expression during *Artemia* development, *FEBS Lett.*, 155, 93, 1983.
11. **Bagshaw, J. C.,** Developmentally regulated gene expression in *Artemia*: histone gene expression in newly hatched larvae, *Develop. Genet.*, 3, 41, 1982.
12. **Díaz-Guerra, M., Quintanilla, M., Palmero, I., Sastre, L., and Renart, J.,** Differential expression of a gene highly homologous to c-*ras* during the development of the brine shrimp, *Biochem. Biophys. Res. Commun.*, 162, 802, 1989.
13. **Squires, S. L. and Acey, R. A.,** A new method for the purification of nuclei from *Artemia*, in *Cell and Molecular Biology of Artemia Development*, Warner, A. H., MacRae, T. H., and Bagshaw, J. C., Eds., Plenum Press, New York, 1987, 221.

14. **Cruces, J., Wonenburger, M. L. G., Díaz-Guerra, M., Sebastián, J., and Renart, J.,** Identification of a satellite in the *Artemia* genome, in *Artemia Research and its Application,* Vol. 2, Decleir, W., Moens, L., Slegers, H., Jaspers, E., and Sorgeloos, P., Eds., Universa Press, Wetteren, Belgium, 1987, 403.
15. **Cruces, J., Wonenburger, M. L. G., Díaz-Guerra, M., Sebastián, J., and Renart, J.,** Satellite DNA in the crustacean, *Artemia, Gene,* 44, 341, 1986.
16. **Barigozzi, C., Badaracco, G., Plevani, P., Baratelli, L., Profeta, S., Ginelli, E., and Meneveri, R.,** Heterochromatin in the genus *Artemia, Chromosoma,* 90, 332, 1984.
17. **Cruces, J.,** Caracterización del genoma y de los genes ribosómicos de *Artemia,* Ph.D. thesis, Universidad Autónoma de Madrid, Spain, 1982.
18. **Warner, A. H. and Bagshaw, J. C.,** Absence of detectable 5-methylcytosine in DNA of embryos of the brine shrimp *Artemia, Develop. Biol.,* 102, 264, 1984.
19. **Vaughn, J. C.,** DNA reassociation kinetic analysis of the brine shrimp *Artemia salina, Biochem. Biophys. Res. Commun.,* 79, 525, 1977.
20. **Vaughn, J. C.,** DNA sequence organization in the genome of the brine shrimp, *Artemia salina,* in *Biochemistry of Artemia Development,* Bagshaw, J. C. and Warner, A. H., Eds., University Microfilm International, Ann Arbor, 1979, 190.
21. **Hedgecock, D., Tracey, M. L., and Nelson, K.,** Genetics, in *The Biology of Crustacea,* Vol. 2, Abele, L. G., Ed., Academic Press, New York, 1982, 288.
22. **Badaracco, G., Baratelli, L., Ginelli, E., Meneveri, R., Plevani, P., Valsasnini, P., and Barigozzi, C.,** Variations in repetitive DNA and heterochromatin in the genus *Artemia, Chromosoma,* 95, 71, 1987.
23. **Clark, C. G.,** On the evolution of ribosomal RNA, *J. Mol. Evol.,* 25, 343, 1987.
24. **Diels, L., De Baere, R., Vandenberghe, A., and De Watcher, R.,** The sequence of the 5S ribosomal RNA of the crustacean *Artemia salina, Nucleic Acids Res.,* 9, 5141, 1981.
25. **Ursi, D., Vandenberghe, A., and De Watcher, R.,** The sequence of the 5.8S ribosomal RNA of the crustacean *Artemia salina,* with a proposal for a general secondary structure model for 5.8S ribosomal RNA, *Nucleic Acids Res.,* 10, 3517, 1982.
26. **Vaughn, J. C., Sperbeck, S. J., and Hughes, M. J.,** Molecular cloning and characterization of ribosomal RNA genes from the brine shrimp. Nucleotide sequence analysis and evolution of the 5.8S rRNA gene region and its flanking nucleotides, *Biochim. Biophys. Acta,* 783, 144, 1984.
27. **Bagshaw, J. C., Skinner, H. B., Burn, T. C., and Perry, B. A.,** Nucleotide sequence of the 5S RNA gene and flanking regions interspersed with histone genes, *Nucleic Acids Res.,* 15, 3628, 1987.
28. **Cruces, J., Díaz-Guerra, M., Gil, I., and Renart, J.,** The 5S rRNA-histone repeat in the crustacean *Artemia.* Structure, polymorphism and variation of the 5S rRNA segment in different populations, *Nucleic Acids Res.,* 17, 6283, 1989.
29. **Díaz-Guerra, M., Cruces, J., Sebastián, J., and Renart, J.,** Cloning of a 5S rRNA gene from *Artemia* larvae, *Biol. Cell.,* 45, 116, 1982.
30. **Cruces, J., Sebastián, J., and Renart, J.,** Ribosomal ribonucleic acids from the crustacean *Artemia, FEBS Lett.,* 137, 108, 1982.
31. **Roberts, M. P. and Vaughn, J. C.,** Ribosomal RNA sequence conservation and gene number in the larval brine shrimp, *Biochim. Biophys. Acta,* 697, 148, 1982.
32. **Nelles, L., Fang, B. L., Volckaert, G., Vandenberghe, A., and De Watcher, R.,** Nucleotide sequence of a crustacean 18S RNA gene and secondary structure of eukaryotic small subunit ribosomal RNAs, *Nucleic Acids Res.,* 12, 8749, 1984.
33. **Ishikawa, H.,** Comparative studies on the thermal stability of ribosomal RNAs. IV. Thermal stability and molecular integrity of ribosomal RNAs from several protostomes (rotifers, round-worms, liver flukes, and brine shrimp), *Comp. Biochem. Physiol.,* 58(B), 1, 1977.
34. **Nelles, L., Van Broeckhoven, C., De Watcher, R., and Vandenberghe, A.,** Location of the hidden break in large subunit ribosomal RNA of *Artemia salina, Naturwissenschaften,* 71, 634, 1984.
35. **Cruces, J., Sebastián, J., and Renart, J.,** Restriction mapping of the rRNA genes from *Artemia* larvae, *Biochem. Biophys. Res. Commun.,* 98, 404, 1981.
36. **Vaughn, J. C., Whitman, D. J., Bagshaw, J. C., and Helder, J. C.,** Molecular cloning and characterization of ribosomal RNA genes from the brine shrimp, *Biochim. Biophys. Acta,* 697, 156, 1982.
37. **Gallego, M. E., Medina, R., Cruces, J., Domínguez, E., Sebastián, J., and Renart, J.,** Ribosomal RNA gene types in *Artemia,* in *Artemia Research and Its Application,* Vol. 2, Decleir, W., Moens, L., Slegers, H., Jaspers, E., and Sorgeloos, P., Eds., Universa Press, Wetteren, Belgium, 1987, 423.
38. **Gallego, M. E., Díaz-Guerra, M., Cruces, J., Sebastián, J., and Renart, J.,** Characterization of two types of rRNA gene repeat units from the crustacean *Artemia, Gene,* 48, 175, 1986.
39. **Piot, E., Bosmans, M.-J., De Chaffoy de Courcelles, D., and Kondo, M.,** Structure of the ribosomal repeat units in *Artemia* DNA, *Arch. Int. Physiol. Biochim.,* 89, (B)128, 1981.
40. **Gil, I., Gallego, M. E., Renart, J., and Cruces, J.,** Identification of the transcriptional initiation site of ribosomal RNA genes in the crustacean *Artemia, Nucleic Acids Res.,* 15, 6007, 1987.

41. **Koller, H. T., Frondorf, K. A., Maschner, P. D., and Vaughn, J. C.,** *In vivo* transcription from multiple spacer rRNA gene promoters during early development and evolution of the intergenic spacer in the brine shrimp *Artemia, Nucleic Acids Res.*, 15, 5391, 1987.
42. **Gil, I. and Cruces, J.,** personal communication, 1988.
43. **Labhart, P. and Reeder, R. H.,** Characterization of three sites of RNA 3′ end formation in the *Xenopus* ribosomal gene spacer, *Cell*, 45, 431, 1986.
44. **Grummt, I., Kuhn, A., Bartsch, I., and Rosenbauer, H.,** A transcription terminator located upstream of the mouse rDNA initiation site affects rRNA synthesis, *Cell*, 47, 901, 1986.
45. **Cross, N. and Dover, G.,** Tse-tse fly rDNA: an analysis of structure and sequence, *Nucleic Acids Res.*, 15, 15, 1987.
46. **Bagshaw, J. C., Andrews, M. T., and Perry, B. A.,** Organization and expression of histone genes in the brine shrimp *Artemia*, in *Histone Genes, Structure, Organization and Regulation*, Stein, G. S., Stein, J. L., and Marzluff, W. F., Eds., John Wiley & Sons, New York, 1984, 181.
47. **Andrews, M. T., Vaughn, J. C., Perry, B. A., and Bagshaw, J. C.,** Interspersion of histone and 5S rRNA genes in *Artemia, Gene*, 51, 61, 1987.
48. **Cruces, J., Díaz-Guerra, M., Sebastián, J., and Renart, J.,** Organization and sequence of 5S rRNA genes in *Artemia*, in *Artemia Research and Its Application*, Vol. 2, Decleir, W., Moens, L., Slegers, H., Jaspers, E., and Sorgeloos, P., Eds., Universa Press, Wetteren, Belgium, 1987, 393.
49. **Renart, J., Cruces, J., Díaz-Guerra, M., Domínguez, E., Franco, E., Gil, I., Macías, M. T., Palmero, I., and Sastre, L.,** Structure and expression of ribosomal RNA genes from *Artemia*, in *Cell and Molecular Biology of Artemia Development*, Warner, A. H., MacRae, T. H., and Bagshaw, J. C., Eds., Plenum Press, New York, 1989, 287.
50. **Artavanis-Tsakonas, S., Schedl, P., Tschudi, C., Pirrota, V., Steward, R., and Gehring, W. J.,** The 5S genes of *Drosophila melanogaster, Cell*, 12, 1057, 1977.
51. **Childs, G., Maxson, R., Cohn, R. H., and Kedes, L.,** Orphons: dispersed genetic elements derived from tandem repetitive genes of eukaryotes, *Cell*, 23, 651, 1981.
52. **Bollon, A. P.,** Organization of fungal ribosomal RNA genes, in *The Cell Nucleus rDNA*, Busch, H. and Rothblum, L., Eds., Academic Press, New York, 1982, 68.
53. **Weiner, A. M. and Emery, H. S.,** The rDNA of *Dictyostelium discoideum* and *Physarum polycephalum*, in *The Cell Nucleus rDNA*, Busch, H. and Rothblum, L., Eds., Academic Press, New York, 1982, 127.
54. **Drouin, G., Hofman, J. D., and Doolittle, W. F.,** Unusual ribosomal RNA gene organization in copepods of the genus *Calanus, J. Mol. Biol.*, 196, 943, 1987.
55. **Ulbrich, N., Kumagi, I., and Erdmann, V. A.,** The number of ribosomal RNA genes in *Thermus thermophilus* HB8, *Nucleic Acids Res.*, 12, 2055, 1984.
56. **Lund, E., Dahlberg, J. E., Lindahl, L., Juskunas, S. R., Dennis, R. P., and Nomura, M.,** Transfer RNA genes between 16S and 23S rRNA genes in rRNA transcription units of *E. coli, Cell*, 7, 165, 1976.
57. **Dover, G. A.,** Molecular drive: a cohesive mode of species evolution, *Nature*, 299, 111, 1982.
58. **Martinez, J. L., Díaz-Guerra, M., and Renart, J.,** unpublished data.
59. **Franco, E.,** Aislamiento y caracterizaci6n de los genes que codifican ubicuitina en el crustáceo *Artemia*. Ph.D. thesis, Universidad Aut6noma de Madrid, Spain, 1988.
60. **Palmero, I. and Sastre, L.,** Complementary DNA cloning of a protein highly homologous to mammalian sarcoplasmic reticulum Ca-ATPase from the crustacean *Artemia*, J. Mol. Biol., 210, 737, 1989.
61. **Sastre, L.,** personal communication, 1988.
62. **Van Hemert, F., Amons, R., Pluijms, W. J. M., Van Ormondt, H., and Möller, W.,** The primary structure of elongation factor EF-1α from the brine shrimp *Artemia, EMBO J.*, 3, 1109, 1984.
63. **Lenstra, J. A., Van Vliet, A., Arnberg, A. C., Van Hemert, F. J., and Möller, W.,** Genes coding for the elongation factor EF-1α in *Artemia, Eur. J. Biochem.*, 155, 475, 1986.
64. **Lee, H., Simon, J. A., and Lis, J. T.,** Structure and expression of ubiquitin genes of *Drosophila melanogaster, Mol. Cell. Biol.*, 8, 4727, 1988.
65. **Neuman-Silberberg, F. S., Schejter, E., Hoffman, F. M., and Shilo, B.,** The *Drosophila ras* oncogenes: structure and nucleoside sequence, *Cell*, 37, 1027, 1984.
66. **Barbacid, M.,** *Ras* genes, *Ann. Rev. Biochem.*, 56, 779, 1987.
67. **Sastre, L., Palmero, I., Macías, M. T., Gil, I., Franco, E., Domínguez, E., Díaz-Guerra, M., Quintanilla, M., Cruces, J., and Renart, J.,** cDNA cloning of developmentally regulated *Artemia* genes, in *Cell and Molecular Biology of Artemia Development*, Warner, A. H., MacRae, T. H., and Bagshaw, J. C., Eds., Plenum, New York, 1989, 319.
68. **Batuecas, B., Marco, R., Calleja, M., and Garesse, R.,** Molecular characterization of *Artemia* mitochondrial DNA: cloning, physical mapping and preliminary gene organization, in *Artemia Research and its Applications*, Decleir, W., Moens, L., Slegers, H., Jaspers, E., and Sorgeloos, P., Eds., Vol. 2, Universa Press, Wetteren, Belgium, 1987, 355.
69. **Batuecas, B., Garesse, R., Calleja, M., Valverde, J. R., and Marco, R.,** Genome organization of *Artemia* mitochondrial DNA, *Nucleic Acids Res.*, 16, 6515, 1988.

70. **Wolstenholme, D. R., MacFarlaine, J. L., Okimoto, R., Clary, D. O., and Wahleithner, J. A.**, Bizarre tRNAs inferred from DNA sequences of mitochondrial genomes of nematode worms, *Proc. Natl. Acad. Sci. U.S.A.*, 84, 1324, 1987.
71. **De Brujin, M. H. L.**, *Drosophila melanogaster* mitochondrial DNA: a novel organization and genetic code, *Nature*, 304, 234, 1983.
72. **Garesse, R.**, *Drosophila melanogaster* mitochondrial DNA: gene organization and evolutionary considerations, *Genetics*, 118, 649, 1988.
73. **Batuecas, B., Marco, R., and Garesse, R.**, unpublished data.
74. **Woltensholme, D. R. and Clary, D. O.**, Sequence evolution of *Drosophila* mitochondrial DNA, *Genetics*, 105, 725, 1985.
75. **Goddard, J. M. and Woltensholme, D. R.**, Origin and direction of replication in mitochondrial DNA molecules from *Drosophila melanogaster*, *Proc. Natl. Acad. Sci. U.S.A.*, 75, 3886, 1978.
76. **Olivo, P. D., Van de Walle, M. J., Laipis, P. J., and Hauswirth, W. W.**, Nucleotide sequence evidence for rapid genotypic shifts in the bovine mitochondrial DNA D-loop, *Nature*, 306, 400, 1983.
77. **Upholt, W. B.**, Estimation of DNA sequence divergence from comparison of restriction endonuclease digests, *Nucleic Acids Res.*, 4, 1257, 1977.
78. **Moritz, C., Dowling, T. E., and Brown, W. M.**, Evolution of animal mitochondrial: relevance for population biology and systematics, *Ann. Rev. Ecol. Syst.*, 18, 262, 1987.
79. **Nel, R.**, *Molecular Evolutionary Genetics*, Columbia University Press, New York, 1987, 86.
80. **Abreu-Grobois, F. A.**, A review of the genetics of *Artemia*, in *Artemia Research and its Application*, Vol. 1, Decleir, W., Moens, L., Slegers, H., Jaspers, E., and Sorgeloos, P., Eds., Universa Press, Wetteren, Belgium, 1987, 61.
81. **Kornberg, A.**, DNA replication, *J. Biol. Chem.*, 236, 1, 1988.
82. **Nakayasu, H. and Berezney, R.**, Mapping replicational sites in the eucaryotic cell nucleus, *J. Cell Biol.*, 108, 1, 1989.
83. **Campos Ortega, J. A. and Hartenstein, V.**, *The Embryonic Development of Drosophila melanogaster*, Springer Verlag, Heidelberg, 1985.
84. **Kornberg, A.**, *DNA Replication*, W. H. Freeman, San Francisco, 1980.
85. **Nakanishi, Y. H., Iwasaki, T., Okigaki, T., and Kato, H.**, Cytological studies of *Artemia salina*. I. Embryonic development without cell multiplication after the blastula stage in encysted dry eggs, *Annot. Zool. Japan.*, 35, 223, 1962.
86. **Olson, C. S. and Clegg, J. S.**, Cell division during the development of *Artemia salina*, *Wilhelm Roux Arch.*, 189, 1, 1978.
87. **Slater, J. M. and McLennan, A. G.**, DNA polymerases α and γ during pre-emergence and early larval development of *Artemia*, *Europ. J. Biochem.*, 129, 415, 1982.
88. **McLennan, A. G. and Miller, D.**, *Artemia* DNA polymerases in *Artemia Research and its Applications*, Decleir, W., Moens, L., Slegers, H., Jaspers, E., and Sorgeloos, P., Eds., Vol. 2, Universa Press, Wetteren, Belgium, 1987, 253.
89. **Blow, J.**, Eucaryotic chromosome replication requires both α and δ DNA polymerases, *Trends Genet.*, 5, 134, 1989.
90. **Domingo, A., Martin, E., Cervera, M., and Marco, R.**, Biochemical characterization and developmental behavior of *Artemia* embryonic and nauplial deoxyribonucleases *Biochemistry* 15, 4125, 1986.
91. **Grummt, F., Waltz, Jantzen, H. M., Hamprecht, K., Huebscher, U., and Kvenzle, C. C.**, Diadenosine 5′, 5′′′-P^1, P^4-tetraphosphate, a ligand of the 57 kilodalton subunit of DNA polymerase α, *Proc. Natl. Acad. Sci. U.S.A.*, 76, 6031, 1979.
92. **Zourgui, L., Tharaud, D., Solari, A., Litvak, S., and Tarrago-Litvak, L.**, Stimulation of DNA synthesis by microinjection of diadenosine 5′, 5′′′-P^1, P^4-tetraphosphate (Ap4A) into *Xenopus laevis* oocytes, *Dev. Biol.*, 103, 409, 1984.
93. **McLennan, A. G. and Prescott, M.**, Diadenosine 5′, 5′′′-P^1, P^4-tetraphosphate in developing embryos of *Artemia*, *Nucl. Acids Res.*, 12, 1609, 1984.
94. **Sillero, A. and Sillero, M. A. G.**, Purine nucleotide metabolism in *Artemia*, in *Biochemistry and Cell Biology of Artemia*, MacRae, T. H., Bagshaw, J. C., and Warner, A. H., Eds., CRC Press, Boca Raton, FL, 1989, 96.
95. **Sillero, A. and Sillero, M. A. G.**, Interconversion of purine nucleotides in *Artemia*: a review, in *Artemia Research and its Applications*, Decleir, W., Moens, L., Slegers, H., Jaspers, E., and Sorgeloos, P., Eds., Vol. 2, Universa Press, Wetteren, Belgium, 1987, 105.
96. **Palmero, I., Renart, J., and Sastre, L.**, Isolation of cDNA clones coding for mitochondrial 16S ribosomal RNA from the crustacean *Artemia*, *Gene*, 68, 239, 1988.
97. **Marco, R., Garesse, A., and Vallejo, C. G.**, Mitochondrial unmasking and yolk platelets metabolization during early development in *Artemia salina*, in *The Brine Shrimp Artemia*, Vol. 2, Persoone, G., Sorgeloos, P., Roels, D., and Jasper, E., Eds., Universa Press, Wetteren, Belgium, 1980, 481.

98. **Marco, R., Garesse, A., and Vallejo, C. G.,** Storage of mitochondria in the yolk platelets of *Artemia* dormant gastrulae, *Cell Mol. Biol.*, 27, 515, 1981.
99. **McCourt, R. P.,** The radiobiology of *Artemia*: a review, in *Artemia Research and its Applications*, Decleir, W., Moens, L., Slegers, H., Jaspers, E., and Sorgeloos, P., Eds., Vol. 1, Universa Press, Wetteren, Belgium, 277, 1987, 227.
100. **Feulgen, R.,** Ueber Nucleo-Gelase, *Hoppe-Seyler's Z. Physiol. Chem.*, 237, 261, 1935.
101. **Urieli-Shoval, S. Y., Gruenbaum, Y., Sedat, J., and Razin, A.,** The absence of detectable methylated bases in *Drosophila melanogaster, FEBS Lett.*, 146, 148, 1982.
102. **Bird, A. P.,** DNA methylation versus gene expression, *J. Embryol. Exp. Morph.*, 83, Suppl. 31, 1984.
103. **Bellini, L.,** Osservazioni sugli acidi nucleici nello sviluppo di Artemia salina Leach, *Ricerca Scient.*, 30, 816, 1960.
104. **James, T. C., Hentschel, C. C., and Tata, J. R.,** Gene expression during resumption of development of encysted *Artemia* gastrulae, in *Biochemistry of Artemia Development*, Bagshaw, J. C. and Warner, A. H., Eds., University Microfilm International, Ann Arbor, MI, 1979, 167.
105. **Maesen, G. D. F. and Möller, W.,** personal communication, 1989.
106. **MacRae, T. H., Campbell, E. J., and Langdon, C. M.,** Tubulin synthesis and microtubule organization in *Artemia*, in *Cell and Molecular Biology of Artemia Development*, Warner, A. H., MacRae, T. H., and Bagshaw, J. C., Eds., Plenum, New York, 1989.
107. **Flytaznis, C. N., Hough-Evans, B. R., Britten, R. J., and Davidson, E. H.,** Gene transfer by microinjection into the sea urchin egg, in *Developmental Genetics of Higher Organisms*, Malacinski, G. M., Ed., Macmillan, New York, 1988, 147.
108. **Hernandorena, A.,** An increased dietary tryptophan requirement induced by interference with purine interconversion in *Artemia*, in *Artemia Research and Its Applications*, Decleir, W., Moens, L., Slegers, H., Jaspers, E., and Sorgeloos, P., Eds., Vol. 2, Universa Press, Wetteren, Belgium, 1987, 233.
109. **Gardiner, M. S.,** *The Biology of Invertebrates*, McGraw-Hill, New York, 1972.
110. **Scagel, R. F., Bandoni, R. J., Maze, J. R., Rouse, G. E., Schofield, W. B., and Stein, J. R.,** *Plants: An Evolutionary Survey*, Wadsworth, Beltmont, CA, 1984.
111. **Cameron, G., Hazledine, D., Jones, S., Stoesser, G., Kalm, P., and Nial, J.,** EMBL Nucleotide Sequence Data Library, European Molecular Biology Laboratory, Heidelberg, W. Germany, 1987.
112. **Davis, B. D., Dulbecco, R., Eisen, H. N., and Ginsberg, H. S.,** *Microbiology*, Harper & Row, New York, 1980, 29.
113. **Powell, R. I. J., Trotman, C. N. A., and Tate, W. P.,** Molecular Biology of *Artemia* haemoglobins, in *Proc. NATO-Advanced Research Workshop on Cell and Molecular Biology of Artemia Development*, Montreal, Canada, 1988, 28.
114. **MacRae, T. H.,** *Artemia* tubulin: a model protein for the study of gene regulation and organelle formation during development, in *Biochemistry and Cell Biology of Artemia*, MacRae, T. H., Bagshaw, J. C., and Warner, A. H., Eds., CRC Press, Boca Raton, FL, 1989, 213.
115. **McGinnis, W.,** Homeo box sequences of the antennapedia class are conserved only in higher animal genomes, *Cold Spring Harbor Symp. Quant. Biol.*, 50, 263, 1985.
116. **Manzanares, M., Marco, R., and Garesse, R.,** unpublished data.
117. **Schuh, R., Aicher, W., Gaul, U., Côté, S., Preiss, A., Maier, D., Seifert, E., Nauber, U., Schröder, C., Kemler, R., and Jäckle, H.,** A conserved family of nuclear proteins containing structural elements of the finger protein encoded by Krüppel, a *Drosophila* segmentation gene, *Cell*, 47, 1025, 1986.
118. **Maesen, G. D. F., Amons, R., Maasen, J. A., and Möller, W.,** Primary structure of elongation factor 1β from *Artemia, FEBS Lett.*, 208, 77, 1986.
119. **Maesen, G. D. F., Amons, R., Zeelen, J. P., and Möller, W.,** Primary structure of elongation factor 1γ from *Artemia, FEBS Lett.*, 223, 181, 1987.
120. **Cruz-Alvarez, M. and Pellicer, A.,** Cloning of a full-length complementary DNA for an *Artemia salina* glycine-rich protein, *J. Biol. Chem.*, 262, 13377, 1987.
121. **Wharton, K. A., Yedvobnick, B., Finnerty, V. G., and Artavanis-Tsakonas, S.,** Opa: a novel family of transcribed repeats shared by the Notch locus and other developmentally regulated loci in *Drosophila melanogaster, Cell*, 40, 55, 1985.
122. **Medina, R., Gallego, M. E., Renart, J., and Cruces, J.,** unpublished results.

Chapter 2

POST-DORMANCY TRANSCRIPTION AND TRANSLATION IN THE BRINE SHRIMP

Warren P. Tate and Craig J. Marshall

TABLE OF CONTENTS

I. INTRODUCTION

Artemia at certain stages in its life cycle has an extraordinary ability to form a quiescent cyst which is able to survive for long periods in extremely harsh conditions. The cyst comprises about 4000 cells enclosed in a number of tough and largely impervious membranes and shells[1-3] and its metabolic activity is scarcely detectable. Upon the addition of water, development of the *Artemia* embryo begins and a fully formed nauplius hatches from the cyst after about 24 h.[4] The mechanisms by which the development of the brine shrimp embryo is halted, and by which the appropriate development of the shrimp occurs upon rehydration, are particularly relevant to eukaryotic cell development. Given this interest in *Artemia* differentiation, what then are some of the things that play a part in the metabolic stasis of the brine shrimp cyst?

The *Artemia* cyst contains components necessary for the expression of genes both at the transcriptional level and the translational level, but their activity seems to be regulated in the maintenance of the cryptobiotic state. For example, although the RNA polymerases necessary to transcribe gene sequences are present, their activity is suppressed by an inhibitor. Considerably more is known about regulators of translation. As we shall see later, this is probably the main point of regulation in maintenance of the dormant and the rehydrated states of *Artemia* embryos. The mRNA of desiccated cysts is found in two forms of messenger, ribonuclear, protein particles (mRNP particles): a cytosolic form and a membrane-bound form. Each of these forms apparently limits the availability of mRNA for translation, at the same time maintaining a large pool of mRNA that can be made available for protein synthesis at the appropriate signal.

Ribosomes themselves are found in the cryptobiotic embryo, but are not associated with mRNA nor are they present as polysomes, the characteristic of cells actively synthesizing proteins. The presence of ribosomes is not surprising, the expenditure of considerable energy is required to manufacture ribosomes, and the ribosomal proteins themselves require ribosomes for their synthesis.[5]

The protein synthetic factors[6-9] are probably the most studied of the effectors of post-dormancy events. Modulation of the amounts of molecules that affect protein synthetic factors is relevant to the repression of translation during dormancy and the coordinated establishment of protein synthesis in the developing embryo. Roles have been proposed for initiation, elongation, and termination factors in the control of metabolism, both on their own and in conjunction with other molecules that modify their specific activity.

II. UNCOUPLED MORPHOGENESIS AND CELL DIVISION

A. COMPARISON WITH SEEDS AND NEWLY FERTILIZED EGGS

Comparisons are sometimes drawn between *Artemia* cysts and seeds or eggs. This is misleading in the sense that an egg or seed contains very little embryonic tissue. In contrast, the *Artemia* embryo is a well-developed collection of about 4,000 cells which are in a state of suspended animation. The addition of water is all that is required to restart development. On the other hand, the cyst is analogous to the seed in that it is dormant, storable, polyploid, and contains a large store of unprogrammed ribosomes and translational components.

The activation of development associated with the fertilization of sea urchin eggs[10] and the germination of seeds[11] depends on the activation of masked mRNA and is insensitive to actinomycin D, an inhibitor of transcription, and is therefore under translational control. The onset of protein synthesis observed upon hydration of *Artemia* cysts suggests that the same is the case with *Artemia* but it is clear that transcription is also important as the embryo develops and the two mechanisms are closely interrelated. The difficulty of getting molecules such as actinomycin into the embryo during the first stages of development to make an assessment of the effects of specific inhibitors of the molecular biological events is a limiting feature.

A rapid mobilization of ribosomes into polysomes observed in *Artemia* with the onset of development is characteristic of many plant seeds upon germination[12] as well as in quiescent animal cells entering a proliferative state.[13] In *Artemia*, however, only a small percentage of the available ribosomes seem committed to polysomes, in contrast to most active mammalian tissues which have a high proportion of their ribosomes in this state.[13]

B. ABSENCE OF CELL DIVISION DURING DEVELOPMENT

The first four hours of development following the hydration of the cyst is reversible, and dormancy can be reestablished by dehydration during this period without loss of viability.[14] This facility can be reconciled with the natural history of *Artemia*. A cyst may be cast on the shores of a receding lake and if the embryo is to stand a reasonable chance of surviving it would be important that the lake is refilled to some extent. A brief shower will not provide enough water for the brine shrimp to go through its normal life cycle; a considerable advantage is likely to be gained therefore by the capacity of the organism to resuspend animation if the water available in the environment is too little to allow successful development.

The cytological events leading from this reversible phase to the nonreversible phase of active morphogenesis and differentiation are unclear. For these first periods of development no cell division occurs in the embryo, the number of nuclei remaining at about 4000 per shrimp[15] despite the large degree of internal differentiation and obvious morphological changes (e.g., the formation of the eye spot) that have accompanied the development of the gastrula to the prenauplius.[16] Development to the motile nauplius results in a 10% increase in nuclei,[15] and newly hatched nauplii can continue to develop normally for up to two days when mitosis is inhibited.[17] It is this uncoupling of cell division from differentiation and morphogenesis which has attracted interest in *Artemia*; the study of other model systems in developmental biology, such as the sea urchin, have established a dogma that development proceeds by the spatially coordinated synthesis of new cell types mediated by variable gene expression.[18]

It is likely that the absence of cell division during early development is closely linked with the capacity to resuspend metabolic activity. Clearly, any cell differentiation is likely to prove irreversible; probably only certain states are resistant to dehydration. Thus a transcriptional and translational pause in a cell adapted for the rigors of dehydration, which still can allow normal development when the "pause" is released, and where the "pause" can be reimposed when appropriate, would seem to be more suitable than a system which relied on cell division for the resumption of normal development. Whatever the origins, this remarkable separation of morphogenesis from new cell synthesis in the initial post-gastrular development of *Artemia* makes it an ideal system for assessing the role of protein synthesis in the initial developmental events.

III. ONSET OF DEVELOPMENT: REGULATION OF PROTEIN SYNTHESIS

The biochemical mechanisms underlying the changes in protein synthesis appear to be highly complex; the pertinent questions are whether the events are regulated by:

1. The physiological state of the ribosomes
2. A limiting supply of a component of the translational machinery
3. The availability of function mRNA
4. A combination of these factors

A. THE RIBOSOMES

1. Store of Ribosomes

The fact that there is a large store of 80S ribosomes devoid of mRNA in the cyst and that the onset of protein synthesis is rapid, suggests that new synthesis of ribosomes is not a factor in the control of early development. The relatively low proportion of this pool appearing in polysomes may be indicative of inhibitory molecules still continuing to have some effect after initiation of development.

2. Physiological State of the Ribosomes

Ribosomes from the cyst and early developing embryo have a lower specific activity of protein synthesis than those from late embryos.[19-21] This was thought to be due to an inhibitory protein on the large subunit. In addition, a small RNA molecule was isolated from the ribosomes by salt washes.[22-23] This molecule was an inhibitor of activity but did not account fully for the observed changes in ribosome function during development. Ribosomes from the cyst and early stages of development (up to 8 h) were able to support polypeptide release factor activity only poorly compared with those isolated from the later stages leading to the emergence of the nauplius.[7] It seemed clear from these studies that ribosomes were less active in cysts than in actively translating stages of development but that the transition was a gradual increase in specific activity rather than a dramatic change. This can be explained by several possible mechanisms such as, for example, inhibitory substances preventing the formation of active ribosomal complexes and modifications to the ribosomes themselves.

a. Inhibitor Proteins

Removal of a heat labile protein of M_r 13,000 from functionally repressed ribosomes results in increased activity.[24] This protein seems to mediate its activity at the elongation step of protein biosynthesis, antagonizing the eEF-2 mediated formation of the 80S-poly(U)-Phe-tRNA ternary complex, and may be an altered form of an elongation factor. There are parallels with the development of sea urchins where a protein associated with the unfertilized but not the fertilized egg inhibits polyphenylalanine synthesis *in vitro*.[25,26]

b. Inhibitor and Activator RNA

The inhibitor of ribosomal function isolated from salt washes by Lee-Huang et al.[22] was an oligonucleotide which inhibits the translation of both artificial and natural mRNAs *in vitro*. This RNA had a complement present at later stages of development which forms a complex with the inhibitor but its appearance has not been correlated with an onset of protein synthesis. The inhibitor RNA had a M_r of 6000 and was rich in U, whereas the so-called activator was somewhat larger (M_r 9000), rich in G, and the two RNAs had differential sensitivities to RNase A and RNase T1.

c. Modification

The difficulties of getting radiolabeled precursor molecules into the cyst makes a study of subtle changes in ribosome structure difficult. Nevertheless, Williamson[21] found that ribosomal protein patterns for ribosomes isolated at different stages of development showed changes in a diagonal series of acidic proteins. The pattern is analogous to that given by the different phosphorylation derivatives of rat liver ribosomal protein S6,[27] a protein whose phosphorylation is modulated by hormones and growth factors.

B. TRANSLATIONAL FACTORS

The requirement that the cyst be able to respond rapidly and reversibly to hydration and be capable of the high rates of protein synthesis that are required in the proper development of the organism strongly suggests that alterations to translation are the most important mechanisms of control in the cryptobiotic and developing *Artemia*. Protein synthesis factors seem to play a major part in the maintenance of dormancy.

1. Initiation Factors

There are differences of opinion as to whether or not the levels of the initiation factors vary in developing *Artemia*. Initially, it was reported that the activity of eukaryotic initiation factor 2 (eIF-2) was much reduced in ametabolic cysts when compared with active nauplii,[28] but it was subsequently shown that the total activities of eIF-2-containing fractions from dormant and active *Artemia* are similar.[29] It is likely that levels of eIF-2 do not change during development and that control of this type is not present in *Artemia*.[30] Phosphorylation of eIF-2 is associated with control of translation in heme-deficient rabbit reticulocyte lysates.[31] More recently De Herdt et al.[32] have found that eIF-2 isolated from cryptobiotic embryos is associated with an mRNP fraction active in protein synthesis, and is absent from the translationally inactive component which is associated with a small RNA and a 64,000 M_r protein.

Limited availability of any initiation factor can reduce the expression of all mRNA, by slowing the rate of the first major step in protein biosynthesis. Moreover, it can also control the expression of individual mRNAs within a total mRNA population. Not all mRNAs are equally competent for initiation and there is likely to be competition among different mRNA species. Some mRNAs may be translated more efficiently than others. This phenomenon was once attributed to the presence of message specific initiation factors,[33-34] but it is now understood that different mRNAs have initiation sites of differing affinities for the ribosome. To date, the importance of mRNA competition or mRNA specific factors in the control of protein synthesis during development in *Artemia* has not been assessed.

2. Elongation Factors

In addition to the inhibitory activities of elongation factors described above, Daum et al.[35] have shown that the amount of EF-Tu (eEF-1) present in the embryo after the beginning of development increases as judged both by two-dimensional, gel electrophoresis and by cDNA hybridization studies using a probe derived from the amino acid sequence of *Artemia* EF-Tu.[36] The *Artemia* eEF-1 is composed of three polypeptides: eEF-1α which is associated with the binding of aminoacyl tRNA to the ribosome, eEF-1β which catalyses the hydrolysis of the eEF-1α-bound GTP, and eEF-1γ that forms complexes with eEF-1β, but whose function is unknown.[37]

The eEF-1 is associated with another aspect of control in the developing nauplius. The gene that codes for eEF-1α has been cloned by Möller's group[38] and shown to be similar to bacterial EF-Tu. Analysis of the production of mRNA of eEF-1α and of the proteins of ribosomal large subunits eL12 and eL12′ demonstrated a coordinate expression. Although there is an intron splicing site common to both of these primary transcripts, it is not clear if expression of these genes is under the same control in the genome of or the expression is coordinated in some other way.[38]

About half of the elongation factor 2 (eEF-2) present in dormant *Artemia* embryos occurs in a complexed form. The presence of the inhibitory oligonucleotide prevents the formation of the 80S-poly(U)-phe-tRNA ternary complex in an *in vitro* assay.[24] During development, the amount of the oligonucleotide inhibitor declines, and there is a disappearance of complexed EF-2 and a concomitant increase in the free form.[39] This mechanism of control is similar to that reported in the muscle of chick embryos[34] and in rabbit reticulocytes.[40]

3. Termination Factors

Extracts from cysts and from late developing embryos are equally capable of mediating termination of the globin subunit from reticulocyte polysomes[41] and there is no evidence to date that the factor in the cyst is complexed to an inhibitory component. It seems to have the typical eukaryotic release-factor structure.[42] As described above, however, ribosomes from dormant cysts do not respond as well to purified release factors and their respective termination codons as do ribosomes from embryos 18 h after the resumption of development.[7] It is suggested that some factor interferes with the interaction of the release factor with the ribosome, and that this disappears as the embryo develops.

C. AVAILABILITY OF FUNCTIONAL mRNA

1. Masked RNA

The mRNA that is present in dormant cysts has a number of unusual states, all of which seem to lack template activity. There are no polysomes present in the encysted embryo[43] and the mRNA is found, instead, in the form of mRNP particles[44-46] or associated with membranes independent of the ribosomes.[47] In these forms the RNA is not available for translation. Although it forms a large store of mRNA, it is not available for the production of protein immediately upon hydration. Membrane-bound mRNA can be isolated by sucrose gradient centrifugation of cyst homogenates.[48] After detergent treatment to release the mRNP particles from the membrane, the majority of the poly(A)$^+$RNA sediment at 40S, and very little poly(A)$^+$RNA is associated with ribosomes. The proteins of the mRNP membrane-bound particles are hydrophobic as expected. In contrast, RNP particles found in the developing embryo associated with polysomes have a markedly different protein complement.[49]

2. P38 and HD40

A 38,000 M_r poly(A) + binding protein (designated P38) was isolated from cryptobiotic cysts of Van Hove et al.[50] and this comprised 60 — 70% of the protein component of nonpolysomal mRNP particles. P38 is found in at least seven forms which reflect the degree of phosphorylation of the protein, a modification that may affect the ability of the protein to bind to mRNA, or perhaps may change the specificity of the binding site. That P38 has also been shown to inhibit protein synthesis in rabbit reticulocyte lysate,[51] and to accumulate in the nucleus of *Xenopus laevis* oocytes,[52] suggests that the function of P38 is not restricted to *Artemia*. The mechanism of action of P38 is not entirely clear. Both P38 and the HD40 (helix-destabilizing protein 40) described by Szer et al.,[53] are able to disrupt double stranded poly(A) – poly(U) complexes. P38 and HD40 are identical as judged by molecular weight, immunoreactivity in double immunodiffusion assays, by comparison of tryptic peptides, by two-dimensional electrophoresis, and by the presence of the unusual amino acid dimethylarginine and the absence of cysteine from both proteins. Van Hove et al.[50] have also shown that P38 declines in abundance in the cytosol of nauplii after the onset of development, suggesting that this protein plays a role in the establishment or maintenance of the cryptobiotic state.

A further level of control is implied by the finding that the degree of phosphorylation is correlated with the ability of the protein to bind to mRNA; thus dephosphorylation of P38 reduces the capacity for mRNA binding, and so would allow the translation of that mRNA. Thoen et al.[54] have isolated a cAMP independent protein kinase from cryptobiotic *Artemia* cysts that is associated with cytoplasmic mRNP particles. This enzyme has the capacity to phosphorylate *Artemia* eIF-2α, and more importantly, P38. It is likely that the presence of this protein kinase maintains the phosphorylation of P38 and thus the capacity of P38 both to bind poly(A)$^+$RNA and to suppress its translation. The phosphorylation of eIF-2α is thought not to be physiologically significant despite the fact that phosphorylation of eIF-2 is normally associated with inhibition of translation.[54]

3. Diguanosine Nucleotides

The earliest work done on control of development was the examination of the nucleotides present in *Artemia*. This work is reviewed in Clegg and Conte.[55] A number of unusual nucleotides were discovered, G(5')pppp(5')G, usually abbreviated as Gp_4G and a number of variants where the number of phosphates differed. The Gp_4G is important as a purine source for nucleic acid synthesis, particularly after emergence of the nauplius. *Artemia* is apparently not capable of purine ring synthesis *de novo* and so are reliant on endogenous stores and the environment. Accordingly, Gp_4G has been shown to be absent from all stages apart from the embryo of *Artemia* and the ovaries of females producing cysts. It has also been shown that an enzyme (GTP:GTP guanylyltransferase) catalyzes the reversible formation of GP_4G from two molecules of GTP. Thus, stores of GP_4G are not only available to provide purine rings but are also a repository of "high energy phosphate" which is then available for use during the time when the developing embryo is metabolizing anaerobically.

4. Methylation and Capping of mRNA

In common with other eukaryotic organisms, the mRNA of *Artemia* are found to be "capped" at the 5'-terminus with one of a number of variations on 7-methylguanosine. Typically, 75% of the mRNA isolated from *Artemia* has a cap of the form m^7G5'ppp5'AmpGp and the remainder has the structure m^7G5'ppp5'GmpGp.[56] Translation of mRNA isolated from brine shrimp embryos was found to be inhibited by the inclusion in the assay of the cap analogue, 7-methylguanosine 5'-monophosphate as well as by other synthetic analogues.[57] The only other form of methylation of mRNA reported by these workers was the presence of N^6-methyladenosine at internal positions of the ribonucleotide chain. The absence of a cap structure, which can suppress the translation of mRNA, does not appear to be a major control mechanism in *Artemia* since the mRNA isolated from both dormant cysts and developing embryos contain the cap and the methylase.[57] However, methylation of the large stores of diguanosine nucleotides may provide a potent cap analogue which could play a role in the control of protein synthesis.[30]

5. Translational Control RNA

A translational inhibitor could be removed from polyA$^+$RNP particles during affinity chromatography with EDTA allowing the RNA to be translated *in vitro*.[58] It was not determined whether this RNA was the same as that isolated from ribosomes. Su[59] observed a profound decrease in the template-inactive mRNP fraction and concomitant increase in a template-active and protein-free mRNA throughout the development of the *Artemia* embryo. A small RNA was isolated from the RNP fractions which acted as translational repressor. The levels of the molecule decreased during development of the embryo coinciding with activation of the template RNA. A possible scheme is shown in Figure 1.

IV. ONSET OF DEVELOPMENT: ROLE OF TRANSCRIPTION

A. RNA POLYMERASES AND THEIR SYNTHESIS

Dormant *Artemia* embryos, in common with most other eukaryotic organisms, contain three DNA-dependent RNA polymerases; RNA polymerases I, II, and III. Of these, RNA polymerase II is the enzyme concerned with the transcription of protein-encoding genes to form a primary transcript, subsequently processed into mature mRNA. RNA polymerase I transcribes ribosomal RNA genes while RNA polymerase III produces transcripts of tRNA and other small RNAs. The amounts of the three enzymes vary independently during embryonic development[60] although all increase in specific activity. Isolation and characterization of RNA polymerases I and II showed that the amount of these enzymes in older nauplii (72 h) was somewhat less than in younger animals (36 h).[61] The activities of the DNA-dependent RNA-polymerases in the desiccated cyst are suppressed by a protein factor, the amount of which declines after hydration and the restart of metabolism.

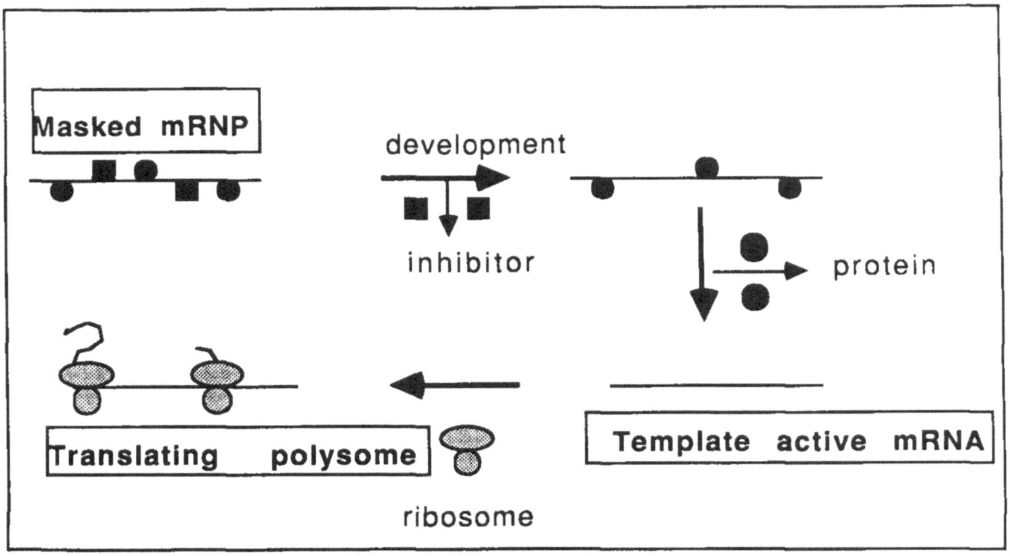

FIGURE 1. Activation of masked mRNA during development. A small RNA inhibitor (block) and protein is lost from the masked RNA during development.

Since protein synthetic activity has been observed within 30 min of rehydration of the cyst[62] and polysome formation after only 5 min[63] the role of transcriptional factors may be longer term in the control of development, rather than the primary effectors of the resumption of development.

B. TRANSCRIPTIONAL "ON" AND "OFF" SWITCHES

Immediately after hydration a burst of RNA synthesis is observed in the cyst. Shortly after the embryo hatches, the rate of RNA synthesis declines to about 20% of the maximal level seen immediately after hydration.[64] Bagshaw has called these changes in RNA synthesis the transcriptional "on" and "off" switches. The activities of the RNA polymerases was also shown to vary, increasing at the time of the "on" switch and declining with the "off" switch. A factor, known as the S protein, is present in dormant and developing embryos until about the time of emergence.[65] The S protein is found associated with a 15S complex, and has the property of stimulating the transcription of RNA by RNA polymerase II. The "on" switch may be a reflection of the stimulation of RNA polymerase conjoint with increased synthesis of the polymerase proteins. The question of whether or not the synthesis of the S protein is under some sort of translation control has not been examined.

The transcriptional "off" switch is not as straight forward as it might appear. Bagshaw et al.[66] have shown that polymerase synthesis is not simply turned off after hatching, as *de novo* synthesis, measured by the incorporation into polymerase protein of radioactivity (in the form of $NaH^{14}CO_3$), remains significant even as the total amount of polymerase declines. Other authors have suggested that the "off" switch is a result of the starvation conditions under which *Artemia* are often cultured.[67,68]

C. TRANSCRIPTIONAL REPROGRAMMING

Amaldi et al.[69] observed only a small increase in the amounts of polyA$^+$mRNA in the first 4 h of development although a significant accumulation of both polyA$^+$ and poly A$^-$RNA occurred by 16 h.[70] It has been suggested that the role of early transcriptional events is to renew the preexisting mRNA during embryogenesis.

James and Tata[71] found that the overall sequence complexity of the RNA remained relatively constant during the first 20 h of development but there were qualitative changes within the high abundancy class poly A^+mRNA which indicated some form of reprogramming of the transcriptional machinery was occurring. For example, Fisher et al.[72] have shown (in connection with the Na^+-K^+ ATPase) that the amount of mRNA in developing *Artemia* embryos rises such that at 28 h after hydration, the level of mRNA is about three times what it was at dormancy, and that the levels decline again so that by 40 h the amount of mRNA is much what it was at dormancy.

V. DEVELOPMENTAL CHANGES IN PROTEINS SYNTHESIZED

A. PATTERN OF POLYPEPTIDES SYNTHESIZED

Several investigators found little difference in the pattern of polypeptides synthesized *in vitro* by mRNA extracted from embryonic *Artemia*, although there is a change in the intracellular distribution of the mRNA. The polypeptides synthesized depends on the subcellular origin of the mRNA suggesting that there was a flow of informational RNA from cytoplasmic RNP particles into the ribosomal fraction.[69,59] These observations were consistent with translational control, there being a stage-specific gene expression controlled by the release of performed "masked" mRNA from the RNP particles to the ribosomes.

The first studies of mRNA translation from dormant cysts were *in vivo* analyses of the latent period (about 30 min) between hydration and detectable protein synthesis.[73] Nillson and Hultin[74] demonstrated that there existed in cryptobiotic *Artemia* a store of mRNA that is translationally blocked during dormancy but which is released from this block upon hydration. Most of the mRNA present in the cyst is stored in the form of mRNP particles that are translationally inactive (see Section III C.1, "Masked RNA") and different mRNA-containing fractions support different rates of protein synthesis.[75] Lake[76] has shown variation in the pattern of proteins synthesized by translation of the different mRNP fractions found in dormant cysts suggesting that different types of mRNA molecules are specifically bound by different RNA-binding proteins. Comparison of the translation products of cytoplasmic and membrane-bound mRNP particles demonstrated considerable differences showing that there were fraction-specific mRNAs,[76] although there were some common proteins.[77] This segregation of mRNA into these two classes of particles probably reflects some sort of control, possibly the different protein components of the cytosolic and membrane-bound mRNP bind to different sequences on the mRNA molecules.

What are some of the types of proteins that are synthesized during development, how do these proteins change, and what is the connection between these proteins and the requirements of the embryo?

B. EXAMPLES OF SPECIFIC PROTEINS

1. The 19S Complex

Present in the dormant cyst is a protein that forms 19S aggregates. The complex, 3.6 mg/g cysts, consist of a 27,000 M_r protein that declines in amount such that none is detectable seven days after hatching.[78] The function of the complex is not definitely known; it is not involved in translation of mRNA. A role as an osmoregulator, maintaining the level of water in the cyst above some critical level during desiccation, has been proposed. Radiation studies have shown that the water in the cysts is not "free" and it is possible that the binding of water to the protein (presumably by Gibbs-Donnan electrostatic interactions) "captures" the water.[79]

2. Ribosomal Proteins

As might be expected, despite the store of ribosomes present in *Artemia* cysts, one of the first classes of proteins translated after the resumption of development is the ribosomal proteins.[80] The mRNAs coding for ribosomal proteins are found in the cytoplasm of dormant cysts as mRNP particles, and these are found to be translated within 30 min of hydration.

It is also likely that many of the other proteins associated with the normal machinery of the cell are translated immediately after hydration, as considerable damage may have occurred to many of the cell components during the time of desiccation and the exposure to the heat that often accompanies the cryptobiotic state (irreversible damage to the cell machinery may explain the relatively low proportion of cysts which hatch, observed in many batches of *Artemia* cysts). Furthermore, there may be some components that cannot tolerate dehydration and so must be synthesized before complete cell function can be restored.

3. The Na^+,K^+ ATPase

Control of the Na^+,K^+ ATPase is exerted with subtlety. The involvement of this protein in a mechanism that serves to remove the sodium that diffuses into the shrimp's blood from the high salt environment is likely to be important. Early in development the permeability of the cyst wall and membranes is such that the embryo will need to excrete little Na^+, as this ion does not freely diffuse into the animal, but once the nauplius has hatched it may be exposed to extremely high concentrations of Na^+ with which it must be able to cope. In the nauplius there is a structure called the salt gland which is very rich in Na^+,K^+ ATPases, whereas in the adult, the salt pumps are found in the gills.

The presence of enzymatically active NA^+,K^+ ATPases was first detected in newly-hatched nauplii,[81,82] and was shown to be first synthesized about 4 h before emergence.[83] The lag between incorporation of radiolabel into the Na^+,K^+ ATPase and the detection of Na^+,K^+ ATPase activity has been attributed to the time taken for the proper association of the ATPase subunits and for their insertion into the membrane. The Na^+,K^+ ATPase of *Artemia* is composed of two subunits, α and β, where the α-subunit is present in two forms α_1 and α_2.[83] The relative amounts of α_1 and α_2 were shown to vary during development leading to the suggestion that perhaps different mRNAs directed their synthesis. However both α-subunits are chemically very similar[84] and are likely related by some form of post-translational processing. Although further work showed that the amount of mRNA for each subunit, α and β, varied during development, this variation seems not to account for the changes in the amounts of the α_1 and α_2 ATPase subunits.[72] Fisher et al.[72] have also demonstrated, by radio-immune assay, the presence of low levels of both the α and β subunits even in dormant cysts, with a significant increase of both subunits occurring at about 6 — 8 h after hydration. The amount of the Na^+,K^+ ATPase present in developing nauplii reached a plateau at about 24 — 30 h. After early development, control of the Na^+,K^+ ATPase is likely to be governed by such factors as the salt content of the brine in which the shrimp are living.

4. Artemia Hemoglobins

Among the most abundant of the proteins present in the nauplius are the hemoglobins. There are three hemoglobins, HbI, HbII, and HbIII, each having different oxygen affinities.[85] The hemoglobins are apparently present in *Artemia* as a mechanism to improve the uptake of oxygen. Highly saline solutions have a much lower oxygen tension than that found in water which makes oxygen more difficult to extract from the environment, and it is likely that the hemoglobin aids in this function. *Artemia* is an effective osmoregulator and maintains the osmolality of its hemolymph in the face of a considerable pressure of Na^+ by an active excretion of salt.[55] Such ion pumping has a considerable metabolic cost and increases the demand for oxygen. In a sense, the control of hemoglobin synthesis is similar to the control of synthesis of the Na^+,K^+ ATPases as similar underlying phenomenon are involved, although hypoxia appears to be the stimulus for synthesis of hemoglobins.[86,87]

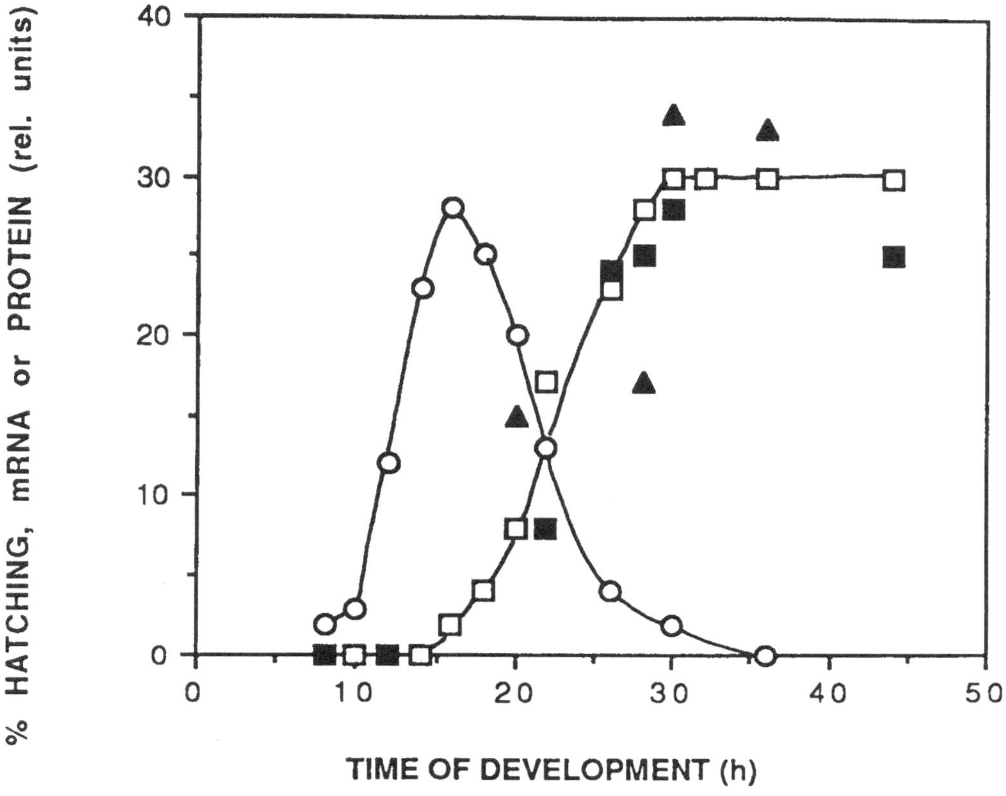

FIGURE 2. Expression of globin mRNA and appearance of hemoglobin protein occurs with hatching of the nauplius. □, Hatching; ▲, Hb mRNA; ■, Hb protein; ○, prenauplius.

In vivo pulse-labeling studies of developing *Artemia* nauplii showed a peak of incorporation of radio-labeled [^{3}H]leucine at 24 — 48 h after the resumption of metabolic activity. As much as 20% of the total protein present in the nauplius was in the form of hemoglobin. Of this, by far the greater part was HbII, some was HbIII, and a little was HbI.[88] The appearance of hemoglobin coincides with hatching of the embryo from the cyst. No hemoglobin is detectable in the dormant cyst but Manning et al.[89] were able to translate mRNA isolated from nauplii in both *Xenopus* oocytes and rabbit reticulocyte lysate and to isolate and quantitate the translated hemoglobin by immunoprecipitation with a hemoglobin specific antibody. Powell[90] using a sensitive dot blot assay found that hemoglobin was detected just at the time that the prenauplius emerged from the hatching membrane and was released as a free swimming nauplius. Hemoglobin mRNA was detected at the same time. Here the gene expression is being controlled at the level of transcription rather than translation, apparently as a result of a physiological switch at the time of hatching (see Figure 2). Whether this is the result of the induction of an oxygen-sensing mechanism, sensitive to the oxygen tension of the saline environment, which can then regulate expression of the hemoglobin genes, or whether a simple "developmental clock" is responsible, will only be resolved by further study.

VI. CONCLUSION

Considerable effort has been made to understand the events that surround the establishment of cryptobiosis and the manner in which metabolism resumes upon hydration. A summary of probable events is shown in Figure 3. An understanding of the mechanisms

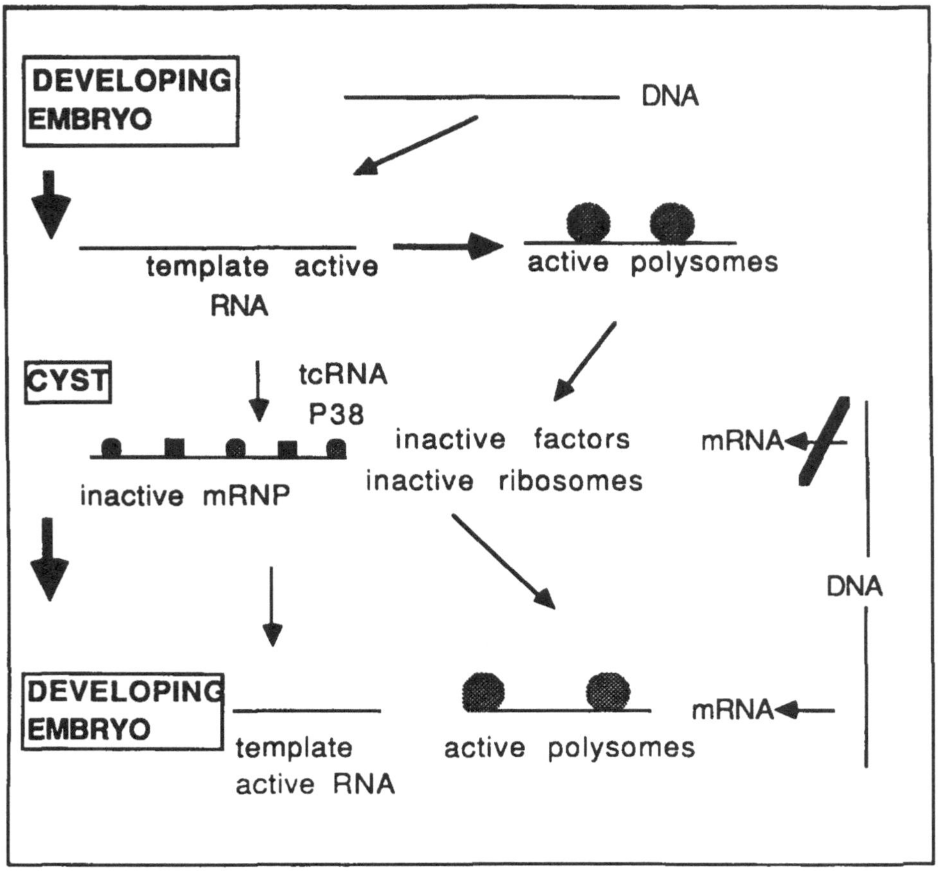

FIGURE 3. Transition from developing embryo to cyst and reactivation of development involves both translational and transcriptional mechanisms. tcRNA, translational control RNA; P38, see text.

which operate in the curious state of cryptobiosis may assist in understanding embryogenesis in general.

What has been learned of the dormancy of *Artemia*? We know that the desiccated cyst has a number of control mechanisms that suppress normal metabolism. These range from mechanisms that suppress transcription, proteins that sequester a store of mRNA in an untranslatable form, to specific inhibitors of many of the ribosomal functions necessary for proper protein synthesis. The primary event in the reestablishment of normal metabolism seems to be a release of the translational blocks leading to a flow of preexisting mRNA into active templates supporting protein synthesis, an activation of ribosomes and the general protein synthesis machinery. That this occurs very rapidly suggests that the lifting of translational control is the first key event, but transcriptional events play an important role subsequently along with translation in regulating the development of the *Artemia* embryo.

REFERENCES

1. **Needham, J. and Needham, D. M.,** On phosphorus metabolism in embryonic life. I. Invertebrate eggs, *J. Exp. Biol.*, 7, 317, 1930.
2. **Gilchrist, B. M. and Green, J.,** The pigments of *Artemia*, *Proc. R. Soc., Ser. B*, 152, 118, 1960.

3. **Anderson, E. D., Lochhead, J. H., Lochhead, M. S., and Huebner, E.,** The origin and structure of the teriary envelope in thick-shelled eggs of the brine shrimp, *Artemia, J. Ultrastruct. Res.*, 32, 497, 1970.
4. **Sorgeloos, P.,** Life history on the brine shrimp, *Artemia*, in *The Brine Shrimp Artemia*, Vol. 1, 2, and 3, Persoone, G., Sorgeloos, P., Roels, O., and Jaspers, E., Eds., Universa Press, Wetteren, Belgium, 1980, xix.
5. **Hultin, T. and Morris, J. E.,** The ribosomes of encysted embryos of *Artemia salina* during cryptobiosis and resumption of development, *Dev. Biol.*, 17, 143, 1968.
6. **MacRae, T. H., Woodley, C. L., and Wahba, A. J.,** Characterization of polypeptide chain initiation factors during *Artemia salina* development, in *Biochemistry of Artemia Development*, Bagshaw, J. C. and Warner, A. H., Eds., University Microfilms International, Ann Arbor, 1979, 2.
7. **Reddington, M. A., Williamson, D., and Tate, W. P.,** A release factor from the cyst of *Artemia salina*; characterization and activity on stage-specific ribosomes, in *Biochemistry of Artemia Development*, Bagshaw, J. C. and Warner, A. H., Eds., University Microfilms International, Ann Arbor, 1979, 42.
8. **Yablonka-Reuvini, Z. and Warner, A. H.,** Characterization of elongation factor 2 in dormant cysts and developing embryos of *Artemia salina*, in *Biochemistry of Artemia Development*, Bagshaw, J. C. and Warner, A. H., Eds., University Microfilms International, Ann Arbor, 1979, 22.
9. **Janssen, G. M. C. and Moller, W.,** Kinetic studies on the role of elongation factors 1β and 1γ in protein synthesis, *J. Biol. Chem.*, 263, 1773, 1988.
10. **Humphreys, T.,** Efficiency of translation of messenger-RNA before and after fertilization in sea urchins, Dev. Biol., 20, 435, 1969.
11. **Marcus, A. and Feeley, J.,** Ribosome activation and polysome formation *in vitro*: requirement for ATP, *Proc. Natl. Acad. Sci. U.S.A.*, 56, 1770, 1966.
12. **Jachymczyk, W. J. and Cherry, J. H.,** Studies on messenger RNA from peanut plants: *in vitro* polyribosome formation and protein synthesis, *Biochim. Biophys. Acta*, 157, 368, 1968.
13. **Thomas, G. and Gordon, J.,** Regulation of protein synthesis during the shift of quiescent animal cells into the proliferative state, *Cell. Biol. Int. Reports*, 3, 307, 1979.
14. **Morris, J. E.,** Hydration, its reversibility, and the beginning development in the brine shrimp, *Artemia salina, Comp. Biochem. Physiol.*, 39(A), 843, 1971.
15. **Olson, C. S. and Clegg, J. S.,** Nuclear numbers in encysted dormant embryos of different *Artemia salina* populations, *Experientia*, 32, 864, 1976.
16. **Benesch, R.,** Zur ontogenie und morphologie von *Artemia salina* L., *Zool. Jordbrukstek. Anat.*, 86, 307, 1969.
17. **Finamore, F. J. and Clegg, J. S.,** Biochemical aspects of morphogenesis in the brine shrimp, *Artemia salina*, in *The Cell Cycle, Gene Enzyme Interactions*, Padilla, G. M., Whitsonn, G. L., and Cameron, I. L., Eds., Academic Press, New York, 1969, 249.
18. **Hentschel, C. C. and Tata, J. R.,** The molecular embryology of the brine shrimp, *Trends Biochem. Sci.*, 1, 97, 1976.
19. **Huang, F. L. and Warner, F. L.,** Control of protein synthesis in brime shrimp embryos by repression of ribosomal activity, *Arch. Biochem. Biophys.*, 163, 716, 1974.
20. **Moens, L. and Kondo, M.,** Polysome-dependent synthesis of embryonic proteins of *Artemia salina* during cell differentiation and analysis of heme-containing protein, *Dev. Biol.*, 49, 457, 1976.
21. **Williamson, D. W.,** The control of ribosome activity during embryonic development in *Artemia salina*, M.S. thesis, University of Otago, New Zealand, 1980.
22. **Lee-Huang, S., Sierra, J. M., Naranjo, R., Filipowicz, W., and Ochoa, S.,** Eukaryotic oligonucleotides affecting messenger RNA translation, *Arch. Biochem. Biophys.*, 180, 276, 1977.
23. **Lee-Huang, S., Sierra, J. M., Filipowicz, W., and Ochoa, S.,** Translation inhibitors and activators in *Artemia salina* embryos, *Fed. Proc., Abstract 1888*, 35, 1728, 1976.
24. **Warner, A. H., Shridhar, V., and Finamore, F. J.,** Isolation and partial characterization of a protein synthesis inhibitor from brine shrimp embryos, *Can. J. Biochem.*, 5, 965, 1977.
25. **Metafora, S., Felicetti, L., and Gambino, R.,** The mechanism of protein synthesis activation after fertilization of sea urchin eggs, *Proc. Natl. Acad. Sci. U.S.A.*, 68, 600, 1971.
26. **Gambino, R., Metafora, S., Felicetti, L., and Raisman, J.,** Properties of the ribosomal salt wash from unfertilised and fertilised sea urchin eggs and its effect on natural mRNA translation, *Biochim. Biophys. Acta*, 312, 377, 1973.
27. **Wool, I. G., Collatz, E., Tsurgi, K., Lin, A., Ulbrich, N., and Lightfoot, H.,** Structure and function of eukaryote ribosomes, *Proc. Aust. Biochem. Soc.*, 10, 6, 1977.
28. **Filipowicz, W., Sierra, J. M., and Ochoa, S.,** Polypeptide chain intitiation in eukaryotes. V. Initiation factor MP in *Artemia salina* embryos, *Proc. Natl. Acad. Sci. U.S.A.*, 72, 3947, 1975.
29. **MacRae, T. H., Royehowdhury, M., Houston, K. J., Woodley, C. L., and Wahba, A. J.,** Protein synthesis in brine shrimp embryos. Dormant and developing embryos of *Artemia* contain equivalent amounts of chain initiation factor, *Eur. J. Biochem.*, 100, 67, 1979.

30. **Hultin, T. and Nilsson, M. O.,** The molecular biology of *Artemia*, in *The Brine Shrimp Artemia*, Vol. 2, *Physiology, Biochemistry, Molecular Biology*, Persoone, G., Sorgeloos, P., Roels, O. A., and Jaspers, E., Eds., Universa Press, Wetteren, Belgium, 1980, 83.
31. **Ochoa, S. and de Haro, C.,** Regulation of protein synthesis in eukaryotes, *Ann. Rev. Biochem.*, 48, 549, 1979.
32. **De Herdt, E., Piot, E., Wahba, A., and Slegers, H.,** Initiation factor eIF2 associated with non-polysomal poly(A)-containing messenger ribonucleoproteins of cryptobiotic gastrulae of *Artemia salina, Eur. J. Biochem.*, 151, 455, 1985.
33. **Wigle, D. T. and Smith, A. E.,** Specificity in initiation of protein synthesis in a fractionated mammalian cell-free system, Nat. New Biol., 242, 136, 1973.
34. **Heywood, S. M., Kennedy, D. S., and Bester, A. J.,** Separation of specific initiation factors involved in the translation of myosin and myoglobin messenger RNAs and the isolation of a new RNA involved in translation, *Proc. Natl. Acad. Sci. U.S.A.*, 71, 2428, 1974.
35. **Daum, III, H. A., Bragg, P. W., Sittman, D. B., Dholakia, J. N., Woodley, C. L., and Wahba, A. J.,** The expression of a gene for eukaryotic elongation factor Tu in *Artemia* during development. Translation of poly(A)$^+$RNA and the use of a synthetic oligonucleotide to detect the presence of eukarotic elongation factor Tu-specific mRNA, *J. Biol. Chem.*, 260, 16347, 1985.
36. **Amons, R., Pluijms, W., Roobol, K., and Moller, W.,** Sequence homology between EF-1α, the "α" chain of elongation factor 1 from *Artemia salina* and elongation factor EF-Tu from *Escherichia coli, FEBS Lett.*, 153, 37, 1983.
37. **Maessen, G. D. F., Amons, R., Zeelen, J. P., and Moller, W.,** Primary structure of elongation factor 1γ from *Artemia, FEBS Lett.*, 223, 181, 1987.
38. **Lenstra, J. A., Van Vliet, A., Arnberg, A. C., Van Hemert, F. J., and Möller, W.,** Genes coding for the elongation factor EF-1α in *Artemia, Eur. J. Biochem.*, 155, 475, 1986.
39. **Yablonka-Reuvini, Z., Fontaine, J. J., and Warner, A. H.,** Distribution of elongation factor 2 between particulate and soluble fractions of the brine shrimp *Artemia* during early development, *Can J. Biochem. Cell. Biol.*, 61, 833, 1983.
40. **Bogdanovsky, D., Herman, W., and Schapira, G.,** Presence of a new RNA species among the initiation protein factors in eukaryotes translation, *Biochem. Biophys. Res. Commun.*, 54, 25, 1973.
41. **Reddington, M. A., Fong, A. P., and Tate, W. P.,** The termination of the synthesis of proteins *in vitro* with extracts from the undeveloped cyst of *Artemia salina, Dev. Biol.*, 63, 402, 1978.
42. **Reddington, M. A. and Tate, W. P.,** A polypeptide chain release factor from the undeveloped cyst of the brine shrimp, *Artemia salina, FEBS Lett.*, 97, 335, 1979.
43. **Nilsson, M. O. and Hultin, T.,** Characteristics and intracellular distribution of messengerlike RNA in encysted embryos of *Artemia salina, Dev. Biol.*, 38, 138, 1974.
44. **Kondo, M., Moens, L., Slegers, H., and Frans, V.,** The poly(A)-containing ribonucleotide complexes in the dormant embryos of *Artemia salina, Hoppe-Seyler's Zeitsch. Physiol. Chem.*, 355, 1218, 1974.
45. **Grosfeld, H. and Littauer, U. Z.,** Cryptic form of mRNA in dormant *Artemia salina* cysts, *Biochem. Biophys. Res. Commun.*, 67, 176, 1975.
46. **Felicetti, L., Pierandrei-Amaldi, O., Noretti, S., Campioni, N., and Urbani, C.,** Intracellular distribution, sedimentation values and template activity of polyadenylic acid-containing RNA stored in *Artemia salina* cysts, *Cell Differ.*, 4, 339, 1975.
47. **Lake, M. and Hultin, T.,** Characterization of poly(A)+RNA-containing structural component directly associated with cytoplasmic membranes in dormant *Artemia* cysts, *Biochim. Biophys. Acta*, 606, 286, 1980.
48. **Hultin, T., Lake, M., Nilsson, M. O., and Nygard, O.,** Role of cytoplasmic membranes in the latency of protein synthesis in *Artemia* embryos, in *The Brine Shrimp Artemia*, Vol. 2, *Physiology, Biochemistry, Molecular Biology*, Persoone, G., Sorgeloos, P., Roels, O. A., and Jaspers, E., Universa Press, Wetteren, Belgium, 1980, 449.
49. **Lake, M. and Hultin, T.,** A comparison between free membrane-associated mRNP particles in dormant cysts and polysomal mRNP particles in nauplii of *Artemia salina* 12th Feb. Meeting, Dresden, F.R.G., Abstract 164, 1978.
50. **Van Hove, L., Thoen, C., Backhovens, H., and Slegers, H.,** The function of messenger RNA associated poly(A)-binding proteins, in *Artemia Research and Its Applications*, Vol. 2, *Physiology, Biochemistry, Molecular Biology*, Decleir, W., Moens, L., Slegers, H., Jaspers, E., and Sorgeloos, P., Eds., Universa Press, Wetteren, Belgium, 1987, 189.
51. **De Herdt, E., Thoen, C., Van Hove, L., Roggen, E., Piot, E., and Slegers, H.,** Identification and properties of the 38000-Mr poly(A)-binding protein of non-polysomal messenger ribonucleoproteins of cryptobiotic gastrulae of *Artemia salina, Eur. J. Biochem.*, 139, 155, 1984.
52. **De Herdt, E., Marbaix, G., Tencer, R., and Slegers, H.,** Subcellular distribution in *Xenopus laevis* oocytes of a microinjected poly(A)-binding protein from *Artemia salina* gastrulae, *Eur. J. Biochem.*, 132, 623, 1984.

53. **Marvil, D. K., Nowak, L., and Szer, W.,** Single-stranded nucleic acid-binding protein from *Artemia salina, J. Biol. Chem.*, 255, 6466, 1980.
54. **Thoen, C., Van Hove, L., and Slegers, H.,** Identification of the substrates of the casein kinase II associated with non-polysomal messenger ribonucleoproteins of *A. salina* cryptobiotic embryos, *Mol. Biol. Rep.*, 11, 69, 1986.
55. **Clegg, J. S. and Conte, F. P.,** A review of the cellular and developmental biology of *Artemia*, in *The Brine Shrimp Artemia*, Vol. 2, *Physiology, Biochemistry, Molecular Biology*, Persoone, G., Sorgeloos, P., Roels, O. A., and Jaspers, E., Eds., Universa Press, Wetteren, Belgium, 1980, 11.
56. **Groner, Y., Grosfeld, H., and Littauer, U. Z.,** 5'-capping structures of *Artemia salina* in RNA and translational inhibition by cap analogs, *Eur. J. Biochem.*, 71, 281, 1976.
57. **Muthukrishnan, S., Filipowcz, W., Sierra, J. M., Both, G. W., Shatkin, A. J., and Ochoa, S.,** mRNA methylation and protein synthesis in extracts from embryos of brine shrimp, *Artemia salina, J. Biol. Chem.*, 250, 9336, 1975.
58. **Slegers, H., Mettrie, R., and Kondo, M.,** Evidence for a cytoplasmic translational inhibitor RNA in *Artemia salina* gastrula embryos, *FEBS Lett.*, 80, 390, 1977.
59. **Su, S. J.,** Translational control during initiation of development of dormant embryos of *Artemia salina*, M.S. thesis, University of Otago, New Zealand, 1980.
60. **Renart, J. and Sebastian, J.,** Characterization and levels of the RNA polymerases during the embryogenesis of *Artemia salina, Cell Differ.*, 5, 97, 1976.
61. **Birndorf, H. C., D'Alessio, J., and Bagshaw, J. C.,** DNA dependent RNA polymerases from *Artemia* embryos. Characterization of polymerases I and II from nauplius larvae, *Dev. Biol.*, 45, 34, 1975.
62. **Clegg, J. S.,** Protein synthesis in the absence of cell division during the development of *Artemia salina* embryos, *Nature*, 212, 517, 1966.
63. **Golub, A. and Clegg, J. S.,** Protein synthesis in *Artemia salina* embryos: I. Study on polyribosomes, *Dev. Biol.*, 17, 644, 1968.
64. **McClean, D. K. and Warner, A. H.,** Aspects of nucleic acid metabolism during development of the brine shrimp, *Artemia salina, Dev. Biol.*, 24, 88, 1971.
65. **D'Alessio, J. M. and Bagshaw, J. C.,** DNA-dependent RNA polymerases from Artemia salina. Characterization of a protein factor from developing embryos that stimulates Artemia RNA polymerase II, *Dev. Biol.*, 70, 71, 1979.
66. **Bagshaw, J. C., Acey, R., and Helder, J. C.,** RNA polymerases and transcriptional switches in developing *Artemia*, in *The Brine Shrimp Artemia*, Vol. 2, *Physiology, Biochemistry, Molecular Biology*, Persoone, G., Sorgeloos, P., Roels, O. A., and Jaspers, E., Eds., Universa Press, Wetteren, Belgium, 1980, 278.
67. **Sebastian, J., Cruces, J., Osuna, C., and Renart, J.,** Role of the RNA polymerases in the regulation of transcription during the early development of *Artemia*, in *The Brine Shrimp Artemia*, Vol. 2, *Physiology, Biochemistry, Molecular Biology*, Persoone, G., Sorgeloos, P., Roels, O. A., and Jaspers, E., Universa Press, Wetteren, Belgium, 1980, 335.
68. **Hentschel, C. C. and Tata, J. R.,** Differential activation of free and template engaged RNA polymerase I and II during the resumption of development of dormant *Artemia* gastrulae, *Dev. Biol.*, 57, 293, 1977.
69. **Amaldi, P. P., Felicetti, L., and Campioni, N.,** Flow of informational RNA from cytoplasmic poly(A)-containing particles to polyribosomes in *Artemia salina* cysts at early stages of development, *Dev. Biol.*, 59, 49, 1977.
70. **Sierra, J. M., Filipowicz, W., and Ochoa, S.,** Messenger RNA in undeveloped and developing *Artemia salina* embryos, *Biochem. Biophys. Res. Commun.*, 69, 181, 1976.
71. **James, T. C. and Tata, J. R.,** Messenger RNA during early embryogenesis in *Artemia salina*. Altered translatability and sequence complexity, *Differentiation*, 16, 9, 1980.
72. **Fisher, J. A., Baxter-Lowe, L. A., and Hokin, L. E.,** Regulation on Na, K-ATPase biosynthesis in developing *Artemia salina, J. Biol. Chem.*, 261, 515, 1986.
73. **Golub, J. S.,** Studies on protein synthesis and polyribosomes during dormancy and development in *Artemia salina* Leach and the dissociation of oxidized 80S ribosomes by sodium deoxycholate, Ph.D. thesis, University of Miami, FL, Dissertation Abstracts, 30, 3494, 1969.
74. **Nilsson, M. O. and Hultin, T.,** Evidence of a store of informational RNA in encysted embryos of *Artemia salina, Exp. Cell Res.*, 72, 145, 1972.
75. **Nilsson, M. O. and Hultin, T.,** Characteristics and intracellular distribution of messenger-like RNA in encysted embryos of *Artemia salina, Dev. Biol.*, 38, 138, 1974.
76. **Lake, M.,** The relationship between polysomal and latent membrane-bound messenger ribonucleoprotein particles in *Artemia salina, Int. J. Biochem.*, 16, 1015, 1984.
77. **Slegers, H., De Herdt, E., and Kondo, M.,** Non-polysomal poly(A)-containing messenger ribonucleoproteins of cryptobiotic gastrulae of *Artemia salina, Eur. J. Biochem.*, 117, 111, 1981.
78. **De Herdt, E., Slegers, H., and Kondo, M.,** The 27000-protein of the 19S cytoplasmic complex of *Artemia* is one of the major RNA-binding proteins, in *The Brine Shrimp Artemia*, Vol. 2, *Physiology, Biochemistry, Molecular Biology*, Persoone, G., Sorgeloos, P., Roels, O. A., and Jaspers, E., Universa Press, Wetteren, Belgium, 1980, 395.

79. **Clegg, J. S., Szwarnowski, S., McClean, V. E. R., Sheppard, R. J., and Grant, E. H.,** Interrelationships between water and cell metabolism in *Artemia* cysts. X. Microwave dielectric studies, *Biochim. Biophys. Acta,* 721, 458, 1982.
80. **Pierandrei-Amaldi, P. and Campioni, N.,** Messenger RNA for ribosomal proteins in dormant and developing *Artemia salina* embryos, *Biochim. Biophys. Acta,* 655, 359, 1981.
81. **Ewing, R. D., Peterson, G. L., and Conte, F. P.,** Larval salt gland of *Artemia salina* nauplii. III. Localization and characterization of the sodium + potassium-activated adenosine triphosphatase, *J. Comp. Physiol.,* 88, 217, 1974.
82. **Conte, F. P., Droukas, P. C., and Ewing, R. D.,** Development of sodium regulation and *de novo* synthesis of Na + K-activated ATPase in larval brine shrimp *Artemia salina, J. Exp. Zool.,* 202, 339, 1977.
83. **Peterson, G. L. and Hokin, L. E.,** Biogenesis of the (Na + K)-activated adenosine triphosphatase during naupliar development in *Artemia,* in *Biochemistry of Artemia Development,* Bagshaw, J. C. and Warner, A. H., Eds., University Microfilms International, Ann Arbor, 1979, 120.
84. **Peterson, G. L., Churchill, L., Fisher, J. A., and Hokin, L. E.,** Structural and biosynthetic studies on the two molecular forms of the (Na^+ + K^+)-activated adenosine triphosphatase large subunit in *Artemia salina* nauplii, *J. Exp. Zool.,* 221, 295, 1982.
85. **D'Hondt, J., Moens, L., Heip, J., D'Hondt, A., and Kondo, M.,** Oxygen-binding characteristics of three extracellular hemoglobins of *Artemia salina, Biochem. J.,* 171, 705, 1978.
86. **Heip, J., Moens, L., and Kondo, M.,** Effect of concentrations of salt and oxygen on the synthesis of extracellular hemoglobins during development of *Artemia salina, Dev. Biol.,* 63, 247, 1978.
87. **Ferry, J. A., Nichols, R. C., Condon, S. J., Stubbs, J. D., and Bowen, S. T.,** Increase in net synthesis of the B-polypeptide (relative to the a-polypeptide) in hypoxia, *Biochim. Biophys. Acta,* 739, 249, 1983.
88. **Heip, J., Moens, L., Herstens, R., Wood, E. J., Heyligen, H., Van Broekhoven, A., Vrints, R., De Chaffoy, D., and Kondo, M.,** *Artemia* extracellular hemoglobins: ontogeny, structure and *in vivo* radiolabeling, in *The Brine Shrimp Artemia,* Vol. 2, *Physiology, Biochemistry, Molecular Biology,* Persoone, G., Sorgeloos, P., Roels, O. A., and Jaspers, E., Universa Press, Wetteren, Belgium, 1980, 427.
89. **Manning, A. M., Ting, G. S., Mansfield, B. C., Trotman, C. N. A., and Tate, W. P.,** The isolation and faithful translation of *Artemia* naupliar haemoglobin mRNA, *Biochem. Int.,* 12, 715, 1986.
90. **Powell, R. J.,** Characterization of the developmental appearance of hemoglobin in *Artemia,* B.S. (Hon) thesis, University of Otago, New Zealand, 1987.

Chapter 3

ENZYME ACTIVITIES THROUGH DEVELOPMENT: A SYNTHESIS OF THE ACTIVITY AND CONTROL OF THE VARIOUS ENZYMES AS THE EMBRYO MATURES

Herman Slegers

TABLE OF CONTENTS

I. INTRODUCTION

Artemia females can either release encysted embryos by the oviparous pathway or give birth to nauplius larvae by the ovoviviparous pathway. The cysts released from the ovisac are developmentally arrested and can remain in this state for long periods. These cysts are dormant and an endogenous mechanism controls the initiation of development. Dormancy can be terminated by a dehydration-rehydration cycle. These rehydrated cysts exist in a quiescent state termed anhydrobiosis and the environment governs the subsequent development.[1]

As some storage procedures result in a transition from dormancy to quiescence, the existing *Artemia* bibliography, from a biochemical point of view, does not allow a distinction between both states. Furthermore, differences in homogenization procedures and assay conditions may result in quantitative and qualitative differences in enzyme activities as well as cellular localization. With this in mind, data reported for dormant cysts have to be interpreted with caution.

Dormant cysts have a very low metabolic rate when the water content is below 0.1 to 0.3 g H_2O/g dry weight and metabolism is drastically changed after rehydration above 0.65 g H_2O/g dry weight.[2] Since the embryonic cuticle layer is impermeable to nonvolatile organic molecules and since protein synthesis is almost absent at the onset of development, all energy sources, enzymes, and substrates involved in the initial metabolic conversions should be present in dormant cysts. Therefore, enzyme regulation by activators, inhibitors, ionic strength, and pH may be of major importance in *Artemia* development.

It is beyond the scope of this review to summarize all the information on enzymes studied in *Artemia*. Clegg and Conte[1] have already collated some of these enzymes and classified them according to their presence or absence in gastrulae and their change in activity upon development. From this compilation it is evident that the majority of enzymes already exist in dormant gastrulae and that many enzymic activities are almost constant during the development of gastrulae to nauplii. The latter enzymes are mainly functional in the homeostasis of the cells and as such have no specific function in the cascade of events which precede emergence and hatching. It is the aim of this review to describe those enzymes which are or have been proposed to be involved in the initiation of development of *Artemia* dormant cysts as well as in the sequential events of preemergence and hatching.

II. THE INITIATION OF DEVELOPMENT: THE ACTIVATION OF TREHALASE

The mechanisms for the initiation of development in dormant cysts are largely unknown. One of the first phenomena which may initiate a cascade of events is the increase in the

TABLE 1
Properties of the Acid and Neutral Trehalases in *Artemia* Dormant Cysts

Property	Acid trehalase[a]	Neutral trehalase[b]	
Cellular localization	Particulate fraction	Cytosol	Cytosol
Molecular mass (M_r)	75,000	110,000 (pH 8.6)	235,000 (pH 6.3)
pI	6.2	n.m.	n.m.
pH optimum	5.6	7.0—7.2	7.6
K_m (trehalose)	4.2 m*M*	16.4 m*M*	34.4 m*M*
Effectors		K_i	K_i
ATP	No inhibition	0.91 m*M*	0.17 m*M*
cAMP	No inhibition	No inhibition	No inhibition

Note: n.m., not measured.

[a] Compiled from Ballario, P., Bergami, M., Cacace, M. G., Scala, F., and Silvestri, L., *Comp. Biochem. Physiol.*, 61, 265, 1978.
[b] Compiled from Hand, S. C. and Carpenter, J. F., *Science*, 232, 1535, 1986.

intracellular pH. The reversible transitions between metabolically inactive and active states are correlated with an increase of the pH from 6.3 to >7.9.[3,4] Initially the change from an acidic to a basic intracellular pH was proposed to be responsible for the activation of dormant cysts. However, the measurement of an internal pH >7.9 in dormant cysts collected in laboratory conditions as well as the existence in these cysts of at least two compartments with a different pH necessitates a revision of previous interpretations and indicates that termination of dormancy involves more than a simple increase in pH.[4]

The intracellular pH is one of the factors involved in the control of the activity of enzymes, ion conductivity, the cell cycle, and the proliferative capacity of cells.[5] The pH increase observed in dormant cysts certainly activates several stored enzymes. One of these enzymes, trehalase or α-α-trehalose 1-D-glycohydrolase (E.C. 3.2.1.28), is involved in the interruption of dormancy and the activation of metabolism. Trehalose, glycerol, and glycogen are the main carbohydrate reserves of dormant cysts. Trehalose represents at least 15% of the dry weight and is synthesized at the expense of glycogen during the formation of the encysted embryos.[6] Trehalose is also the major free carbohydrate of yolk platelets.[7] Preemergent development is mainly dependent on the degradation of trehalose which is rapidly metabolized into glycerol and glycogen in order to provide energy.[8-9] In anhydrobiotic organisms cleavage of trehalose is correlated with the interruption of dormancy.[4]

Acidic and neutral trehalase activities have been described for *Artemia*. An acidic trehalase activity has been purified by Ballario et al.,[10] from the 1,000 g sediment and seems to be analogous to the particle-bound trehalases purified from cell membranes, cell walls, vacuoles, and mitochondria.[10-12] In *Artemia* the cellular localization of this enzyme is unknown. Although yolk platelets contain a large amount of trehalose, the assumption that the acidic trehalase activity is confined to the latter is speculative.[1]

The properties of the acidic trehalase are listed in Table 1. In dormant cysts the pH is close to the optimum pH of this enzyme. The trehalase becomes activated upon cyst formation by internal acidification, as also observed in yeast.[13] A sugar transport function has often been proposed for particle-bound trehalase.

A neutral trehalase activity is located in the cytosol. The presence of soluble trehalase was first reported by Boulton and Huggins[14] but the enzyme was only recently characterized by Hand and Carpenter.[15] The enzyme exists in two interconvertible forms with a M_r of 110,000 and 235,000. The equilibrium between both forms is strongly dependent on pH.

The low M_r form exists at pH 8.6 and the high M_r form at pH 6.3. Both forms have approximately the same maximal activity at pH 7.4 but significantly differ in activity below this optimum pH. After termination of dormancy the increase in pH mobilizes active trehalase by dissociation of almost inactive multimers. Both forms are inhibited by ATP, the multimer being more sensitive to ATP inhibition. The properties of the neutral trehalase are compared with those of the acidic trehalase in Table 1. The pH-dependent association-dissociation of neutral trehalase and the inhibition of its activity by ATP is a mechanism which is involved in the control of energy supply. Concomitantly, metabolism is shifted from a low metabolic rate sustained by acidic trehalase to a high metabolic rate caused by the activation of neutral trehalase.

As already suggested by Clegg and Conte[1] the initial function of α-α-trehalose may not be its use as a carbohydrate reserve. A phenomenon in anhydrobiosis is the preservation of membranes necessary for the integrity of the cell. Water is an essential component for the maintenance of the morphology of bilayers which undergo phase transitions upon dehydration. The disaccharide α-α-trehalose is one of the most effective saccharides in the preservation of the morphology and activity of membranes in the absence of water. From studies on phospholipid model systems, Chandrasekhar and Gaber[16] demonstrated that α-α-trehalose mimics the interaction of water with the phospholipid monolayer. Membrane phospholipids are hydrated to 0.25 g H_2O/g dry weight. Dormant cysts resume metabolism above 0.1 to 0.3 g H_2O/g dry weight, and trehalose is mainly metabolized when sufficient water is already present to replace the membrane-associated trehalose. The imbibition of water is further facilitated by glycerol synthesized at the expense of trehalose. The accumulated glycerol decreases the osmotic pressure difference and increases water uptake.[9] The enzymes involved in glycerol synthesis are unknown.

III. DEGRADATION OF YOLK PLATELETS

A. NONENZYMIC AND ENZYMIC DEGRADATION OF YOLK PLATELETS

Yolk platelets contain all necessary metabolites for the preemergent development of the cyst[7] and at least 80% of the total cyst protein is contained in these organelles.[17] Preemergent development takes place in the absence of external nutrient supply and 25% of the yolk proteins are metabolized before hatching.[18] Yolk degradation proceeds in two phases. In the first apparently nonenzymic phase, a 50% reduction in dry weight is correlated with a pH increase from 6.3 to 8.2.[19] At pH 8.0 more than 80% of the protein is released in a lipid to protein ratio characteristic for lipovitellin. Conversely, acidification blocks this mechanism of yolk solubilization. Release of free amino acids, nucleoside triphosphates, and carbohydrates is not detected.[19]

A fusion of yolk platelets with lysosomes initiates the second phase of yolk degradation.[20] Biochemical and morphological data demonstrate the presence of lysosomal structures in yolk platelets.[18,20] Yolk structures in different states of degradation have been separated by isopycnic sucrose density gradient centrifugation and found to be associated with lysosomal hydrolytic enzymes. Several hydrolytic activities have been assayed in the yolk containing particulate fraction, e.g., β-N-acetyl galactosaminidase (EC 3.2.1.53), acid lipase (EC 3.1.1.3), β-glucosidase (EC 3.2.1.21), acid phosphatase (EC 3.1.3.2), acid phosphodiesterase (EC 3.1.4.-), DNAase (EC 3.1.4.-), RNAase (EC 3.1.4.-), and acid protease (EC 3.4.21.-).[21] These enzymes have maximal activities at different times of development. Before hatching, acid proteases and acid phosphatases are the main lysosomal activities detected in partially degraded yolk platelets.[18] Yolk degradation by lysosomes is a prerequisite for development which is arrested in the presence of lysosomotrophic agents. These basic agents increase the internal pH and lower the activity of hydrolytic enzymes involved in yolk degradation. The degradation of yolk platelets is not continuous with time. The existence

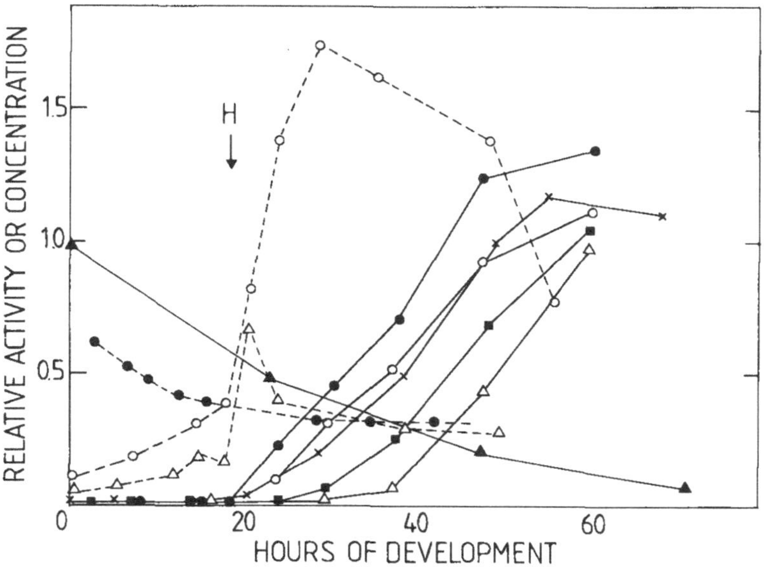

FIGURE 1. Developmental changes in proteolytic activities in relation to yolk degradation. Yolk degradation (▲—▲, 1.0 = 100% yolk) is compared with the activity of lysosomal thiol protease (0---0), (1.0 = 73 × 10^{-3} U/10^6 animals); cytosolic thiol protease (●---●, 1.0 = 530 × 10^{-3} U/g wet weight cysts); acid protease (Δ---Δ, 1.0 = 15 × 10^{-3} U/10^6 animals); trypsin-like protease (X—X, 1.0 = 563 × 10^{-3}/10^6 animals); protease A (Δ—Δ; 1.0 = 9.0 s.a.); protease B (0—0, 1.0 = 22.9 s.a.); protease C (●—●, 1.0 = 22.9 s.a.); and protease D (■—■, 1.0 = 3.8 s.a.). The use of different enzyme assays and enzyme unit definitions does not allow a quantitative comparison of the enzyme activities. For the assay procedures and definitions of the used units the reader has to consult the original publications. (The data are collated from Sillero et al.,[198] Perona and Vallejo,[21] Warner and Shridhar,[26] Marco et al.,[24] Ezquieta and Vallejo,[33] and Osuna et al.[47]) Based on the time of hatching a correction is made for the effect of the incubation temperature on cyst development. H, hatching; s.a., specific activity.

of at least five cycles of degradation has been observed. It is not known if these cycles are correlated with the pH-dependent or the enzyme-dependent degradation or a combination of both.

A lysosomal and cytosolic thiol protease,[20-22,25-26] a trypsin-like protease,[31-34] an acid protease,[24] and an alkaline protease[46-48] have been implicated in yolk degradation. The developmental changes of these enzyme activities are compared in Figure 1. Only the thiol proteases and the acid protease are active in the initial hours of cyst development and may be involved in yolk degradation before hatching.

The lysosomal acid protease involved in yolk degradation is a thiol protease.[20,22] The cathepsin B-like enzyme (EC 3.4.22.-) is a monomeric M_r 68,000 glycoprotein. The properties of the enzyme are presented in Table 2. The enzyme has a broad substrate specificity and can use protamine, bovine serum albumin, cytosolic proteins, and lipovitellin as substrates but preferentially hydrolyses an Arg-Arg peptide bond. Lipovitellin is degraded into a sequence of intermediate polypeptides similar to the *in vitro* degradation pattern.[23]

The cytosolic thiol protease has properties apparently similar to the lysosomal enzyme.[25-26] This cathepsin-B is a dimeric glycoprotein with a M_r of 55,000 and composed of subunits with M_r 31,500 and 25,900. The properties of the enzyme are compared with those of the lysosomal thiol protease (Table 2). The enzyme preferentially uses protein substrates and bovine serum albumin is a better substrate than Arg-Arg containing synthetic peptides. The

TABLE 2
Properties of Thiol Proteases Involved in Yolk Degradation

Property	Lysosomal thiol protease[a]		Cytosolic thiol protease[b]	
Molecular mass (M_r)	68,000		55,000	
Subunit composition	Monomer		Dimer M_r 31,500 M_r 25,900	
pH	pH 3.5—4.0		pH 3.5	
pH stability	pH 7.0		pH 6.0	
		K_m (μM)		K_m (μM)
Substrates	Protamine	20	Protamine	n.m.
	cb-A-A-OMe-Nap	20	BSA	12
	Pr-Ph-A-Nan	130	b-A-Nap	130
	BSA	n.m.	cb-A-A-Nap	240
	Lipovitellin	n.m.	EF2	n.m.
	Cytosolic proteins	n.m.	Hemoglobin	n.m.
			Lipovitellin	n.m.
			ts-A-me	n.a.
			b-A-nan	n.a.
		% Activity		% Activity
Effectors (using protamine[a] and *N,N*-dimethyl BSA[b] as substrates)	Antipain (1 μg/ml)	0	Antipain (0.4 μM)	50
	Leupeptin (1 μg/ml)	0	Leupeptin (0.7 m*M*)	50
	KI (1 m*M*)	0	Iodoacetate (1 m*M*)	13
	Cu^+Zn^+ (1 m*M*)	10	1, 10 phenanthroline (1 m*M*)	157
	STI (10 μg/ml)	100	STI (1 mg/ml)	98
	PMSF (10 μM)	100	PMSF (130 m*M*)	100
			D.T.T. (1 m*M*)	117

Note: The substrates are arranged according to their affinity; n.m., not measured K_m; n.a., no affinity; b-A-Nap, *N*-α-benzoyl-DL-arginine-β-naphthylamide; cb-A-A-OMe-Nap, carbobenzoxy-arginyl-arginyl-4 methoxy β-naphthylamide; cb-A-A-Nap, carbobenzoxy-arginyl-arginine-β-naphthylamide; Pr-Ph-A-Nan, D-prolyl-L-phenylalanyl-L-arginine-p-nitro anilide; STI, soybean trypsin inhibitor; ts-A-me, *p*-toluene sulfonyl-L-arginine methylester.

[a] Compiled from Perona and Vallejo;[21,22] Perona et al.[20].
[b] Compiled from Warner and Shridhar;[26] Warner.[27].

M_r, composition, stability, activity range, and substrate specificity are different for the cytosolic and lysosomal cathepsin-B. Otherwise, both enzymes have comparable properties.[27] The cytosolic enzyme displays a high specificity for *Artemia* proteins at pH 6.0 and hydrolyses elongation factor EF2 and lipovitellin. The pattern of lipovitellin hydrolysis is very similar to the degradation pattern observed *in vivo* suggesting a function in yolk degradation. Moreover, the activity of this thiol protease parallels yolk degradation (Figure 1). The cytosolic thiol protease activity decreases very rapidly up to emergence. Between emergence and hatching the activity slightly increases due to activation of enzyme previously complexed with inhibitors.[26-27] These inhibitors are only partially characterized. Besides its proposed function in yolk degradation the cytosolic thiol protease may also be functional in the cessation of protein synthesis by hydrolysis of EF2 upon acidification of the intracellular pH to 6.8 in anaerobic conditions.

Marco et al.[24] have shown that in dormant cysts an acid protease activity which is mainly localized in yolk platelets shifts to the cytosol upon development. Just before hatching the cytosolic acid protease attains its maximal activity. After hatching, the activity decreases to

70% of its maximal value. The optimum pH of this acid protease shifts from 3.5 to 6.5 upon purification. This enzyme may be a precursor of alkaline proteases and appears in the cytosol as a result of an unmasking by the acid pH and may be involved in yolk degradation (Figure 1).[24,46-47]

B. ORGANELLES, ENZYMES, AND ENZYME EFFECTORS RELEASED UPON YOLK DEGRADATION

Yolk degradation is more than a simple mechanism of energy and nutrient supply for the development of dormant cysts. Yolk degradation is an essential step in the cascade of events leading to emergence and proceeds in two phases. In the first phase yolk proteins are solubilized by the internal pH increase of cysts upon development. In the second phase enzymic degradation of yolk takes place. In this phase enzymes stored in yolk become activated. However, further research is needed to evaluate the contribution of these activated enzymes to the process of yolk degradation and to the subsequent developmental program of the dormant cyst. It is unlikely that yolk degradation is not accompanied by the release of small metabolites affecting the activity of the cytosolic enzymes. Although there is no direct evidence for this event, enzyme effectors such as polyphosphates, present in yolk platelets in high concentrations, are implicated in the onset of protein synthesis. This step in development will be discussed in Section IV. Hentschel and Tata[28] have pointed to the importance of yolk substances in the regulation of cyst development.

Before discussing the organelles and enzymes stored in yolk in more detail two considerations must be emphasized. First, the localization of organelles and enzymes is deduced from assays on crude particulate fractions prepared in conditions destructive and nondestructive to yolk platelets. To our knowledge a straightforward technique such as immune electron microscopy has not been applied. Second, the activities proposed to be stored in the yolk platelets of dormant cysts are present in very low concentrations and show a tremendous increase after hatching, strongly arguing for a function in larval development. Due to the lack of knowledge of the *in vivo* substrates of these enzymes, their function remains unknown. From these two considerations it is evident that the involvement in cyst development of enzymes stored in yolk platelets is far from resolved.

1. Mitochondrial Unmasking

Smith et al.[30] have shown that 50% of the respiratory capability is gained within one hour of development. The initial oxygen consumption is closely related to changes in the structure of mitochondria and to the increase in specific activity of cytochrome oxidase (EC 1.9.3.1), cytochrome b and c. The onset of mitochondrial activity is unrelated to yolk degradation and is due to the maturation of free mitochondria. On the contrary, the increase in mitochondrial activity between emergence and hatching is related to yolk degradation.[29] A number of mitochondrial enzymes have been located in the yolk platelet-containing particulate fraction. All these enzymes show an increase in activity around hatching coincident with yolk metabolism. From electron-microscopic studies and density-gradient centrifugation, mitochondria apparently remain associated with yolk platelets while the latter are metabolized.[29]

2. Enzymes Activated upon Yolk Degradation

Several hydrolytic enzymes localized in yolk platelets become activated upon yolk degradation, e.g., a trypsin-like protease (EC 3.4.21.-),[31-34] endoribonuclease VI (EC 3.1.4.-),[35-39] and DNase A-I (EC 3.1.4.-).[40-41]

a. Trypsin-Like Protease

A latent, trypsin-like, proteolytic activity has been detected in yolk platelets.[31] The masking of the enzyme is due to a very tight association with lipovitellin, the main storage protein of yolk platelets.[32,34] The cleavage of α-lipovitellin to α-10 by limited proteolysis is required for enzyme inhibition. After hatching, the enzyme appears in the cytosol in parallel with the degradation of yolk platelets (Figure 1). The cytosolic proteolytic activity is still masked in a precursor complex by its association with degraded lipovitellin.[31,33] The enzyme is partially active and is completely inhibited by the trypsin inhibitors, soybean trypsin inhibitor and leupeptin. The trypsin-like protease is further processed in the cytosol and activation is accompanied by a decrease in M_r to 29,000. The cytosolic thiol protease which degrades lipovitellin is probably involved in this activation process.[27] The enzyme preferentially uses protamine and α-N-benzoyl-DL-arginine-p-nitroanilide as substrates and requires divalent ions to stabilize its activity after extensive purification.[33] Chaotrophic agents (chloroquine, KSCN, and KI) reduce the molecular mass of the enzyme from 29,000 to 21,000 without loss of activity but with a decrease in divalent ion requirement. The enzyme is also inhibited by Gp_4G, the main polyphosphate present in yolk platelets.[42-43] The latter inhibition is dependent on the substrates used. In the cytosol of dormant cysts the Gp_4G concentration is 10-fold lower than in yolk platelets and decreases from 1.2 mM to 0.1 mM upon preemergent development.[44,45] The cytosolic decrease of Gp_4G not only activates the trypsin-like enzyme but also modulates its substrate specificity towards arginine-rich substrates.[43] The physiological substrates of the enzyme have not been identified but the enzyme may be involved in the hydrolysis of proteins in early development.[33] However, the optimum activity between pH 7.5—9.0 and the steep increase in activity after hatching also suggests a function in larval development (Figure 1). The activation of the trypsin-like protease resembles the induction after hatching of four alkaline proteases already present in a masked, high molecular-weight complex in early development.[46-48] In Section VII the properties of the trypsin-like protease are compared with those of the alkaline proteases (Table 7). The enzyme has similar properties to the alkaline protease B and closely parallels its activity after hatching (Figure 1).

b. Endoribonuclease VI

Dormant cysts and developing embryos contain very low levels of ribonuclease activity. A ribonuclease is induced in early larval development.[35] The enzyme is also detected in the yolk platelet containing fraction of dormant cysts.[36] The enzyme is a 5'-endoribonuclease, has an optimal activity at pH 7.5 and requires Mg^{2+} or Mn^{2+} ions. Ca^{2+}-ions are very efficient in the restoration of enzyme activity after EDTA treatment.[35] The metal-ion dependence is at variance with those reported by Quintanilla and Renat.[37] The endoribonuclease has a high specificity for poly(U) and poly(A) but does not hydrolyse poly G and DNA, tRNA, which is only slowly degraded, is a potent inhibitor of poly(U) hydrolysis, indicating a high affinity of the enzyme for tRNA. The enzyme has been purified to homogeneity and consists of a single polypeptide chain with a M_r of 38,000. Using dinucleotide substrates it is shown that the enzyme is very specific for UpN phosphodiester bonds. UpA is the best substrate of the enzyme which is named endoribonuclease VI.[37] The activity of this endoribonuclease is modulated by several purine nucleotides.[38-39] Gp_4G inhibits the enzyme at physiological concentrations (K_i = 75 μM). This inhibition explains the low ribonuclease activity found in dormant cysts and the unmasking of the enzyme upon purification. Endoribonuclease VI has been implicated in mRNA processing and may be involved in the activation of stored messenger ribonucleoproteins (mRNP) in the early stages of development.[49] One of the activation mechanisms of mRNP may be degradation of the associated poly(U)-rich inhibitor RNA.[50-51]

c. Deoxyribonuclease A-I

DNAase activity in dormant cysts is detected only with very sensitive methods.[40] The

DNAase activity as observed for proteases and ribonucleases significantly increases after hatching. Before hatching, the activity is mainly associated with yolk platelets and only 14% of the activity is recovered from the cytosol.[41] In dormant cysts the DNAase activity is exclusively DNAase A-I/A-I*. The activity has been purified and exists in two molecular forms with a M_r of 1.1×10^6 (A-I*) and 46,000 (A-I). The high M_r form continuously gives rise to the low M_r form. DNAase A-I/A-I* has an optimal activity at pH 8.8, requires Mg^{2+} ions, is slightly inhibited by Ca^{2+} ions, and preferentially degrades double stranded DNA. The enzyme becomes activated upon yolk metabolization and may be involved in the degradation of DNA associated with yolk platelets.[41]

IV. THE ONSET OF PROTEIN SYNTHESIS

Although the presence of stored mRNA has been demonstrated in several laboratories, no endogenous protein synthesis has been observed in dormant cysts.[52-57] Protein synthesis is detectable after hydration of cysts and rapidly increases upon development (Figure 2B). The onset of protein synthesis takes place in the absence of transcription and stored mRNA becomes actively translated at the onset of development. mRNA from dormant cysts is not efficiently translated in a cell-free protein synthesizing system.[50,55,58-59] Ninety percent of the translatable mRNA is located in the poly(A)-lacking RNA fraction. The poly(A)-containing mRNA is primarily inactive, with the exception of the membrane-associated poly(A)-containing mRNA.[55,58] Kallin and Hultin[60] reported that the poly(A)-containing mRNA of the membrane fraction is also associated with an inhibitor of mRNA translation. In agreement with our observations, some isolation conditions may dissociate an inhibitor from mRNA or mRNP.[50,63] The association of mRNA with proteins into mRNP does not significantly alter the translational activity of cytosolic poly(A)-lacking and poly(A)-containing mRNA.[61,62]

Several mechanisms, acting on different steps of protein synthesis, have been proposed to inhibit protein synthesis in dormant cysts, e.g., repression of stored messenger ribonucleoproteins by association with an inhibitor RNA,[49-51] phosphorylation of initiation factor eIF2 by a cAMP-dependent protein kinase,[64] inhibition of elongation by an EF2-like factor;[65,66] inhibition of mRNA-ribosome complex formation by methylated and unmethylated dinucleoside polyphosphates,[67-69] ribosomes inactive in poly(U)-directed polyphenylalanine synthesis by association of loosely bound soluble inhibitors,[70] and phosphorylation of the poly(A)-binding protein by the mRNP-associated protein kinase.[71] The disaggregation of elongation factor EF1, the dissociation of elongation factor EF2 from multiprotein complexes, control of the level of aminoacyl-tRNA by aminoacyl-tRNA hydrolases, and the redistribution of poly(A)-polymerase have been implicated in the resumption or increase in the rate of protein synthesis.[72-78] In a review on the molecular biology of *Artemia* several of these mechanisms have been collated.[79] The mechanisms of translation are discussed in more detail by Tate and Marshall (this volume); discussion in this chapter will be only of those repression or activation mechanisms which are due to the action of enzymes.

A. REGULATION OF mRNA TRANSLATION BY METHYLATED AND UNMETHYLATED POLYPHOSPHATES

Three enzymes are involved in the formation of the cap structure in mRNA. These enzymes are mRNA-guanylyltransferase (EC 2.7.7.50), mRNA-(guanine-7-) methyl transferase (EC 2.1.1.56) and mRNA-(nucleoside-$O^{2'}$-) methyl transferase (EC 2.1.1.57). The RNA-guanylyltransferase has been purified from the cytosol of dormant cysts.[80] The enzyme is composed of a single polypeptide of M_r 73,000 and transfers the GMP moiety of GTP to the 5′-diphosphate terminus of mRNA. A covalent GMP-enzyme intermediate is formed by reaction with GTP and displacement of PPi. The formed bond is a phosphoamide linkage with a ϵ-NH_2 group of a yet unidentified amino acid. The formation of the enzyme inter-

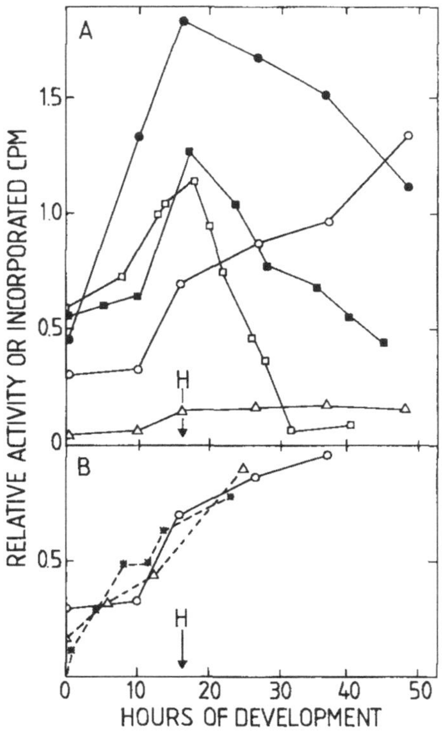

FIGURE 2. A. Developmental changes in enzyme activities involved in RNA transcription and polyadenylation. Enzyme activity of DNA-dependent RNA polymerases I (●—●), II (0—0), and III (Δ—Δ) (1.0 = 33 × 10³ U/mg protein); histone acetyltransferase using endogenous histones (□—□, 1.0 = 42.4 pmol ^{14}C-acetyl) and poly(A) polymerase (■---■ 1.0 = 130 U/ 10^5 animals). The use of different enzyme assays and enzyme unit definitions does not allow a quantitative comparison. For the assay procedures and the definitions of the used units the reader has to consult the original publications. (The data are collated from Renart and Sebastian,[155] Sastre and Sebastian,[123] Cano et al.,[199] and Cano and Pestana.[168]) Based on the time of hatching a correction is made for the effect of the incubation temperature on cyst development (H, hatching).

B. Comparison of RNA synthesis and protein synthesis upon development. The activity of DNA-dependent RNA polymerase II (0—0, 1.0 = 33 × 10³ U/mg protein) is compared with the *in vivo* ^{14}C-uracil incorporation in *de novo* synthesized RNA (*---*, 1.0 = 5200 cpm/mg cysts). Proteins synthesized by total RNA extracted at different developmental times is measured by *in vitro* incorporation of ^{35}S-methionine in a cell-free translational system (Δ---Δ, 1.0 = 1520 cpm/μg RNA). For the assay procedures and enzyme unit definition the reader has to consult the original publications. (The data are collated from Renart and Sebastian,[155] Cano et al.,[199] Susheela and Jayaraman,[128] and James and Tata.[59]) H, hatching.

mediate requires Mg^{2+}-ions and is inhibited by PPi. The transguanylylation reaction requires an RNA chain with a 5′-diphosphate end and the γ-phosphate of the primary transcript has to be removed. An RNA-5′-triphosphatase activity is associated with the enzyme. Limited trypsin digestion results in the generation of two domains with a M_r of 44,000 and 20,000 with a guanylyltransferase and RNA-5′-triphosphatase activity respectively.[80,81] Both domains can be separated by ion-exchange chromatography on CM-sephadex.

Extracts of dormant and developing cysts are able to methylate reovirus unmethylated mRNA at the 5′-end to form 5′-terminal structures of the type m^7Gp_3G and $m^7Gp_3G^m$.[67] The same observation has been made by Warner using dinucleoside polyphosphates.[45] Gp_2G and Gp_3A were efficiently methylated and Gp_3G and Gp_4G were methylated to a lesser extent. Moreover, Mizumoto and Kaziro[81] demonstrated that *Artemia* extracts can also methylate ribose-O^2-polyphosphates at the 7-position of guanosine using S-adenosylmethionine as methyl donor. These methylating enzymes have not been purified. From the foregoing it is evident that in *Artemia* dormant cysts, all enzymes which are required for the capping of mRNA are present. In dormant and developing cysts mRNA is methylated and contain the 5′-terminal $m^7Gp_3G^m$ cap structure and lack of mRNA methylation is not responsible for the absence of protein synthesis in dormant cysts.

The cytosol of dormant cysts contains high concentrations of Gp_2G, Gp_3G, Gp_3A, and Gp_4G.[45] Little is known about the function of these polyphosphates in *Artemia*. To date, only the synthesis of Gp_4G by GTP:GTP guanylyltransferase (EC 2.7.7.45) isolated from yolk platelets and its hydrolysis into GTP and GMP by the cytosolic *bis*-(5′-guanosyl) tetraphosphatase (EC 3.6.1.17) is well documented.[42,82-86] The synthesis and function of the other dinucleoside polyphosphates is still unknown. From the knowledge of Gp_4G metabolism, it is not unlikely that some of these polyphosphates are synthesized in yolk platelets and appear in the cytosol after yolk degradation. With the exception of Ap_4A, a decrease in the levels of polyphosphates in the cytosol is observed upon development and a function in the repression of protein synthesis has been suggested. Gp_2G, Gp_3G, and Gp_4G inhibit the binding of m^7 Gp_3GpC to the cap-binding protein and the extent of inhibition increases with the number of phosphates.[68] Moreover, Gp_3G and Gp_3A as well as $m^7Gp_3G^m$ and m^7Gp_3Am inhibit *in vitro* mRNA translation by preventing the formation of a ribosome-mRNA initiation complex.[68,69]

B. INITIATION OF PROTEIN SYNTHESIS

1. Regulation of eIF2 Activity

The absence of protein synthesis in dormant cysts was initially attributed to a depletion of protein synthesis initiation factors.[87-89] However, dormant and developing cysts contain comparable amounts of eIF2.[90,91] The purification and characterization of *Artemia* eIF2 has been extensively reviewed.[90-95] eIF2 is composed of three subunits of M_r 42,000 (α), 43,000 (β) and 52,000 (γ). The factor is functional in the GTP-dependent binding of Met-$tRNA_f$ and its transfer to a 40S-ribosome-initiation complex.

The cytosol of dormant cysts contains high levels of protein kinase activity. A cAMP-dependent protein kinase has been partially purified by ion-exchange chromatography on DEAE-cellulose and phosphocellulose using histones as a substrate.[64] As observed in other systems the catalytic subunit of this cAMP-dependent protein kinase is a potent inhibitor of translation and acts at the level of initiation. The enzyme requires ATP and cAMP and is inhibited by hemin which interferes with the binding of cAMP. This enzyme activates an inactive eIF2-kinase, the heme-controlled repressor (HCR) or heme-regulated inhibitor (HRI), which has not been characterized in *Artemia*. In several mammalian systems the phosphorylation of eIF2α by HCR inhibits translation by impairment of the ability of GDP displacement from eIF2. GDP in a guanine nucleotide exchange reaction by a guanine exchange

TABLE 3
Comparison of Ribosome- and mRNP-Associated Casein Kinase II

Property	mRNP-Associated casein kinase II[a]		Ribosome-associated casein kinase II[b]		Ribosome-associated eIF2α-kinase[c]	
Purification	mRNP Salt wash		Ribosome dissociation		Ribosome salt wash	
Molecular mass (M_r)	136,000		127,000		150,000—180,000	
Subunits (M_r)	36,500(α)		35,500(α)		n.m.	
	33,000(α′)		32,500(α′)			
	28,000(β)		29,000(β)			
Structure	$\alpha_2\beta_2$, $\alpha\alpha'\beta_2$		$\alpha_2\beta_2$, $\alpha\alpha'\beta_2$		n.m.	
pH	pH 7—8		pH 7—8		pH 7.5	
Ion requirement	$MgCl_2$, $MnCl_2$ 10—12 m*M*		$MgCl_2$ 5—6 m*M*		$MgCl_2$ 10 m*M*	
	Inhibited by KCl		Inhibited by $MnCl_2$, $CaCl_2$, KCl		KCl 200 m*M*	
		K_m		K_m		K_m
Substrates	ATP (casein)	7.1 μ*M*	ATP (casein)	3.1 μ*M*	ATP (casein)	13 μ*M*
	GTP (casein)	3.9 μ*M*	GTP (casein)	15.6 μ*M*	GTP (casein)	10 μ*M*
	in situ		*in situ*			
	M_r 38,000 poly(A)-binding protein	n.m.	S_6, S_8, L_{18}, L_{13}, L_9 ribosomal proteins	n.m.		
	in vitro		*in vitro*			
	eIF2(α) with ATP	n.m.	eIF2(α) with GTP	n.m.	eIF2(α) (GTP)	2 μ*M*
	eIF2(α) and eIF2(β) with GTP	n.m.	eIF2(α) and eIF2(β) with ATP	n.m.	eIF2(α) (ATP)	7 μ*M*
	Casein (ATP)	7 μ*M*	Casein	n.m.	Casein	68 μ*M*
	Poly(A)-binding proteins	n.m.	Poly(A)-binding proteins	n.m.		
		K_i		K_i		K_i
Inhibitors	Heparin	0.01 μ*M*	Heparin	0.002 μ*M*	Heparin	0.03 μ*M*
	Poly(L)-glutamic acids	10 μ*M*	Poly(L)-glutamic acids	100 μ*M*		
	Caffeine	0.7 m*M*	Caffeine	5 m*M*		
	Ap_4A	150 μ*M*			Gp_2	37 μ*M*
					Gp_5G	65 μ*M*
					Gp_4G	140 μ*M*
					Gp_3G	200 μ*M*

	2,3-Bisphosphoglycerate	5 m*M*	
	Hemin	7 μ*M*	Hemin slightly inhibitory
	Poly A · poly U	16 μg	
	Poly G · poly C	18 μg	

Note: The substrates are arranged according to their affinity; n.m., not measured K_m.

[a] Compiled from Thoen et al.;[71,104] Slegers and Aerden.[109]
[b] Compiled from Thoen et al.[98]
[c] Compiled from Woodley et al.;[94] Wahba and Woodley.[95]

factor (GEF).[96] In contrast, *Artemia* eIF2α retains its activity after phosphorylation by HCR of rabbit reticulocytes although the rate of nucleotide exchange is lowered to less than 50%.[97] A casein kinase II is isolated from the high salt wash of ribosomes and from the cytosol.[94] This kinase phosphorylates eIF2α from *Artemia*, yeast, and wheat germ but eIF2β, from rabbit reticulocytes. The purified enzyme has a M_r of 150,000 to 180,000 and utilizes ATP and GTP. The properties of this enzyme are presented in Table 3. The enzyme does not alter the GDP/GTP exchange or Met-tRNA$_f$ binding of eIF2 and diguanosine polyphosphates inhibit its activity. Dinucleoside polyphosphates present in the cytosol may be involved in the regulation of eIF2 phosphorylation *in vivo*.

In our laboratory a ribosome-associated casein kinase II has been purified with very similar properties.[98] A comparison is made in Table 3. This enzyme efficiently phosphorylates eIF2. eIF2α is mainly phosphorylated when GTP is used as phosphoryl donor and eIF2α and eIF2β are phosphorylated when ATP is used. The enzyme is tightly associated with 80S ribosomes and is only released after dissociation of the latter into 40S and 60S subunits. Ribosomal proteins which become phosphorylated *in situ* are S_6 and S_8 of the 40S subunit and L18/L13 and L9 of the 60S subunit.[99] S_6 and S_8 are proteins involved in poly(U) binding and are localized on both sides of the mRNA-binding site.[100-101] L18 is localized in the EF1 binding site.[102] In comparison with the phosphorylation of eIF2, ribosomal proteins are not significantly phosphorylated. However, these proteins are already phosphorylated *in vivo* and may be the physiological substrates of the ribosome-associated protein kinase.

The identity between the two ribosome-associated casein kinases II is not demonstrated and although both enzymes efficiently phosphorylate eIF2 their function in translational regulation remains unknown.

2. Activation of Stored mRNP

A discrete number of proteins is associated with nonpolysomal mRNP of dormant cysts. Several of these proteins have been identified, e.g., the poly(A)-binding proteins, eIF2 present on a small translatable mRNP fraction, and a protein kinase.[49] At least two mechanisms regulate the translational activity of mRNP. One mechanism is inhibition of translation by a mRNP-associated inhibitor ribonucleoprotein (iRNP) which interferes with the ternary complex formation between eIF2, GTP, and Met-tRNA$_f$. This mechanism has been reviewed elsewhere.[49] The other mechanism is phosphorylation-dephosphorylation of mRNP-proteins by an mRNP-associated protein kinase and two, cytosolic, polycation-stimulated, protein phosphatases.[103]

a. mRNP-Associated Protein Kinase

Evidence for a function of phosphorylation in the repression of mRNP comes from the observed translational inactivation of mRNP after its incubation with ATP or GTP before addition to a cell-free translational system.[71] The mRNP-associated protein kinase is purified and identified as a casein kinase II. The enzyme has a M_r of 136,000 and is composed of two subunits of M_r 36,500 (α) and 28,000 (β). The α-subunit is the catalytic subunit and is easily converted to α′ (M_r 33,000) by limited proteolysis without loss of activity. The enzyme has a $\alpha_2\beta_2$ or $\alpha\alpha'\beta_2$ structure. The properties of the enzyme are compared with those of the ribosome-associated protein kinases in Table 3. In mRNP the enzyme mainly phosphorylates the M_r 38,000 poly(A)-binding protein.[104] Phosphorylation of the poly(A)-binding protein increases its interaction with the poly(A)-sequence of mRNA and results in the repression of mRNP translation.[105] Although exogenous eIF2 is very efficiently phosphorylated by the mRNP-associated protein kinase, *in situ* phosphorylation of eIF2 has not been observed.

The initial pH increase observed after termination of dormancy does not inactivate the mRNP-associated protein kinase. The activation of stored mRNP requires the inhibition of this enzyme by specific protein kinase effectors or the action of protein phosphatases. Potential protein kinase inhibitors present in *Artemia* are dinucleoside polyphosphates and 2,3-bisphosphoglycerate.

In the initial hours of *Artemia* development an increase in Ap_4A is observed in the cytosol.[106] This dinucleoside tetraphosphate is formed in the backreaction of aminoacyl-tRNA ligases (EC 6.1.1.1-6.1.1.22) with ATP.[107] Ap_4A has been shown to bind to a protein kinase without affecting its activity. The latter enzyme is unrelated to the mRNP-associated protein kinase.[108] In contrast AP_4A strongly inhibited the mRNP-associated protein kinase.[109] The inhibition of the ribosome- and mRNP-associated casein kinase II by diguanosine polyphosphates and diadenosine polyphosphates, respectively, strongly argues for a regulation function of these compounds during mRNP translation.

In erythroid cells 2,3-bisphosphoglycerate has been identified as a regulator of casein kinase II.[110] This glycolytic intermediate is quickly formed in *Artemia* yolk platelets after termination of dormancy.[111] Since yolk degradation is one of the initial events observed after initiation of development, 2,3-bisphosphoglycerate may be released into the cytosol. 2,3-bisphosphoglycerate inhibits the mRNP-associated protein kinase but to a lesser extent than Ap_4A. Inhibition by both effectors is noncompetitive with respect to ATP.

b. Protein Phosphatases Involved in mRNP-Protein Dephosphorylation

Protein phosphatases active in the dephosphorylation of mRNP-proteins are detected in the postribosomal supernatant of dormant embryos.[112] At least five protein phosphatases were resolved by ion-exchange chromatography on DEAE-cellulose using phosphorylase-a as a substrate. Only two polycation stimulated phosphatases were active in the dephosphorylation of mRNP-proteins. The main activity, designated PCS-X, has a M_r of 225,000. The other phosphatase, designated PCS-Y, has a M_r of 346,000.[103] PCS-X has been purified to homogeneity and is composed of two subunits of M_r 75,000 (α) and 40,000 (β). The β-subunit is the catalytic subunit which is released from the enzyme after treatment with ethanol or upon freeze-thawing in the presence of 1.7% (v/v) 2-mercaptoethanol.[113] The enzyme has a $\alpha_2\beta_2$ structure. PCS-Y has a more complex composition and is composed of several proteins of M_r 100,000, 91,000, 70,000, 60,000, 56,000, and 37,000. The M_r 70,000 and 37,000 have a comparable M_r to the α and β subunits of PCS-X. Both enzymes are composed of the same core subunits and it is suggested that the additional proteins may modulate the enzymic activity.[103] Both phosphatases are active on basic proteins, e.g., protamine, histones, and polylysine, and are inhibited by NaF, ATP and PPi. These properties are consistent with a classification as polycation stimulated protein phosphatases with properties very similar to protein phosphatase 2A from rabbit skeletal muscle.[114]

Purified PCS-X and PCS-Y dephosphorylate mRNP-proteins forty times more efficiently than phosphorylase-a. Protein substrates of these protein phosphatases were identified by sequential phosphorylation-dephosphorylation and dephosphorylation-phosphorylation of mRNP-proteins.[105] Both enzymes dephosphorylate the M_r 38,000 poly(A)-binding protein. A substrate of M_r 65,000 is only detected after dephosphorylation of mRNP prior to phosphorylation by the mRNP-associated protein kinase. The latter substrate has a similar molecular mass as the protein component of the inhibitor RNP and is one of the major phosphoproteins of mRNP. It is not unlikely that phosphorylation-dephosphorylation of this protein regulates the association of iRNP to mRNP.

3. Polyadenylation of mRNP

Polyadenylation is a general step in the nuclear and cytoplasmic processing of mRNA.[115-116] The cytoplasmic polyadenylation has been proposed to function in RNA stability and in the activation of stored mRNA.[117-120] Drastic change in polyadenylation of mRNA concomitant with an increase in protein synthesis has been observed in several systems after fertilization or resumption of development.[121-122] In *Artemia* dormant cysts poly(A)-polymerase (EC 2.7.7.19) is mainly localized in the cytosol but polyadenylating activities gradually appear in the nuclear and membrane fractions upon development.[76,77,123-125] The cytosolic enzyme is not associated with cytoplasmic stored nonpolysomal mRNP in contrast to the membrane-bound enzyme which is associated with mRNP.[125-127] The latter enzyme is involved in the enrichment of poly(A)-containing RNA in the membrane fraction in the initial hours of development.[128] The total polyadenylating activity is maximal at hatching and at this stage is correlated with the processing of *de novo* synthesized RNA (Figure 2).

Poly(A)-polymerase has been purified to near homogeneity from the cytosol of dormant cysts and from the membrane fraction of four hours-old activated embryos.[124,125,129] The properties of these enzymes are compared in Table 4. Both enzymes have a different M_r and differ in the binding to DEAE-cellulose and phosphocellulose. However, the cytosolic enzyme has properties inherent for membrane proteins. It is a glycoprotein and easily forms aggregates. A decrease in M_r has been observed from M_r 70,000 to 60,000 without loss in activity but further degradation to M_r 35,000 inactivates the enzyme.[131] The membrane-associated enzyme is already present in dormant cysts and becomes activated upon development by a yet, unidentified mechanism. Phosphorylation of the enzyme has been suggested to induce activation.[125] Neither of these properties unequivocally determines the relationship between both polyadenylating activities, and transfer of the cytosolic poly(A)-polymerase to the membrane fraction upon development is speculative.

The cytosolic poly(A)-polymerase can be separated into two enzyme fractions by affinity chromatography on poly(A)-Sepharose in stringent binding conditions.[131] The unbound fraction, poly(A)-polymerase I, elicited a broad substrate specificity but did not elongate mRNP. The bound fraction, poly(A)-polymerase II, very efficiently elongated 17S mRNP.[130] The specificity of substrate polyadenylation is determined by a small RNP present in the poly(A)-polymerase II fraction. The latter RNP is composed of a small RNA with a length of 83 nucleotides and a M_r 38,000 protein resembling the poly(A)-binding protein.

The function of the cytosolic and membrane bound poly(A)-polymerases is unknown. The efficient *in vitro* polyadenylation of stored mRNP does not activate mRNP for translation in a cell-free translational system. Moreover, the absence of poly(A)-sequences in mRNP actively engaged in protein synthesis suggests that these enzymes are more likely involved in the selection of mRNA for translation. The maintenance of a poly(A) steady-state length is required for the association of poly(A)-binding proteins implicated in the control of mRNA translation.

C. THE REGULATION OF THE RATE OF PROTEIN SYNTHESIS

1. Disaggregation of Elongation Factor $EF1_H$

EF1 catalyzes the binding of aminoacyl-tRNA to the A-site of the ribosome. Two forms of elongation factor 1 are present in *Artemia*. The heavy form ($EF1_H$) decreases upon cyst development and the light form ($EF1_L$) becomes the dominant form in nauplii.[72-74,132,133] The change in the form of EF1 is correlated with the level of protein synthesis and the heavy form is proposed as a storage form of EF1.

$EF1_H$ has a M_r of 2 to 5 $\times$ 10^5 and is composed of three polypeptides of M_r 51,000 (α), 26,000 (β) and 46,000 (γ) in stoichiometric ratio.[133] $EF1_L$ only contains the α-polypeptide. Significant amounts of phosphatidylcholine and neutral lipids are associated with $EF1_H$. The heavy enzyme form is disaggregated by elastase, phospholipase, GTP, and GDP.[74] In developing cysts and naupliar extracts a protease converts $EF1_H$ to $EF1_L$.[73] This protease is inhibited by PMSF but not by the carboxypeptidase inhibitor *O*-phenanthroline. The serine protease proposed to mobilize $EF1_L$ has not been characterized. $EF1_H$ consists of an EF1α.EF1βγ complex functionally analogous to bacterial EFTu.EFTs.[134] EF1β is easily

TABLE 4
Properties of Cytosolic and Membrane Associated Poly(A)-Polymerase

Property	Cytosolic poly(A)-polymerase[a]			Cytosolic poly(A)-polymerase[b]			Membrane-associated poly(A)-polymerase[c]	
Elution DEAE-cellulose	80 m*M* KCl			80 m*M* NH_4Cl			Unbound	
Elution phospho-cellulose	—			100—160 m*M* KCl			210—260 m*M* KCl	
Elution poly(A)-sepharose	Unbound			Unbound (PAP I) 750 m*M* KCl (PAP II)			1 M KCl	
Molecular mass M_r	70,000			70,000			48,000	
pI	6.0			6.6—7.6			n.m.	
pH	8.0—8.5			8.0—8.5			8.0	
Mn^{2+}	2—4 m*M*			1 m*M*			1m*M*	
Substrates	ATP	K_m = 0.04 m*M*		ATP	K_m = 0.1 m*M*			n.m.
		%	K_m		PAPI %	PAPII %		%
	Artemia rRNA	100		$(A)_4$	100	100	tRNA	100
	Torula RNA	45	30 μ*M*	$(A)_{54}$	2.5	76.4	rRNA	93
	Poly(A)	42	10 μ*M*	rRNA	n.d.	44	Poly(U)	182
	Poly(AG)	40		Poly(A)	2.5	n.d.	Poly(A)	108
				tRNA	22.5	25		
	Cytosolic poly A^+ mRNP		0.017 μ*M*	17S mRNP	n.d.	500	—	—
	Membrane poly A^+ mRNP		0.014 μ*M*	—	—	—	Membrane mRNP	High affinity
Inhibitors	Monovalent and divalent ions			Monovalent and divalent ions				n.m.

Note: n.d., not detected; n.m., not measured.

[a] Compiled from Sastro and Sebastian.[76,123,126,129]
[b] Compiled from Roggen and Slegers;[124] Roggen et al.[130,131]
[c] Compiled from Jeyaraj et al.;[77] Raja et al.[125]

dissociated from the complex and may explain the disaggregation of the heavy form by limited proteolysis or guanosine nucleotides. The heavy form of the enzyme shows little GTP-binding activity nor is a stable complex detected between $EF1_H$, GTP, and aminoacyl-tRNA.[135] On the contrary, EF1α forms a stable GTP independent complex with aminoacyl-tRNA. This binary complex binds GTP and EF1α catalyzes the GTP dependent binding of aminoacyl-tRNA to the A-site of ribosomes concomitant with the hydrolysis of GTP.[136] EF1βγ is a guanine exchange factor active on EF1α.GDP. The function of EF1γ is still unknown but has a specific affinity for tubulin and membranes and has been proposed to bind EF1 to the cytoskeleton of the cell. The latter factor also accelerates the nucleotide exchange of EF1β suggesting that EF1γ regulates the function of EF1β.[137] A casein kinase II activity is associated or copurified with EF1βγ which phosphorylates the β-chain of the enzyme and decreases the GDP/GTP exchange reaction. The phosphorylation site in EF1β is Ser^{89} located in the N-terminal region of an acidic sequence characteristic for casein kinase II substrates. The kinase is inhibited by heparin and diphosphoglycerate.[138] The inhibition of EF1 mediated binding of aminoacyl-tRNA to ribosomes by heparin has been described by Slobin.[139] The EF1-associated protein kinase may be related to the ribosome- and mRNP-associated protein kinases (IV.B.1., IV.B.2.a).

More detailed information on the characterization and function of *Artemia* elongation factor EF1 can be found in the reviews of Hultin and Nilsson,[79] Möller et al.,[140] and Tate and Marshall (this volume).

2. Dissociation of the Multiprotein-EF2 Complex

EF2, the analogue of elongation factor G in procaryotes, catalyzes the translocation of the ribosome to 3 bases closer to the 3′-end of mRNA with ejection of the deacylated tRNA from the P-site of the ribosome. In developing cysts the amount of EF2 is regulated by at least two mechanisms. The first mechanism is very similar to the one reported for $EF1_H$ (Section IV.C.1). Analogous to EF1, several forms of EF2 are identified in dormant cysts. Beside the free form sedimenting at 5.2S (M_r 95,000) two heavy forms are detected sedimenting at 16S to 20S and at 40S to 50S.[75,141] Only a small fraction of EF2 is associated with ribosomes. A redistribution of EF2 between the multiprotein-EF2 complex and the free form takes place upon development. The former decreases by more than 30% before hatching coincident with the increase in protein synthesis. After 30 h of development, the complexed EF2 is reduced to 20% of its initial value and is proposed as a storage form of EF2. The latter enzyme form is 25 to 75 times more active in EF2-mediated ribosome translocation by protection of EF2 against proteolytic degradation.[75]

Degradation of EF2 is the second control mechanism of EF2 activity in developing cysts and nauplii. A very specific hydrolysis of EF2 is observed which degrades EF2 in two polypeptides of M_r 41,000 and 36,000. This degradation is very likely due to a thiol protease which very specifically degrades EF2 into fragments of M_r 54,000 and 40,000 and subsequently into fragments of M_r 45,000 and 38,000.[26,27]

An EF2-like inhibitor of protein synthesis has been detected in EF2 containing fractions.[66] The inhibitor coelutes with EF2 in gel filtration and affinity chromatography on GTP-agarose and is only separated from EF2 by chromatography on hydroxyapatite. The inhibitor is a GTP-binding protein, easily lost upon purification and may be EF2 nicked by the thiol protease.

D. REGULATION OF PROTEIN SYNTHESIS AT THE LEVEL OF AMINOACYLATED tRNA

In addition to ribosomes, mRNAs, and elongation factors, polypeptide chain elongation requires aminoacylated tRNA. The latter forms a ternary complex with GTP and EF1α (Section IV.C.1). This complex reacts with the ribosome where the aminoacylated tRNA attaches to the ribosomal A-site. The cellular amount of aminoacylated tRNA has been implicated in the regulation of protein synthesis.[142]

In *Artemia* several enzymes have been characterized which are involved in tRNA processing and in the developmental change of the amount of aminoacylated tRNA, e.g., ATP(CTP)tRNA nucleotidyltransferase (EC 2.7.7.21-25) and aminoacyl-tRNA hydrolase (EC 3.1.1.29).

1. tRNA Nucleotidyltransferases

Eucaryotic tRNA genes are transcribed as long precursors which are processed by endonucleases to mature tRNA species. The processing requires the removal of 5′- and 3′-end sequences and in a subset of tRNA gene transcripts the excision of intervening sequences.[143] The processed tRNA has no 3′-terminal -CCA sequence and cannot participate in protein synthesis. In the nucleus tRNA is further processed by nucleotide modifications and -CCA addition. The latter sequence has a specific function in the interaction of tRNA with amino acid-tRNA ligases (EC 6.1.1.1-6.1.1.22), in the selection mechanism for the correct amino acid and in the attachment to the ribosome.[144]

Dormant cysts contain tRNA nucleotidyltransferases which catalyze the *in vitro* incorporation of AMP or CMP into the 3′-ends of incomplete tRNA chains.[145] These enzymes require Mg^{2+}-ions and are distinct from poly(A)-polymerase. The activity increases fourfold upon development to early larval stages and is impaired in nondialyzed extracts by an unidentified low molecular-weight inhibitor.

2. Aminoacyl-tRNA Hydrolases

Several amino acid-tRNA hydrolases have been identified in *Artemia*. Dormant cysts contain a specific N-acetylphenylalanine-tRNA hydrolase which does not hydrolyze other free or N-substituted aminoacylated tRNAs.[146] The enzyme has a pH optimum at pH 7-7.5 and requires divalent ions. At least two forms of the enzyme can be separated by ion-exchange chromatography on DEAE-sephadex.[147] The form eluting at pH 7.0 and at 250 m*M* Tris-HCl corresponds to the N-acetylphenylalanine-tRNA hydrolase. The other form elutes at 500 m*M* Tris HCl, which is the position of an unspecific aminoacyl-tRNA hydrolase, but differs from the latter by its 5- to 10-fold more efficient use of N-acetylphenylalanine-tRNA as substrate. The enzyme forms have a M_r of 35,000 and 70,000. The total N-acetylphenylalanine-tRNA hydrolase activity does not change significantly upon development. A transition from the light form to the heavy form is observed.[147] The light form decreases upon development and attains 25% of its initial amount at hatching. Conversely the heavy form increases 5- to 6-fold after hatching.

An unspecific hydrolase is induced after hatching.[148] This enzyme uses free and N-substituted aminoacyl-tRNAs and differs from peptidyl-tRNA hydrolases detected in other systems which do not hydrolyze N-substituted aminoacyl-tRNAs. The enzyme has a M_r of 55,000, has a very high affinity for aminoacyl-tRNA (K_m = 1 μM), and is inhibited by deacylated tRNA. The expression of this enzyme is blocked by protein synthesis inhibitors indicating that it is synthesized *de novo* before hatching. The physiological function of these tRNA hydrolases is still unknown.

V. ENZYMES IN PREEMERGENT DEVELOPMENT: THE REGULATION OF DNA TRANSCRIPTION

Preemergent development takes place without DNA synthesis and cell division. Up to the prenauplius stage, gene replication is not required and differentiation occurs primarily by rearrangement of preexisting cells.[149] Differentiation requires the translation of stored mRNA as well as *de novo* RNA synthesis.[57,128,150,151] RNA synthesis becomes already detectable *in vivo* after 15 min of development (Figure 2B) and a considerable divergence in sequence complexity of three abundancy classes of mRNA from dormant cysts and nauplii indicate extensive DNA transcription upon development.[52-59] Some of the mRNAs synthesized *de novo* in preemergent development code for enzymes or transport proteins functional in cyst development and emergence, e.g., EF1α,[95,140] (Na + K)ATPase,[152] and the oxygen-carrier hemoglobin.[153] The enzymes involved in the regulation of DNA transcription are DNA-dependent RNA polymerases (EC 2.7.7.6), and histone acetyltransferases (EC 2.3.1.48). In Figure 2 the developmental changes of these enzymes are compared with RNA synthesis and protein synthesis.

A. DNA-DEPENDENT RNA POLYMERASES

The presence of DNA-dependent RNA polymerases in dormant cysts, developing cysts, and larvae has been shown in several laboratories.[154-156] Due to differences in homogenization, purification, and assay methods, published results are slightly at variance. However, from these studies one may conclude that in dormant cysts detectable levels of RNA polymerases I, II, and III are present (Figure 2A). RNA polymerase III is very sensitive to proteolytic degradation and is only detected in stringent conditions. The RNA polymerase activity in dormant cysts is less than 10% of the activity in nauplii. More than 90% of polymerase I and II is unbound to chromatin and present in nuclei in a free form and becomes actively engaged in transcription upon development. According to Bagshaw et al.[157] 70 to 80% of the RNA polymerase activity is located in the cytosol. The properties of DNA-dependent RNA polymerases I and II are collated in Table 5. In dormant and developing cysts the activity of RNA polymerases is controlled by cytosolic inhibitors.[159,160] One of these inhibitors, designated I_1, is active on RNA polymerase I and acts at the level of initiation. The inhibitor has been partially purified from the cytosol. The inhibitor may be a ribonucleoprotein and binds to double stranded DNA before formation of the initiation complex but is inactive on the elongation process.[160]

A stimulator of RNA polymerase II activity has been isolated from the cytosol of dormant and developing cysts.[161] The factor, designated S protein, sediments at 15S and contains a core of two polypeptides with M_r 183,000 and 67,000 in a 1:2 ratio. The protein acts on the enzyme and not on the DNA template. The amount of protein S decreases in the period of preemergence and early larval development. The mechanism of activation is unknown.

In larval development RNA polymerase activity seems to be regulated by proteases.[162] After hatching RNA polymerase I decreases to a very low level and after 40 h of development at 30°C RNA polymerase II represents 80 to 90% of the total activity[154,156,162] (Figure 2). Protease B, an enzyme induced after hatching, has been identified as the proteolytic activity involved in the *in vitro* inactivation of RNA polymerase I (Section VII).[47,48,162] RNA polymerase II has been purified from different stages of *Artemia* development by Cruces et al.,[163] and has a more complex composition than reported by Bagshaw.[164] The enzyme has a complex composition and consists of 12 subunits. The largest subunit of M_r 205,000 is converted into a M_r 172,000 polypeptide by limited proteolysis by soluble extracts of nauplii. Proteolysis is inhibited by low concentrations of soybean trypsin inhibitor. The latter inhibits proteases A, B, and C which are induced in early larval development (Section VII.B). The M_r 172,000 component is the highest M_r subunit of RNA polymerase II in larvae. The proteolytic modification of the M_r 205,000 subunit has been proposed as the regulation mechanism of RNA polymerase II activity upon larval development.[163]

B. HISTONE ACETYLTRANSFERASES

Histone acetylation has been mentioned as a mechanism for the regulation of gene expression. Although the biological function of histone acetylation is unknown, histone acetyltransferase has been implicated in the regulation of chromatin structure and function.[166]

TABLE 5
Properties of DNA-dependent RNA Polymerases

Property	RNA polymerase I (a)	RNA polymerase I (b)	RNA polymerase I (c)	RNA polymerase II (a)	RNA polymerase II (b)	RNA polymerase II (c)	RNA polymerase III (a)
DEAE							
Elution $(NH_4)_2SO_4$	150 m*M*	100—150 m*M*	100—150 m*M*	250 m*M*	200—250 m*M*	200—250 m*M*	350 m*M*
Ion requirements							
Mn^{2+} optimum	1.5—2.0 m*M*	0.5 m*M*	1 m*M*	1.5—2.0 m*M*	0.5 m*M*	1 m*M*	1.5—2.0 m*M*
Mg^{2+} optimum	5 m*M*	0.5—12 m*M*	2—3 m*M*	5 m*M*	5—12 m*M*	4—10 m*M*	5 m*M*
$(NH_4)_2SO_4$ optimum	60 m*M*	40 m*M*	60 m*M*	60—140 m*M*	120 m*M*	80—100 m*M*	Inhibition
pH optimum	n.m.	pH 7.5—8.5	n.m.	n.m.	pH 7.5—8.5	n.m.	n.m.
Molecular mass (M_p)	n.m.	n.m.	n.m.	205,000 (d)	(l)	n.m.	n.m.
				172,000 (l)	170,000		
				140,000	130,000		
				32,500	36,000		
				25,600	24,000		
				22,000			
				19,400			
				18,000			
				16,800			
				14,700			
				14,000			
				12,700			
				11,500			
Effectors							
α-Amanitin sensitivity	I_{50} = 1000 μg ml^{-1}	Insensitive	I_{50} = 1000 μg ml^{-1}	I_{50} = 0.005 μg ml	I_{50} = 0.01 μg ml^{-1}	I_{50} = 0.01 μg ml^{-1}	Insensitive
Cycloheximide		No inhibition			No inhibition		
Rifampicin		No inhibition			No inhibition		
Gp_4G			No effect			No effect	
Gp_3G			No effect			No effect	
Actinomycin D			I_{50} = 1μg ml^{-1}		I_{50} = 1 μg ml^{-1}		
Aurintricarboxylic acid			Inhibition		Stimulation optimum 400 μ*M*		

Note: n.m., not measured; d, dormant cyst; l, larvae.

Compiled from:
(a) Renart and Sebastian,[155] Osuna and Sebastian,[162] Cruces et al.[163]
(b) Birndorf et al.,[154] Bagshaw.[164]
(c) Swennen,[165] Swennen et al.[158]

In *Artemia*, nuclear endogenous acetylation increases upon cyst development but decreases to low levels after hatching[167] (Figure 2). Soluble chromatin shows the same pattern of acetylating activity. Initially the changes in activity have been reported to be due to regulation of histone acetyltransferase by a developmentally controlled inhibitor of the enzyme.[168] Additional results, however, indicate that the latter inhibitor was a small DNA sequence arising from endonucleolytic attack of DNA in nuclear extracts. Most of the inhibitor activity is suppressed by DNAase I.

In several systems the existence of cytosolic and nuclear histone acetyltransferases has been reported. Histones H_1, H_{2A}, and H_4 become irreversibly acetylated by cytosolic enzymes before termination of polypeptide chain synthesis, and in interphase, chromatin histones are reversibly acetylated by nuclear enzymes.

In *Artemia* three isoenzymes, designated histone acetyltransferase I, II, and III have been characterized.[169-171] Histone acetyltransferase I has been purified from larvae. The enzyme is chromatin bound and shows a salt-dependent dissociation from a M_r 280,000 to a M_r 170,000 form. The reaction velocity-substrate saturation dependence is sigmoid which is a characteristic for allosteric enzymes with association-dissociation characteristics. The Lys-rich histone H_1 is preferentially used as substrate and the Arg-rich histone H_4, an *in vivo* substrate of the enzyme, is poorly acetylated. In contrast, acetylation of unfractionated histones is mainly confined to the Arg-rich histones H_3 and H_4. H_4 specifically inhibits the acetylation of H_1.[169,170] Polyamines increase the rate of transfer of acetyl groups from acetyl CoA to endogenous histones. Of the polyamines, spermine, is the best effector and activates acetyltransferase I severalfold. Concomitantly, the H_3 substrate saturation curve shifts from a sigmoid to a hyperbolic shape.[172] Histone acetyltransferases II and III are purified from the cytosol of larvae.[171] Histone acetyltransferase II specifically uses histone H_4 as substrate and acetylation is enhanced by histone H_1. The change in the shape of the reaction velocity-substrate saturation curve from a sigmoid to a hyperbolic shape suggests that H_1 acts as an effector of histone acetyltransferase II. Histone acetyltransferase III is the major acetyltransferase activity in *Artemia* and changes from a balanced acetylation of histones H_1, H_3, and H_4 at low substrate concentration to a preferential acetylation of H_3 and H_4 at high concentrations of unfractionated histones. Up to now the function of these enzymes in gene expression is unknown.

VI. ENZYMES IMPLICATED IN EMERGENCE AND HATCHING

During hydration of dormant cysts and preemergent development, free glycerol synthesized at the expense of stored trehalose, produces a high internal osmotic pressure.[9] As already mentioned in Section II, the glycerol gradient established between the inside and the outside of the cyst controls the water uptake by passive diffusion. When sufficient water is imbibed, differentiation can take place by rearrangement of preexisting cells. Sato[173] suggested the participation of a "hatching" enzyme in the process of emergence. This enzyme has been proposed to degrade the outer cuticular membrane of the cyst after the first emergent stage E_1. The enzyme has never been found.

In the early development, the hatching membrane and the exoskeleton are formed between the inner cuticular membrane and the outer membrane of the embryo.[174] Chitin has been identified as a component of the exoskeleton and is degraded by at least two enzymes: chitinase (EC 3.2.1.14) and N-acetyl glucosaminidase (EC 3.2.1.30, EC 3.2.1.50).[175] Constant levels of these enzymes are already present in dormant and developing cysts. At the onset of emergence the amount steadily increases up to the naupliar stage (Figure 3). The degradation of chitin facilitates the rupture of the cyst shell by the pressure built up by glycerol in the extra embryonic space. The rupture of the shell at E_1 is accompanied by an increase in the level of sodium and a transition takes place from a hyperosmotic to a

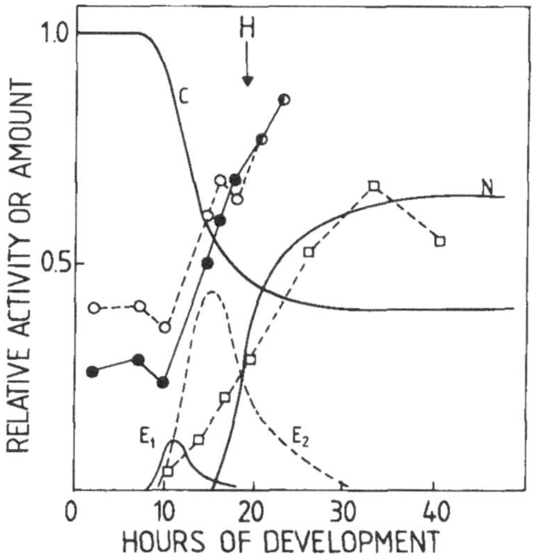

FIGURE 3. Developmental changes in enzyme activities in emergence. Comparison of the activity of chitinase (●—●, 1.0 = 0.45 $^{E}585$/mg dry weight); N-acetyl glucosaminidase (0---0, 1.0 = 2.3 U/mg dry weight); and (Na+K)-ATPase (□---□, 1.0 = 5.25 μmol P_i/h/mg protein) with the composition of a growing population of developing cysts (C), emerged embryos (E_i), fully emerged embryos (E_2), and nauplii (N). The use of different enzyme units does not allow a quantitative comparison of enzyme activities. For the assay procedures and unit definitions the reader has to consult the original publications. (The data are collated from Clegg and Conte,[1] Peterson et al.,[180] and Funke and Spindler.[175])

hypoosmotic regulation.[176,177] The morphological changes which accompany this transition have been reviewed.[1,178] A model for the bioenergetic regulation of ion transport has been presented by Conte.[178] In this model a coupling between glycolysis and ion transport is proposed, based on the observed intimate contact between plasma membranes and mitochondria, the presence in the chloride cells of the salt gland of a high number of yolk platelets, and enzymes involved in the C_4 dicarboxylic acid pathway.

The functional enzyme in the regulation of sodium levels in the emerging nauplius is the (Na+K)-activated adenosine triphosphatase (EC 3.6.1.3).[177] It has been implicated in emergence by creating a salt gradient between the embryo and the inner cuticular membrane.[179] (Na+K)-ATPase activity is undetectable in the early stages of cyst development and the enzyme activity appears in the first emergence stage E_1 although the subunits were synthesized several hours in advance (Figure 3).[180] The appearance of (Na+K)-ATPase is not correlated with the developmental changes of other membrane-associated enzymes. 5′-Nucleotidase (EC 3.1.3.5) and glucose-6-phosphatase (EC 3.1.3.9), a plasma membrane, and endoplasmic reticulum associated enzyme, respectively, show a constant level of activity in preemergent development and the activity of both enzymes strongly increases after hatching.[180]

The (Na+K)-ATPase has been purified from the membrane fraction of *Artemia* at several developing stages.[181-183] The enzyme is composed of two subunits. The larger subunit exists in two forms of M_r 100,000 (α_1) and 94,000 (α_2) and is the catalytic subunit as deduced from its potassium sensitivity and Na^+-Mg^{2+} dependent phosphorylation. The small subunit has a M_r of 40,000 (β) and is the subunit exposed to the external surface.[177,181,182] The α and β subunits are glycoproteins and contain 5.2% and 9 to 23% carbohydrate, respectively. In comparison with (Na+K)-ATPases in other systems, an $(\alpha)_2(\beta)_2$ structure is proposed for the enzyme.[184] The subunit α_2 is not a degradation product of α_1 but both forms are synthesized by different mRNAs. The relative proportions of mRNAα_1 and mRNAα_2 remain constant upon development and the decline in the amount of α_2 from a α_1/α_2 ratio of 1:1 to 4:1 suggests the existence of an additional regulatory mechanism. The *de novo* synthesis of the α subunits takes place on membrane-bound polysomes and precedes the synthesis of the β subunit which is synthesized on cytosolic polysomes.[152,185] The N-terminal sequences of the α subunits have been determined. The third position in α_1 is a Gly and is replaced by a Lys in α_2. The sequences are very hydrophobic and are quite different from the sequences reported for other vertebrate (Na+K)-ATPases.[186]

VII. ENZYMES IN LARVAL DEVELOPMENT

Cell division is almost absent in preemergent development but is initiated after hatching. The number of cells slowly increases from approximately 3,900—4,000 cells to 4,500—5,000 cells in the instar I nauplius. Up to this stage, cell differentiation and morphogenesis is still remarkably independent from cell division.[187] Upon further development, the DNA content increases approximately twofold and is correlated with considerable changes in morphogenesis.[187,188] These morphological changes have been summarized in several reviews.[1,189,190]

Concomitantly with cell division, cell differentiation, and morphogenesis, still 50 to 55% of the initial number of yolk platelets are degraded in larval development. In the subsequent sections there will be a description of the enzymes involved in DNA replication and the enzymes correlated with cell differentiation. Due to the lack of histochemical data, enzyme activities implicated in differentiation-related morphogenesis are very poorly documented in contrast to the enzymes involved in protein, RNA, and DNA synthesis.

A. THE REPLICATION OF DNA

DNA polymerases or DNA nucleotidyltransferases (EC 2.7.7.7.) are present in the particulate fractions of dormant cysts.[191,192] Ion-exchange chromatography on Cibracon blue 3-GA Matrex separated cytosolic DNA polymerase α and γ.[192] DNA polymerase α is responsible for nuclear DNA replication and γ, for mitochondrial replication. DNA polymerase β, a chromatin-bound enzyme involved in DNA repair, is not detected in dormant cysts and larvae although a cytosolic protein of M_r 48,000 cross-reacts with antichick DNA polymerase β.[193,194] The complete inhibition of DNA polymerase activity by N-ethylmaleimide, an inhibitor of DNA polymerase α and γ, further supports the absence of DNA polymerase β activity.[191,192] The absence of DNA polymerase β is proposed to be due to the extreme lability of this enzyme or its presence in an inactive form.

The properties of DNA polymerase α and β are collated in Table 6. Both enzymes are very sensitive to proteolytic degradation and the change in DNA polymerase activity upon development may not reflect their presence *in vivo*.[193] Maximal activity for DNA polymerase γ is observed in preemergent development and for DNA polymerase α at emergence. From the results of Ficoll density gradient centrifugation it is proposed that the enzymes shift from a sequestered state to a transient soluble state before they become associated with their respective replication sites.[192] In Figure 4 the enzyme activities are compared with DNA synthesis and with cytochrome oxidase activity taken as a measure for mitochondrial activity.[180,195,196] A time difference of approximately 15 to 20 h is observed between the appearance of DNA polymerase α and its use in nuclear DNA replication. If the same lag phase exists for DNA polymerase γ the estimated developmental time for mitochondrial replication is 23 to 30 h or just before the second activity maximum of cytochrome oxidase. This agrees with the observation that respiratory activity in preemergent results from the unmasking or maturation of stored mitochondria.[29,30]

TABLE 6
Properties of DNA Polymerases

Property	DNA Polymerase α[a]	DNA Polymerase γ[a]	DNA Polymerase α and γ[b]
Sedimentation	7.6S	8.5S	7—8S
M_r	110,000	118,000	n.m.
pH optimum	8.0	8.5	7.5 and 9.0
Ion requirement			
Mg^{2+}	1 m*M*	10 m*M*	8 m*M*
Mn^{2+}	n.m.	n.m.	1.6 m*M*
K^+	50 m*M*	80 m*M*	Inhibition I_{50} = 100 m*M*
Inhibitors	K_i	K_i	
NEM	1 m*M*	1.5 m*M*	Inhibition
Aphidicolin	2 m*M*	No inhibition	n.m.
ddTTP	Inhibition	No inhibition	n.m.

Note: NEM, N-ethylmaleimide; ddTTP, 2′-3′-dideoxythymidine-5′-triphosphate; n.m., not measured.

[a] Compiled from Slater and McLennan;[192] McLennan and Miller.[193]
[b] Compiled from Cacace et al.[191]

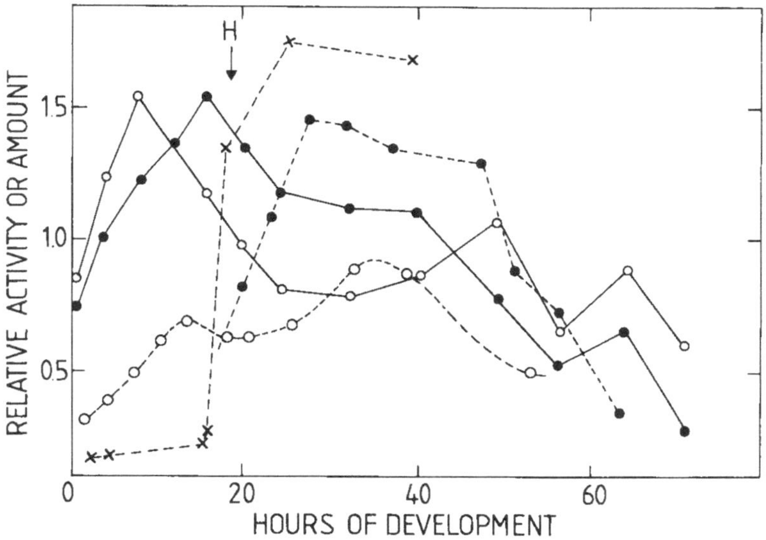

FIGURE 4. Developmental changes of DNA polymerase activities and DNA replication. The activity of DNA polymerase α (●—●, 1.0 = 5.55 U/10^6 animals) and DNA polymerase γ (0—0, 1.0 = 5.55 U/10^7 animals) is compared with nuclear DNA synthesis (●---●, 1.0 = 3630 cpm mg^{-1} DNA), cytochrome oxidase (0---0, 1.0 = 10.8 μmoles oxidized/h/mg protein) and mitochondrial activity (X---X, 1.0 = 65 μatoms of oxygen/min/10^6 animals). For the assay procedures and unit definitions the reader has to consult the original publication. (The data are collated from Slater and McLennan,[192] Bagshaw,[196] Peterson et al.,[180] and Vallejo et al.[195]) Based on the time of hatching a correction is made for the effect of the incubation temperature on cyst development (H, hatching).

In mammalian systems Ap_4A has been implicated in the regulation of cellular proliferation.[107] This dinucleoside polyphosphate is an effector of DNA replication and binds to DNA polymerase α.[197] In the development of *Artemia* the Ap_4A concentration increases 125-fold up to 3.3 μ*M* at emergence.[106] However, no binding to DNA-polymerase α could be demonstrated. The Ap_4A-binding protein is a protein kinase of M_r 93,000 with a catalytic subunit of M_r 72,000. Minor polypeptides of M_r 35,000, 40,000, and 42,000 are also present in the kinase containing fractions. The Ap_4A-binding protein kinase sediments at 4.8S well separated from DNA polymerases α and γ.[108] The enzyme phosphorylates histone 2B but its activity is not affected by Ap_4A binding. On the contrary, cyclic nucleotides did not bind to the protein kinase but significantly affected the protein kinase activity. The enzyme is not related to the casein kinases II involved in the regulation of protein synthesis (Section IV.B, Table 3).

B. ENZYMES CORRELATED WITH CELL DIFFERENTIATION

As already mentioned, few studies relate enzyme activities with cell differentiation and morphogenesis. Only the differentiation of cells involved in active transport functions has been correlated with the appearance of enzymic activities, e.g., cells of the digestive tract and of the excretory and osmoregulatory system. The enzymes induced upon differentiation of these cells are alkaline proteases (EC 3.4.21.-), (Na + K)-ATPase (EC 3.6.13) and alkaline phosphatases (EC 3.1.3.1).

The properties of (Na + K)-ATPase are described in Section VI. This enzyme is a marker for the differentiation of the salt gland.

In several systems the appearance of proteolytic activities is correlated with the differentiation of the digestive tract. In *Artemia* the alimentary tube differentiates around hatching and is fully functional at the first larval molt.[200] Several protease activities increase in the period of gut formation, e.g., proteases A, B, C, D, and the trypsin-like protease activated upon yolk degradation[31-34,46-48] (Figure 1 and Section III.B.2a). The properties of these enzymes have been collated in Table 7. Protease B and the trypsin-like protease have very similar properties but their identity has not been demonstrated. As a result of *in vitro* experiments, a function in yolk and RNA polymerase degradation has been proposed for these alkaline proteases.[47] *In vivo*, these enzymes may also be functional as digestive enzymes or in the destruction of the larval salt gland.[47,201] Samain et al.[202] studied the digestive enzymes, trypsin and amylase. These enzymes are independently synthesized upon development and are induced 4 to 7 days after hatching, well separated from the alkaline proteases A, B, C, and D. Apparently the alkaline proteases and the digestive enzyme trypsin are unrelated and immunochemical and immunohistochemical data are required for the elucidation of the localization and function of the alkaline proteases.

Alkaline phosphatase is a more suitable marker of differentiation. This enzyme is mainly bound to membranes of cells with a high metabolic activity. The function of this enzyme is unclear but seems to be implicated in active transmembrane transport or in contact digestion.[204] The work of Raineri[203] clearly shows the usefulness of this enzyme in differentiation studies.

In *Artemia* embryonic development, a low alkaline phosphatase activity is located in yolk platelets. Upon development, the enzyme first appears in differentiating endodermal cells and is associated with the basolateral plasma membranes. At emergence the enzyme activity increases rapidly in the differentiating midgut and the antennal glands concomitantly with the induction of a new isoenzyme. Different isospecies are located in the midgut, antennal, and maxillary glands. In the second-instar nauplius when the alimentary tract is completely formed, most intestinal-like alkaline phosphatase activity is associated with microvilli and participates in the breakdown and adsorption of nutrients.[203]

VIII. CONCLUSIONS

From the collated data it is obvious that many enzymes have been purified from *Artemia*. From a biochemical point of view the mechanisms of yolk degradation, protein-, RNA-, and DNA-synthesis are well documented, in comparison with the phenomena which take

TABLE 7
Properties of Alkaline Proteases

Property	Trypsin- like protease[a]		Protease A[b]		Protease B[c]		Protease C[d]		Protease D[e]	
Molecular mass M_r										
Masked form	650,000		100,000—125,000		100,000—125,000		100,000—125,000		100,000—125,000	
Purified	29,000		38,000		33,000		34,000		n.m.	
KI treatment	21,000		21,000		21,000		21,000		n.m.	
pH	pH 7.5—9.0		pH 7.5—7.8		pH 7.5—7.8		pH 7.5—7.8		n.m.	
Stability	Inactive at acidic pH		Inactive below pH 6.3		Inactive below pH 6.3		Inactive below pH 6.3		n.m.	
		K_m		K_m		K_m		K_m		K_m
Substrates	Lipovitellin	6.6 μ*M*	bc-leu-p	53 μ*M*	b-A-Nan	20 μ*M*	b-A-Nan	11 μ*M*	Casein	n.m.
	Protamine	n.m.	b-A-Nan	n.a.	Casein	n.m.	ts-A-me	n.m.	Azocasein	n.m.
	b-A-Nan	n.m.			Azocasein	n.m.	Azocasein	n.m.		
	Polylysine	n.m.			Volk proteins	n.m.	RNA polymerases	n.m.		
	BSA	n.m.			bc-leu-p	n.a.	Casein	n.a.		
	Histones	n.m.					bc-leu-p	n.a.		
	% Activity									
Effectors (Using protamine[a], bc-leu-p[b] and b-A-Nan[c,d] as substrates)	$MgCl_2$, $CaCl_2$ (10 m*M*)	100%	No divalent ions required		No divalent ions required		No divalent ions required			
	EDTA (5 m*M*)	9%								
	KCl, NaCl (10 m*M*)	10%								
	Mn^{2+}, Ca^{2+}, Zn^{2+}, Cu^{2+}, Fe^{2+}	Inhibition								
				K_i		K_i		K_i		
	STI (100 μg)	4%	STI	0.05 μ*M*	STI	0.02 μ*M*	STI	0.013 μ*M*	STI	n.i.
			OTI	0.096 μ*M*	OTI	0.02 μ*M*	OTI	0.046 μ*M*	OTI	n.i.
			TLCK	n.i.	TLCK	10 μ*M*	TLCK	80 μ*M*	TLCK	n.i.
			PMSF	1 μ*M*	PMSF	2 m*M*	PMSF	n.i.	PMSF	n.i.
	Leupeptin (10 μg)	10%								
	Lipovitellin	38%								
	Gp_4	G (K_i = 0.3 μ*M*)								
	Ap_4A	n.i.								

TABLE 7 (continued)
Properties of Alkaline Proteases

Note: The substrates are arranged according to their affinity; n.m., not measured K_m; n.a., no affinity; n.i., no inhibiton. Abbreviations: BSA, bovine serum albumin; b-A-Nan, α-*N*-benzoyl-DL-arginine *p*-nitroanilide; bc-leu-p, *N*-carbobenzoxy-L-leucine *p*-nitrophenyl ester; OTI, ovomucoid trypsin inhibitor; STI, soybean trypsin inhibitor; ts-A-Me, *p*-toluene-sulfonyl-L-arginine methyl ester.

Compiled from:

[a] Ezquieta and Vallejo.[31-34,43]

[b,c,d,e] Garesse et al.,[46] Osuna et al.,[47] Olalla et al.[48]

place upon differentiation. The main deficiency in several studies is the lack of an unequivocal demonstration of the functions of these enzymes which are mainly deduced from *in vitro* experiments or which are still unknown. Furthermore, straightforward identification of apparently similar enzymes by immunochemical methods has not been applied.

In the absence of enzyme synthesis and proteolytic degradation, enzyme activity is the result of the action of inhibitors and activators. Enzyme effectors in *Artemia* development are poorly characterized and warrant further research. Several unanswered questions may be solved if more straightforward techniques such as immunohistochemical localization, mRNA localization by *in situ* hybridization with cDNA probes, and cell culturing are adopted in future research.

REFERENCES

1. **Clegg, J. S. and Conte, F. P.,** A review of the cellular and developmental biology of *Artemia*, in *The Brine Shrimp Artemia*, Vol. 2, Persoone, G., Sorgeloos, P., Roels, O., and Jaspers, E., Eds., Universa Press, Wetteren, Belgium, 1980, 11.
2. **Clegg, J. S.,** Hydration dependent metabolic transitions and the state of cellular water in *Artemia* cysts, in *Dry Biological Systems*, Crowe, J. and Clegg, J. S., Eds., Academic Press, New York, 1978, 117.
3. **Busa, W. B. and Crowe, J. H.,** Intracellular pH regulates transitions between dormancy and development of brine shrimp (*Artemia salina*) embryos, *Science*, 221, 366, 1983.
4. **Crowe, J. H., Crowe, L. M., Drinkwater, L., and Busa, W. B.,** Intracellular pH and anhydrobiosis in *Artemia* cysts, in *Artemia Research and Its Applications*, Vol. 2, Decleir, W., Moens, L., Slegers, H., Jaspers, E., and Sorgeloos, P., Eds., Universa Press, Wetteren, Belgium, 1987, 19.
5. **Madshus, I. H.,** Regulation of intracellular pH in eukaryotic cells, *Biochem. J.*, 250, 1, 1988.
6. **Clegg, J. S.,** The origin of trehalase and its significance during the formation of encysted dormant embryos of *Artemia salina*, *Comp. Biochem. Physiol.*, 14, 135, 1965.
7. **Warner, A. H., Puodziukas, J. G., and Finamore, F. J.,** Yolk platelets in brine shrimp embryos. Site of biosynthesis and storage of diguanosine nucleotides, *Exp. Cell Res.*, 70, 365, 1972.
8. **Emerson, D. N.,** The metabolism of hatching embryos of the brine shrimp, *Artemia salina*, *Proc. Natl. Acad. Sci. U.S.A.*, 42, 131, 1963.
9. **Clegg, J. S.,** The control of emergence and metabolism by external osmotic pressure and the role of the free glycerol in developing cysts of *Artemia salina*, *J. Exp. Biol.*, 41, 879, 1964.
10. **Ballario, P., Bergami, M., Cacace, M. G., Scala, F., and Silvestri, L.,** α,α-trehalase from the brine shrimp *Artemia salina*. Purification and properties, *Comp. Biochem. Physiol.*, 61, 265, 1978.
11. **Chen, C. C., Guo, W. J., and Isselbacher, K. J.,** Rat intestinal trehalase. Studies of the active site, *Biochem. J.*, 247, 715, 1987.
12. **Londesborough, J. and Varimo, K.,** Characterization of two trehalases in baker's yeast, *Biochem. J.*, 219, 511, 1984.
13. **Valle, E., Bergillos, L., Gascon, S., Parra, F., and Ramos, S.,** Trehalase activation in yeasts is mediated by an internal acidification, *Eur. J. Biochem.*, 154, 247, 1986.
14. **Boulton, A. P. and Huggins, A. K.,** Glycolytic activity in crustaceans, *Comp. Biochem. Physiol.*, 33, 491, 1970.
15. **Hand, S. C. and Carpenter, J. F.,** pH-Induced metabolic transitions in *Artemia* embryos mediated by a novel hysteretic trehalase, *Science*, 232, 1535, 1986.
16. **Chandrasekhar, I. and Gaber, B. P.,** Stabilization of the bio-membrane by small molecules: Interaction of trehalaose with the phospholipid bilayer, *J. Biomol. Struct. Dyn.*, 5, 1163, 1988.
17. **Vallejo, C. G., Perona, R., Garesse, R., and Marco, R.,** The stability of the yolk granules of *Artemia*. An improved method for their isolation and study, *Cell Diff.*, 10, 343, 1981.
18. **Perona, R., Bes, J. C., and Vallejo, C. G.,** Degradation of yolk in the brine shrimp *Artemia*. Biochemical and morphological studies on the involvement of the lysosomal system, *Biol. Cell*, 63, 361, 1988.
19. **Utterback, P. J. and Hand, S. C.,** Yolk platelet degradation in preemergence *Artemia* embryos: response to protons *in vivo* and *in vitro*, *Am. J. Physiol.*, 252, R774, 1987.
20. **Perona, R., Ezquieta, B., and Vallejo, C. G.,** The degradation of yolk in *Artemia*, in *Artemia Research and Its Applications*, Vol. 2, Decleir, W., Moens, L., Slegers, H., Jaspers, E., and Sorgeloos, P., Eds., Universa Press, Wetteren, Belgium, 1987, 105.

21. **Perona, R. and Vallejo, C. G.,** Acid hydrolases during *Artemia* development: a role in yolk degradation, *Comp. Biochem. Physiol.*, 81(B), 993, 1985.
22. **Perona, R. and Vallejo, C. G.,** The lysosomal proteinase of *Artemia*. Purification and characterization, *Eur. J. Biochem.*, 124, 357, 1982.
23. **De Chaffoy de Courcelles, D. and Kondo, M.,** Lipovitellin from the crustacean, *Artemia salina*. Biochemical analysis of lipovitellin complex from the yolk granules, *J. Biol. Chem.*, 255, 6727, 1980.
24. **Marco, R., Vallejo, C. G., Garesse, R., and Perona, R.,** A hypothesis on the activation process of proteolytic activities during *Artemia* early development, in *The Brine Shrimp Artemia,* Vol. 2, Persoone, G., Sorgeloos, P., Roels, O., and Jaspers, E., Eds., Universa Press, Wetteren, Belgium, 1980, 315.
25. **Nagainis, P. A. and Warner, A. H.,** Evidence for the presence of an acidic protease and protease inhibitors in dormant embryos of *Artemia salina, Dev. Biol.*, 68, 259, 1979.
26. **Warner, A. H. and Shridhar, V.,** Purification and characterization of a cytosol protease from dormant cysts of the brine shrimp *Artemia, J. Biol. Chem.*, 260, 7008, 1985.
27. **Warner, A. H.,** The role of proteases and their control in *Artemia* development, in *Artemia Research and Its Applications,* Vol. 2, Decleir, W., Moens, L., Slegers, H., Jaspers, E., and Sorgeloos, P., Eds., Universa Press, Wetteren, Belgium, 1987, 203.
28. **Hentschel, C. C. and Tata, J. R.,** The molecular embryology of the brine shrimp, *Trends Biochem. Sci.*, 1, 97, 1976.
29. **Marco, R., Garesse, R., and Vallejo, C. G.,** Mitochondrial unmasking and yolk platelets metabolization during early development in *Artemia*, in *The Brine Shrimp Artemia,* Vol. 2, Persoone, G., Sorgeloos, P., Roels, O., and Jaspers, E., Eds., Universa Press, Wetteren, Belgium, 1980, 481.
30. **Schmitt, H., Grossfeld, H., and Littauer, U. Z.,** Mitochondrial biogenesis during differentiation of *Artemia salina* cysts, *J. Cell Biol.*, 58, 643, 1973.
31. **Ezquieta, B. and Vallejo, C. G.,** The trypsin-like proteinase of *Artemia*. Yolk localization and developmental activation, *Comp. Biochem. Physiol.*, 82(B), 731, 1985.
32. **Ezquieta, B. and Vallejo, C. G.,** Lipovitellin inhibition of *Artemia* trypsin-like proteinase: a role for storage protein in regulating proteinase activity during development, *Arch. Biochem. Biophys.*, 250, 410, 1986.
33. **Ezquieta, B. and Vallejo, C. G.,** *Artemia* trypsin-like proteinase a developmentally regulated proteinase, in *Artemia Research and Its Applications*, Vol. 2, Decleir, W., Moens, L., Slegers, H., Jaspers, E., and Sorgeloos, P., Eds., Universa Press, Wetteren, Belgium, 1987, 53.
34. **Ezquieta, B. and Vallejo, C. G.,** *Artemia* trypsin-like proteinase: developmental activation is inhibited by a lysosomal agent, *Biol. Cell,* 58, 227, 1986.
35. **Sebastian, J. and Heredia, C. F.,** Purification and properties of a ribonuclease induced during the early larval development of *Artemia salina, Eur. J. Biochem.*, 90, 405, 1978.
36. **Heredia, C. F.,** Developmental changes of *Artemia* ribonuclease, *Comp. Biochem. Physiol.*, 78(B), 407, 1984.
37. **Quintanilla, M. and Renart, J.,** Purification and characterization of a novel UpN-specific endoribonuclease VI from *Artemia* larvae, *J. Biol. Chem.*, 257, 12594, 1982.
38. **Heredia, C. F.,** Inhibition of endoribonuclease VI from *Artemia* larvae by cytidine-2-phosphate, *Biochem. Biophys. Res. Commun.*, 121, 408, 1984.
39. **Grau, P. P. and Heredia, C. F.,** Diguanosine 5′,5‴-P^1,P^4-tetraphosphate and other purine nucleotides inhibit endoribonuclease VI from *Artemia, FEBS Lett.*, 236, 291, 1988.
40. **Cervera, M., Martin, E., Domingo, A., Vallejo, C. G., and Marco, R.,** DNase activity during early development of *Artemia*, in *The Brine Shrimp Artemia,* Vol. 2, Persoone, G., Sorgeloos, P., Roels, O., and Jaspers, E., Eds., Universa Press, Wetteren, Belgium, 1980, 293.
41. **Domingo, A., Cervera, M., Martin, E., and Marco, R.,** Biochemical characterization and developmental behavior of *Artemia* embryonic and nauplial deoxyribonucleases, *Biochemistry*, 25, 4125, 1986.
42. **Warner, A. H.,** The biosynthesis, metabolism and function of dinucleoside polyphosphates in *Artemia* embryos: a compendium, in *The Brine Shrimp Artemia,* Vol. 2, Persoone, G., Sorgeloos, P., Roels, O., and Jaspers, E., Eds., Universa Press, Wetteren, Belgium, 1980, 105.
43. **Ezquieta, B. and Vallejo, C. G.,** Diguanosine 5′,5‴-P^1,P^4-tetraphosphate causes specific inhibition and desensitization of *Artemia* trypsin-like proteinase in the hydrolysis of a high affinity, arginine rich substrate, *Biochim. Biophys. Acta,* 883, 380, 1986.
44. **Sillero, A. and Ochoa, S.,** Nuclear localization of diguanosine polyphosphates in *Artemia* embryos, *Arch. Biochem. Biophys.*, 143, 548, 1971.
45. **Warner, A. H.,** Studies on the biosynthesis and function of dinucleoside polyphosphates in *Artemia* embryos, in *Regulation of Macromolecular Synthesis by Low Molecular Weight Mediators,* Koch, G. and Richter, D., Eds., Academic Press, New York, 1979, 161.
46. **Garesse, R., Perona, R., Marco, R., and Vallejo, C. G.,** The unmasking of proteolytic activity during the early development of *Artemia salina*. Identification of a precursor after hatching, *Eur. J. Biochem.*, 106, 225, 1980.

47. **Osuna, C., Olalla, A., Sillero, A., Sillero, M. A. G., and Sebastian, J.,** Induction of multiple proteases during the early larval development of *Artemia salina, Dev. Biol.*, 61, 94, 1977.
48. **Olalla, A., Osuna, C., Sebastian, J., Sillero, A., and Sillero, M. A. G.,** Purification and properties of three proteases from the larvae of the brine shrimp *Artemia salina, Biochim. Biophys. Acta*, 523, 181, 1978.
49. **Slegers, H., De Herdt, E., Piot, E., Backhovens, H., Thoen, C., Van Hove, L., Roggen, E., and Aerden, M.,** Activation of stored messenger ribonucleoproteins: Identification and function of the proteins associated with non-polysomal poly(A)-containing messenger ribonucleoproteins of cryptobiotic gastrulae of *Artemia* sp., in *Biochemistry and Cell Biology of Artemia*, MacRae, T., Bagshaw, J., and Warner, A. H., Eds., CRC Press, Boca Raton, FL, 1989, 21.
50. **Slegers, H., Mettrie, R., and Kondo, M.,** Evidence for a cytoplasmic translational inhibitor RNA in *Artemia salina* gastrulae embryos, *FEBS Lett.*, 80, 390, 1977.
51. **Lee-Huang, S., Sierra, J. M., Naranjo, R., Filipowicz, W., and Ochoa, S.,** Eucaryotic oligonucleotides affecting mRNA translation, *Arch. Biochem. Biophys.*, 180, 276, 1977.
52. **Clegg, J. S. and Golub, A. L.,** Protein synthesis in *Artemia salina* embryos. II. Resumption of RNA and protein synthesis upon cessation of dormancy in the encysted gastrulae, *Dev. Biol.*, 19, 178, 1969.
53. **Nilsson, M. O. and Hultin, T.,** Evidence for a store of informational RNA in encysted embryos of *Artemia salina, Exp. Cell Res.*, 72, 145, 1972.
54. **Grosfeld, H. and Littauer, U. Z.,** Cryptic form of mRNA in dormant *Artemia salina* cysts, *Biochem. Biophys. Res. Commun.*, 67, 176, 1975.
55. **Sierra, J. M., Filipowicz, W., and Ochoa, S.,** Messenger RNA in undeveloped and developing *Artemia salina* embryos, *Biochem. Biophys. Res. Commun.*, 69, 181, 1976.
56. **Slegers, H. and Kondo, M.,** Messenger ribonucleoprotein complexes of cryptobiotic embryos of *Artemia salina, Nucl. Acids Res.*, 4, 625, 1977.
57. **Amaldi, P. P., Felicetti, L., and Campioni, N.,** Flow of informational RNA from cytoplasmic poly(A)-containing particles to polyribosomes in *Artemia* cysts at early stages of development, *Dev. Biol.*, 59, 49, 1977.
58. **Nilsson, M. O. and Hultin, T.,** Characteristics and intracellular distribution of messenger-like RNA in encysted embryos of *Artemia salina, Dev. Biol.*, 38, 138, 1974.
59. **James, T. C. and Tata, J. R.,** Messenger RNA during early embryogenesis in *Artemia salina*. Altered translatability and sequence complexity, *Differentiation*, 16, 9, 1980.
60. **Kallin, B. and Hultin, T.,** The inhibited protein synthesis in dormant *Artemia salina* cysts, *10th Internat. Congress of Biochemistry*, Hamburg, Abstr. 03-6-109, 1976.
61. **De Herdt, E., Slegers, H., Piot, E., and Kondo, M.,** Non-polyadenylated 22S ribonucleoprotein particle is insensitive to translational inhibitor RNA of cryptobiotic gastrulae of *Artemia salina, Nucl. Acids Res.*, 7, 1363, 1979.
62. **De Herdt, E., Piot, E., Wahba, A., and Slegers, H.,** Initiation factor eIF2 associated with non-polysomal poly(A)-containing messenger ribonucleoproteins of cryptobiotic gastrulae of *Artemia salina, Eur. J. Biochem.*, 151, 455, 1985.
63. **Slegers, H., De Herdt, E., Mettrie, R., Piot, E., and Kondo, M.,** Characterization of repressed poly(A)-containing ribonucleoproteins in cryptobiotic gastrulae of *Artemia salina*, in *Biochemistry of Artemia Development*, Bagshaw, J. C. and Warner, A. H., Eds., University Microfilms International, Ann Arbor, MI, 1979, 84.
64. **Sierra, J. M., De Haro, C., Datta, A., and Ochoa, S.,** Translational control by protein kinases in *Artemia salina* and wheat germ, *Proc. Natl. Acad. Sci. U.S.A.*, 74, 4356, 1977.
65. **Warner, A. H., Shridhar, V., and Finamore, F. J.,** Isolation and partial characterization of a protein synthesis inhibitor from brine shrimp embryos, *Can. J. Biochem.*, 55, 965, 1977.
66. **Warner, A. H., Shridhar, V., and Yablonka-Reuveni, Z.,** Partial characterization of a protein synthesis inhibitor from *Artemia* cysts, in *Biochemistry of Artemia Development*, Bagshaw, J. C. and Warner, A. H., Eds., University Microfilms International, Ann Arbor, MI, 1979, 58.
67. **Muthukrishnan, S., Filipowicz, W., Sierra, J. M., Both, G. W., Shatkin, A. J., and Ochoa, S.,** mRNA methylation and protein synthesis in extracts from embryos of brine shrimp *Artemia salina, J. Biol. Chem.*, 250, 9336, 1975.
68. **Filipowicz, W., Furuichi, Y., Sierra, J. M., Muthukrishnan, S., Shatkin, A. J., and Ochoa, S.,** A protein binding the methylated 5′-terminal sequence, m^7 GpppN, of eukaryotic messenger RNA, *Proc. Natl. Acad. Sci. U.S.A.*, 73, 1559, 1976.
69. **Groner, Y., Grosfeld, H., and Littauer, U. Z.,** 5′-capping structures of *Artemia salina* mRNA and the translational inhibition by cap analogs, *Eur. J. Biochem.*, 71, 281, 1976.
70. **Huang, F. L. and Warner, A. H.,** Control of protein synthesis in brine shrimp embryos by repression of ribosomal activity, *Arch. Biochem. Biophys.*, 163, 716, 1974.
71. **Thoen, C., Van Hove, L., Piot, E., and Slegers, H.,** Purification and characterization of the messenger ribonucleoprotein-associated casein kinase II of *Artemia salina* cryptobiotic gastrulae, *Biochim. Biophys. Acta*, 783, 105, 1984.

72. **Slobin, L. I. and Möller, W.,** Changes in form of elongation factor during development of *Artemia salina, Nature,* 258, 452, 1975.
73. **Twardowski, T., Hill, J. M., and Weissbach, H.,** Disaggregation of elongation factor 1 by extracts of *Artemia salina, Biochem. Biophys. Res. Commun.,* 71, 826, 1976.
74. **Nombela, C., Redfield, B., Ochoa, S., and Weissbach, H.,** Elongation factor 1 from *Artemia salina*: Properties and disaggregation of the enzyme, *Eur. J. Biochem.,* 65, 395, 1976.
75. **Yablonka-Reuveni, Z., Fontaine, J. J., and Warner, A. H.,** Distribution of elongation factor 2 between particulate and soluble fractions of the brine shrimp *Artemia* during early development, *Can. J. Biochem. Cell Biol.,* 61, 833, 1982.
76. **Sastre, L. and Sebastian, J.,** Poly(A) polymerase activity during the early development of *Artemia*, in *The Brine Shrimp Artemia,* Vol. 2, Persoone, G., Sorgeloos, P., Roels, O., and Jaspers, E., Eds., Universa Press, Wetteren, Belgium, 1980, 325.
77. **Jeyaraj, N., Talib, S., Louis, J., Susheela, C., and Jayaraman, K.,** Occurrence of poly(A) polymerase in particles rich in poly(A) RNA in the developing embryos of *Artemia*, in *The Brine Shrimp Artemia,* Vol. 2, Persoone, G., Sorgeloos, P., Roels, O., and Jaspers, E., Eds., Universa Press, Wetteren, Belgium, 1980, 305.
78. **Gallego, M. E. and Heredia, C. F.,** Purification and properties of an inducible aminoacyl-tRNA hydrolase from *Artemia* larvae, *Biochim. Biophys. Acta,* 696, 57, 1982.
79. **Hultin, T. and Nilsson, M. O.,** The molecular biology of *Artemia*, in *The Brine Shrimp Artemia,* Vol. 2, Perssone, G., Sorgeloos, P., Roels, O., and Jaspers, E., Eds., Universa Press, Wetteren, Belgium, 1980, 83.
80. **Yagi, Y., Mizumoto, K., and Kaziro, Y.,** Limited tryptic digestion of messenger RNA capping enzyme from *Artemia salina*. Isolation of domains for guanylyltransferase and RNA 5′-triphosphatase, *J. Biol. Chem.,* 259, 4695, 1983.
81. **Mizumoto, K. and Kaziro, Y.,** Messenger RNA capping enzyme from eukaryotic cells, *Progr. Nucl. Acids Res. Mol. Biol.,* 34, 1, 1987.
82. **Warner, A. H. and Finamore, E. J.,** Isolation, purification and characterization of P^1,P^4-asymmetrical pyrophosphohydrolase from brine shrimp eggs, *Biochemistry,* 4, 1568, 1965.
83. **Warner, A. H., Beers, P. C., and Huang, F. L.,** Biosynthesis of the diguanosine nucleotides. I. Purification and properties of an enzyme from yolk platelets of brine shrimp embryos, *Can. J. Biochem.,* 52, 231, 1974.
84. **Warner, A. H. and Huang, F. L.,** Biosynthesis of the diguanosine nucleotides. II. Mechanism of action of GTP:GTP guanylyltransferase on nucleotide metabolism in brine shrimp embryos, *Can. J. Biochem.,* 52, 241, 1974.
85. **Vallejo, C. G., Lobaton, C. D., Quintanilla, M., Sillero, A., and Sillero, M. A. G.,** Dinucleoside tetraphosphatase in rat liver and *Artemia salina, Biochim. Biophys. Acta,* 438, 304, 1976.
86. **Sillero, A. and Sillero, M. A. G.,** Purine nucleotide metabolism in *Artemia*, in *Biochemistry and Cell Biology of Artemia,* MacRae, T. H., Bagshaw, J. C., and Warner, A. H., Eds., CRC Press, Boca Raton, FL, 1989, 95.
87. **Sierra, J. M., Meier, D., and Ochoa, S.,** Effect of development on the translation of messenger RNA in *Artemia salina* embryos, *Proc. Natl. Acad. Sci. U.S.A.,* 71, 2693, 1974.
88. **Filipowicz, W., Sierra, J. M., and Ochoa, S.,** Polypeptide chain initiation in eukaryotes. Initiation factor M.P. in *Artemia salina* embryos, *Proc. Natl. Acad. Sci. U.S.A.,* 72, 3947, 1975.
89. **Filipowicz, W., Sierra, J. M., Nombela, C., Ochoa, S., Merrick, W. C., and Anderson, W. F.,** Polypeptide chain initiation in eukaryotes: initiation factor requirements for translation of natural messengers, *Proc. Natl. Acad. Sci. U.S.A.,* 73, 44, 1976.
90. **MacRae, T. H., Woodley, C. L., and Wahba, A. J.,** Characterization of polypeptide chain initiation factors during *Artemia salina* development, in *Biochemistry of Artemia Development,* Bagshaw, J. C. and Warner, A. H., Eds., University Microfilms International, Ann Arbor, MI, 1979, 2.
91. **MacRae, T. H., Roychowdhury, M., Houston, K. J., Woodley, C. L., and Wahba, A. J.,** Protein synthesis in brine shrimp embryos, dormant, and developing embryos of *Artemia salina* contain equivalent amounts of chain initiation factor 2, *Eur. J. Biochem.,* 100, 67, 1979.
92. **Warner, A. H., MacRae, J. H., and Wahba, A. J.,** The use of *Artemia salina* for developmental studies. Preparation of embryos, tRNA, ribosomes, and initiation factor 2, *Methods Enzymol.,* 60, 298, 1979.
93. **Wahba, A. J. and Woodley, C. L.,** Molecular aspects of development in the brine shrimp *Artemia, Prog. Nucl. Acid Res. Mol. Biol.,* 31, 221, 1984.
94. **Woodley, C. L., Dholakia, J. N., and Wahba, A. J.,** Polypeptide chain initiation in *Artemia* embryos: functional and structural studies on chain initiation factor in *Artemia*, in *Artemia Research and Its Applications,* Vol. 2, Decleir, W., Moens, L., Slegers, H., Jaspers, E., and Sorgeloos, P., Eds., Universa Press, Wetteren, Belgium, 1987, 499.

95. **Wahba, A. J. and Woodley, C. L.,** Molecular aspects of development in the brine shrimp *Artemia*: structural and functional studies on the eukaryotic polypeptide initiation factor 2 and expression of the eukaryotic elongation factor Tu during development, in *Biochemistry and Cell Biology of Artemia*, MacRae, T. H., Bagshaw, J. C., and Warner, A. H., Eds., CRC Press, Boca Raton, FL, 1989, 57.
96. **Pain, V. M.,** Initiation of protein synthesis in mammalian cells, *Biochem. J.*, 235, 625, 1986.
97. **Mehta, H. B., Dholakia, J. N., Roth, W. W., Parekh, B. S., Montelaro, R. C., Woodley, C. L., and Wahba, A. J.,** Structural studies on the eukaryotic chain initiation factor 2 from rabbit reticulocytes and brine shrimp *Artemia* embryos. Phosphorylation by the heme-controlled repressor and casein kinase II, *J. Biol. Chem.*, 261, 6705, 1986.
98. **Thoen, C. and Slegers, H.,** Ribosome-associated cyclic nucleotide-independent protein kinase of *Artemia salina* cryptobiotic gastrulae, *Biochim. Biophys. Acta*, 825, 268, 1985.
99. **Thoen, C., De Herdt, E., and Slegers, H.,** Identification of the ribosomal proteins phosphorylated by the ribosome-associated casein kinase II from cryptobiotic gastrulae of the brine shrimp *Artemia* sp., *Biochem. Biophys. Res. Commun.*, 135, 347, 1986.
100. **Tolan, D. R. and Traut, R. R.,** Protein topography of the 40S ribosomal subunit from rabbit reticulocytes shown by cross-linking with 2-imino-thiolane, *J. Biol. Chem.*, 256, 10129, 1981.
101. **Westermann, P. and Nygard, O.,** The spatial arrangement of the complex between eukaryotic initiation factor eIF-3 and 40S ribosomal subunit, *Biochim. Biophys. Acta*, 741, 103, 1983.
102. **Bielka, H.,** The eukaryotic ribosome, *Trends Biochem. Sci.*, 3, 156, 1978.
103. **Thoen, C., Van Hove, L., and Slegers, H.,** The function of phosphorylation-dephosphorylation in messenger ribonucleoprotein translation, *Adv. Prot. Phosphatases*, 3, 121, 1986.
104. **Thoen, C., Van Hove, L., and Slegers, H.,** Identification of the substrates of the casein kinase II associated with non-polysomal messenger ribonucleoproteins of *A. salina* cryptobiotic embryos, *Molec. Biol. Rep.*, 11, 69, 1986.
105. **Van Hove, L., Thoen, C., Cohen, P., and Slegers, H.,** Dephosphorylation of cytoplasmic non-polysomal messenger ribonucleoproteins from cryptobiotic gastrulae of *Artemia salina*, *Biochem. Biophys. Res. Commun.*, 131, 1241, 1985.
106. **McLennan, A. G. and Prescott, M.,** Diadenosine 5′,5‴-P^1,P^4-tetraphosphate in developing embryos of *Artemia*, *Nucl. Acids Res.*, 12, 1609, 1984.
107. **Zamecnik, P.,** Diadenosine 5′,5‴-P^1,P^4-tetraphosphate (Ap_4A): its role in cellular metabolism, *Anal. Biochem.*, 134, 1, 1983.
108. **Prescott, M. and McLennan, A. G.,** The protein kinase activity associated with the major (5′-adenosyl)tetraphosphate binding protein of *Artemia*, in *Artemia Research and Its Applications*, Vol. 2, Decleir, W., Moens, L., Slegers, H., Jaspers, E., Eds., Universa Press, Wetteren, Belgium, 1987, 125.
109. **Slegers, H. and Aerden, M.,** Messenger ribonucleoproteins of cryptobiotic gastrulae of *Artemia* sp.: mechanisms of activation and repression of non-polysomal messenger ribonucleoproteins, in *Cell and Molecular Biology of Artemia Development*, Warner, A. H., Bagshaw, J. C., and MacRae, T. H., Eds., Plenum Press, New York, 1989, 259.
110. **Hathaway, G. M. and Traugh, J. A.,** Regulation of casein kinase II by 2,3-bisphosphoglycerate in erythroid cells, *J. Biol. Chem.*, 259, 2850, 1984.
111. **Raja, L. M. V., Louis, J., and Jayaraman, K.,** Demonstration of the formation of 1,3 diphosphoglyceric acid as an intermediate in high energy phosphate metabolism during reinitiation of development in *Artemia*, *Biochim. Biophys. Acta*, 723, 410, 1983.
112. **Thoen, C., Van Hove, L., Cohen, P., and Slegers, H.,** Identification of protein phosphatases dephosphorylating mRNP proteins from cryptobiotic gastrulae of the brine shrimp *A. salina*, *Biochem. Biophys. Res. Commun.*, 131, 84, 1985.
113. **Thoen, C., Van Hove, L., and Slegers, H.,** Purification, subunit structure and properties of a high-molecular-mass protein phosphatase capable of dephosphorylating mRNP of the brine shrimp *Artemia* sp., *Eur. J. Biochem.*, 163, 503, 1987.
114. **Merlevede, W.,** Protein phosphates and protein phosphatases. Landmarks in an eventful century, *Adv. Prot. Phosphatases*, 1, 1, 1985.
115. **Brawerman, G.,** The role of the poly(A) sequence in mammalian messenger RNA, *Crit. Rev. Biochem.*, 10, 1, 1981.
116. **Darnell, J. E., Jr.,** Variety in the level of gene control in eukaryotic cells, *Nature*, 297, 365, 1982.
117. **Slater, I. and Slater, D. W.,** Polyadenylation and transcription following fertilization, *Proc. Natl. Acad. Sci. U.S.A.*, 71, 1103, 1974.
118. **Wilt, F. H.,** The dynamics of maternal poly(A)-containing mRNA in fertilized sea urchin eggs, *Cell*, 11, 673, 1977.
119. **Egrie, J. C. and Wilt, F. H.,** Changes in poly(adenylic acid) polymerase activity during sea urchin embryogenesis, *Biochemistry*, 18, 269, 1979.
120. **Clegg, K. B. and Piko, L.,** RNA synthesis and cytoplasmic polyadenylation in the one-cell mouse embryo, *Nature*, 295, 342, 1982.

121. **Rosenthal, E. T., Tansey, T. R., and Ruderman, J. V.,** Sequence-specific adenylations deadenylations accompany changes in the translation of maternal messenger RNA after fertilization of *Spisula* oocytes, *J. Mol. Biol.*, 166, 309, 1983.
122. **James, T. C., Hentschel, C. C., and Tata, J. R.,** Gene expression during resumption of development of encysted *Artemia* gastrulae, in *Biochemistry of Artemia Development*, Bagshaw, J. C. and Warner, A. H., Eds., University Microfilms International, Ann Arbor, MI, 1979, 167.
123. **Sastre, L. and Sebastian, J.,** Developmental changes in poly(A) polymerase activity in *Artemia, Eur. J. Biochem.*, 135, 69, 1983.
124. **Roggen, E. and Slegers, H.,** Isolation and characterization of cytoplasmic poly(A) polymerase from cryptobiotic gastrulae of *Artemia salina, Eur. J. Biochem.*, 147, 225, 1985.
125. **Raja, L. M. V., Louis, J., and Jayaraman, K.,** Periodicity in elongation of oligo(A) tracts of RNA in developing embryos of *Artemia*: the role of RNP associated poly(A) polymerase, in *Artemia Research and Its Applications*, Vol. 2, Decleir, W., Moens, L., Slegers, H., Jaspers, E., and Sorgeloos, P., Eds., Universa Press, Wetteren, Belgium, 1987, 333.
126. **Sastre, L. and Sebastian, J.,** In vitro polyadenylation of messenger ribonucleoprotein particles stored in *Artemia* dormant embryos, *Biochem. Biophys. Res. Commun.*, 108, 1609, 1982.
127. **De Herdt, E., Piot, E., Kondo, M., and Slegers, H.,** The 38000-M_r poly(A)-binding protein of non-polysomal messenger ribonucleoproteins of cryptobiotic gastrulae of *Artemia salina, Eur. J. Biochem.*, 122, 453, 1982.
128. **Susheela, C. and Jayaraman, K.,** On the mode of activation of sequestered messengers in *Artemia salina, Differentiation*, 5, 29, 1976.
129. **Sastre, L. and Sebastian, J.,** Purification and properties of a polyadenylate polymerase from *Artemia* dormant embryos, *Biochim. Biophys. Acta*, 661, 54, 1981.
130. **Roggen, E., Dams, E., and Slegers, H.,** The primer specificity of cytoplasmic poly(A) polymerase from cryptobiotic gastrulae of *Artemia salina, Biochim. Biophys. Acta*, 825, 21, 1985.
131. **Roggen, E., Aerden, M., and Slegers, H.,** The regulation of cytoplasmic polyadenylation in *Artemia* sp. cryptobiotic gastrulae, in *Artemia Research and Its Applications*, Vol. 2, Decleir, W., Moens, L., Slegers, H., Jaspers, E., and Sorgeloos, P., Eds., Universa Press, Wetteren, Belgium, 1987, 471.
132. **Nagata, S., Iwasaki, K., and Kaziro, Y.,** Distribution of the low molecular weight form of eukaryotic elongation factor 1 in various tissues, *J. Biochem.*, 80, 73, 1976.
133. **Slobin, L. I. and Möller, W.,** Characterization of developmentally regulated forms of elongation factor 1 in *Artemia salina*. 1. Purification and structural properties of the enzymes, *Eur. J. Biochem.*, 69, 351, 1976.
134. **Slobin, L. I. and Möller, W.,** The heavy form of elongation factor 1 in *Artemia salina* embryos is functionally analogous to a complex of bacterial factors EF-Tu and EF-Ts, *Biochem. Biophys. Res. Commun.*, 74, 356, 1977.
135. **Slobin, L. I. and Möller, W.,** Characterization of developmentally regulated forms of elongation factor 1 in *Artemia salina*. 2. Functional properties of the enzyme, *Eur. J. Biochem.*, 69, 367, 1976.
136. **Roobol, K. and Möller, W.,** The role of guanine nucleotides in the interaction between aminoacyl-tRNA and elongation factor 1 of *Artemia salina, Eur. J. Biochem.*, 90, 471, 1978.
137. **Janssen, G. M. C. and Möller, W.,** Elongation factor 1$\beta\gamma$ from *Artemia*. Purification and properties of its subunits, *Eur. J. Biochem.*, 171, 119, 1988.
138. **Janssen, G. M. C., Maessen, G. D. F., Amons, R., and Möller, W.,** Phosphorylation of elongation factor 1β by an endogenous kinase affects its catalytic nucleotide exchange activity, *J. Biol. Chem.*, 263, 11063, 1988.
139. **Slobin, L. I.,** The inhibition of elongation factor 1 activity by heparin, *Biochem. Biophys. Res. Commun.*, 73, 539, 1976.
140. **Möller, W., Amons, R., Janssen, G., Lenstra, J. A., and Maassen, J. A.,** Studies on elongation factor 1 and ribosomes of *Artemia*, in *Artemia Research and Its Applications*, Vol. 2, Decleir, W., Moens, L., Slegers, H., Jaspers, E., and Sorgeloos, P., Eds., Universa Press, Wetteren, Belgium, 1987, 451.
141. **Yablonka-Reuveni, Z. and Warner, A. H.,** Characterization of elongation factor 2 in dormant cysts and developing embryos of *Artemia salina*, in *Biochemistry of Artemia Development*, Bagshaw, J. C. and Warner, A. H., Eds., University Microfilms International, Ann Arbor, MI, 1979, 22.
142. **Ogilvie, A., Huschka, U., and Kersten, W.,** Control of protein synthesis in mammalian cells by aminoacylation of transfer ribonucleic acid, *Biochim. Biophys. Acta*, 565, 293, 1979.
143. **Deutscher, M. P.,** Processing of tRNA in prokaryotes and eukaryotes, *Crit. Rev. Biochem.*, 17, 45, 1984.
144. **Sprinzl, M. and Cramer, F.,** The -C-C-A end of tRNA and its role in protein biosynthesis, *Progr. Nucl. Acid Res. Mol. Biol.*, 22, 1, 1979.
145. **Del Rio, J. M. and Heredia, C. F.,** Transfer RNA nucleotidyltransferase from *Artemia*. Developmental changes and integrity of transfer RNA, *Comp. Biochem. Physiol.*, 86(B), 291, 1987.

146. **Miralles, J., Sebastian, J., and Heredia, C. F.**, Independent temporal expression of two N-substituted aminoacyl-tRNA hydrolases during the development of *Artemia salina*, *Biochim. Biophys. Acta*, 518, 326, 1978.
147. **Leon, J. and Heredia, C. F.**, N-acetylphenylalanyl-tRNA hydrolase from *Artemia*. Identification of two molecular forms and their evolution during early differentiation, *Biochim. Biophys. Acta*, 653, 350, 1981.
148. **Gallego, M. E. and Heredia, C. F.**, Purification and properties of an inducible aminoacyl-tRNA hydrolase from *Artemia* larvae, *Biochim. Biophys. Acta*, 696, 57, 1982.
149. **Golub, A. and Clegg, J. S.**, Protein synthesis in *Artemia salina* embryos. I. Studies on polyribosomes, *Dev. Biol.*, 17, 644, 1968.
150. **Grosfeld, H. and Littauer, U. Z.**, The translation in vitro of mRNA from developing cysts of *Artemia salina*, *Eur. J. Biochem.*, 70, 589, 1976.
151. **Grosfeld, H., Soreq, H., and Littauer, U. Z.**, Membrane associated cytoplasmic mRNA in *Artemia salina*. Functional and physical changes during development, *Nucl. Acids Res.*, 4, 2109, 1977.
152. **Fisher, J. A., Baxter-Lowe, L. A., and Hokin, L. E.**, Site of synthesis of the α and β subunits of the Na,K-ATPase in brine shrimp nauplii, *J. Biol. Chem.*, 259, 14217, 1984.
153. **Manning, A. M., Ting, G. S., Mansfield, B. C., Trotman, C. N. A., and Tate, W. P.**, The isolation and faithful translation of *Artemia* naupliar hemoglobin mRNA, *Biochem. Int.*, 12, 715, 1986.
154. **Birndorf, H. C., D'Allesio, J., and Bagshaw, J. C.**, DNA-dependent RNA-poly-merases from *Artemia* embryos. Characterization of polymerases I and II from nauplius larvae, *Dev. Biol.*, 45, 34, 1975.
155. **Renart, J. and Sebastian, J.**, Characterization and levels of the RNA-polymerases during the embryogenesis of *Artemia salina*, *Cell Diff.*, 5, 97, 1976.
156. **Hentschel, C. and Tata, J. R.**, Differential activation of free and template engaged RNA polymerase I and II during resumption of development of dormant *Artemia* gastrulae, *Dev. Biol.*, 57, 283, 1977.
157. **Bagshaw, J. C., Acey, R., and Helder, J. C.**, RNA polymerases and transcriptional switches in developing *Artemia*, in *The Brine Shrimp Artemia*, Vol. 2, Persoone, G., Sorgeloos, P., Roels, O., and Jaspers, E., Eds., Universa Press, Wetteren, Belgium, 277, 1980.
158. **Swennen, L., Moens, L., Heip, J., and Kondo, M.**, Preferential stimulation of aurintricarboxylic acid of DNA-dependent RNA polymerase II activity in isolated larval nuclei of *Artemia salina*, *Hoppe-Seyler's Physiol. Chem.*, 362, 133, 1981.
159. **D'Alessio, J. M. and Bagshaw, J. C.**, DNA-dependent RNA polymerases from *Artemia salina*. IV. Appearance of nuclear RNA polymerase activity during pre-emergence development of encysted embryos, *Differentiation*, 8, 53, 1977.
160. **Sebastian, J., Cruces, J., Osuna, C., and Renart, J.**, Role of the RNA polymerases in the regulation of transcription during the early development of *Artemia*, in *The Brine Shrimp Artemia*, Vol. 2, Persoone, G., Sorgeloos, P., Roels, O., and Jaspers, E., Eds., Universa Press, Wetteren, Belgium, 335, 1980.
161. **D'Alessio, J. M. and Bagshaw, J. C.**, DNA-dependent RNA polymerases from *Artemia salina*. Characterization of a protein factor from developing embryos that stimulates *Artemia* RNA polymerase II, *Dev. Biol.*, 70, 71, 1979.
162. **Osuna, C. and Sebastian, J.**, Levels of the RNA polymerases during the early larval development of *Artemia*, *Eur. J. Biochem.*, 109, 383, 1980.
163. **Cruces, J., Diaz, V., Quintanilla, M., Renart, J., and Sebastian, J.**, Purification and subunit structure of RNA polymerase II from different stages of *Artemia* development, *Eur. J. Biochem.*, 141, 279, 1984.
164. **Bagshaw, J. C.**, DNA-dependent RNA polymerases from *Artemia salina*. Subunit structure of polymerase II, *Nucl. Acids Res.*, 3, 1449, 1976.
165. **Swennen, L.**, The DNA-dependent RNA polymerases of *Artemia salina*, Ph.D. thesis, University of Antwerp, Belgium, 1979.
166. **Wu, R. S., Panusz, H. T., Hatch, C. L., and Bonner, W. M.**, Histones and their modifications, *CRC Crit. Rev. Biochem.*, 20, 201, 1986.
167. **Cano, A., Estepa, I., and Pestana, A.**, Regulation of histone acetylation in *Artemia*, in *The Brine Shrimp Artemia*, Vol. 2, Persoone, G., Sorgeloos, P., Roels, O., and Jaspers, E., Eds., Universa Press, Wetteren, Belgium, 1980, 367.
168. **Cano, A. and Pestana, A.**, Regulation of histone acetyltransferase activity during the development of *Artemia salina*, *Dev. Biol.*, 54, 276, 1976.
169. **Cano, A. and Pestana, A.**, Purification and properties of a histone acetyltransferase from *Artemia salina*, highly efficient with H1 histone, *Eur. J. Biochem.*, 97, 65, 1979.
170. **Cano, A., Estepa, I., and Pestana, A.**, Substrate regulation of histone acetyltransferase from *Artemia salina*, *Biochem. Biophys. Res. Commun.*, 92, 163, 1980.
171. **Estepa, I. and Pestana, A.**, Isolation and partial characterization of three histone-specific acetyltransferases from *Artemia*, *Eur. J. Biochem.*, 132, 249, 1983.
172. **Estepa, I. and Pestana, A.**, Activation of polyamines of the acetylation of endogenous histones in isolated chromatin and nuclei from *Artemia*, *Eur. J. Biochem.*, 119, 431, 1981.

173. **Sato, N. L.**, Enzymatic contribution to the encystment of *Artemia salina, Sci. Rep. Tohoku Univ.*, 33, 319, 1967.
174. **Morris, J. and Afzelius, B. A.**, The structure of the shell and outer membranes in encysted *Artemia salina* embryos during cyrptobiosis and development, *J. Ultrastr. Res.*, 20, 244, 1967.
175. **Funke, B. and Spindler, K. D.**, Developmental changes of chitinolytic enzymes and ecdysteroid levels during the early development of the brine shrimp *Artemia*, in *Artemia Research and Its Applications*, Vol. 2, Decleir, W., Moens, L., Slegers, H., Jaspers, E., and Sorgeloos, P., Eds., Universa Press, Wetteren, Belgium, 1987, 67.
176. **Geldiay, R., Koray, T., and Buyukisik**, Effect of different glycerol concentrations upon the hatching of *Artemia*, in *The Brine Shrimp Artemia*, Vol. 2, Persoone, G., Sorgeloos, P., Roels, O., and Jaspers, E., Eds., Universa Press, Wetteren, Belgium, 1980, 147.
177. **Conte, F. P., Droukas, P. C., and Ewing, R. D.**, Development of sodium regulation and de novo synthesis of Na+K-activated ATPase in larval brine shrimp, *Artemia salina, J. Exp. Zool.*, 202, 339, 1977.
178. **Conte, F. P.**, Role of C-4 pathway in crustacean chloride cell function, *J. Am. Physiol.*, 238, R269, 1980.
179. **Trotman, C. N. A., Mansfield, B. C., and Tate, W. P.**, Inhibition of emergence, hatching, and protein biosynthesis in embryonic *Artemia salina, Dev. Biol.*, 80, 167, 1980.
180. **Peterson, G. L., Ewing, R. D., and Conte, F. P.**, Membrane differentiation and *de novo* synthesis of the ($Na^+ + K^+$)-activated adenosine triphosphatase during development of *Artemia salina* nauplii, *Dev. Biol.*, 67, 90, 1978.
181. **Peterson, G. L., Ewing, R. D., Hootman, S. R., and Conte, F. P.**, Large scale purification and molecular and kinetic properties of the (Na+K)-activated adenosine triphosphatase from *Artemia salina* nauplii, *J. Biol. Chem.*, 253, 4762, 1978.
182. **Peterson, G. L. and Hokin, L. E.**, Biogenesis of the (Na+K)-activated adenosine triphosphatase during naupliar development in *Artemia*, in *Biochemistry of Artemia Development*, Bagshaw, J. C. and Warner, A. H., Eds., University Microfilms International, Ann Arbor, MI, 1979, 120.
183. **Peterson, G. L. and Hokin, L. E.**, Improved purification of brine shrimp (*Artemia salina*) ($Na^+ + K^+$)-activated adenosine triphosphatase and amino acid and carbohydrate analysis of the isolated subunits, *Biochem. J.*, 192, 107, 1980.
184. **Peterson, G. L. and Hokin, L. E.**, Molecular weight and stoichiometry of the sodium- and potassium-activated adenosine triphosphatase subunits, *J. Biol. Chem.*, 256, 3751, 1981.
185. **Fischer, J. A., Baxter-Lowe, L. A., and Hokin, L. E.**, Site of synthesis of the α and β subunits of the Na,K-ATPase in brine shrimp nauplii, *J. Biol. Chem.*, 259, 14217, 1984.
186. **Morohashi, M. and Kawamura, M.**, Solubilization and purification of *Artemia salina* (Na,K)-activated ATPase and NH_2-terminal amino acid sequence of its larger subunit, *J. Biol. Chem.*, 259, 14928, 1984.
187. **Nakanishi, Y. H., Iwasaki, T., Okigaki, T., and Kato, H.**, Cytological studies of *Artemia salina*. I. Embryonic development without cell multiplication after the blastula stage in encysted dry eggs, *Annot. Zool. Jpn.*, 35, 223, 1962.
188. **Benesch, R.**, Zur ontogenie und morphologie von *Artemia salina* L., *Zool. Jordbrukstek. Anat.*, 86, 307, 1969.
189. **Freeman, J. A.**, The integument of *Artemia* during early development, in *Biochemistry and Cell Biology of Artemia*, MacRae, T. H., Bagshaw, J. C., and Warner, A. H., Eds., CRC Press, Boca Raton, FL, 1989, 233.
190. **Criel, L.**, Morphology of *Artemia*, in *Artemia Biology*, Browne, R. A., Sorgeloos, P., and Trotman, C. N. A., Eds. (in this volume).
191. **Cacace, M. G., Ballario, P., Colapicchioni, C., and Bergami, M.**, DNA polymerase in *Artemia* embryos, in *The Brine Shrimp Artemia*, Vol. 2, Persoone, G., Sorgeloos, P., Roels, O., and Jaspers, E., Eds., Universa Press, Wetteren, Belgium, 1980, 285.
192. **Slater, J. M. and McLennan, A. G.**, DNA polymerases α and γ during pre-emergence and early development of *Artemia, Eur. J. Biochem.*, 129, 415, 1982.
193. **McLennan, A. G. and Miller, D.**, *Artemia* DNA polymerases: activity gel analysis of DNA polymerase-α and -γ polypeptides and immunological detection of a polypeptide related to DNA polymerase-β, in *Artemia Research and Its Applications*, Vol. 2, Decleir, W., Moens, L., Slegers, H., Jaspers, E., and Sorgeloos, P., Eds., Universa Press, Wetteren, Belgium, 1987, 253.
194. **McLennan, A. G.**, Immunological detection of a DNA polymerase-β polypeptide in embryonic cysts and larvae of the brine shrimp *Artemia, Comp. Biochem. Physiol.*, 86(B), 315, 1987.
195. **Vallejo, C. G., Sillero, M. A. G., and Marco, R.**, Mitochondrial maturation during *Artemia salina* embryogenesis. General description of the process, *Cell. Mol. Biol.*, 25, 113, 1979.
196. **Bagshaw, J. C.**, Developmentally regulated gene expression in *Artemia*: Histone gene expression in newly hatched larvae, *Dev. Genet.*, 3, 41, 1982.
197. **Weinmann-Dorsch, C., Hedl, A., Grummt, I., Albert, W., Ferdinand, F. J., Friis, R. R., Pierron, G., Moll, W., and Grummt, F.**, Drastic rise of intracellular adenosine (5′)tetraphospho(5′)adenosine correlates with onset of DNA synthesis in eukaryotic cells, *Eur. J. Biochem.*, 138, 179, 1984.

198. **Sillero, M. A., Burillo, S. L., Dominguez, E., Olalla, A., Osuna, C., Renart, J., Sebastian, J., and Sillero, A.,** Multiple proteolytic enzymes in *Artemia*, in *The Brine Shrimp Artemia,* Vol. 2, Persoone, G., Sorgeloos, P., Roels, O., and Jaspers, E., Eds., Universa Press, Wetteren, Belgium, 1980, 343.
199. **Cano, A., Cruces, J., Estepa, I., Gallego, M. E., Sillero, M. A. G., Heredia, C. F., Llorente, P., Olalla, A., Osuna, C., Pestana, A., Renart, J., Ruiz, A., Sastre, L., Sebastian, J., and Sillero, A.,** Developmental changes of enzyme levels during *Artemia salina* differentiation, in *Biochemistry of Artemia Development,* Bagshaw, J. C. and Warner, A. H., Eds., University Microfilms International, Ann Arbor, MI, 1979, 150.
200. **Hootman, S. R. and Conte, F. P.,** Fine structure and function of the alimentary epithelium in *Artemia salina* nauplii, *Cell Tissue Res.,* 155, 423, 1974.
201. **Warner, A. H.,** Proteases and protease inhibitors in *Artemia* and their role in the developmental process, in *Biochemistry and Cell Biology of Artemia,* MacRae, T. H., Bagshaw, J. C., and Warner, A. H., Eds., CRC Press, Boca Raton, FL, 1989, 113.
202. **Samain, J. F., Moal, J., Daniel, J. Y., Lecoz, J. R., and Jezequel, M.,** The digestive enzymes amylase and trypsin during the development of *Artemia*: effect of food conditions, in *The Brine Shrimp Artemia,* Vol. 2, Persoone, G., Sorgeloos, P., Roels, O., and Jaspers, E., Eds., Universa Press, Wetteren, Belgium, 1980, 239.
203. **Raineri, M.,** Histochemical and biochemical study of alkaline phosphatase (ALP) activity in developing embryos and larvae of *Artemia*, in *Artemia Research and Its Applications,* Vol. 2, Decleir, W., Moens, L., Slegers, H., Jaspers, E., and Sorgeloos, P., Eds., Universa Press, Wetteren, Belgium, 1987, 139.
204. **McComb, R. B., Bowers, G. J., and Posen, S.,** *Alkaline Phosphatases,* Plenum Press, New York, 1979.

Chapter 4

NORMALITY AND ABNORMALITY IN EARLY DEVELOPMENT

Clive N. A. Trotman

TABLE OF CONTENTS

I. INTRODUCTION

In the phraseology of Needham and Needham[1] nearly sixty years ago, "...if an *Artemia* egg before development is crushed, nothing but a transparent jelly is visible, together with a few orange coloured oil drops." Yet if the "egg", or encysted embryo, is crushed after a few hours of incubation in sea water, an entirely different entity is expelled; an apparently intact prenauplius larva complete with the appendages of locomotion and feeding, antennae, a light sensitive eyespot, and a digestive tract. At the stage of approximately 16 h, depending on conditions, a prenauplius displays unmistakable signs of life with rhythmic movements of its appendages and streaming of the hemolymph. It escapes from its hatching envelope and swims away.

In that period of a few hours the embryo has switched from a condition of cryptobiosis and virtually total internal dehydration that may have lasted for decades. A kit of about 4000 cells having no recognizable gross morphology has remustered and differentiated into a complex free swimming crustacean without undergoing any cell division[2] or DNA synthesis.[3] More remarkably, the animal has ruptured and escaped from an extremely thick, tough, multilayered, and impermeable cyst wall that has afforded protection from the adverse environment.

The seemingly simple event of emergence from the cyst, in which the cyst wall splits and the prenauplius surrounded by a thin hatching membrane is expelled, appears progressively more complex as we identify more prerequisites. A complex mechanism has correspondingly more points of vulnerability to disruption. Conversely, the critical examination of abnormalities in the emergence mechanism is providing extra clues about the sequence of events in normal development and has contributed to the identification of some of the factors involved.

It is intended here to review deviations from the normally smooth path of early development in *Artemia* and to discuss the inferences from those deviations in terms of a model of the emergence mechanism. The review is intentionally selective and, bearing in mind the highly international interest in *Artemia* development, the most accessible references have been cited where possible.

II. NORMAL EARLY DEVELOPMENT

A. OUTLINE MODEL OF NORMAL EMERGENCE

A preliminary model (Figure 1) is proposed to review and assess the evidence for how *Artemia* emerges. Figure 2 shows the relationship between the component parts of the model and the corresponding microanatomical structures. The model takes into account the need for the developing prenauplius to split the extremely thick and evidently strong wall of the cyst and later to escape in a swift, clean movement. Internal hydrostatic pressure of osmotic origin is the logical force to rupture the cyst wall.

The question is, how is this osmotic potential generated? The chief prerequisites for emergence to occur are water, oxygen, and a temperature that supports metabolism. It is proposed in Figure 1 that biochemical pumps generate a high concentration of ions or solutes at the most logical place, namely between the innermost permeability barrier of the cyst wall and the embryo itself. The pumps depend on a supply of energy derived from the oxidative metabolism of stored substrates.

B. IS OSMOTIC PRESSURE A FEASIBLE MECHANISM OF CYST RUPTURE?

For an osmotic potential to develop, a semipermeable barrier has to exist that will at least impede the egress of some dissolved molecules while permitting the entry of water. The structure of the cyst wall at the electron microscopical level has been described in

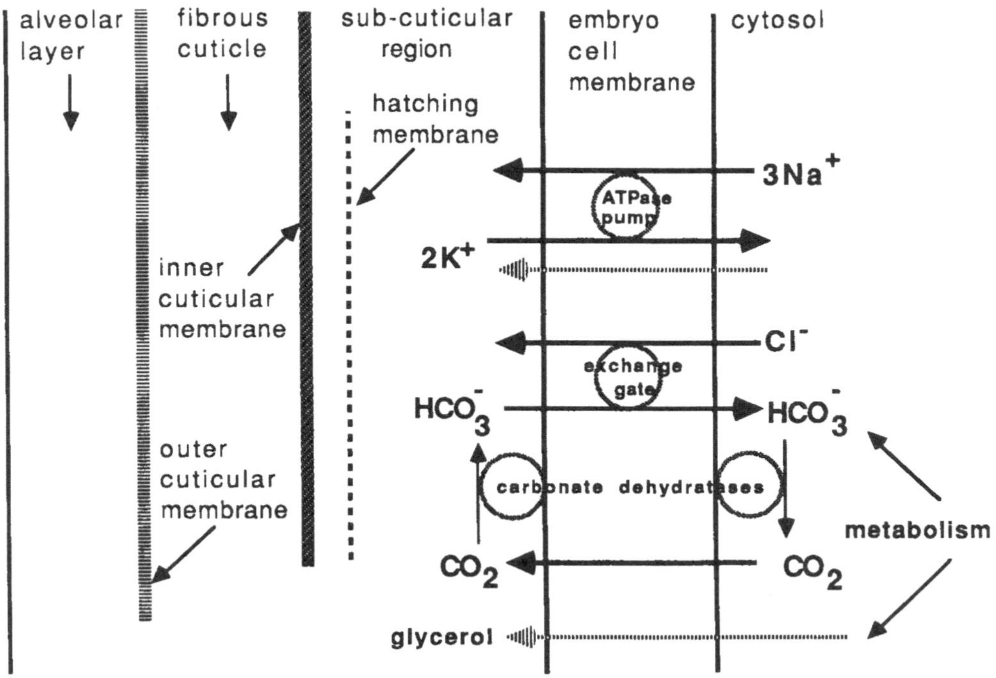

FIGURE 1. Outline model of a mechanism for the generation of an osmotic potential within the permeability barrier of the *Artemia* cyst. Compare with the microanatomical features illustrated in Figure 2.

detail.[4,5] Permeation with lead nitrate clearly showed the outer cuticular membrane to be the outer permeability barrier to external lead ions.[5] This membrane is believed to be impermeable to virtually all ions and solutes with the exception of water and dissolved gases, which appear to exchange freely between the embryo and the environment. Some volatile solvents including ethanol and chloroform might penetrate.[6]

An unstudied aspect is whether the several membranes of the *Artemia* cyst change their permeabilities through emergence and hatching. Permeability could be modified according to the presence of adjacent membranes, in response to internal hydrostatic pressure changes, or in consequence of the metabolism of any associated trehalose.[7] The steady rise in chitinase and *N*-acetyl-glucosaminidase activities in the period after 9 h cyst hydration through hatching[8] offers further potential for alteration of the properties of cuticular structures.

It is not known whether ions and solutes diffusing outward from the embryo are able to traverse the membrane bounding the inner surface of the cuticular fibrous layer, i.e., the inner cuticular membrane. The model accommodates either an impermeable or a partially permeable inner cuticular membrane.

Presumably the incipient exoskeleton of the nauplius also develops selective permeability prior to excystment; however, it would be less logical for the organism itself to suffer rapid osmotic expansion, and the observation is that it is principally the fluid surrounding the organism which increases in volume. Furthermore, the osmotic pressure attainable would be far higher by confining the potential to a small volume between the embryo and the cyst wall.

Morphometry of newly emerged prenauplii indicates the total volume enclosed by the hatching membrane to average 2.4 times the volume of the cyst.[9] Some water-filled space within the hatching membrane enclosure is evident, as is mild turgor compared with punctured sacs, e.g., those from which the nauplius has escaped. These observations are compatible

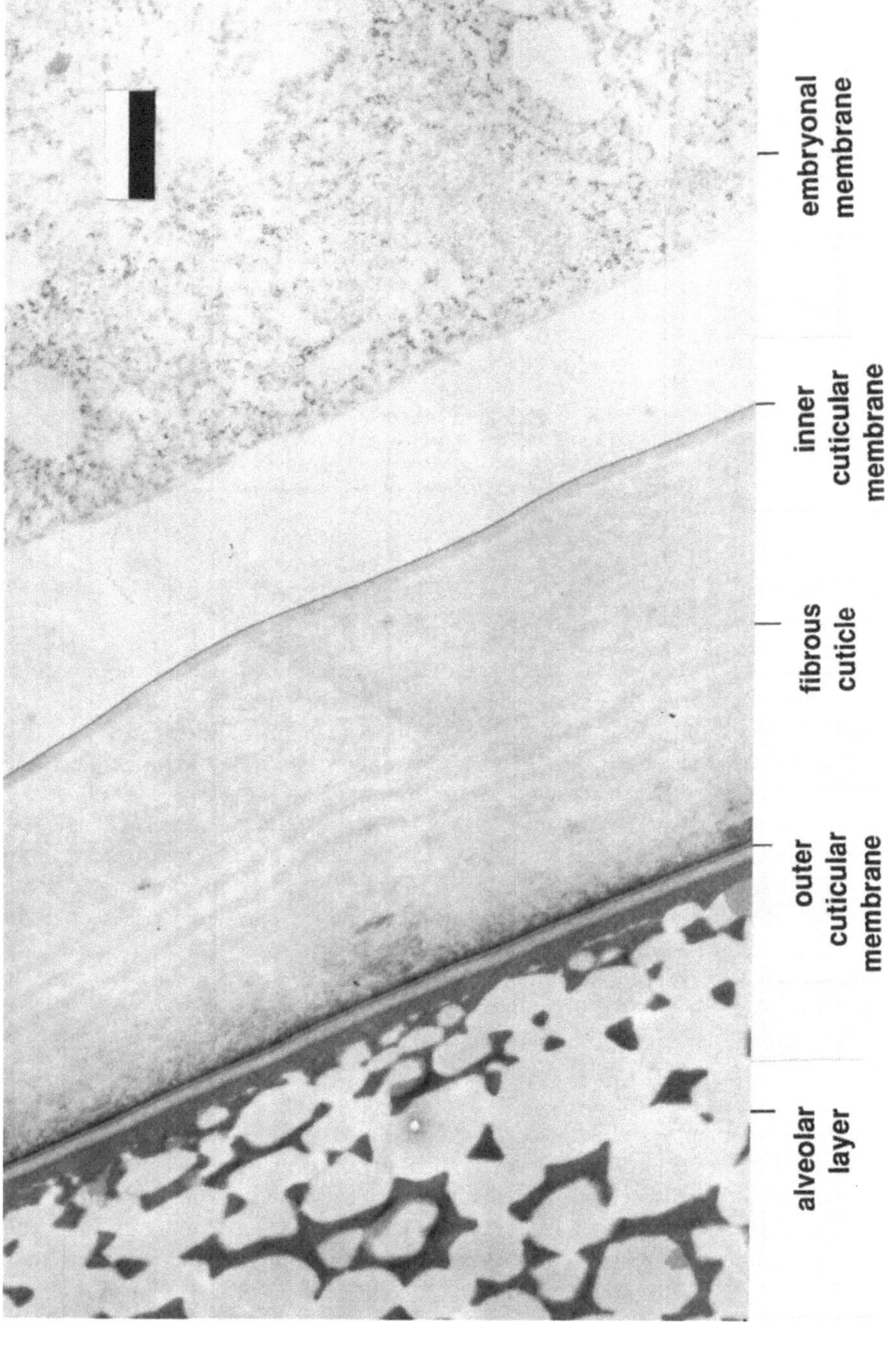

FIGURE 2. Transverse section of cyst with enclosed prenauplius at 18 h stage. The hatching membrane has not separated from the inner cuticular membrane. Electron microscopy; glutaraldehyde, uranyl acetate, lead citrate used in all transmission figures. Bar = 1 μm.

with the emergence of the prenauplius being facilitated by a marked expansion of the hatching membrane sac in response to an influx of water.

C. EMERGENCE IS CORRELATED WITH AEROBIC METABOLISM

Cysts incubated in water at room temperature commence respiration within minutes,[10] provided they have been activated after diapause by certain stimuli including dehydration and transient exposure to light.[11-14] The mechanism of release from diapause has been reviewed[15] and is considered in detail by Drinkwater and Clegg in this volume.[16] Respiration is inhibited by incubation at 4°C, by inadequate hydration,[17,18] by incubation in salt or carbohydrate media exceeding 60 to 70 atmospheres of osmotic pressure (about 2 molal NaCl),[19] by depression of the internal pH,[20-22] or of course if oxygen is withheld. Inhibition of aerobic respiration always prevents emergence, even if continued in hydrated cysts for a period of months, but it does not destroy the capacity to emerge when respiration is restored.[9,23,24] A restricted anaerobic metabolism involving nucleotides occurs[25] but does not involve glycolysis[26] and does not support emergence.

The osmotic expulsion mechanism evidently is dependent on aerobic metabolism. The inhibitory effect of hyperosmolality on respiration appears to operate through consequential hypohydration directly inhibiting trehalase and hexokinase (and possibly phosphofructokinase).[27] The same enzymes are implicated in the inhibitory effect of low internal pH[28,29] but the hypohydration effect is independent of internal pH.[27] Mitochondria would be deprived of their main substrate, which appears to derive from the glycolysis of trehalose since respiratory quotients measured in emerging cysts give a value close to one.[19]

One common factor in these inhibitory mechanisms is cessation of mitochondrial ATP generation. Recent measurements[27] confirm that ATP production is inhibited on transfer of cysts after 4 h of incubation to a dehydrating 5 *M* NaCl medium. Interestingly, the adenylate energy charge falls only slightly, suggesting incomplete inhibition or the utilization of nonglycolytic substrates, possibly from degradation of the yolk granules in which many mitochondria are embedded.[30,31] By contrast, transfer to anoxic incubation, which would inhibit cytochrome oxidase, causes the adenylate energy charge to fall considerably.[27]

1. How Might ATP Be Implicated In Emergence?

Clegg[19] and Conte et al.[32] have demonstrated that glycerol, which should be an acceptable candidate for the osmotic potential, accumulates in the preemergent cyst and escapes into the medium when the cyst ruptures. Conversion of trehalose to glycerol is an ATP requiring pathway, specifically at the phosphofructokinase reaction. It would be necessary for half of the trehalose to be metabolised to pyruvate or alternatively for oxidative phosphorylation to be operational for there to be sufficient ATP to energize the diversion of the remainder of the trehalose to glycerol.

Glycerol can be invoked as the primary component of the osmotic potential, or as a secondary consequence of the need to metabolize trehalose, either to restore membranes to the hydrated state or to generate ATP. Possibly all three mechanisms are involved and the net result is that glycerol provides initially a convenient osmotic gradient for water imbibition, as Clegg[19] and Conte[32] have suggested, until an ATP-energized ionic gradient supervenes.

It is possible to rationalize the appearance of glycerol for other reasons than primarily for osmotic purposes. Glycolysis requires a supply of NAD^+ for the oxidation of glyceraldehyde-3-phosphate. If glycolysis initially exceeds the capacity of the nascent mitochondria to dispose of pyruvate, then NAD^+ could be regenerated from NADH by the reduction of glyceraldehyde-3-phosphate of glycerol phosphate (subsequently hydrolyzed by a phosphatase), at a cost of half of the glycolytic substrate. The benefit is a net gain of 2 ATP molecules per glucose unit glycolyzed. Thus the diversion of substrate to glycerol might be dictated more by the demand for ATP in the period of around 7 to 10 h taken by mitochondria to become fully structured and functional.[10,33] Despite their consumption of oxygen in this period, the mitochondria may be less than fully coupled and may in fact be net importers of ATP. Furthermore, the metabolism of trehalose may have more to do with a reversal of a mechanism for the preservation of membranes in the desiccated state by complexation with trehalose[7,22] than with energy production per se.

Glycerol would accumulate throughout the embryo and cyst in the absence of a specific secretory mechanism. The desirability of the osmotic potential being localized outside the embryo implies the existence of an extrusion mechanism. It is not known whether glycerol is actively transported in *Artemia*, but the model (Figure 1) incorporates a conventional ATPase pump exchanging Na^+ for K^+ and the concept can be broadened to envisage the extrusion of other solutes.

Na,K-ATPases certainly exist in preemergent *Artemia*; the activity is detectable by 8 h.[32] The mRNA of the ATPase subunits is differentially expressed,[34] that of the beta subunit changing little but the alpha mRNA rising from zero prior to 6 h to a peak at 16 h, corresponding approximately to the period between early rupture and completed emergence. The location of most of the ATPase is in the salt gland,[32] which has an osmoregulatory role after hatching but there is no reason why it should not function as the generator of the osmotic potential before emergence.

The time courses published by Conte et al.[32] show a Na^+ influx over the 8 to 15 h period which, in a nonsynchronous culture, may represent a sudden influx to each individual as the outer cuticular membrane ruptures. Interestingly, Na,K-ATPase is activated at about the same time.[32] Conversely, Geldiay et al.[35] measured an increased efflux of Na^+ into sodium-free glycerol media at times corresponding to the emergent phase. These two groups' findings are not contradictory since both support the conclusion that Na^+ becomes exchangeable between embryo and medium at the time of emergence. Rupture of the outer cuticular membrane appears to coincide with a time of changeover from a glycerol passive gradient to an ionic gradient maintained by active pumping.

D. ULTRASTRUCTURAL ASPECTS OF EMERGENCE

Dissolution of the alveolar layer[36] (dechorionation, decapsulation) by treatment of cysts with hypochlorite does not interfere with the subsequent emergence and hatching of viable nauplii. The protective role of the alveolar layer can thereby be distinguished from the functional role of interior membranes in emergence.

Scanning electron microscopy of dechorionated cysts (Figure 3) reveals a prominent network of structural elements, which were first identified by light microscopy and in thin sections.[5] When dechorionated cysts undergo emergence the rupture spreads preferentially along the lines of these structural elements (Figure 4). At higher magnification the center line of the element appears to be the zone of separation, possibly indicative of the fibrous layer of the cuticle having been assembled by edgewise fusion of polygonal plates. In thin sections (Figure 5) the structural element displays a bilateral structure with stained components corresponding to the plate edges, and a central, unstained zone. The fortuitous view of a fractured structural element in Figure 6 shows, however, that separation extending through the outer cuticular membrane and fibrous region is not necessarily accompanied immediately by breakage of the inner cuticular membrane.

After preparation of developing dechorionated cysts for scanning electron microscopy by dehydration, the bulk of each polygonal plate often appears to be substantially thinner than the structural element bounding it (Figure 3), although this difference is rarely prominent in thin sections where dehydration has been achieved by replacement with embedding resin (Figure 5). This suggests that the fibrous cuticle could act as a liquid reservoir and absorb part of the osmotic potential. As the fibrous region swells, the outer cuticular membrane may perform the dual functions of sealing the osmotic potential and mechanically preventing premature rupture of the fibrous region at the plate boundaries.

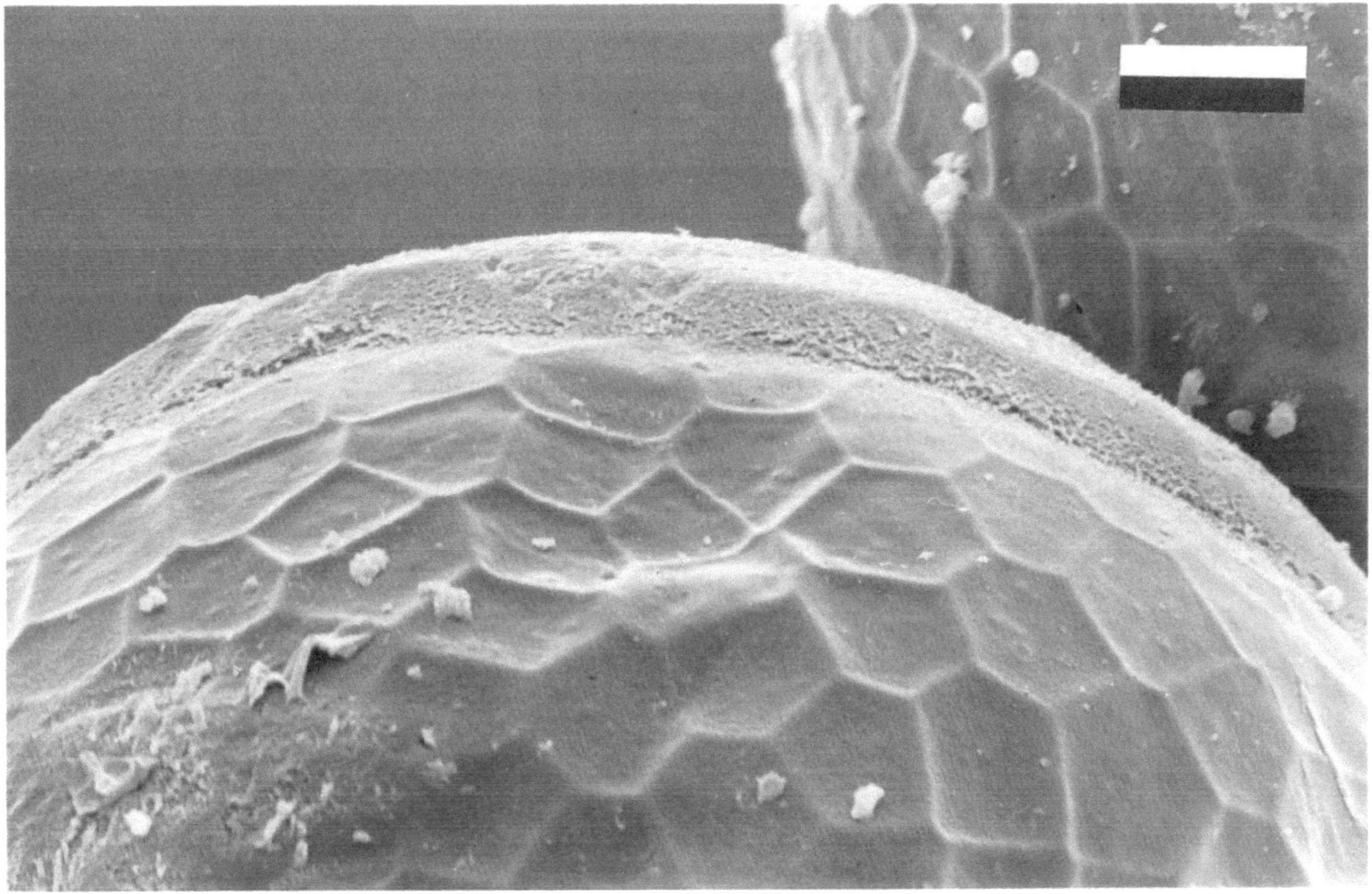

FIGURE 3. Scanning electron microscopy of a cyst dechorionated with hypochlorite and incubated 18 h, showing the polygonal pattern of structural elements within the cuticle. The surface of the outer cuticular membrane is exposed except at the top of the image where a remnant of alveolar layer has survived. Bar = 20 μm.

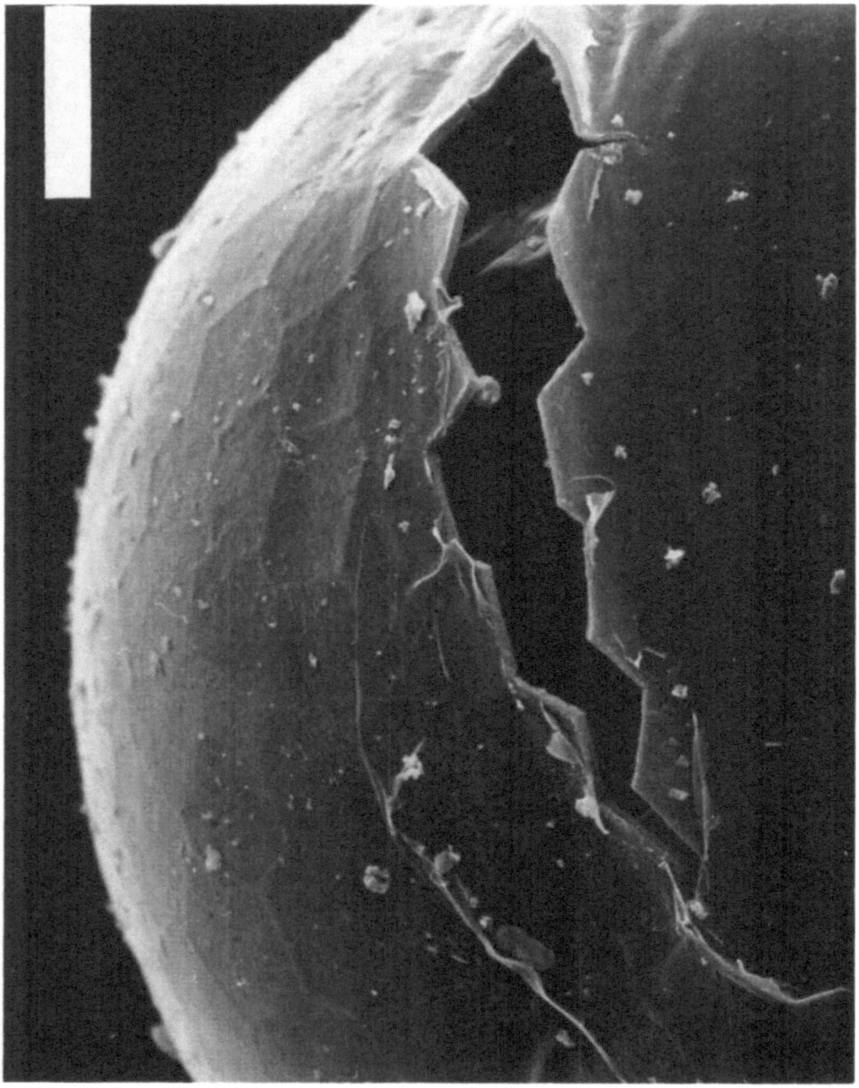

FIGURE 4. A dechorionated cyst commencing emergence after 16 h incubation. The cuticle has fractured along the structural elements between polygonal plates. Bar = 20 μm.

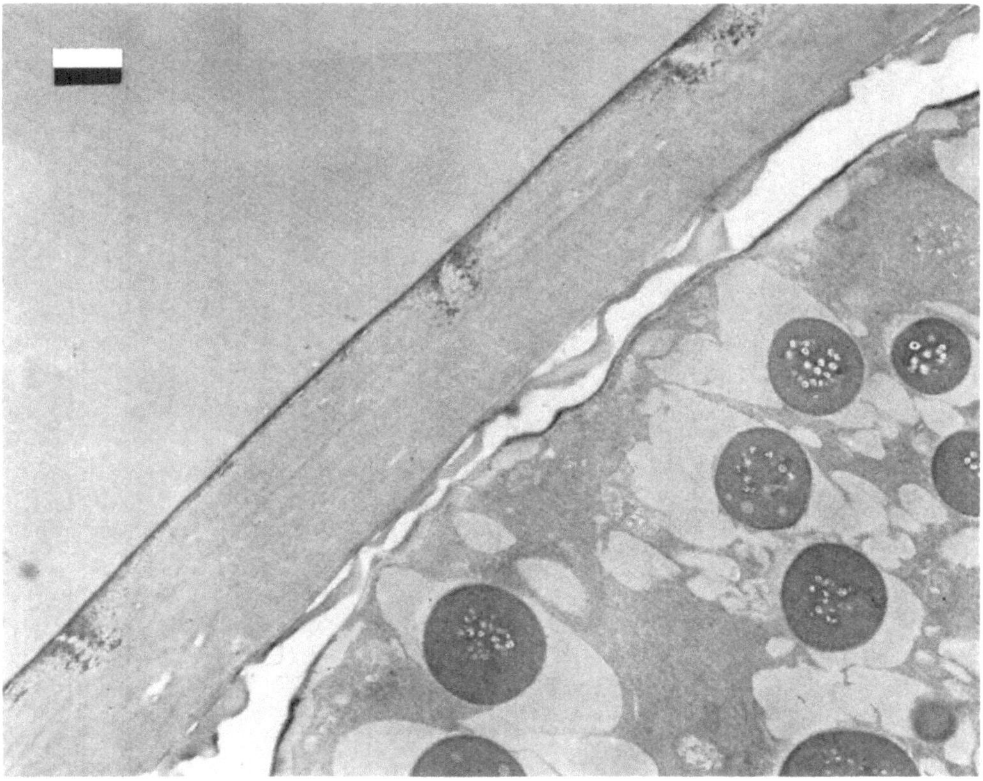

FIGURE 5. Dechorionated cyst. In the cuticle three structural elements are seen in section. Bar = 2 μm.

III. ABNORMAL EARLY DEVELOPMENT

A. DEFECTIVE EMERGENCE

The crucial aspect of the osmotic expansion mechanism is that if it is not sufficiently vigorous, the prenauplius is not propelled clear of the cyst. Certain distinct abnormalities of emergence can be traced to inadequate expansion.

The most conspicuous version of the abnormality is the oval-shaped prenauplius (Figure 7) which, we have reasoned,[37] results from retention of the inner cuticular membrane. Another abnormality at the late-E1 stage (Figure 8) is related to the oval abnormality, but with the broken shell retained. The E1 stage (terminology of Nakanishi et al.[2]) is a normal but transient stage of emergence; the abnormality is in the arrest of the organism in this or a late, almost-E2 condition as in Figure 8.

The appearance of these abnormalities in cysts incubated with cyanide[9] is compatible with a deficient ATP supply attributable to inhibition of oxidative phosphorylation. In terms of the model, the generation of an inadequate osmotic potential means that the influx of water, although adequate to crack the alveolar layer, is insufficient to rupture the inner cuticular membrane. Thus the hatching membrane sac which propels the organism from the cyst is prevented from inflating. The detachment of the cyst wall segments leading to the oval stage is fortuitous, but the prenauplius nevertheless starves because the inner cuticular and hatching membranes surrounding it obstruct feeding.

The observation that a deficiency of bicarbonate ion in the incubation medium provokes the same abnormalities as are attributed to ATP deficiency,[37] suggest a possible involvement

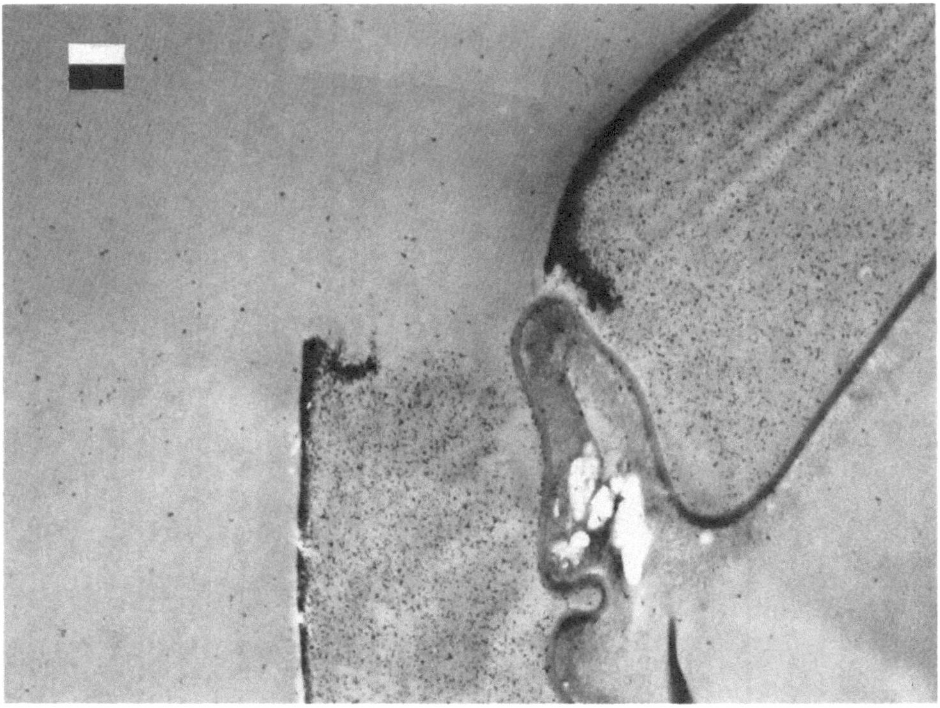

FIGURE 6. Section of cyst cuticle fractured through the center of a structural element and throughout the thickness of the cuticle except for the inner cuticular membrane. (The colloidal gold stain is related to a different study.) Bar = 1 μm.

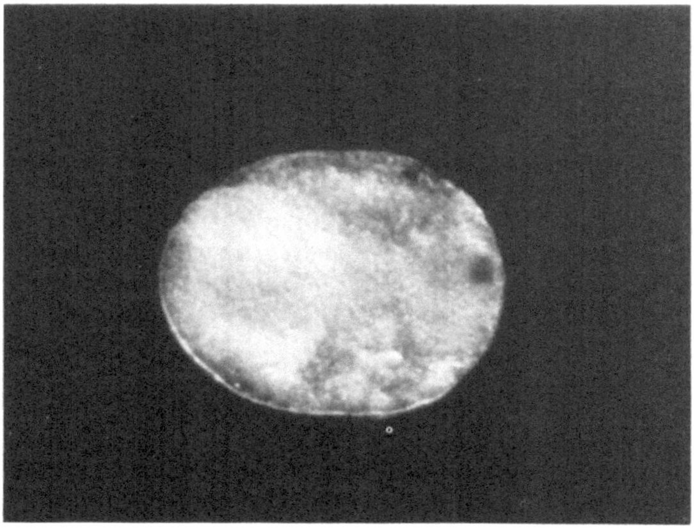

FIGURE 7. Abnormal oval prenauplius; eyespot at right; optical, dark field.

FIGURE 8. Abnormal encysted prenauplius arrested at late E1 stage.

of HCO_3^- in generation of the osmotic potential. The model in Figure 1 therefore incorporates two classical HCO_3^- or CO_2 utilizing reactions.

The first is a Cl^- for HCO_3^- exchange gate, a logical adjunct to an ATPase pumping out Na^+ in exchange for K^+. Significantly, external HCO_3^- is found not to be required until shortly before emergence.[37] The inference is that the embryo may be self-sufficient in HCO_3^- derived from decarboxylation reactions, diffusing out of the embryo as CO_2 but largely retained within the cyst as HCO_3^- and returning to the embryo in exchange for Cl^-.

The second is carbonate dehydratase. By favoring within the cyst the existence of CO_2 inside the embryo and HCO_3^- on the outer surface of the exoskeleton, assuming those to be the predominant equilibrium species at the respective locations, carbonate dehydratase could facilitate transport both of HCO_3^- into and of CO_2 out of the embryo while also minimizing outward loss through the cyst wall of dissolved CO_2. The abnormal emergence that follows cyst-wall rupture in HCO_3^- deficient incubation medium may result from an efflux of accumulated metabolic HCO_3^- into the medium after rupture of the outer cuticular membrane and additionally provides a clue that the inner cuticular membrane may be partially permeable.

Rafiee et al.[38] observed the same abnormalities in response to cadmium toxicity as we reported for HCO_3^- deficiency, and suggested a possible effect through microtubule disruption. It is significant that cadmium forms an insoluble carbonate, however, and an alternative mechanism could involve the formation of an insoluble complex at carbonate binding sites.

If carbonate dehydratase is present, it ought to be measurable. The enzyme is known in crustaceans.[39] Efforts by Conte[40] and efforts in our laboratory to detect it directly in prenauplii have not been successful beyond possible traces. Nevertheless, 18 h nauplii treated with acetazolamide and sulphanilamide, recognized as carbonate dehydratase inhibitors,[41] suffer marked toxic effects including a swelling compatible with impaired osmoregulation.[37] This tantalizing indirect suggestion of a carbonate dehydratase is deserving of further exploration and possible extrapolation to earlier stages with the aid of molecular probes of gene expression.

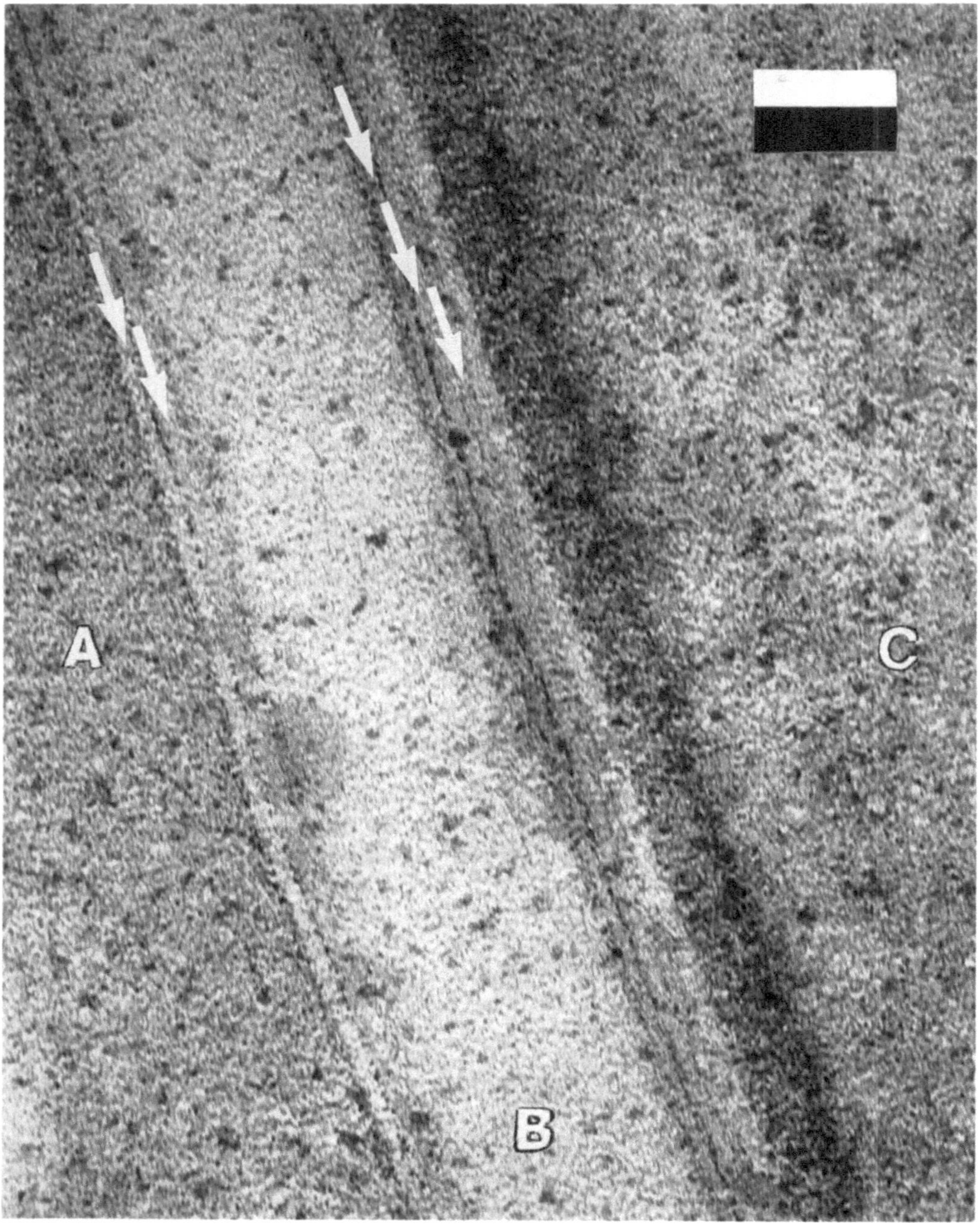

FIGURE 9. Outer cuticular membrane located between alveolar layer (A) and fibrous cuticle (C). Amorphous zone (B). Arrows indicate structures approximately 7—8 nm wide, i.e., having the dimensions of potential bilayer membranes. (Original magnification 100,000.) Bar = 50 nm.

De Chaffoy et al.[42] noted that embryos excised from the uterus before the second gastrula stage and then cultured *in vitro* develop into a defective and nonviable naupliar stage. These explants also fail to establish a permeability barrier within their outer cuticular membrane. Reference to the high magnification micrograph of outer cuticular membrane in Figure 9 reveals a central amorphous zone of about 80 nm wide, flanked on its alveolar surface by

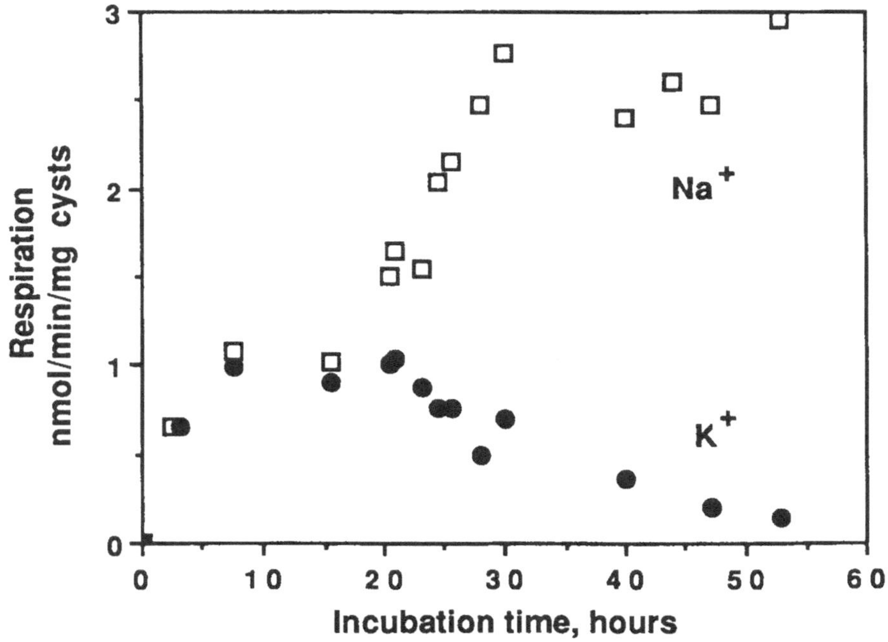

FIGURE 10. Respiration of cysts incubated in 600 m*M* NaCl or 600 m*M* KCl media, each containing 10 m*M* Tris-Cl, pH 8. Bacterial activity was suppressed with gentamicin 100 μg/ml. Emergence to the E2 stage was normal in both media but only cysts incubated in Na^+ medium progressed from E2 to hatching.

two juxtaposed structures, each matching the classical 7 to 8 nm dimension of a lipid bilayer membrane, and on its cuticular surface by at least a triple barrier of such membranes. It is the amorphous zone that apparently fails to develop properly *ex utero*.[42] If, however, the embryo is removed from the uterus after the second gastrula stage, then an impermeable outer cuticular membrane is already in place and development proceeds normally. This important observation is consistent with a requirement for the developing late embryo to retain certain metabolites but leaves open the question of whether these are required for development of the embryo or for its emergence, or for both functions. It may be significant that the *ex utero* embryos in these experiments were cultured in HCO_3^--free medium.[42]

B. DEFECTIVE HATCHING

The response of the prenauplius to molecules applied outside the hatching membrane indicates that the membrane is at least partially permeable.

Prenauplii treated with 500 μ*M* cycloheximide have shown impaired locomotor activity associated with a marked retardation of hatching (40% hatching after 37 h; controls 90%).[9] Whatever the consequence of action of this protein synthesis inhibitor in *Artemia*, the hatching membrane is evidently permeable to a molecule of M_r 281.

The hatching membrane appears also to be permeable to certain ions. Incubation of cysts in media based on KCl in place of NaCl results in full emergence of respiring and apparently normal prenauplii at the conventional time (Figure 10). These then fail to commence limbic movements and remain arrested at the E2 stage with dwindling respiration (Figure 10) unless the KCl is replaced with NaCl within a few hours of emergence (Figure 11), showing that the prenauplius is sensitive to ionic species external to the hatching membrane.

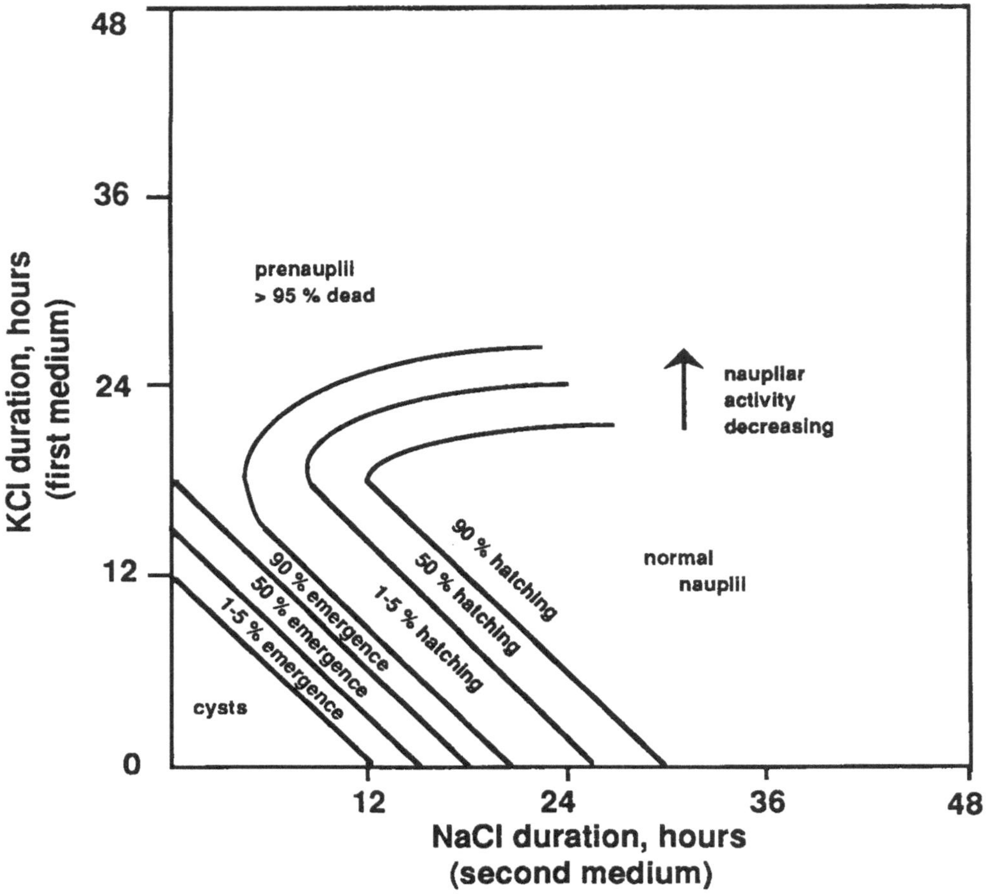

FIGURE 11. Emergence and hatching of cysts incubated in 600 m*M* KCl for different periods of time, followed by transfer to 600 m*M* NaCl, 10 m*M* KCl, 10 m*M* $MgCl_2$; both media contained 10 m*M* Tris, pH 8, and gentamicin 10 μg/ml. The contours interpolate points of equal emergence or hatching percentage; e.g., 18 h KCl followed by 12 h NaCl gave 90% hatching, whereas 24 h KCl, 9 h NaCl gave 1—5% hatching.

IV. DISCUSSION

A. NORMAL EMERGENCE MECHANISM

A detailed scenario of the emergence mechanism can now be pieced together from the evidence described in the earlier sections. The dual sources of osmotic potential that are evident, namely glycerol and ionic gradients, can be reconciled by incorporating both into a biphasic model which proposes that the former is augmented by the latter.

What are the initial events of this scenario? Hydration of the cyst allows water to traverse the alveolar and cuticular regions of the cyst wall to rehydrate the embryo. If the embryo has been activated,[16] also dissolved oxygen is available, the temperature supports metabolism and the internal pH is within limits,[22] then metabolism and respiration start. Trehalose is metabolized initially by glycolysis and glycerol production, later increasingly via the citric acid cycle and oxidative phosphorylation to generate ATP, CO_2, and H_2O.

At this stage two critical components determining emergence are becoming available, namely glycerol and ATP. Accumulating glycerol constitutes an osmotic gradient in and around the embryo, extending possibly as far as the outer cuticular membrane to permeate the fibrous cuticle. The resulting internal hydrostatic pressure fractures the inelastic alveolar layer and the osmotic reservoir expands slightly, revealing the E1 stage, after nominally 8 to 12 h under laboratory conditions.

ATPase expression or activation is programmed to commence shortly before this time, between 6 h and 8 h. The combined requirement for respiration and HCO_3^- suggests that the embryo extrudes NaCl into the osmotic reservoir, as Na^+ exchanged for K^+ by an ATP-energized pump and as Cl^- exchanged for HCO_3^- passively. The K^+ and the HCO_3^- are each recycled; classically K^+ leaks back out across the membrane, while HCO_3^- equilibrates with CO_2 (catalyzed by carbonate dehydratase) and diffuses out as the dissolved gas. There may also be a tendency for the osmotic potential to be reabsorbed by the embryo, for instance via the alimentary tract,[40] necessitating continual extrusion.

What are the critical structures mediating emergence? Hydrostatic pressure builds beneath the hatching and cuticular membranes. The cuticle is liable to failure at the boundaries of the hexagon plates but is constrained by the outer cuticular membrane. The alveolar layer cannot be regarded as a crucial constraint since emergence proceeds normally in de-chorionated cysts. Ultimately, the outer cuticular membrane fails catastrophically and the rupture is propagated extensively, partly along the structural elements and throughout the depths of the fibrous layer. The hatching membrane sac then inflates, finally rupturing the inner cuticular membrane surrounding it, forcing apart the cuticular and alveolar remnants and propelling the enclosed prenauplius from the cyst as the E2 stage.

The evidence[37] of pathological failure to progress from E1 to E2 suggests that it is possible for the osmotic pressure to be adequate to rupture the alveolar layer, outer cuticular membrane, and fibrous cuticle, but inadequate to complete the process by rupturing the inner cuticular membrane. Once the outer cuticular membrane is breached the osmotic potential will begin to dissipate to the environment through the partially permeable inner cuticular membrane and hatching membrane and, unless the hatching membrane sac inflates promptly, the opportunity is lost.

Conversely, the hatching membrane never bursts during normal emergence but fills to a fixed capacity. Nor is there evidence of special strength in the hatching membrane, which, on the contrary, is the most difficult membrane to discern in sections. The partial permeability of the hatching membrane appears to provide the necessary mechanism to limit the degree of inflation in response to any disparity in osmotic potential at emergence time.

There is no compelling evidence that the point of rupture of the cyst wall is other than random in relation to the orientation of the embryo, as expected for the failure of a pressurized sphere. The common appearance of the eyespot in the rupture window, followed by head-first emergence, is explicable by a passive rotation of the embryo and hatching membrane to match the slightly conical shape acquired by the cyst as the alveolar rupture widens. The fit of the hatching membrane to the shape of the prenauplius, and the fusion of the caudal part of the hatching membrane to the cuticle, are unremarkable considering the ecdytic origin of the cuticular and hatching membranes.[8,43]

1. Unsolved Problems

The area of greatest uncertainty in the above model is the source of the proposed ionic gradient. Is an internal embryonic source available? An order of magnitude estimate of average Na^+ concentration throughout a 200 μm diameter cyst would have an upper limit of about 70 m*M* taking Conte's[32] measurements of 0.8 μmol Na^+/mg dry weight and cyst weight of 3.5 μg, and a lower limit a quarter of that in the case[44] of cysts less endowed with Na^+ and with a weight of 1.6 μg.[45] Such a Na^+ complement ought to be ample for the generation of a much higher concentration if confined to a small osmotic reservoir. The reservation to this calculation is that it is not known how much of the measured Na^+ was entrapped within the alveolar spaces outside the permeability barrier and unavailable to the embryo.

The alternative is an external ion source, but enhanced detail of the rupture mechanism and of membrane permeabilities is required in order to refine the model. Much has been based on the valuable but quantal result of Morris and Afzelius[5] in the identification of one specific permeability barrier at the outer cuticular membrane. It is possible that the combined permeabilities of the other membranes of the cyst, which might also undergo changes during development or in response to pressure or tension, could permit the enveloped prenauplius to act as a net importer of environmental ions in the late emergent phase. The apparent ability of the system to exchange HCO_3^- with the environment in this phase supports such a concept. The complexity of the outer cuticular membrane illustrated in Figure 9 and the precedent of the variation in permeability with developmental conditions identified by De Chaffoy et al. in experiments with embryos *ex utero*[42] suggest considerable scope for the modulation of cyst permeability during development.

The ability of cysts to reach the E2 stage in non-ionic media such as mannitol[19] or glycerol[35] superficially argues against the concept of an external source for an ionic gradient, except that adventitious ions would be available from both alveolar and extraneous sources and from buffers unless rigorously excluded.

We have deduced that the emergent phase in a mixed culture is accompanied by either an efflux or an influx, depending on the concentration difference prevailing, of glycerol, HCO_3^-, Na^+, and plausibly other solutes and ions, but the key unanswered question is the precise timing in an individual cyst of the onset of this access to the environment. It makes a considerable difference to the interpretation according to whether accessibility has to await the exposure to the medium of the outer cuticular membrane, the inner cuticular membrane, or the hatching membrane. Considerable scope remains for further work to refine the detailed timescale of a complex process of emergence.

ACKNOWLEDGMENTS

The author is grateful for grants and support from the New Zealand Lottery Grants Board and University Grants Committee; the Belgian National Science Foundation; the Corday Morgan Fellowship of the Royal Society of Chemistry; the E. G. Johnstone Trust; and the J. D. S. Roberts Foundation.

REFERENCES

1. **Needham, J. and Needham, D. M.,** On phosphorous metabolism in embryonic life. I. Invertebrate eggs, *J. Exp. Biol.*, 7, 317, 1930.
2. **Nakanishi, Y. H., Iwasaki, T., Okigaki, T., and Kato, H.,** Cytological studies of *Artemia salina*. I. Embryonic development without cell multiplication after the blastula stage in encysted dry eggs, *Annot. Zool. Jpn.*, 35, 223, 1962.
3. **Nakanishi, Y. H., Okigaki, T., Kato, H., and Iwasaki, T.,** Cytological studies of *Artemia salina*. II. Deoxyribonucleic acid (DNA) content and the chromosomes in encysted dry eggs and nauplii, *Proc. Jpn. Acad.*, 39, 306, 1963.
4. **Morris, J. E.,** Dehydrated cysts of *Artemia salina* prepared for electron microscopy by totally anhydrous techniques, *J. Ultrastuct. Res.*, 25, 64, 1968.
5. **Morris, J. E. and Afzelius, B. A.,** The structure of the shell and outer membranes in encysted *Artemia salina* embryos during cryptobiosis and development, *J. Ultrastruct. Res.*, 20, 244, 1967.
6. **Iwanami, Y.,** Retaining the viability of the resting egg of brine shrimp (*Artemia salina*) in organic solvent, *Exp. Cell Res.*, 78, 470, 1973.
7. **Crowe, J. H., Crowe, L. M., and Chapman, D.,** Preservation of membranes in anhydrobiotic organisms: the role of trehalose, *Science*, 223, 701, 1984.

8. **Funke, B. and Spindler, K.-D.**, Developmental changes of chitinolytic enzymes and ecdysteroid levels during the early development of the brine shrimp *Artemia*, in *Artemia Research and Its Applications*, Vol. 2, *Physiology, Biochemistry, Molecular Biology*, Decleir, W., Moens, L., Slegers, H., Sorgeloos, P., and Jaspers, E., Eds., Universa Press, Wetteren, Belgium, 1987, 67.
9. **Trotman, C. N. A., Mansfield, B. C., and Tate, W. P.**, Inhibition of emergence, hatching and protein biosynthesis in embryonic *Artemia salina*, *Devel. Biol.*, 80, 167, 1980.
10. **Schmitt, H., Grossfeld, H., and Littauer, U. Z.**, Mitochondrial biogenesis during differentiation of *Artemia salina* cysts, *J. Cell Biol.*, 58, 643, 1973.
11. **Sorgeloos, P.**, First report on the triggering effect of light on the hatching mechanism of *Artemia salina* dry cysts, *Mar. Biol.*, 22, 75, 1973.
12. **Van der Linden, A., Blust, R., and Decleir, W.**, The influence of light on the hatching of *Artemia* cysts (Anostraca: Branchiopoda: Crustacea), *J. Exp. Mar. Biol. Ecol.*, 92, 207, 1985.
13. **Van der Linden, A., Vankerckhoven, I., Caubergs, R., and Decleir, W.**, Action spectroscopy of light-induced hatching of *Artemia* cysts (Branchiopoda: Crustacea), *Mar. Biol.*, 91, 239, 1986.
14. **Van der Linden, A., Blust, R., Cuypers, K., Thoeye, C., and Bernaerts, F.**, An action spectrum for light-induced hatching of *Artemia* cysts, in *Artemia Research and Its Applications*, Vol. 2, *Physiology, Biochemistry, Molecular Biology*, Decleir, W., Moens, L., Slegers, H., Sorgeloos, P., and Jaspers, E., Eds., Universa Press, Wetteren, Belgium, 1987, 181.
15. **Lavens, P. and Sorgeloos, P.**, The cryptobiotic state of *Artemia* cysts, its diapause deactivation and hatching: a review, in *Artemia Research and Its Applications*, Vol. 3, *Ecology, Culturing, Use in Aquaculture*, Sorgeloos, P., Bengtson, D. A., Decleir, W., and Jaspers, E., Eds., Universa Press, Wetteren, Belgium, 1987, 27.
16. **Drinkwater, L. and Clegg, J. S.**, Experimental biology of cyst diapause (in this volume), Chapter 5, 1991, 93.
17. **Clegg, J. S.**, Interrelationships between water and cellular metabolism in *Artemia* cysts. II. Carbohydrates, *Comp. Biochem. Physiol.*, 53(A), 83, 1976.
18. **Clegg, J. S.**, Interrelationships between water and cellular metabolism in *Artemia* cysts. III. Respiration, *Comp. Biochem. Physiol.*, 53(A), 89, 1976.
19. **Clegg, J. S.**, The control of emergence and metabolism by external osmotic pressure and the role of free glycerol in developing cysts of *Artemia salina*, *J. Exp. Biol.*, 41, 879, 1964.
20. **Busa, W. B., Crowe, J. H., and Matson, G. B.**, Intracellular pH and the metabolic status of dormant and developing *Artemia* embryos, *Arch. Biochem. Biophys.*, 216, 711, 1982.
21. **Busa, W. B. and Crowe, J. H.**, Intracellular pH regulates transitions between dormancy and development of the brine shrimp (*Artemia salina*) embryos, *Science*, 221, 366, 1983.
22. **Crowe, J. H., Crowe, L. M., Drinkwater, L., and Busa, W. B.**, Intracellular pH and anhydrobiosis in *Artemia* cysts, in *Artemia Research and Its Applications*, Vol. 2, *Physiology, Biochemistry, Molecular Biology*, Decleir, W., Moens, L., Slegers, H., Sorgeloos, P., and Jaspers, E., Eds., Universa Press, Wetteren, Belgium, 1987, 19.
23. **Dutrieu, J. and Crestia-Blanchine, D.**, Resistance des oeufs durables hydrates d'*Artemia salina* a l'anoxie, *C. R. Acad. Sci. Paris*, 263, 998, 1966.
24. **Ewing, R. D. and Clegg, J. S.**, Evidence for a single form of lactate dehydrogenase in *Artemia salina*, *Fed. Proc.*, 27 (Abstr. 3199), 785, 1968.
25. **Stocco, D. M., Beers, P. C., and Warner, A. H.**, Effect of anoxia on nucleotide metabolism in encysted embryos of the brine shrimp, *Dev. Biol.*, 27, 479, 1972.
26. **Ewing, R. D. and Clegg, J. S.**, Lactate dehydrogenase activity and anaerobic metabolism during embryonic development in *Artemia salina*, *Comp. Biochem. Physiol.*, 31, 297, 1969.
27. **Glasheen, H. S. and Hand, S. C.**, Anhydrobiosis in embryos of the brine shrimp *Artemia*: characterization of metabolic arrest during reductions in cell-associated water, *J. Exp. Biol.*, 135, 363, 1988.
28. **Carpenter, J. F. and Hand, S. C.**, pH-sensitive control points of carbohydrate metabolism in encysted embryos of *Artemia*, *Am. Zool.*, 23 (Abstr. 63), 891, 1983.
29. **Carpenter, J. F. and Hand, S. C.**, pH-dependent arrestment of carbohydrate metabolism during anaerobic dormancy and aerobic acidosis in *Artemia* cysts, *Am. Zool.*, 25 (Abstr. 263), 50(A), 1985.
30. **Batuecas, B., Garesse, R., and Marco, R.**, The isolation of the yolk granules DNA in the crustacean *Artemia* and the storage of mitochondria in the yolk granules of the dormant gastrulae, *Biol. Cell*, 45 (Abstr. 340), 160, 1982.
31. **Vallejo, C. G. and Perona, R.**, Microscopical study of the mechanisms of yolk degradation in *Artemia* and the concomitant release of mitochondria and other stored informational material, *Biol. Cell*, 45 (Abstr. 156), 97, 1982.
32. **Conte, F. P., Droukas, P. C., and Ewing, R. D.**, Development of sodium regulation and *de novo* synthesis of Na+K-activated ATPase in larval brine shrimp, *Artemia salina*, *J. Exp. Zool.*, 202, 339, 1977.
33. **Jardel, J.-P. and Nayudu, P. R. V.**, Mitochondrial maturation of *Artemia salina*, *Micron*, 13, 365, 1982.

34. **Guo, J. Z. and Hokin, L. E.**, Differential gene expression of sodium/potassium ATPase alpha- and beta-subunits in developing brine shrimp, *Book of Abstracts, NATO Advanced Research Workshop on Cell and Molecular Biology of Artemia Development*, 20 (Abstr. A16), 1988.
35. **Geldiay, R., Koray, T., and Buyukisik, B.**, Effect of different glycerol concentrations upon the hatching of *Artemia*, in *The Brine Shrimp Artemia*, Vol. 2, *Physiology, Biochemistry, Molecular Biology*, Persoone, G., Sorgeloos, P., Roels, O., and Jaspers, E., Eds., Universa Press, Wetteren, Belgium, 1980, 147.
36. **Bruggeman, E., Sorgeloos, P., and Vanhaecke, P.**, Improvements in the decapsulation technique of *Artemia* cysts, in *The Brine Shrimp Artemia*, Vol. 3, *Ecology, Culturing, Use in Aquaculture*, Persoone, G., Sorgeloos, P., Roels, O., and Jaspers, E., Eds., Universa Press, Wetteren, Belgium, 1980, 261.
37. **Trotman, C. N. A., Gieseg, S. P., Pirie, R. S., and Tate, W. P.**, Abnormal development in *Artemia*: defective emergence of the prenauplius with bicarbonate deficiency, *J. Exp. Zool.*, 243, 225, 1987.
38. **Rafiee, P., Matthews, C. O., Bagshaw, J. C., and MacRae, T. H.**, Reversible arrest of *Artemia* development by cadmium, *Can. J. Zool.*, 64, 1633, 1986.
39. **Henry, R. P.**, The function of invertebrate carbonic anhydrase in ion transport, *Ann. N.Y. Acad. Sci.*, 429, 544, 1984.
40. **Conte, F. P.**, Structure and function of the crustacean larval salt gland, *Int. Rev. Cytol.*, 91, 45, 1984.
41. **Pocker, Y. and Sarkenen, S.**, Carbonic anhydrase: structure, catalytic versatility, and inhibition, *Adv. Enzymol.*, 47, 149, 1978.
42. **De Chaffoy, D., De Maeyer-Criel, G., and Kondo, M.**, On the permeability and formation of the embryonic cuticle during development *in vivo* and *in vitro* of *Artemia salina* embryos, *Differentiation*, 12, 99, 1978.
43. **Belk, D.**, Embryonic cuticles of *Artemia* during diapause and hatching: insights from comparison with other branchiopoda, *J. Crustacean Biol.*, 7, 691, 1987.
44. **Cowgill, U. M., Emmel, H. W., Boggs, G. U., Murphy, P. G., Gersich, F. M., Takahashi, I. T., and Bengtson, D. A.**, Variations in chemical composition of *Artemia* cysts from three geographical locations, in *Artemia Research and Its Applications*, Vol. 1, *Morphology, Genetics, Strain Characterization, Toxicology*, Sorgeloos, P., Bengtson, D. A., Decleir, W., and Jaspers, E., Eds., Universa Press, Wetteren, Belgium, 1987, 173.
45. **Trotman, C. N. A.**, unpublished data.

Chapter 5

EXPERIMENTAL BIOLOGY OF CYST DIAPAUSE

Laurie E. Drinkwater and James S. Clegg

TABLE OF CONTENTS

I. INTRODUCTION

During the normal course of reproductive events in *Artemia*, the diapause embryo, also called a "cyst", experiences diverse environmental extremes and thus exhibits a variety of corresponding adaptations at all levels of organization. These responses to environmental conditions occur in conjunction with endogenously controlled metabolic transitions, of which perhaps the most striking is the shift in embryos from the diapause state to the resumption of development. Because exogenously and endogenously imposed metabolic and developmental arrests often occur simultaneously, it is necessary to consider both in order to understand diapause. Consequently, while the focus of this chapter is on the control of diapause in *Artemia*, we will also cover relevant aspects of environmentally imposed dormancy.

We have not attempted to review all of the literature on the effects of environmental conditions on diapause and hatching nor have we addressed the biogeography of diapause. Readers are referred to the recent paper by Berthelemy-Okazaki[1] and the thorough review by Lavens and Sorgeloos[2] for more extensive information on these subjects. We are concerned mainly with *Artemia franciscana* since the most detailed studies of diapause regulation have been conducted on two natural populations of this bisexual species: one from San Francisco Bay, California (referred to as SFB), the other from the Great Salt Lake, Utah (referred to as GSL).

A. TERMINOLOGY

Given the capacity for two fundamentally different dormant states in these cysts, it is necessary to define the terms we will use in the ensuing discussion. Embryos in *diapause* are kept dormant by some endogenous, physiological, and/or structural mechanisms since they will not resume development even under favorable conditions.[3-5] Many organisms enter diapause as a result of environmental cues that can be interpreted to forecast the coming of adverse conditions and thus become dormant prior to experiencing unfavorable conditions. Thus, diapause can serve as an important timing mechanism to synchronize the appearance of actively feeding stages with environmental food sources. Release from diapause, which we will term "activation", usually requires transient exposure to a specific environmental stimulus (or stimuli), and resumption of development and metabolism associated with it are possible only after this treatment. The word "transient" is important here in order to distinguish between activation cues and the environmental conditions which are simply required for development.

In contrast, we use the term "quiescence" to refer to a hypometabolic state (which in this case includes developmental arrest) imposed by unfavorable external conditions such as insufficient water or oxygen, or low temperature.[4] In this case, the embryo is responding directly to an unfavorable environment; as soon as favorable conditions are restored, normal metabolism and development are resumed. Thus, in this review we will refer to any cyst that has not been subjected to diapause breaking cues as a *diapause cyst*, while the term *activated cyst* will be used to refer to cysts whose diapause has been terminated. Using these definitions, diapause in *Artemia* is equivalent to dormancy in seeds. While the term "dormancy" generally has this specific meaning to botanists, zoologists tend to use it in a more general sense, referring to any hypometabolic state whether intrinsic or extrinsic.[36] This is certainly the case in the *Artemia* literature. Figure 1 summarizes these terms as we will apply them to *Artemia*.

B. THE DIAPAUSE EMBRYO

Two modes of reproduction have been described in *A. franciscana*: for a given clutch of eggs, females can produce either encysted embryos which are released in diapause (ovi-

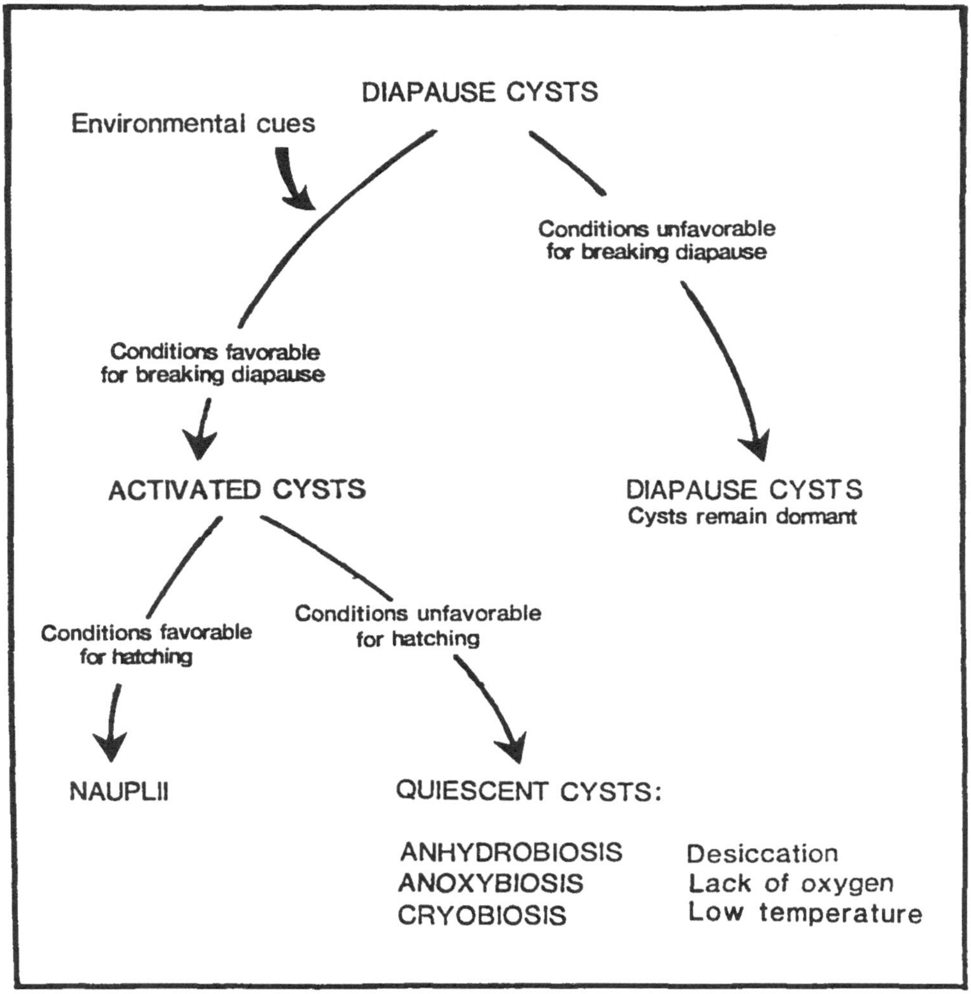

FIGURE 1. Terminology used in this review to refer to cysts in various metabolic states. (From Drinkwater, L. E. and Crowe, J. H., *J. Exp. Zool.*, 241, 297, 1987. With permission.)

parous route) or embryos which continue to develop in the ovisac, resulting in the release of nauplius larvae (ovoviviparous route) (reviewed by Clegg and Conte[8]). The diapause embryos appear to differ biochemically from those destined for the ovoviviparous route early in development.[9] At ten h of development, presumptive diapause embryos have large stores of trehalose, a nonreducing disaccharide, which is apparently synthesized by the embryo while still in the ovisac. Trehalose ultimately comprises at least 15% of the cyst's dry weight.[10,11] It has been proposed that the use of trehalose rather than glycogen as the storage carbohydrate in diapause embryos has adaptive significance since trehalose has the ability to preserve the integrity of membranes and other cellular constituents during desiccation, and is found in other cryptobiotic organisms which are resistant to desiccation.[12-15]

Diapause cysts are released from the ovisac as late gastrulae composed of about 4,000 cells (partly syncytial) surrounded by a complex, noncellular shell[16-19] (Figure 2). Two striking characteristics of the shell are of concern here because they have a major impact on experimental procedures: (1) the inner layer is impermeable to all nonvolatile solutes examined thus far[18,21-24] and (2) the outer layer is very tough and must be removed in many

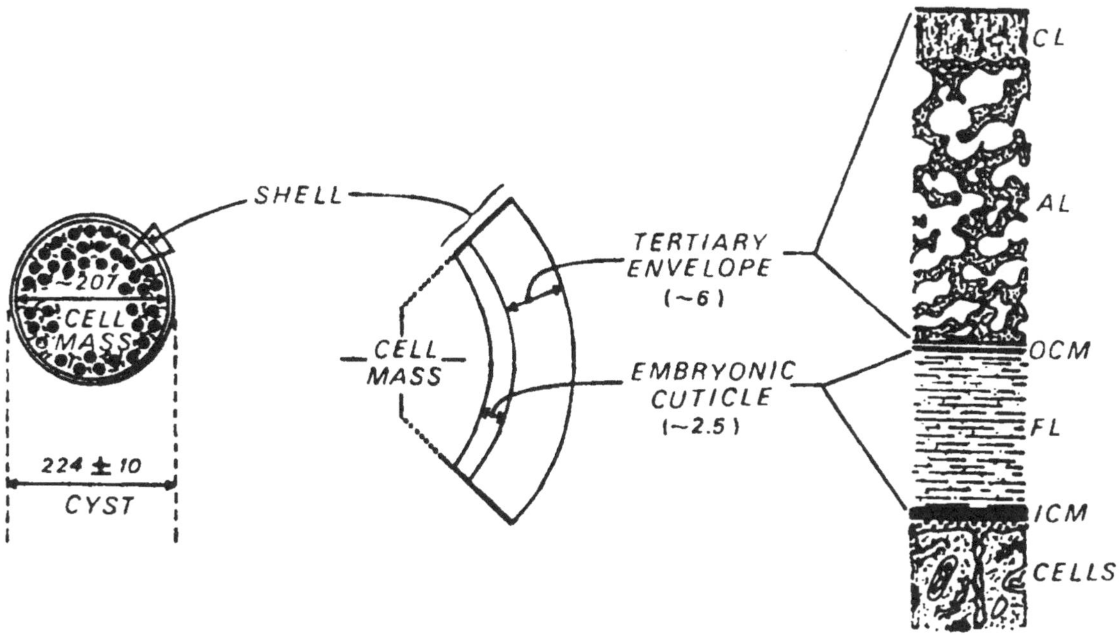

FIGURE 2. Diagram of an *Artemia* cyst at maximum water content. The entire cyst is shown at the left in cross-section; the numbers are in units of μm (± S.D.). Shell ultrastructure is shown to the right: CL (cuticular layer); AL (alveolar layer); OCM (outer cuticular membrane); FL (fibrous layer); ICM (inner cuticular membrane). The OCM appears to be the permeability barrier. Removal of the outer shell by decapsulation removes the CL and AL. (From Clegg, J. S., *Cell Biophys.*, 6, 153, 1984. With permission.)

instances in order to "gently" homogenize the embryos and recover cell organelles. Removal of the outer shell is now referred to as "decapsulation" and is accomplished by immersion of the embryos in solutions containing various concentrations of hypochlorite which oxidize and dissolve the outer layer of the shell leaving the permeability barrier intact.[17] The necessity of this treatment is particularly troublesome in experiments relating to diapause regulation since hypochlorite can, under some circumstances, activate diapause embryos.[25] One must be careful when using hypochlorite since it appears to adsorb to the shell, followed by slow release during incubation or homogenization.

II. METABOLISM IN ACTIVATED CYSTS

A very large amount of information is available on the metabolism that accompanies both the resumption of development and its subsequent suppression by adverse environmental conditions in cysts which have had their diapause terminated.[8,26-28] It is not our intent to repeat a review of this copious literature but rather to provide a brief summary of these metabolic states and some selected recent findings.

A. METABOLISM DURING THE RESUMPTION OF DEVELOPMENT

Once embryos have broken diapause, they will initiate what is termed *preemergence development* (PED) when placed under favorable conditions of sufficient water and oxygen at a permissible temperature.[10,11] The earliest events of PED entail mobilization of energy stores and restoration of RNA and protein synthesis (see Clegg and Conte[8] and recent books).[26-28] Available evidence indicates that the primary substrate of respiration is trehalose which is completely oxidized to CO_2 and H_2O, as well as used as a substrate for glycerol and glycogen synthesis.[10,11] The essential role of trehalose degradation in the resumption of development is perhaps best illustrated by studies on environmentally imposed developmental arrest; in all cases studied to date, the shutdown of trehalose catabolism brings the developmental process to a standstill.[29-32]

Diapause embryos also contain large numbers of yolk platelets which account for about 12% of the dry weight and are the storage site of two-thirds of the protein in the form of a lipovitelline complex.[33-35] Significant portions of the yolk platelets are metabolized during PED.[33] The yolk platelets of *Artemia* embryos have proven to be remarkable organelles. It appears that they may play a role in determining the sequence of events during PED by acting as storage sites for enzymes which are released in metabolically active forms as the yolk platelets are degraded.[36-39] These studies suggest that several hydrolytic enzymes which are present but not active in embryos during the early stages of PED may be associated with the yolk platelets. Later in development the enzymes are released as the yolk platelets are degraded and thus become active. The proteases are the most extensively studied of these enzymes and their role in development and the potential regulatory function of the yolk platelets have recently been reviewed by Warner.[40]

A unique characteristic of the yolk platelets should be emphasized here. It appears that immature mitochondria are also stored in the yolk platelets, a phenomenon discovered by Marco et al.[41] More-or-less mature mitochondria are released, chiefly after hatching as the yolk platelets disintegrate. Although the initial evidence for this was met with skepticism, these findings have been verified by Bes et al.,[42] and can now be considered to be well established.

A synopsis of the known events occurring during development of the diapause embryos is presented in Figure 3. Although highly simplified (a great deal is known about molecular genetics and the protein synthesizing system for example), this chart gives an overview of the sequence of the various categories of metabolic events leading to hatching. Note that termination of diapause does not initiate the resumption of all metabolic faculties, i.e., DNA

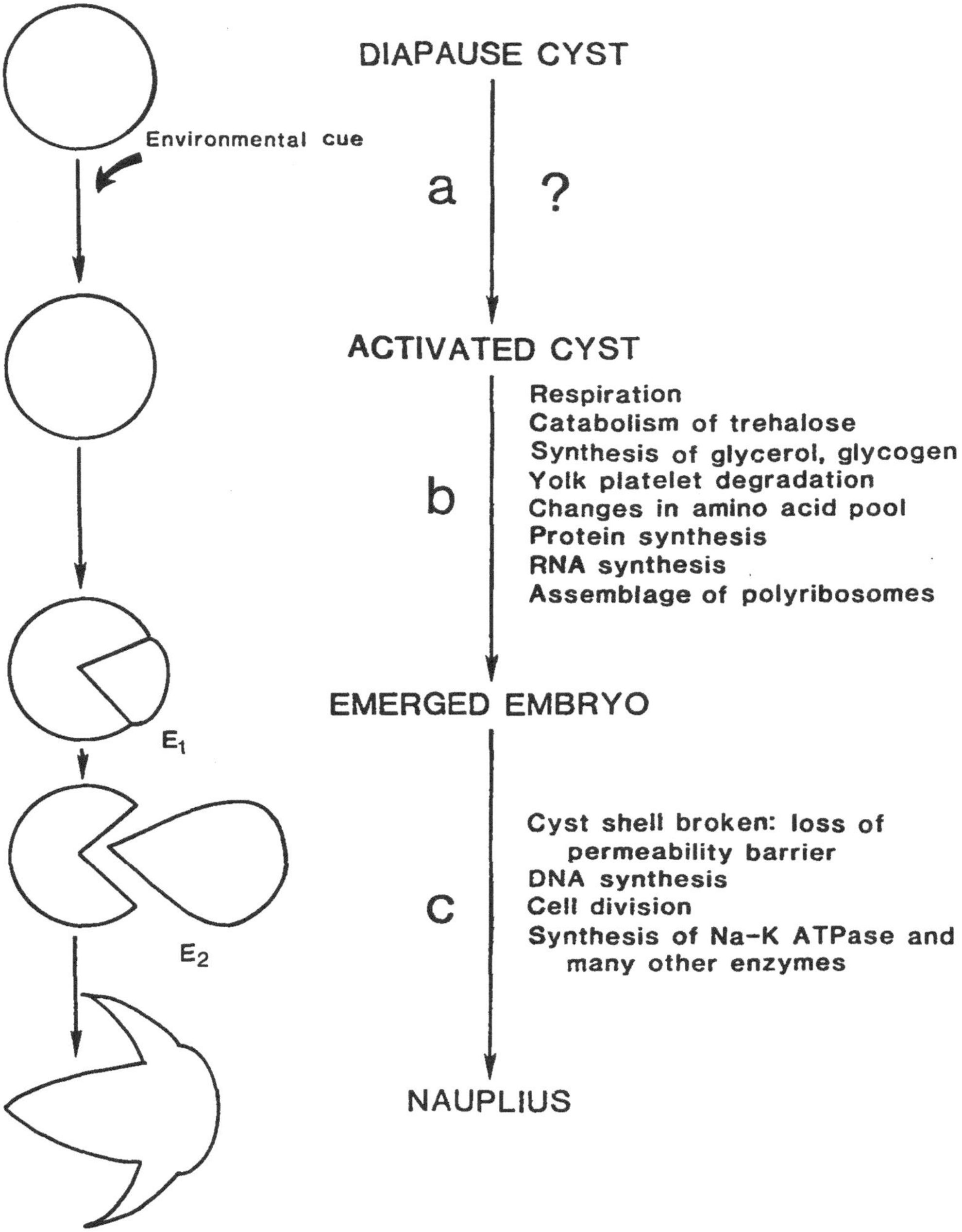

FIGURE 3. Selected metabolic events initiated during the various stages of *Artemia* development. Three notable transitions are included: (a) activation, (b) emergence, and (c) hatching. Emergence occurs in two stages: E_1 and E_2.

synthesis does not occur during PED. Instead, some metabolic functions appear to be mobilized at the time of emergence; however, the regulatory mechanisms controlling this particular transition have not been described in any detail. Furthermore, since all these studies have been conducted with cysts that have been processed commercially and stored in the dried but activated state, PED metabolism actually represents the resumption of metabolic activity and development after anhydrobiosis and not necessarily that associated with the termination of diapause. This makes it difficult to separate those metabolic events which relate solely to resumption of development from those which are connected with recovery from anhydrobiosis.

B. METABOLIC STATUS DURING QUIESCENCE

A. franciscana cysts can withstand essentially complete desiccation, long-term anoxia, even when fully hydrated, and prolonged, low temperatures without loss of viability, either while in diapause or the activated state.[2,8] From the standpoint of metabolic regulation, these astonishing abilities pose a variety of interesting questions. The cells must be able to rapidly halt metabolism in an orderly fashion in response to several very different environmental conditions, and yet be capable of reversing all those events equally quickly. Furthermore, in the case of anoxia, it would seem that an alternate anaerobic metabolism must be activated to allow the cells to perform maintenance functions; however, as we shall see there is reason to believe that anoxia may lead to a complete shutdown in metabolism. Finally, being embryonic, this system must retain, intact, all of its developmental capabilities and potentialities.

1. Anhydrobiosis

One of us (J. S. C.) has previously reviewed, in considerable detail, the metabolic status of activated cysts as a function of water content.[43-45] Glasheen and Hand[46] have contributed most recently to the study of this relationship in an excellent paper. Because all these results have little direct bearing on the diapause condition we simply note that this information is available. It is of interest that the lowest salinity reported to activate cysts (via dehydration) coincides with water contents low enough to inhibit conventional metabolism (0.6 g H_2O)/g dried cysts). An additional feature emerging from this analysis is that cysts containing water levels of about 0.1 g H_2O/g dried cysts or less cannot possibly be metabolizing (in the conventional use of that term). Thus, the severely dried embryo is truly ametabolic, and its properties differ markedly and qualitatively from actively living organisms (and "dead ones", for that matter). We end this brief account by pointing out another interesting feature of the dried and hydrating cyst: as water is added, and metabolism is reinitiated, we are observing the conversion of static nonliving cells into dynamic living ones, at least in those dried cysts that have retained the potential to make this transition. For those cysts that have lost this potential we can only conclude that at some point, they crossed the border into death.

2. Anoxybiosis

As mentioned, both diapause and activated hydrated embryos can survive exposure to anoxia for very long periods of time. The metabolic status of these anoxic embryos is, therefore, of some interest, particularly since it appears that a similar metabolic state can be endogenously imposed in embryos in the presence of oxygen.[47] It has been proposed that both carbohydrate and lipid metabolism are reversibly halted during anoxybiosis in hydrated, activated cysts.[29,48] Subsequent work indicated that the much-reduced, energetic requirements of these cysts might be supplied by the nucleotide, Gp4G (diguanosine-tetraphosphate), which decreased during anoxia.[49] However, no indication of what these energy-requiring processes might be was offered, a point to which we will return. More recently, Hand and Gnaiger[50] using microcalorimetry showed that heat dissipation decreased to about 2% of aerobic values after four h of anoxia, and pointed out that this was the most extreme suppression of metabolism under anoxia observed to date in euryoxic animals. Noting that heat dissipation was continuing to decrease when their studies were terminated, Hand and colleagues subsequently extended their studies to 6 d of anoxia, observing that heat dissipation leveled off at extremely low but measurable levels.[51] Questions arise from this: What is generating this heat? Are the cysts "metabolizing" in the usual sense of that term — coupled enzymatic reactions (pathways) under control of the cell (and organism) which contribute directly to its maintenance and integrity? Hand et al.[51] have proposed that anoxic cysts are indeed metabolizing, based on their observation that decreases in trehalose and glycogen occurred during an anoxic incubation period of six d. We found no measurable change in trehalose or glycerol contents during periods up to 6 months of anoxia, although a small decrease in glycogen was observed prior to, but not after, 3 months of anoxia.[52,53] We believe there are two potential explanations for these differences: (1) different sources of cysts were involved (Hand and colleagues used GSL populations; we used SFB cysts), and (2) different procedures were employed for anoxia (Hand and colleagues used continuous perfusion; we used sealed vials). Thus, at present there is uncertainty about the utilization of cyst carbohydrates for moderate duration anoxia (several days), and the possibility arises that the long-term anoxic cysts reversibly brings its metabolism to a standstill.

If the long-term anoxic cyst carries out some sort of metabolism, it does not appear to be based on carbohydrate utilization; furthermore, that metabolism does not seem to result in the production and release of volatile end products since no change in cyst dry weight was observed after 6 months of anoxia.[53] Because the accumulation of metabolic end products of anoxic metabolism might result in repayment by way of an oxygen debt we (J. S. C.) measured the initial rate of consumption of oxygen by cysts that had previously been incubated under anoxic conditions. Figure 4 is a representative result of such a study for control cysts (hydrated at 0°C for 18 h) and for cysts that had previously experienced 4 months of incubation at 22 to 24°C in anoxic sea water (sealed vials). Not only did anoxic cysts fail to show an oxygen debt, their initial respiration rate was markedly depressed compared to controls. This result has been obtained several times. One possible interpretation for this outcome is that some of the anoxic cysts have been induced to reenter diapause (see Figure 8); however, the situation is more complicated since that would not explain the lag in the increase in rate that occurs when anoxia is terminated (first 15 min of aerobic incubation in Figure 4). We have not yet examined the effects of redrying anoxic cysts on their subsequent respiration after rehydration.

We continue these and other approaches to understanding the metabolic status of anoxic cysts. The question has potential, general significance, in addition to the matter of anoxia in *Artemia* cysts: if the fully hydrated anoxic cyst does bring its metabolism to a standstill then we may ask how it maintains its integrity over periods of months since the teachings of modern biology tell us that living cells rapidly disintegrate in the absence of a constant supply of free energy (i.e., transferable phosphoryl groups). Even utilization of the nucleotide, Gp4G, appears to level off after 2 to 4 months of anoxia,[49] so this candidate seems to be an unlikely free energy source for long-term anoxia. We take up a potential means by which anoxic cysts are "stabilized" in the next section of this paper.

C. REGULATION OF ANOXYBIOSIS

It is currently well established that the shutdown of metabolism during these transitions from aerobic to anoxic conditions is mediated by intracellular pH (pH_i).[50,54,55] Using ^{31}P-nuclear magnetic resonance (NMR), Busa et al.[54] demonstrated that induction of anoxybiosis is accompanied by a decline in pH_i from ≥ 7.9 to about 6.3. The role of pH_i in regulation of this metabolic transition has been supported by manipulations of pH_i in the presence of oxygen.[50,55] Lowering pH_i under aerobic conditions using elevated partial pressures of CO_2 inhibits respiration and development in a manner similar to anoxia. Further studies suggest that the mobilization of energy sources and carbon skeletons (trehalose and the yolk platelets) in the activated cysts is inhibited by the lowered pH resulting from anoxybiosis.[35,56,57] Carpenter and Hand[56] have proposed that trehalase and hexokinase are major control points

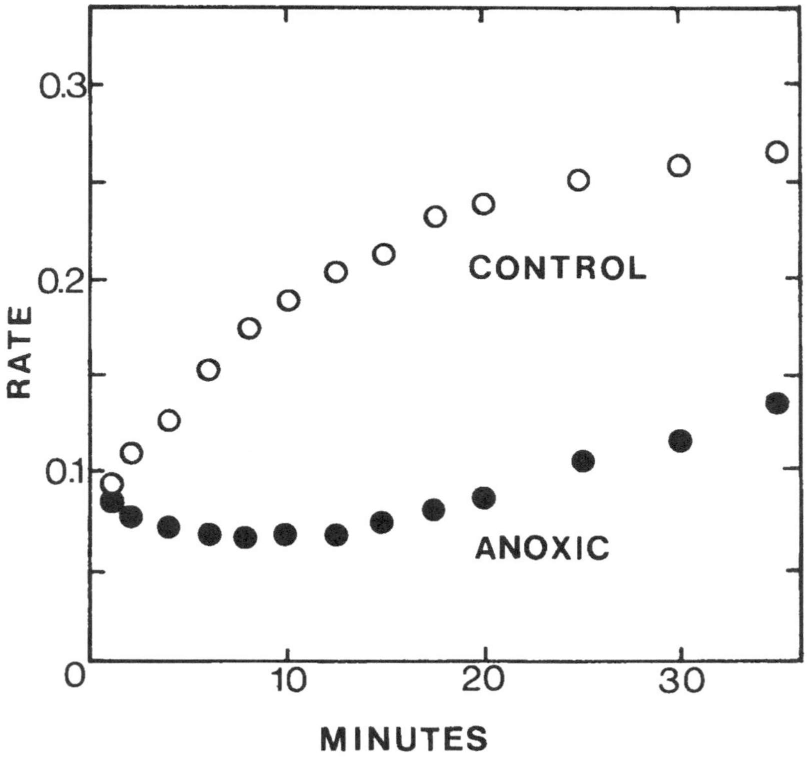

FIGURE 4. Respiration rates (change in % saturation/min/mg dry cysts) of control cysts (open circles) and cysts previously incubated under anoxic conditions (filled circles) in sea water for 4 months. A Strathkelvin oxygen electrode was used to continuously record the change in saturation. The cysts (about 5 mg dry weight) were incubated in 2 ml of 0.2 micron filtered sea water at 25°C.

for carbohydrate metabolism. Studies of trehalase *in vitro* demonstrated the enzyme exhibits reversible, hysteretic interconversion in response to changes in pH of the same magnitude as those occurring *in vivo* resulting in significant decreases in activity at lower pHs.[54,57] Phosphofructokinase is also inhibited *in vitro* at acid pHs.[58] Inhibition of specific pathways of protein and lipid catabolism has not yet been studied; however, it is also possible that mobilization of some of these energy-yielding substrates may be blocked during anoxia by means of the inhibition of their release from the yolk platelets where they are stored.[35] *In vitro*, yolk platelets disintegrate when exposed to alkaline pHs and release large quantities of lipoprotein.[34,35] This process is inhibited upon acidification to pH values similar to those reported *in vivo*. Utterback and Hand[35] reported that during anoxybiosis, release of lipoproteins from the platelets was inhibited, and suggested that a biophysical mechanism may be responsible for this regulation of release of substrates from the yolk platelets under anoxybiosis.

An extremely interesting hypothesis was very recently discussed, and tested by Nelis et al.[59] Starting from results in the unpublished M.S. thesis of Karmiol,[97] Nelis et al.[59] tested the idea that cyst carotenoids acted as clamps to protect macromolecules, and possibly membranes, from hydrolytic attack during prolonged anoxia and desiccation. While that general possibility is extremely attractive, notably in view of our preceding discussion that anoxic cysts may not have active pathways for trehalose, glycerol, and glycogen, the results of Karmiol were not reproduced in detail. Nevertheless, this provocative idea deserves further study because of the potential involvement of carotenoids in cyst stability.

III. INDUCTION OF DIAPAUSE: DETERMINATION OF REPRODUCTION MODE

The production of diapause embryos entails at least two discrete physiological events: determination of reproductive mode in the female and induction of diapause in the embryo.[60] Based on rather scanty data in prediapause embryos,[9] one may speculate that the induction of diapause occurs during the early stages of development. However, we know practically nothing about this matter. In other organisms, it is well established that the induction of embryonic diapause is under maternal control since it is the female's response to environmental conditions that dictates the metabolic state of the mature embryo.[61-64] This appears to be the case in *Artemia*.[1,10,60,65]

The determination of a reproductive mode occurs at some unknown point in the reproductive cycle of the female, possibly even before the eggs have entered the ovisac.[60] The physiological mechanisms controlling a reproductive mode and diapause induction are not known.[1,8] Researchers have noted a correlation between increases in salinity, pigmentation of the hemolymph and shell glands, and a switch to production of cysts in *A. franciscana*,[10,65,66] but the nature of this relationship remains unclear. Comparative studies on eggs and pregastrulae undergoing the two routes of development should prove informative and are badly needed.

The influence of some environmental factors on a reproductive mode has been demonstrated in laboratory experiments. In mass cultures, Versichele and Sorgeloos[65] were able to increase cyst production significantly in the SFB population by subjecting females to cyclic low oxygen levels or a medium containing chelated iron. Based on these results and the observations of others, they suggested that the decreasing levels of oxygen that parallel increasing salinities in the natural habitat induced females to switch to oviparous reproduction. Recently, Berthelemy-Okazaki[1] found that a number of factors, both intrinsic and extrinsic, affect reproductive mode in SFB *Artemia* including brood number, density, photoperiod, salinity, and temperature. Brood number, which is roughly equivalent to maternal age,[67] showed the most consistent correlation with oviparity.[1,60] In one experiment the proportion of oviparous broods increased from 30% of the first brood to 70% of the second brood and then remained about the same through the seventh, the last brood for which data were collected. This trend occurred to varying degrees under all treatments indicating that a reproductive mode is to some extent under endogenous control. Browne[68] has also reported similar correlations between brood number and oviparity in several *A. franciscana* populations.

The effect of hypoxia was not as obvious in these more recent experiments, but the culture conditions and experimental design are significantly different between these and other studies so the results cannot easily be compared. Surprisingly, Berthelemy-Okazaki[1,60] found that salinities of $\geq$ 120 g/l inhibited the switch to oviparity (Figure 5). This finding disagrees with prevailing views on *A. franciscana* life-cycle regulation and illustrates the need for much more research on the physiological ecology of this population. We do not believe this effect of higher salinities on reproductive mode is limited to laboratory conditions. One of us (L. E. D.) has observed situations in San Francisco Bay salterns where essentially all of the females were producing embryos via the ovoviviparous route in ponds with salinities of 150 g/l and noticeable algal blooms.

Why should females continue producing nauplii at higher salinities? It is known that all stages of the life cycle of *A. franciscana* can thrive at these salinities which, in the SFB salterns, may not always indicate the eminent disappearance of the habitat. However, cysts cannot hatch at $\geq$ 120 g/l.[12,31,32,43] Thus, the only way to take advantage of otherwise favorable conditions during higher salinity is through ovoviviparous reproduction. In fact, the ovoviviparous route may be interpreted as an adaptation to the hypersaline conditions

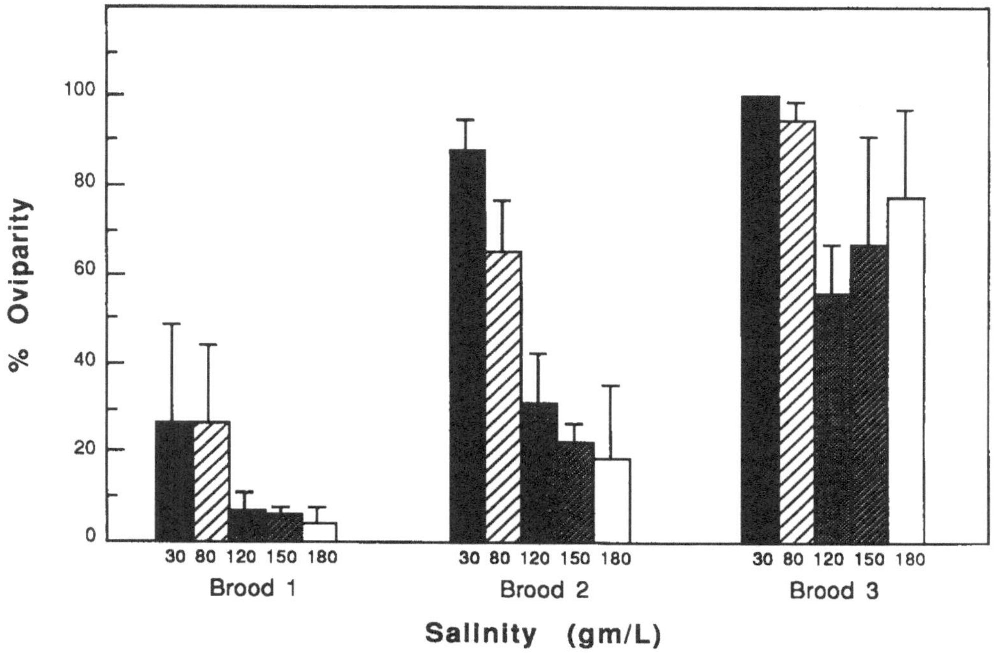

FIGURE 5. Percent oviparity in the first three broods of SFB *A. franciscana* females at five different salinities. Females in salinities of ≥120 g/l show a lower percentage of cyst production in the first two broods. Based on data from Berthelemy-Okazaki.[60]

Artemia inhabit since this mode of reproduction has not been reported in the freshwater Anostrocans.[69] Finally, if one considers the number of cues that terminate diapause in this population,[7] the potential exists for at least some of the cysts produced at the low salinities to be activated and hatch even if salinities remained low.

The few studies that have been conducted on the ecology of other crustaceans and the extensive literature on insects indicate that induction of diapause is not usually dependent upon a single cue and can be very habitat-specific. For example, in an interesting series of reciprocal transfer experiments Hairston[70] found that for the copepod, *Diaptomus sanguineus*, females in isolated ponds were adapted to their specific pond in terms of the switch to production of diapause cysts. The ponds were as close as 100 m from one another but the factors limiting the population were very different: some ponds were temporary and contained no predators while others were permanent and populated by fish which could essentially eliminate the copepod from the pond. The females in different ponds appeared to rely on different signals to determine when to produce diapause cysts. We expect the abiotic factors which influence a reproductive mode in *Artemia* are specific for each population and may be numerous for those populations inhabiting unpredictable environments.

IV. ENVIRONMENTAL CONTROL OF DIAPAUSE TERMINATION

Laboratory experiments have provided evidence that *Artemia* exhibit very similar, general characteristics compared to other organisms with respect to environmental control of diapause. There has been a tendency to conduct research on cyst diapause with the assumption that all *Artemia* cysts will respond to the same environmental cues, particularly desiccation. In the following section we discuss several properties of diapause systems which deserve particular attention within the context of further research in diapause regulation. We use examples from *Artemia* when possible, and also draw briefly from the extensive literature on endogenous dormancy in seeds and insect diapause.

A. INTERSPECIFIC AND INTERPOPULATION VARIATION

As with induction of diapause, the cues which can act as a signal for cyst activation are also habitat specific. This is understandable since the abiotic factors that activate cysts are useful as regulatory cues precisely because they forecast the coming of favorable conditions (both abiotic and biotic). Differences in the response of *Artemia* species to environmental cues have been well documented.[2] SFB *Artemia*, which inhabit a variable, inconsistent habitat, break diapause in response to several cues: dehydration, cold temperatures, and elevated CO_2.[7] Although temperature does fluctuate seasonally in these relatively small shallow ponds, field data indicate that temperatures cold enough to terminate diapause (based on laboratory experiments) do not occur every winter. Salinity can vary in an extremely unpredictable manner, depending on the size of the pond and its proximity to the bay. Consequently, salinities do not always reach the concentrations required to activate the cysts by dehydration.[71] Under these conditions, it is advantageous for diapause embryos to be able to respond to either temperature or salinity (or possibly their interaction as well as other cues). In contrast, *A. monica* (Mono Lake, California) inhabits a large, alpine lake and only breaks diapause after a cold treatment (4°C) of about 90 d.[72] Neither dehydration nor CO_2 activate these cysts.[7] In this environment, temperature is a more reliable cue since the cysts always experience temperatures of 4°C or lower during the winter and do not encounter significant salinity changes.[72-74]

Even in populations of the same species, such as SFB and GSL which are often used interchangeably, highly significant differences exist. While SFB cysts normally break diapause after dehydration,[2,7,65,68] GSL cysts appear to have a requirement for a longterm, cold incubation period while in the desiccated state in order to consistently achieve activation of the majority of the cysts.[2,75] The Great Salt Lake has a much more severe winter than San Francisco Bay, so cold temperature may be more reliable than dehydration as an activation cue. These few examples illustrate how the activation cues correspond with the actual conditions experienced by *Artemia* in its habitat. Clearly, studies on diapause regulation must not only identify species, but should also specify the precise geographical source of the cysts.

B. INTRAPOPULATION VARIATION

Large differences have also been reported in the response of individual cysts within a particular population to any given environmental cue[2,7,72,76] and many contradictions exist in the literature; e.g., significant percentages of GSL diapause cysts are activated by dehydration sometimes[75] while in other experiments only 4% hatch after the same dehydration treatment.[2] Also, typically, the proportion of cysts released from diapause in response to an environmental cue increases over time.[2,7,65,72] In Figure 6 both types of variation are shown for SFB cysts. In this particular case, the field-collected cysts were more easily activated by dehydration, though this is not always the case. In both batches, some individuals are more resistant to activation than others and require a longer dehydration period.

While genetic diversity certainly accounts for some of this variability, differences in the conditions experienced by the female prior to and during embryo formation probably also play a significant role. The effects of environmental conditions during embryo formation on the "depth" of developmental arrest and ease of activation have been documented in other organisms.[3-5,61,62] This relationship has been extensively studied in plants and, in a few cases, the mechanism of the control of environmental conditions on the depth of endogenous dormancy is understood.[61] One case will serve to illustrate the significance. For

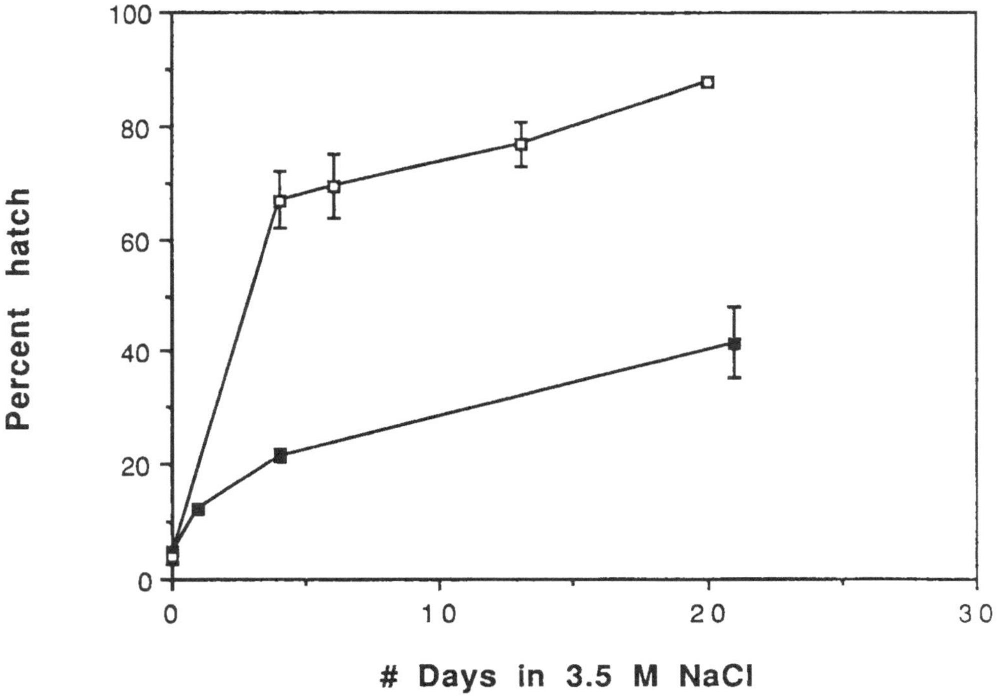

FIGURE 6. Percent hatch in two different batches of SFB *A. franciscana* cysts after preincubation in 3.5 *M* NaCl. In this case, the laboratory-reared cysts (closed squares) are more resistant to activation by this dehydration treatment than the field-collected cysts (open squares). Both batches show an increase in percent activation as time of dehydration increased. Based on data from Drinkwater and Crowe,[7] and Drinkwater, unpublished.[25]

the species *Ononis sicula*, a leguminous desert annual, day length during seed development has a significant effect on seed dormancy: seeds produced under long day-length conditions have very low germination rates while those produced under short days have high germination rates initially.[77,78] This response to photoperiod appears to be mediated by the seed coat. Under short days seeds have a thin coat which allows them to imbibe water and germinate. In contrast, long days result in the production of thick impermeable coats which prevent water uptake and thus, germination.

Although the mechanism regulating the production of diapause in *Artemia* embryos is not known, the available data suggest that in *A. franciscana* the conditions experienced during cyst production have significant influence on the resistance of embryos to activation.[2,7,65] Therefore, it is essential that cysts used for research on diapause regulation are those for which some information on prior history is available.

C. LIGHT AS AN ACTIVATION CUE

There has been a significant amount of interesting work conducted on the effect of light on cyst hatchability.[47,76,79,80] Prior to the work of Van der Linden, many researchers had observed that the hatch rate of *A. franciscana* cysts was often higher when the cysts were hatched in the presence of light and thus illumination was considered to be one of a number of favorable conditions for maximum hatchability.[2]

Recently, Van der Linden[47,76,80] has demonstrated that desiccated *A. franciscana* cyst populations often contain a specific fraction that requires exposure to light in order to hatch. The light-requiring fraction of these populations can hatch in the dark once they have received an adequate dose of light. A brief exposure (30 min) to white light in the presence of oxygen permits hatching of previously desiccated GSL and SFB cysts which would not hatch under dark, otherwise favorable hatching conditions. In other words, rather than being a condition required throughout the resumption of development, such as the presence of oxygen, transient exposure to light permits these cysts to hatch in darkness. These data indicate that light can act as a cue terminating diapause.[47]

Van der Linden and colleagues have characterized this light activation in detail. The actual quanta of light energy required for maximal hatching varies with different batches of cysts and is significantly lower for cysts which have been decapsulated, presumably because the removal of the pigmented outer shell allows the light to reach the photoreceptor.[76] Based on action spectra of hatching induction in decapsulated cysts, they suggest that the photoreceptor is either a phytochrome or haempigment; however they also emphasize that irregularities in the action spectra indicate interference from other pigments and make these experiments difficult to interpret. Considering the number of pigments present in the intact embryo, and the diverse forms these pigments can take, i.e., canthaxanthin is present in two isomeric forms (cis-trans), bound to lipovitellin complexes and may also be bound to other proteins and free in the cytoplasm,[34,59] we must conclude that it is too early to exclude the possibility that other pigments, including the carotenoids may be involved.

Unfortunately, the majority of this excellent work was conducted on commercial GSL cysts which had been previously desiccated so one must be cautious in making conclusions as to the precise role of light in diapause termination. Exposure of SFB cysts which have never been desiccated to ambient light did not result in activation after 50 d[7] and Van der Linden[80] states that light does not activate cysts that have not been desiccated but the data leading to this statement are not provided. Van der Linden has raised the possibility of two phases of diapause in *A. franciscana*, a photorefractory phase (cysts which have not been exposed to diapause breaking conditions) and a photosensitive phase (cysts that have been exposed to known cues but still do not hatch).[47] In view of the data this possibility seems reasonable; however, it is not clear whether exposure to light is a mandatory step in activation. It is well known in other organisms that some signals must be received in a particular sequence in order to result in activation.[61,62] For example, many seeds can only be activated by light following exposure to cold temperatures.[61] Often, a number of interchangeable factors can activate seeds and the sequence of exposure to these cues determines the response of the seed. In some instances, *Daphnia* embryos require activation by light, but this can only occur following cold incubation.[81]

In much of the previous work with *Artemia*, the activation treatments were conducted in the dark but cysts were moved to conditions of ambient light for hatching,[7] or the light conditions during activation are not specified.[65] Some preliminary data suggest that light is not mandatory in diapause termination in SFB cysts. In recent unpublished experiments by one of us (L. E. D.), cysts were activated by dehydration in 3.5 *M* NaCl and hatched under dark and light conditions. The percentage of cysts hatching in the light conditions is always higher but the actual proportion requiring light decreases over time (Figure 7). While it is possible that a segment of these populations is light requiring, no specific cue, including dehydration, has yet been found to be a mandatory requirement, at least for SFB cysts. More research is needed in this area in order to determine the precise requirements for activation.

D. REVERSAL OF ACTIVATION: DIAPAUSE RESTORED?

Although Dutrieu and Chrestia-Blanchine[29] reported no decline in nauplius production from cysts incubated for 5 months under anoxia, occasional observations of a decline in hatch rate have been reported for activated cysts that have been hydrated and incubated for periods of 3 to 5 months under anoxic conditions.[49,52,53] This decline in the percentage of cysts able to hatch after anoxia has always been interpreted as a result of loss of cyst viability.

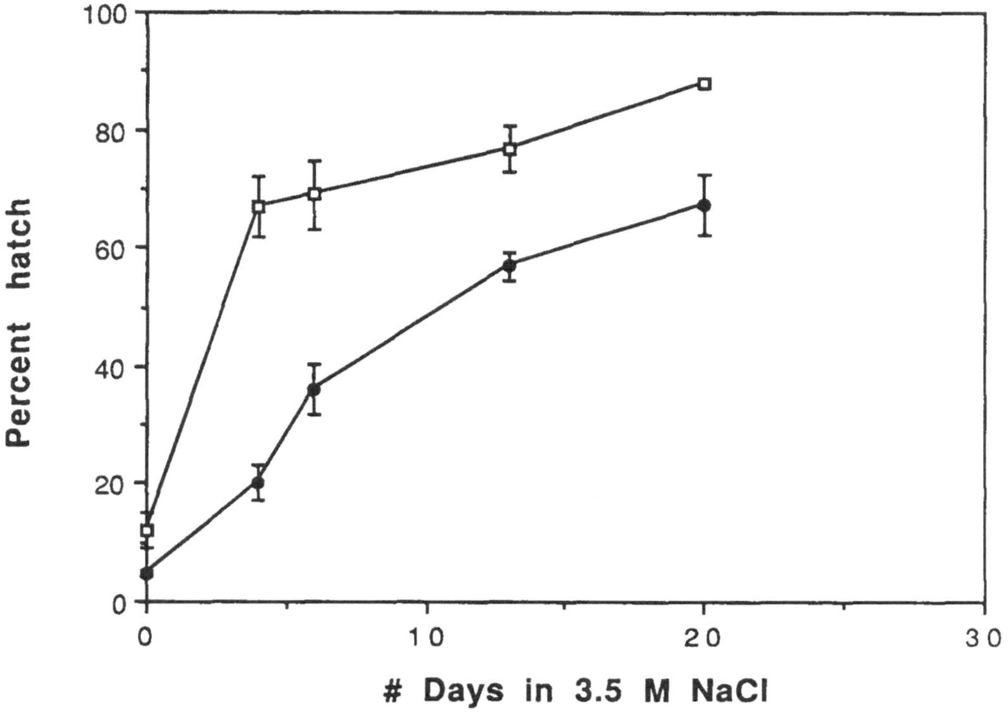

FIGURE 7. Percent hatch after dehydration in 3.5 *M* NaCl under both light (open squares) and dark (closed circles) conditions. The percentage of activated cysts increases over time. While after 4 d dehydration >40% of the cysts required light in order to hatch, after 13 d in the NaCl solution only about 15% more hatching occurs in the light.

More recent unpublished work by one of us (J. S. C.) suggests that at least some of these cysts that do not produce nauplii have not lost their viability but, instead, have had the endogenous suppression of development reinstated. Figure 8 shows the effect of the duration of anoxic incubation (see Clegg and Jackson,[53] for conditions) on the hatch of SFB cysts after their return to aerobic conditions. The open circles in the main body of this graph indicate the percentage of nauplii obtained when anoxic cysts are transferred directly to sea water and incubated aerobically at 25°C for 5 days (hatching conditions). However, a markedly different result is obtained if the anoxic cysts are dried under laboratory conditions (21 to 23°C, R.H. 65%) for 24 h and then incubated under the same hatching conditions (Figure 8, filled circles). A substantial increase in percent hatch occurs, following the dehydration/rehydration cycle. The inset in Figure 8 shows the percentage of anoxic cysts that initially failed to hatch, but did so after dehydration/rehydration. Although this interesting observation requires (and deserves) more detailed study, our current hypothesis is that extended anoxia of hydrated, activated, cyst populations causes some of the cysts to reenter diapause which, in turn, can be terminated by dehydration. We will return to the question of the metabolic status of anoxic cysts later in this paper.

These observations raise the more general question of ''nonviable'' cysts in all *Artemia* populations. The preceding discussion of Figure 8 suggests that we must be very careful about referring to cysts that do not hatch (given optimal hatching conditions) as ''nonviable'' or (worse) ''dead''. In the absence of more detailed confirmation we may only conclude that they do not hatch under those conditions. No doubt, some of these cysts really *are* dead, in the sense that they have lost the ability to resume the metabolism necessary to restore normal development into nauplii, including events of the critical emergence process.

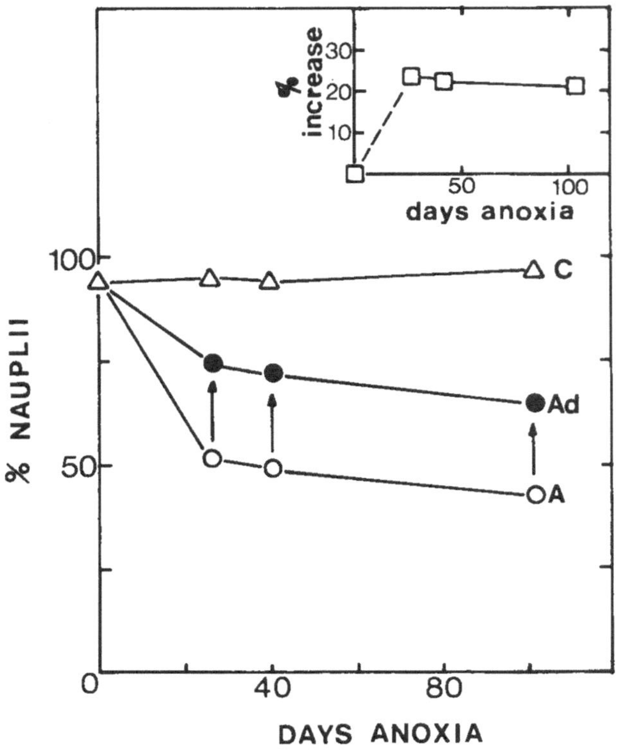

FIGURE 8. Hatching (% nauplii) of cysts incubated either dried, C, (triangles) or in sea water under anoxic conditions at 21—23°C (circles) for the times shown. Open circles (A) are for anoxic cysts transferred directly to aerobic hatching conditions (sea water, 5 d) and filled circles (Ad) are for samples from this same population that were dried (18 h, 21—23°C, relative humidity about 65%) and then exposed to aerobic hatching conditions. The inset shows increases in the yield of nauplii due to this drying-rehydration treatment.

However, the possibility is also entertained that some of these nonhatching cysts will indeed give rise to nauplii, provided that they are given the conditions required to set metabolism and development in motion. It also seems reasonable to suppose that events taking place in the cysts prior to/during the activation process (desiccation, for example) may determine these conditions for restoration of hatching ability. Thus, while most of the cysts in populations exposed to activation processes are indeed activated, others (the nonhatching proportion) may well consist of cysts that have resisted this particular activation cue and remain in diapause (for the same or different reasons) as well as those that are, simply and totally, dead.

To our knowledge, the restoration of hatching ability by dehydration/rehydration of anoxic *Artemia* cysts has not been observed previously. However, a similar phenomenon has been extensively studied in seeds and is termed *secondary dormancy*.[61] Secondary dormancy in seeds can occur in response to changes in temperature, anoxia, and exposure to light of particular wavelengths.[61] Secondary diapause has been reported to occur in *Daphnia* where exposure to blue light results in the reversal of activation.[82]

E. MANIPULATION OF CYST METABOLIC STATE IN THE LABORATORY

We hope the preceding discussion has served to illustrate the complexity of diapause termination. It is not yet possible to study a single cyst; thus, one must always use a population

TABLE 1
Effect of Various Treatments on Diapause Termination for SFB and GSL Cysts

Treatment	Population	
	SFB	**GSL**
Dehydrated cysts		
≥ 2.0 M NaCl	Yes[7]	No data
5.0 *M* NaCl	Yes[7,65]	Sometimes[2,75]
Complete desiccation[a]	Yes[65]	Sometimes[2,75]
Cold treatment during desiccation (−25°C)	No data	Yes[75]
Hydrated cysts		
Cold incubation (8°C)	Yes[7]	No data
Increased levels of CO_2 (up to 20% atmosphere)	Yes[7]	No data
Exposure to light following desiccation	Yes[47]	Yes[47]
Anoxia	No[7]	No data
Alkaline buffer with NH_4^+	No[7]	No data
Oxidants (H_2O_2 and hypochlorite)	Yes[25,86]	Yes[2,86]

Note: Yes, results in activation. No, does not result in activation. Sometimes, reports in literature extremely variable. Unless specified, treatments were applied at temperatures of 23 to 27°C. Since the cysts used in these experiments were produced under varying conditions, this chart should not be interpreted as a comparison but rather as a summary of the environmental conditions which can affect the metabolic status of diapause cysts in these two populations.

[a] Numerous desiccation treatments have been used: air drying, oven drying at 30 to 40°C, and desiccation over $CaCl_2$.

of cysts for metabolic studies, and it is often assumed that all individuals are in the same metabolic state. In reality, the heterogeneity of the requirements for resumption of development indicate that a batch of cysts may represent a continuum of metabolic states from completely activated to extremely dormant.[83] This diversity presumably reflects the uncertainty of the environment and guarantees a portion of the population will remain in diapause, analogous to the seed bank of "weed" species of plants. Hairston and De Stasio[84] have shown that this is the case for the copepod, *Diaptomus sanguineus* where nonhatching diapause embryos remained viable in the mud for at least three years. These dormant embryos were able to hatch and repopulate the pond even during two years of completely unsuccessful reproduction.

For metabolic studies, it is desirable to use cysts which show minimal variation with respect to activation treatments and percent hatch. Interpretation of data from batches with low hatch rates (say, less than 75%) can be very difficult, especially if the metabolic status of the nonhatchers is unknown (i.e., are they in diapause or dead?). While our knowledge of the precise effects of various environmental factors on diapause remains limited, some useful information is available.

The conditions that activate SFB and GSL cysts are summarized in Table 1. Because they so are numerous, precautions must be taken in order to maintain cysts in the diapause state. Indeed, one wonders how the state of diapause is ever maintained in nature! If diapause cysts are not exposed to activating cues, they can be held in the laboratory for months without breaking diapause while maintaining excellent viability.[7] For SFB *Artemia* this involves keeping crowded populations of cysts in a stoppered flask (they eventually become anoxic) at room temperature in 1.25 *M* NaCl in darkness. SFB cysts have also been kept aerobically for up to 50 d in more dilute solutions (usually 0.25 *M* NaCl) under both ambient light and in darkness with no increase in activation.[25] Since dehydration breaks diapause in many species and strains of *Artemia*,[2,68] it does not seem prudent to store cysts in brine solutions or under other conditions allowing dehydration prior to experiments on diapause regulation unless the intention is to activate the cysts. Until the exact nature of light activation becomes clear it seems advisable to avoid exposure to light as much as possible. Furthermore, considering that commercially available cysts have undergone unknown and variable treatments, it is risky to make conclusions about diapause regulation from studies using these cysts.

It is sometimes useful to activate cysts by immersion in H_2O_2 solutions. Dilute H_2O_2 (3 to 5%) is known to activate embryos from many populations without impairing viability[2,85] and brief exposures (20 min) usually result in very high hatch rates (85 to 90%). While this is convenient, it would seem wise to verify that results obtained in this manner can also be achieved with more natural activation treatments. Exposure to hypochlorite, another oxidant, has long been known to increase hatching in activated cyst populations[86] and there has been speculation as to whether this increase in hatch is due to some reaction in the embryo or simply due to removal of part of the shell.[2] One of us (L. E. D.) has found that exposure of diapause embryos to 2.5% hypochlorite (the concentration commonly used to decapsulate cysts) for 15 min at 22°C, increased the hatch rate to 87% compared to a control hatch rate of 5%. Embryos in hypochlorite at 4°C showed a smaller increase in hatching (hatch rate 12%). Thus, it appears that hypochlorite can activate diapause cysts. In view of the effect of H_2O_2 and hypochlorite on diapause embryos, extreme caution should be used during decapsulation, and if possible, this procedure should be avoided when diapause cysts are studied.

V. METABOLIC REGULATION OF DIAPAUSE

Until recently the only tangible evidence of developmental arrest, and thus an endogenously imposed dormancy in the cysts, was the lack of hatching. It is now known that metabolism is severely suppressed in diapause embryos, yet the mechanism of this suppression is unknown.

A. METABOLIC STATUS OF THE DIAPAUSE CYSTS

While diapause cysts do not hatch, they do show signs of active metabolism.[87] Significant respiration rates have been detected in SFB diapause cysts. However, these respiratory rates decline rapidly if the cysts are maintained under aerobic conditions and after five days respiration falls to a low of 0.6 μg O_2/gdw cysts where it levels off. This is about 1% of the initial respiratory rate measured in diapause cysts. Examination of trehalose and glycerol within the same time frame revealed that glycerol content decreased while trehalose increased from 14.6 ± .4% to 17.0 ± 0.4% of the cyst dry weight. Significant differences in glycogen were not detected. The identity of the substrate for this metabolic activity is unknown, as net decline in glycerol is insufficient to supply carbon skeletons for the net increase in trehalose. These observed changes in net carbohydrate levels is the reverse of the rapid mobilization of trehalose and synthesis of glycerol seen during PED. The continued synthesis of trehalose may reflect a prediapause metabolic state. It is conceivable that at least some cysts have not completed prediapause metabolism that presumably occurs in the ovisac, i.e., synthesis of trehalose. When cysts are released from the ovisac, they float on the surface of the pond, sometimes in large aggregations several inches thick. One would suspect that conditions in these aggregations are hypoxic. Thus, many embryos probably shut down aerobic metabolism upon release from the ovisac. It appears that once this prediapause metabolism is completed, metabolic activity is extremely reduced.

During these first few hours of aerobic conditions when respiratory rate is highest in diapause cysts, utilization of high energy phosphates occurs at rates comparable to those seen in activated cysts.[87] ^{31}P-NMR studies have shown that, as in activated cysts, high levels of ribonucleoside di- and triphosphates (NDP, NTP) are present and pH_i is alkaline. It is also interesting to note that diapause cysts respond to anaerobiosis in a manner indistinguishable from the response of activated cysts.[87] When oxygen is removed, pH falls rapidly and NTP/NDP levels decline while inorganic phosphate increases fivefold. Other anabolic processes in addition to trehalose synthesis could be responsible for utilization of the high-energy phosphate compounds and further research in this area is vital to understanding diapause regulation.

B. TREHALASE ACTIVITY

Preliminary evidence suggests trehalase is present in the diapause cysts, though not in an active form.[87] Negligible amounts of trehalase activity have been detected in crude homogenates of diapause SFB *A. franciscana* embryos. In these experiments, if cysts were subjected to conditions known to terminate diapause prior to being homogenized, trehalase activity in the crude homogenates increased as the proportion of activated embryos in the sample increased. Fractionation by centrifugation demonstrated that the trehalase activity was associated with the particulate fraction; the majority of the activity ($\geq$80%) was found in the 500 $\times$ g fraction which contains the yolk platelets. Finally, resuspension of the yolk platelets from diapause embryos in a Monol buffer known to disrupt yolk platelets[88] increased trehalase activity to levels comparable to extracts from activated cysts.

The finding of a trehalase associated with the particulate fraction agrees with the work of Ballario et al.[89] who, using desiccated commercial activated cysts found most of the trehalase activity associated with the yolk platelets. They did not specify which population of *Artemia* was used in these studies; however, judging from the source of the cysts (''Metaframe''), it was probably the SFB population. Carpenter and Hand[90] found only 30% of the trehalase activity associated with the yolk platelets in commercial GSL cysts with a reported hatch rate of 60%. These discrepancies are difficult to interpret since the cysts are probably from two populations and the experimental protocols are quite dissimilar. While this could be simply a population difference it is also possible that trehalase, in functioning as a regulatory enzyme, changes its nature and location during the course of diapause termination and development.

C. METABOLIC STATUS OF LIGHT-ACTIVATED CYSTS

We have covered this work in a separate section because commercial, desiccated cysts were used and the exact nature of the dormant state described below is unclear. For the sake of clarity, until more is known about embryonic diapause in *Artemia*, we reserve the term ''diapause'' for cysts which have not been exposed to diapause-breaking treatments. We recognize, in view of the findings summarized below, it is entirely plausible that a segment of the population could be unaffected by desiccation and remain in diapause.

In these studies with the nonhatching fraction of commercial GSL cysts, Van der Linden et al.[47] also found very low respiratory rates and a shutdown of carbohydrate metabolism. In preparation for these experiments, cysts were first exposed to 5 d of aerobic, dark incubation to allow any completely activated cysts to hatch. Forty percent were able to hatch under these conditions and of the remaining 60%, about half (45%) would resume development and hatch in the dark only after exposure to light. These cysts also had low ATP levels and an acidic pH_i (6.0).[80] Electron microscopy showed no polyribosomes present and the mitochondria did not appear to be metabolically active.[80] It is striking that the metabolic status of these embryos shows significant parallels to that of anoxic activated cysts: acidic pH, low levels of NTP, NDP, high levels of inorganic phosphate, and inhibition of carbohydrate metabolism.

Upon brief exposure to light, activation takes place and the expected events of early PED are initiated: respiratory rate increases, trehalose is degraded, ATP and glycerol levels increase, and pH_i becomes alkaline, mitochondria enlarge and rough endoplasmic reticulum is present.[47,80,91] The earliest event detected by VDL is an increase in cAMP which occurs within a couple of minutes. It is possible that detection of actual changes in carbohydrates, ATP, and pH_i in the light-requiring hatchers could be confounded by the presence of 55% nonhatching cysts. However, the supportive evidence provided by the changes in morphology do suggest that while the metabolic data may be quantitatively imprecise they are qualitatively accurate.[80] The lack of net trehalose degradation combined with the fact that PED is initiated after illumination suggests the endogenous mechanism responsible for suppression of PED is still intact even though the embryos may be in a different phase of diapause, i.e., photosensitive phase. In other words, while the state of the mechanism imposing diapause may have changed, the resulting suppression of the necessary metabolism for development appears to be the same as in diapause (never desiccated) cysts.

D. NATURE OF THE REGULATORY MECHANISM

When one examines the literature on dormancy/diapause it becomes clear that the similarity of *Artemia* to other anhydrobiotes extends beyond the presence of large amounts of trehalose since many of these organisms also experience an endogenously imposed dormancy. These similarities are particularly striking when the cysts are compared to the spores of yeast and fungi, which, like *Artemia* cysts, depend on the rapid mobilization of trehalose during the resumption of activity.[92,93] Interestingly, endogenous dormancy in many of these organisms involves the inhibition of trehalose breakdown by two different mechanisms, depending on the type of trehalase present. In some cases, separation of trehalase and its substrate is responsible for the inhibition of germination of fungal spores.[92] In other species, trehalase is present in the cytoplasm in an inactive form which can be activated via cAMP-induced phosphorylation cascade.[93] While a considerable amount of work has been done on the enzymes activated by the cAMP phosphorylation cascade, the event which actually translates environmental cues that trigger activation into mobilization of this cascade remains unclear.[93]

In fact, this initial event, (i.e., translation of an abiotic signal into a metabolic signal) continues to be elusive in practically all cases of termination of endogenously imposed dormancy studied to date, with the possible exception of the phytochrome system in seeds. This latter system has been extensively studied and it is known that the phytochrome pigment undergoes reversible changes in conformation upon absorption of specific wavelengths of light and the absorption of thermal energy.[61,94] However, the connection between these changes in conformation and termination/reinduction of seed dormancy is not completely understood.

In the case of *Artemia*, research on diapause regulation is in the initial stages so one can only speculate about the nature of the regulatory mechanism. The conditions that terminate diapause can give us some clues about the nature of the initial event required for activation. Whatever the regulatory mechanism is, it must have a biophysical basis that may either break diapause directly or permit some "diapause-breaking metabolism" to occur. While metabolic processes could be occurring under some of the conditions which activate the cysts, such as cold incubation or elevated Co_2, the cysts are also able to break diapause under more extreme conditions. The embryos break diapause in salinities as high as 5.0 *M* NaCl and when completely desiccated. Also, the activation of the cysts apparently increases over time, even at this low-water content indicating that activation is proceeding. The fact that activation apparently continues in cysts which are ametabolic in the normal sense of the word supports the idea that a metabolic process is not responsible for the initial step terminating diapause. Instead, this phase, at least in *A. franciscana*, is probably ametabolic and based on some biophysical (or chemical) event which becomes thermodynamically favorable under the particular circumstances that terminate diapause. Along these lines, it is interesting to note that light activation requires the presence of oxygen, and cannot occur under nitrogen.[80] Yet it is not inhibited by cyanide, suggesting that activation by this particular cue is also not a metabolic process but rather a photochemical event. The evidence for an oxidative process in diapause termination is certainly compelling; however, there is also evidence which suggests that activation by cold temperature can occur under anaerobic conditions.[7] The two observations are not necessarily contradictive because after cold treatment the cysts were moved to hatching conditions of ambient light; thus, the oxidative process could occur at this point. Therefore, light activation may indeed be a different (secondary) process.

The suppression of metabolism and development during diapause seems to stem to some extent from trehalase inhibition. If all the information available is considered, it appears that the metabolic status of diapause embryos is not unlike that of an activated cyst under anoxia: catabolic pathways are shutdown, high energy phosphate compounds are extremely low, pH_i is acidic, and overall metabolic activity is drastically suppressed. It may be that in the diapause cyst, with trehalase activity inhibited, no energy source is available once the prediapause metabolism is completed. If trehalose catabolism is the primary source of energy during PED, its inhibition would be sufficient to result in an acidic pH_i which in turn would shut down other metabolic pathways; i.e., the yolk platelets would be inaccessible.

The possibility that pH_i is the primary regulatory mechanism has been discussed extensively;[7,55,80,86] however, the present data do not support this idea. If depressed pH_i were the regulatory mechanism, then alkalinization should activate diapause embryos. So far, attempts to activate diapause embryos by exposure to NH_4^+ which elevates pH_i have proven to be unsuccessful except in cases where the cysts had already been exposed to a known activation treatment prior to NH_4^+ incubation.[7,55] Thus, depressed pH_i which accompanies the shutdown of metabolism may simply result from suppression of the ATP-generating metabolism. The acidification of pH_i is a common outcome of suppressed metabolism and is thought to be the result of the hydrolysis of NTP, and NDP which yields a net increase in H^+.[95] This drop in pH with lowered energy states has been proposed as a mechanism for orchestrating the suppression of pathways which have no metabolites in common allowing the entire metabolic apparatus of cells to shut down in an orderly fashion.[95,96] While diapause cysts do not appear to be activated by alkalinization, it would be interesting to investigate metabolism in diapause cysts where pH_i has been artificially elevated.

Another potential mechanism which has been proposed involves the carotenoids, specifically canthaxanthin.[10] The literature on carotenoids was recently reviewed by Nelis et al.[59] and although, as these authors note, much of the data is puzzling, it seems premature to eliminate the possibility that the cis-trans isomerization may somehow be involved in regulation of induction/termination of diapause. Several recent findings make this possibility particularly attractive at this juncture: (1) the apparent reversibility of activation; (2) the possibility that trehalase may be associated with the yolk platelets in an inactive form; (3) the observation that light can activate cysts and the potential requirement for an oxidative process; and (4) cysts continue to be activated while dehydrated and the conversion of cis-canthaxanthin to trans-canthaxanthin may continue under these conditions. Thus, examining the content of the cis-trans isomers in diapause cysts and in cysts after activation would be worthwhile. Although the change in cis-trans ratio in desiccated cysts after hydration has been studied, to our knowledge the isomeric conformation of canthaxanthin in the never-desiccated cysts has not yet been investigated.

VI. CONCLUDING COMMENT

The mechanism(s) by which an abiotic signal can be translated into spectacular changes in the metabolic state of an organism continues to be a basic phenomenon which has not been thoroughly described in any organism. We think that the diapause embryo of *Artemia* is an ideal system for such studies, especially because the sequence of events accompanying the resumption of development has been studied extensively. Furthermore, the mechanisms of metabolic arrest during environmentally imposed dormancy are also fairly well understood. Thus, the regulation of diapause in *Artemia* should continue to be an intriguing field of research as many questions remain unanswered in all areas. We hope that future research will focus on cysts which have never been desiccated so that characteristics of the diapause state in these embryos can be clarified.

ACKNOWLEDGMENTS

We would like to thank Dr. Annemie Van der Linden for generously providing us with a copy of her dissertation and submitted manuscript. The production of this manuscript was supported in part by NSF grant DCB-8820347 to J. S. C.

REFERENCES

1. **Berthelemy-Okazaki, N. J. and Hedgecock, D.,** Effect of environmental factors on cyst formation in the brine shrimp *Artemia*, in *Artemia Research and Its Applications,* Vol. 3, Sorgeloos, P., Bengtson, D. A., Decleir, W., and Jaspers, E., Eds., Universa Press, Wetteren, Belgium, 1987, 167.
2. **Lavens, P. and Sorgeloos, P.,** The cryptobiotic state of *Artemia* cysts and its diapause deactivation and hatching: a review, in *Artemia Research and Its Applications,* Vol. 3, Sorgeloos, P., Bengtson, D. A., Decleir, W., and Jaspers, E., Eds., Universa Press, Wetteren, Belgium, 1987, 27.
3. **Amen, R. D.,** The concept of seed dormancy, *Am. Sci.*, 51, 408, 1963.
4. **Mansingh, A.,** Physiological classification of dormancies in insects, *Can. Entomol.*, 103, 983, 1971.
5. **Clutter, M. E., Ed.,** *Dormancy and Developmental Arrest,* Academic Press, New York, 1978.
6. **Keilin, D.,** The problem of anabiosis or latent life: history and current concept, *Proc. R. Soc. London, Ser. B.*, 150, 149, 1959.
7. **Drinkwater, L. E. and Crowe, J. H.,** Regulation of embryonic diapause in *Artemia*: environmental and physiological signals, *J. Exp. Zool.*, 241, 297, 1987.
8. **Clegg, J. S. and Conte, F. P.,** A review of the cellular and developmental biology of *Artemia*, in *The Brine Shrimp, Artemia,* Vol. 2, Persoone, G., Sorgeloos, P., Roels, O., and Jaspers, E., Eds., Universa Press, Wetteren, Belgium, 1980, 11.
9. **Clegg, J. S.,** The origin of trehalose and its significance during the formation of encysted dormant embryos of *Artemia salina, Comp. Biochem. Physiol.*, 14, 135, 1965.
10. **Dutrieu, J.,** Observations biochimiques et physiologiques sur le developpement d'*Artemia salina, Arch. Zool. Exp. Gen.*, 99, 1, 1960.
11. **Clegg, J. S.,** The control of emergence and metabolism by external osmotic pressure and the role of free glycerol in developing cysts of *Artemia salina, J. Exp. Biol.*, 41, 879, 1964.
12. **Clegg, J. S.,** Biochemical adaptations associated with the embryonic dormancy of *Artemia salina, Trans. Am. Microsc. Soc.*, 93, 481, 1974.
13. **Crowe, J. H. and Clegg, J. S., Eds.,** *Anhydrobiosis,* Dowden, Hutchinson, & Ross, Inc., Stroudsburg, PA, 1973.
14. **Crowe, J. H. and Clegg, J. S., Eds.,** *Dry Biological Systems,* Academic Press, New York, 1978.
15. **Crowe, J. H., Crowe, L. M., Carpenter, J. F., Rudolph, A. S., Wistom, C. A., Spargo, B. J., and Anchordoguy, T. J.,** Interactions of sugars with membranes, *Biochim. Biophys. Acta,* 947, 367, 1988.
16. **Benesch, R.,** Zur ontogenie und morphologie von *Artemia salina*, L., *Zool. Jordbrukstek. Anat.*, 86, 307, 1969.
17. **Nakanishi, Y. H., Iwasaki, T., Okigaki, T., and Kato, H.,** Cytological studies of *Artemia salina*. I. Embryonic development without cell multiplication after the blastula stage in encysted dry eggs, *Annot. Zool. Jpn.*, 35, 223, 1962.
18. **Morris, J. G. and Afzelius, B. A.,** The structure of the shell and outer membranes in encysted *Artemia salina* embryos during cryptobiosis and development, *J. Ultrastruct. Res.*, 20, 244, 1967.
19. **Anderson, E., Lochhead, J. H., Lochhead, M. S., and Huebner, E.,** The origin and structure of the tertiary envelope in thick shelled eggs of the brine shrimp, *Artemia, J. Ultrastruct. Res.*, 32, 497, 1971.

20. **Clegg, J. S.,** Interrelationships between water and cellular metabolism in *Artemia* cysts. XI. Density measurements, *Cell Biophys.*, 6, 153, 1984.
21. **Clegg, J. S.,** Protein synthesis in the absence of cell division during the development of *Artemia salina* embryos, *Nature*, 212, 517, 1966.
22. **Clegg, J. S.,** Metabolic studies of cryptobiosis in encysted embryos of *Artemia salina, Comp. Biochem. Physiol.*, 20, 801, 1967.
23. **de Chaffoy, D., De Maeyer-Criel, G., and Kondo, M.,** On the permeability and formation of the embryonic cuticle during development *in vivo* and *in vitro* of *Artemia salina* embryos, *Differentiation* , 12, 99, 1978.
24. **Finamore, F. J. and Clegg, J. S.,** Biochemical aspects of morphogenesis in the brine shrimp, *Artemia salina*, in *The Cell-Cycle: Gene-Enzyme Interactions*, Padilla, G. M., Whitson, G. L., and Cameron, I. L., Eds., Academic Press, New York, 1969, 249.
25. **Drinkwater, L. E.,** unpublished data, 1987.
26. **Sorgeloos, P., Bengtson, D. A., Decleir, W., and Jaspers, E., Eds.,** *Artemia Research and Its Applications*, Vols. 1—3, Universa Press, Wetteren, Belgium, 1987.
27. **MacRae, T. H., Bagshaw, J. C., and Warner, A. H., Eds.,** *Biochemistry and Cell Biology of Artemia*, CRC Press, Boca Raton, FL, 1989, 264.
28. **Warner, A. H., MacRae, T. H., and Bagshaw, J. C.,** *Cellular and Molecular Biology of Artemia Development*, Plenum Press, New York, 1989.
29. **Dutrieu, J. and Chrestia-Blanchine, D.,** Resistance des oeufs durables hedrates *d'Artemia salina* a'l'anoxie, *C. R. Acad. Sci. D.*, 263, 998, 1966.
30. **Clegg, J. S.,** Interrelationships between water and cellular metabolism in cysts. I. Hydration-dehydration from liquid and vapor phases, *J. Exp. Biol.*, 61, 291, 1974.
31. **Clegg, J. S.,** Interrelationships between water and cellular metabolism in *Artemia* cysts. II. Carbohydrates, *Comp. Biochem. Physiol.*, 53a, 83, 1976.
32. **Clegg, J. S.,** Interrelationships between water and cellular metabolism in cysts. III. Respiration, *Comp. Biochem. Physiol.*, 53a, 89, 1976.
33. **Warner, A. H., Beers, P. C., and Huang, F. L.,** Biosynthesis of the diguanosine nucleotides. I. Purification and properties of an enzyme from yolk platelets of brine shrimp embryos, *Can. J. Biochem.*, 52, 231, 1974.
34. **de Chaffoy, D. and Kondo, M.,** Lipovitellin from the crustacean, *Artemia salina*: biochemical analysis of lipovitellin complex from the yolk granules, *J. Biol. Chem.*, 255, 6727, 1980.
35. **Utterback, P. J. and Hand, S. C.,** Yolk platelet degradation in pre-emergence *Artemia* embryos: response to protons *in vivo* and *in vitro, Am. J. Physiol.*, 252, 774, 1987.
36. **Marco, R., Vallejo, C. G., Garesse, R., and Perona, R.,** A hypothesis on the activation process of proteolytic activities during *Artemia* early development, in *The Brine Shrimp, Artemia*, Vol. 2, Persoone, G., Sorgeloos, P., Roels, O., and Jaspers, E., Eds., Wetteren, Belgium, Universa, 1980, 315.
37. **Ezquieta, B. and Vallejo, C. G.,** The trypsin-like proteinase of *Artemia*: yolk localization and developmental activation, *Comp. Biochem. Physiol.*, 82, 731, 1985.
38. **Ezquieta, B. and Vallejo, C. G.,** Lipovitellin inhibition of *Artemia* trypsin-like proteinase: a role for a storage protein in regulating proteinase activity during development, *Arch. Biochem. Biophys.*, 250, 410, 1986.
39. **Domingo, A., Cervera, M., Martin, E., and Marco, R.,** Biochemical characterization and developmental behavior of *Artemia* embryonic and nauplial deoxyribonucleases, *Biochemistry*, 25, 4125, 1986.
40. **Warner, A. H.,** Proteases and protease inhibitors in *Artemia* and their role in the developmental processes, in *Biochemistry and Cell Biology of Artemia*, MacRae, T. H., Bagshaw, J. C., and Warner, A. H., Eds., CRC Press, Boca Raton, FL, 1989, 113.
41. **Marco, R., Garesse, R., and Vallejo, C. G.,** Mitochondrial unmasking and yolk platelets metabolization during early development in *Artemia*, in *The Brine Shrimp, Artemia*, Vol. 2, Persoone, G., Sorgeloos, P., Roels, O., and Jaspers, E., Eds., Wetteren, Belgium, Universa, 1980, 481.
42. **Bes, C., Caratero, C., Caratero, A., Gaubin, Y., and Planel, H.,** Aspects ultrastructuraux de l'embryon dormant *Artemia*, *Bull. Assoc. Anat.*, 67, 255, 1983.
43. **Clegg, J. S.,** Hydration-dependent metabolism transitions and the state of cellular water in *Artemia*, in *Dry Biological Systems*, Crowe, J. H. and Clegg, J. S., Eds., Academic Press, New York, 1978, 117.
44. **Clegg, J. S.,** Metabolism and intracellular environment: the vicinal-water network model, in *Cell-Associated Water*, Drost-Hansen, W. and Clegg, J. S., Eds., Academic Press, New York, 363, 1979.
45. **Clegg, J. S.,** The physical properties and metabolic status of *Artemia* cysts at low water contents: the "water replacement" hypothesis, in *Membranes, Metabolism and Dry Organisms*, Leopold, A. C., Ed., Cornell University Press, Ithaca, 1986, 169.
46. **Glasheen, J. S. and Hand, S. C.,** Anhydrobiosis in embryos of the brine shrimp *Artemia*: characterization of metabolic arrest during reductions in cell-associated water, *J. Exp. Biol.*, 135, 363, 1988.

47. **Van der Linden, A., Blust, R., Van Laere, A. J., and Decleir, W.,** Light-induced release of *Artemia* embryos from diapause: analysis of metabolic status, *J. Exp. Zool.*, 247, 131, 1988.
48. **Ewing, R. D. and Clegg, J. S.,** Lactate dehydrogenase activity and anaerobic metabolism during embryonic development in *Artemia salina, Comp. Biochem. Physiol.*, 31, 297, 1969.
49. **Stocco, D. M., Beers, P. C., and Warner, A. H.,** Effects of anoxia on nucleotide metabolism in encysted embryos of the brine shrimp, *Dev. Biol.*, 27, 479, 1972.
50. **Hand, S. C. and Gnaiger, E.,** Anaerobic dormancy quantified in *Artemia* embryos: a calorimetric test of the control mechanism, *Science*, 239, 1425, 1988.
51. **Hand, S. C.,** Personal communication to J. S. C., 1989.
52. **Clegg, J. S. and Jackson, S. A.,** Aspects of the anaerobic metabolism of *Artemia* cysts, in *Cellular and Molecular Biology of Artemia Development*, Warner, A. H., Bagshaw, J., and MacRae, T., Eds., Plenum Press, New York, 1989, 1.
53. **Clegg, J. S. and Jackson, S. A.,** Long-term anoxia in *Artemia* cysts, *J. Exp. Biol.*, 1989, in press.
54. **Busa, W. B., Crowe, J. H., and Matson, G. B.,** Intracellular pH and the metabolic status of dormant and developing *Artemia* embryos, *Arch. Biochem. Biophys.*, 216, 711, 1982.
55. **Busa, W. B. and Crowe, J. H.,** Intracellular pH regulates the dormancy/development transition of brine shrimp (*Artemia salina*) embryos, *Science*, 221, 366, 1983.
56. **Carpenter, J. F. and Hand, S. C.,** Arrestment of carbohydrate metabolism during anaerobic dormancy and aerobic acidosis in *Artemia* embryos: determination of pH-sensitive control points, *J. Comp. Physiol.*, B, 156, 451, 1986.
57. **Hand, S. C. and Carpenter, J. C.,** pH-Induced metabolic transitions in *Artemia* embryos mediated by a novel hysteretic trehalase, *Science*, 232, 1535, 1986.
58. **Carpenter, J. F. and Hand, S. C.,** Comparison of pH-dependent allostery and dissociation for phosphofructokinases from *Artemia* embryos and rabbit muscle: nature of the enzymes acylated with diethylpyrocarbonate, *Arch-Biochem. Biophys.*, 248, 1, 1986.
59. **Nelis, H. J., Lavens, P., Moens, L., Sorgeloos, P., and DeLeenheer, A. P.,** Carotenoids in relation to *Artemia* development, in *Biochemistry and Cell Biology of Artemia*, MacRae, T. H., Bagshaw, J. C., and Warner, A. H., Eds., CRC Press, Boca Raton, FL, 1989, 159.
60. **Berthelemy-Okazaki, N. J.,** Environmental, biochemical, and genetic factors regulating oviparity in *Artemia*, Ph.D. dissertation, University of California, Davis, 1986.
61. **Bewley, J. D. and Black, M.,** *Physiology and Biochemistry of Seeds*, Vol. 2, Springer-Verlag, New York, 1982, 375.
62. **Tauber, M. J., Tauber, C. A., and Masaki, S.,** Adaptations to hazardous seasonal conditions: dormancy, migration, and polyphenism, in *Ecological Entomology*, Huffaker, C. B. and Rabb, R. L., Eds., John Wiley & Sons, New York, 1984, 149.
63. **Marcus, N. H.,** Photoperiodic and temperature regulation of diapause in *Labidocera aestiva* (Copepoda: Calanoida), *Biol. Bull.*, 162, 45, 1982.
64. **Marcus, N. H.,** Variation in the diapause response of *Labidocera aestiva* (Copepoda: Calanoida) from different latitudes and its importance in the evolutionary process, *Biol. Bull.*, 166, 127, 1984.
65. **Versichele, D. and Sorgeloos, P.,** Controlled production of *Artemia* cysts in batch cultures, in *The Brine Shrimp, Artemia*, Vol. 3, Persoone, G., Sorgeloos, P., Roels, O., and Jaspers, E., Eds., Wetteren, Belgium, Universa, 1980, 246.
66. **Lochhead, J. H.,** Oviparity versus ovoviviparity in the brine shrimp, *Artemia, Bio. Bull.*, 121, 396, 1961.
67. **Belk, D.,** Anostracan mating behavior: a case of scramble-competition polygyny, in *Crustacean Sexual Biology*, Bauer, R. T. and Martin, J. W., Eds., Columbia University Press, in press.
68. **Browne, R. A.,** Reproductive pattern and mode in the brine shrimp, *Ecology*, 61, 466, 1980.
69. **Belk, D.,** personal communication to L. E. D., 1989.
70. **Hairston, N. G., Jr. and Olds, E. J.,** Population differences in the timing of diapause: Adaptation in a spatially heterogeneous environment, *Oecologia*, 61, 42, 1984.
71. **Newman, H.,** personal communication to L. E. D., 1985.
72. **Dana, G.,** Comparative population ecology of the brine shrimp, *Artemia*, M.A. thesis, San Francisco State University, 1981.
73. **Lenz, P. H.,** Ecology of an alkali-adapted variety of *Artemia* from Mono Lake, California, U.S.A., in *The Brine Shrimp, Artemia*, Vol. 3, Persoone, G., Sorgeloos, P., Roels, O., and Jaspers, E., Eds., Wetteren, Belgium, Universa, 1980, 79.
74. **Lenz, P. H.,** Population studies on *Artemia* in Mono Lake, California, Ph.D. thesis, University of California, Santa Barbara, 1983.
75. **Lavens, P., Tackaert, W., and Sorgeloos, P.,** International study on *Artemia*. XLI. Influence of culture conditions and specific diapause deactivation methods on the hatchability of *Artemia* cysts produced in a standard culture system, *Mar. Ecol. Prog. Ser.*, 31, 179, 1986.
76. **Van der Linden, A., Vankerckhoven, I., Caubergs, R., and Decleir, W.,** Action spectrum for light-induced hatching of *Artemia* cysts, (Branchiopoda, Crustacea), *Mar. Biol.*, 1, 239, 1986.

77. **Gutterman, Y. and Heydecker, W.**, Studies of the surfaces of desert plant seeds. I. Effect of day length upon maturation of the seedcoat of *Ononis sicula*, Guss., *Ann. Bot.*, 37, 1049, 1973.
78. **Gutterman, Y. and Evenari, M.**, Effects of the environment of the mother plant on germination by control of seed-coat colour, an index of water permeability, of the desert annual *Ononis sicula* Guss., *J. Ecol.*, 60, 713, 1972.
79. **Sorgeloos, P.**, First report on the triggering effect of light on the hatching mechanism of *Artemia salina* dry cysts, *Mar. Biol.*, 22, 75, 1973.
80. **Van der Linden, A.**, Light-induced development of *Artemia* cysts, Ph.D. dissertation, Universiteit Antwerpen; Universitaire Instelling Antwerpen, Belgium, 1989.
81. **Stross, R. G.**, Photoperiod control of diapause in *Daphnia*. IV. Light and CO_2-sensitive phases within the cycle of activation, *Biol. Bull.*, 140, 137, 1971.
82. **Davison, E. A., Jr. and Stross, R. G.**, A blue-light reversible reaction in an animal system *(Daphnia pulex), Experientia*, 42, 620, 1986.
83. **Come, D.**, Problems of embryonal dormancy as exemplified by apple embryo, *Israel J. Bot.*, 29, 145, 1980/81.
84. **Hairston, N. G., Jr. and De Stasio, B. T., Jr.**, Rate of evolution slowed by a dormant propagule pool, *Nature*, 336, 239, 1988.
85. **Bogatova, I. B. and Schmakova, Z. I.**, Activation of diapausing eggs of *Artemia salina, Gidrogiol. Zh.*, 16, 108, 1980.
86. **Vanhaecke, P. and Sorgeloos, P.**, International study on *Artemia*. XIX. Hatching data on 10 commercial sources of brine shrimp cysts and re-evaluation of the "hatching efficiency" concept, *Aquaculture*, 30, 43, 1983.
87. **Drinkwater, L. E. and Crowe, J. H.**, Metabolic status of *Artemia* cysts during embryonic diapause, *J. Exp. Biol.*, Submitted.
88. **Vallejo, C. G., Perona, R., Garesse, R., and Marco, R.**, The stability of the yolk granules of *Artemia*: an improved method for their isolation and study, *Cell Differ.*, 10, 343, 1981.
89. **Ballario, P., Bergami, M., Cacace, M. G., Scala, F., and Silvestri, S.**, Trehalase from the brine shrimp, *Artemia salina*, purification and properties, *Com. Biochem. Physiol.*, 61(B), 265, 1978.
90. **Hand, S. C. and Carpenter, J. C.**, Multiple forms of trehalase from *Artemia* cysts: pH-dependent kinetic properties and relevance to carbohydrate metabolism, *Am. Zool.*, 25, 51, 1984.
91. **Van der Linden, A., Gadeyne, J., Van Onchelen, H., Van Laere, A., and Decleir, W.**, Involvement of cyclic nucleotides in light induced resumption of development of *Artemia* embryos, *J. Exp. Zool.*, Submitted.
92. **Mandels, G. R.**, Compartmentation of metabolic systems in the regulation of dormancy in fungal spores, *Exp. Mycol.*, 5, 278, 1981.
93. **Thevelein, J. M.**, Regulation of trehalase activity by phosphorylation-dephosphorylation during developmental transitions in fungi, *Exp. Mycol.*, 12, 1, 1988.
94. **Taylorson, R. B.**, Interaction of the phytochrome and other factors in seed germination, in *The Physiology and Biochemistry of Seed Development, Dormancy, and Germination*, Kahn, A. A., Ed., Elsevier Biomedical Press, New York, 323, 1982.
95. **Hochachka, P. W. and Mommsen, T. P.**, Protons and anaerobiosis, *Science*, 219, 1391, 1983.
96. **Busa, W. B. and Nuccitelli, R.**, Metabolic regulation via intracellular pH, *Am. J. Physiol.*, 246, 409, 1984.
97. **Karmiol**, M.S. thesis, University of Windsor, Canada, 1981.

Chapter 6

MORPHOLOGY OF *ARTEMIA*

Godelieve R. J. Criel

TABLE OF CONTENTS

I. EXTERNAL MORPHOLOGY

Scanning electron microscopy provided much details to the external morphology of *Artemia*. Presumed sites of sensory reception were investigated and novel receptors were discovered. Tyson and Sullivan[1,2] and Wolfe[3] both studied the frontal knobs of the protopodite of the male clasper. They found two types of processes projecting from it: numerous small noncellular spines and rarer but larger setae protruding from a broad basal elevation (Figure 1). Only these last ones are innervated.[3] The antenullae of male, female, and nauplius bear two types of innervated sensillae.[4-6] Cuticular setae are found on the adult trunk segments;[7] however, no information is available on their innervation or on their anatomical connection with the central nervous system. The work of Schrehardt[8] improved the understanding of the filter-feeding apparatus. Larval *Artemia* use antennae and mandibles as filtering organs whereas in adult *Artemia* the thoracic limbs delimit a mid-ventral canal called the food groove, where the setae and setulae of the protopodites and endites take over the filter-feeding function; i.e., the particle flow passes the setal filter and suspended bacteria and algae eventually adhere to the setulae. The frontal delimitation of the food groove is marked by the maxillae and maxillulae whose feathered setae carry the food particles through the paragnath channel to the mandibles covered by the labrum. Each mandible possesses peripheral incisors and blunt-crowned teeth used for the mechanical treatment of the filtered food particles. The molar surfaces of the mandibles were described by Tyson and Sullivan.[9,10]

II. INTEGUMENT

The whole body of *Artemia* is covered by an extremely thin exoskeleton. In the larval stages the cuticle is 0.3 to 1.0 μm thin and probably has a similar appearance in all regions of the integument.[11] It is divided into an outer epicuticle and an inner fibrous procuticle without a differentiated exo- and endocuticle. A less uniform cuticle covers the adults[12] (Figure 2); its thickness varies with the region: in the male clasper it reaches 7 μm whereas in the trunk and in the thoracopod exopodite it only attains 1 to 1.5 μm. In most regions it is composed of a thin three-layered epicuticle, a laminated pre-exuvial exocuticle and a fibrous postexuvial endocuticle. The postexuvial endocuticle is lacking on the medially oriented inner surface of the exopodite. The structure of the cuticle covering the gills or metepipodites was not described by Copeland[13] but in his Figure 4 no endocuticle is seen. The differences between the larval and adult cuticle probably indicate further modifications in the cuticle synthesizing mechanism during the larval period.[11] The molting pattern changes as well, i.e., nauplii have a diecdysic type of molt cycle with a predominant premolt stage and a greatly shortened intermolt period.[11] In adults, the intermolt period predominates.[12] Horst[14] studied some ultrastructural aspects of the larval epidermis; the adult epidermal cells have not been fully investigated.

The only tegumental glands that Dornesco and Steopoe[15] discern in *Artemia* are proximal thoracopodal glands, but Benesch[16] claims to have found a similar pair of glands in the abdominal region of the first thoracic segment. However, his description is not in complete

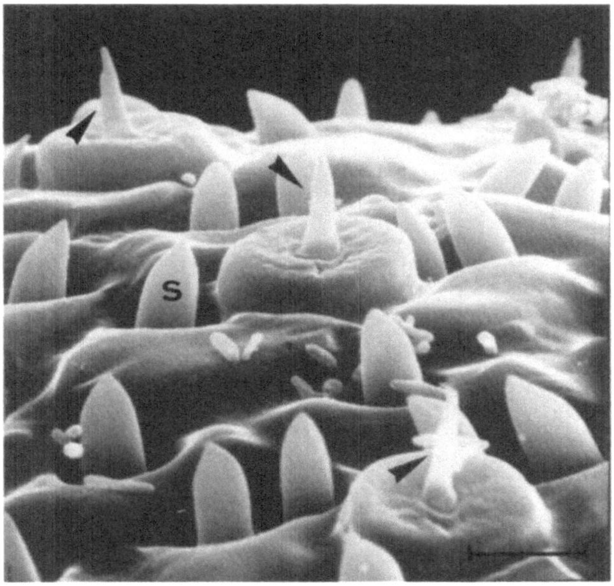

FIGURE 1. Scanning electron micrograph of male frontal knob showing unenervated spines (s) arising from depressions in the exoskeleton and sensory setae (arrows) arising from a dome-shaped supporting cell. (Bar = 5 μm.)

agreement with Dornesco and Steopoe.[15] All authors find one large and two small gland cells but they do not agree which one of these contain the concretions, which made Claus[17] conclude that these glands had a sensory function. Three to four duct cells link those presumed gland cells to the outside. As in *Branchipus*[18] the duct cells of the thoracopodal glands of *Artemia* open at the base of a spine of the protoendite.[16] Dornesco and Steopoe[15] stress the structural differences between the tegumental glands of the saltwater adapted *Artemia* and those of the other freshwater *Anostraca*.

III. CONNECTIVE TISSUE

As soon as differentiating muscle cells and coelom appear in the segments the presumptive, connective tissue can be discerned by its chromatin loaded nuclei.[16] Most of the cells lie dorsally and caudally in the mesoderm and close to the coelom; further on they lie near to the dorsoventral muscles; finally, a compact group of cells can be found in the caudal region of the segment. These cells disperse when the segment further differentiates. From those presumptive connective tissue cells, two types of cells both differentiate with chromatin-rich nuclei. Claus[17] was the first to describe both types. The first type has small nuclei and surrounds muscles, nerve cells, and the gonads with their ducts. The second type (Figure 3 a,b) consists of cells (up to 200 μm in length) with large nuclei, that are connected by long elongated processes to each other, to other organs, and to the exoskeleton by long elongated processes. Benesch[16] calls them fat-storing cells and traces back their origin to stage Na-0 which is the first developmental stage after gastrulation. In the nauplius they are found in the antennal segment, on both sides of the gut diverticles, under the brain and in the labrum. In the post-larval stages they appear in different organs, i.e., in the modified antenna, in the thorax between the dorsal longitudinal muscles and the thoracobrachial limb muscles, in the thoracopods, and in the abdomen. They are most abundant in the labrum, the second antenna, and the mandible. According to Benesch[16] the fat-storing cells of the labrum

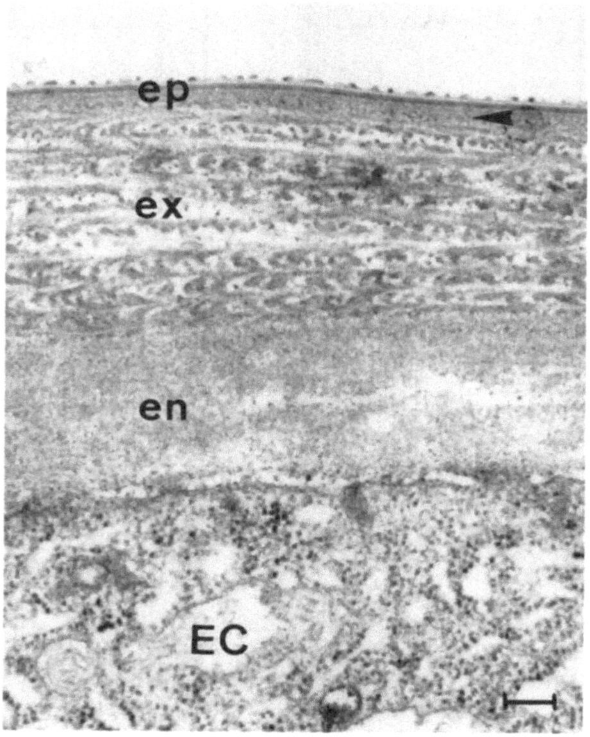

FIGURE 2. Electron micrograph of the cuticle in the trunk region showing the three-layered epicuticle (ep); the exocuticle (ex), with a thin homogeneous outer layer (arrow), and a broad laminated inner layer; the endocuticle (en); and epidermal cell (EC). (Bar = 200 nm.)

originate from the preantennal mesoderm; those of the second antenna and of the mandible, from their respective mesoderm; those of the thorax, from the dorsal mesoderm. According to Claus[17] these cells are the origin of the tendonal sheets whereupon the muscles attach. Benesch[16] only ascribes a mesodermal origin to the large sclerotized tendons of the mandibles and second antennae and presumes the other tendons are from ectodermal origin. The large connective tissue cells were thoroughly studied by Lochhead and Lochhead.[19] From their literature study they conclude that the various terms used for these cells such as fat cells,[17] nephrocytes and "cellules phagocytaires,"[20] and phagocytes[21] are all used in reference to one type of cell and therefore they propose the term "phagocytic storage cells". Functionally these cells seem to combine the characteristics of the nephrocytes and of the fat body in insects. The cytoplasm of these cells exhibits a loose texture and includes a number of large and small vacuoles representing fat and glycogen which has been dissolved. The fat is always in the form of droplets while the distribution of glycogen is diffused. Within each nucleus numerous nucleoli are irregularly distributed. Barigozzi[22] states that the peculiar structure of these nuclei makes counting of chromosomes impossible. These cells are much more numerous and considerably larger in size in females than in males. Using antibodies Van Beek et al.[23] succeeded in localizing the presence of lipovitellin. However, at the electron microscopic level this technique was not successful (personal observations). Another observation linking the function of the phagocytic storage cells with vitellogenesis is the appearance of brown granules when normal vitellogenesis is experimentally inhibited.[19] The storage cells are phagocytic as they sometimes ingest whole blood cells.[19] Moreover Tyson[24]

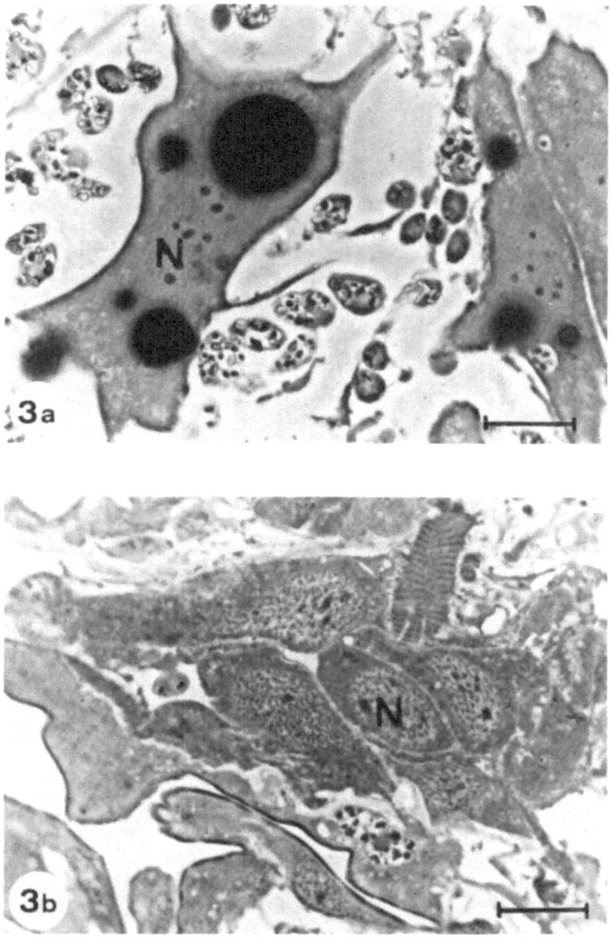

FIGURE 3. Light micrographs of phagocytic storage cells. N, nucleus; bar = 20 μm; 3a is loaded with fat in a well-fed animal; 3b, in a poorly fed animal or under other physiological condition.

has shown spirochetes in vacuoles of amebocytes which are presumably homologous to phagocytic storage cells.

IV. MUSCULAR SYSTEM

When the somites separate, the strands of elongated muscle cells have already attached to the ectoderm.[16] Their location and insertion are temporary, i.e., muscle fibers as well as insertion places which lie close together initially will separate when the appendages develop. The somatic muscle cells become polynucleated before the yolk is resorbed and before the striated fibrils appear.

The segmental origin of the muscular system is still apparent in the adults. The large striated somatic muscles do not insert on the exoskeleton but on the endoskeleton, a chitinous membrane variable in thickness and in distance from the cuticle.[16,17,25] Large, partially branched tendons arise from this membrane and link it to the cuticle. Epidermal cells, modified to tendonal cells, are dispersed throughout the tonofibrils. It is not clear whether these tendonal sheets are from a purely ectodermal origin or whether the mesodermal muscle

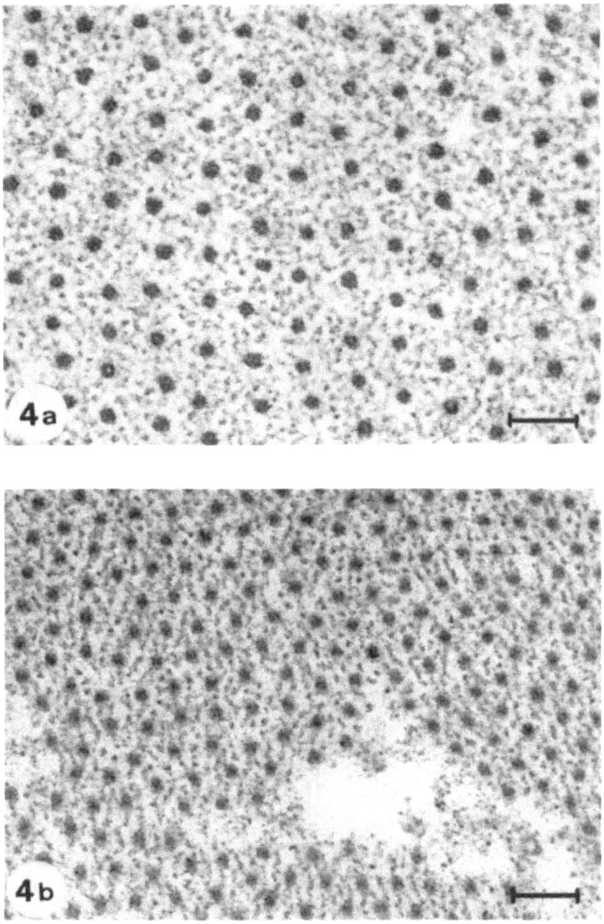

FIGURE 4. Electron micrographs of a transverse section through a somatic muscle (4a), and a visceral muscle (4b). In somatic muscle less thin filaments surround each thick filament than in visceral muscle. (Bar = 100 nm.)

cells participate in their formation.[16,17] The smaller bundles of muscle fibers attach to a barely reinforced basement membrane, the epidermal cells are hardly altered and are connected to the cuticle by few tonofibrils.

Benesch[16] extensively describes both the larval and adult muscular system and endoskeleton, and compares it to the description of the adult *Artemia* by Claus,[17] and to *Chirocephalus*.[26,27] Also, Cassel[25] sketches the adult muscular system.

A summary description of the fine structure of the limb muscle fibers is given by Reger:[28] each thick filament is surrounded with six thin filaments (Figure 4a), and triads of the reticulum are formed at the level of the A-band.

V. CIRCULATORY SYSTEM

A. HEART

As early as 1840, Joly[29] gave the fundamental plan of *Artemia*'s blood circulation as: an open lacunar system characteristic of crustaceans; the heart being a simple longitudinal tube suspended in the body cavity dorsal to the alimentary tract and its anterior end opening

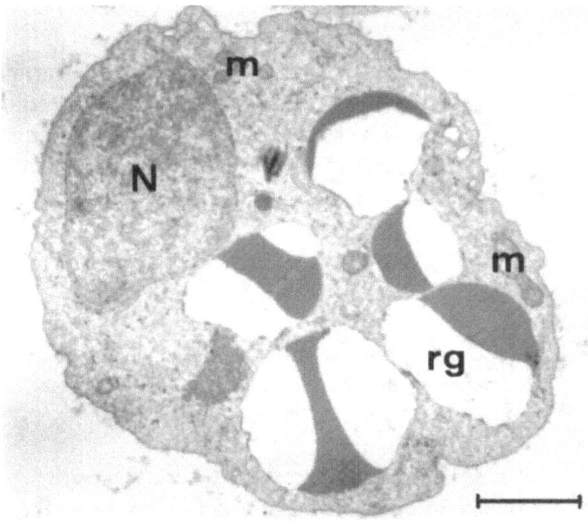

FIGURE 5. Electron micrograph of a blood cell. N, nucleus; m, mitochondria; rg, refringent granules; bar = 2 μm.

freely at the base of the antennae; at the posterior end is a slit, the caudal ostium, and segmentally arranged ostia open over its entire length along both sides. Later, several other descriptions of the circulatory system were published.[16,17,25,30-33] The most comprehensive study however was carried out by Vehsted.[32] He demonstrated how the pericardial septum and its side openings channel the blood circulation in a steady stream in the thorax and abdomen while in the head region connective tissue fibers, muscles, and nerves fulfill the same function. Some controversies exist however about the cranial and caudal opening of the heart, the number of ostia, and the cranial extent of the pericardial septum.

The ultrastructural study of Okland et al.[33] ended the discussion on the number of layers forming the heart. These authors revealed that the heart wall consists of a single layered *myocardium* which is only closed into a tube in its posterior part; in its anterior part it is not closed dorsally but forms a kind of trough with dorsal edges firmly attached to the basement membrane of the epidermis. The cells giving rise to the heart muscle and the pericardial septum are separated from the dorsal longitudinal muscles in the somites as soon as they arise.[16] Weisz[34] considers the cleft arising between the dorsal longitudinal muscles, the heart, and pericardial septum as a dorsal coelom. The ostia and the side openings of the pericardial septum are considered to be secondary structures.

B. BLOOD CELLS AND BLOOD-FORMING ORGANS

The circulating blood cells of *Artemia* are amoeboid cells with a small nucleus and a cytoplasm filled with reflecting granules. All blood cells are of the same type, but vary considerably in size.[17,19,25] The paper by Lochhead and Lochhead[19] is the most relevant on this topic. Although they used several cytochemical techniques, they did not succeed in identifying the chemical composition of the refringent granules. Electron microphotographs show these granules as an electron dense crescent deposited against the surrounding membrane (Figure 5).

Blood cells have a major function in blood clotting and wound healing;[19,35] moreover, in females, they seem to play a role in vitellogenesis[19] which, however, needs further study. The phagocytosing activity postulated for these cells may originate from a confusion with the phagocytic storage cells that we attribute to the connective tissue as does Claus,[17]

Lochhead and Lochhead,[19] and Benesch,[16] but which are classified by other authors as blood cells.[20-21,24,35]

Lochhead and Lochhead[19] extensively described the blood-cell-forming organs. In his doctoral thesis, however, Cassel[25] already presumed that some cell groups in thin, walled pockets on both sides of the leg muscles produced the blood cells. The blood-cell-forming organs are small nodules at the base of each trunk limb. They differ in size and are surrounded by a thin discontinuous membrane. Each nodule consists of a group of rounded cells differing from each other in size and in activity of their nuclei as a result of different stages in mitosis. The resting cells show a uniformity in size of the nucleus and cytoplasm. At all times every cell in the nodule displays a high degree of independence. Toward the periphery of the nodule, the cells show an increase in size of the cytoplasm and a decrease in size of the nucleus. The circulating blood cells have a more extensive cytoplasm and usually carry numerous inclusions. In the spaces adjacent to the blood-forming organs, intermediate stages between the fully developed blood cells and the undeveloped nodular cell are found. Unlike in Malacostraca, no stroma cells are found.

According to Benesch[16] the first blood cells can be distinguished posteriorly and dorsally from the first antenna mesoderm in nauplius-stage Na-5. They look like fat-storaging cells and contain yolk droplets. They are the primary hemocytes before the mandibular and thoracal blood organs, supplying the secondary and tertiary hemocytes, respectively. Benesch[16] considers these organs as vestigial, segmental organs, but Anderson[36] questions this assumption. The mandibular gland differentiates from the mandible mesoderm in the last nauplius stage, Na-6. At the emergence, stage-0, the primary blood cells have become rounded and free of yolk. The segmentally arranged thoracal blood organs differentiate from the coelom and therefore they are not considered to be connective tissue.

VI. DIGESTIVE SYSTEM

The alimentary tract of *Artemia* lies freely in the hemocoel and is bathed directly in the hemolymph. It lacks multicellular digestive glands and can be divided histologically into three regions: a short vertical stomodeum, foregut, or esophagus leading up from the mouth to open at right angles into the mesenteron or midgut which posteriorly joins a short proctodeum, rectum, or hindgut. In front of the point of entrance of the esophagus the midgut expands into a pair of pouches, the gastric caeca.

The histological differences between the three regions have an embryological background: the midgut is from entodermal origin whereas the esophagus and hindgut are from ectodermal origin.[16] Esophagus and hindgut are surrounded by a circular and a longitudinal muscle layer and a dilatator muscle. The midgut is only surrounded by a circular muscle layer. Its epithelium consists of cylindrical to cubical cells with a conspicuous brush border and is separated by a basement membrane from the underlying, circular muscle layer. At the light microscopical level, it was considered that secretion predominated the anterior part of the midgut and resorption predominated the posterior part.[37] The secretion is considered to be holocrine by Frenzel[37] whereas Kuenen[38] finds it to be apocrine up to the third, thoracic segment. Electron microscopy revealed no differentiation into secretory or absorptive cells, but a regional specialization of one cell type.[8,39] Throughout the entire length of the midgut the cells show characteristics of absorptive cells although they also exhibit some secretory characteristics (Figure 6). They are lined with long microvilli which diminish in length in cranio-caudal direction and which are coated with a fine fibrous material, probably mucopolysaccharides. Their apical membranes show apical pits and vesicles. The basal cell membrane of the posterior part of the midgut shows typical infoldings associated with mitochondria (Figure 7). In every cell a phagolysosomal cytoplasmic body suggests an intracellular digestion.[39] Lipid is the main storage product of the anterior part whereas

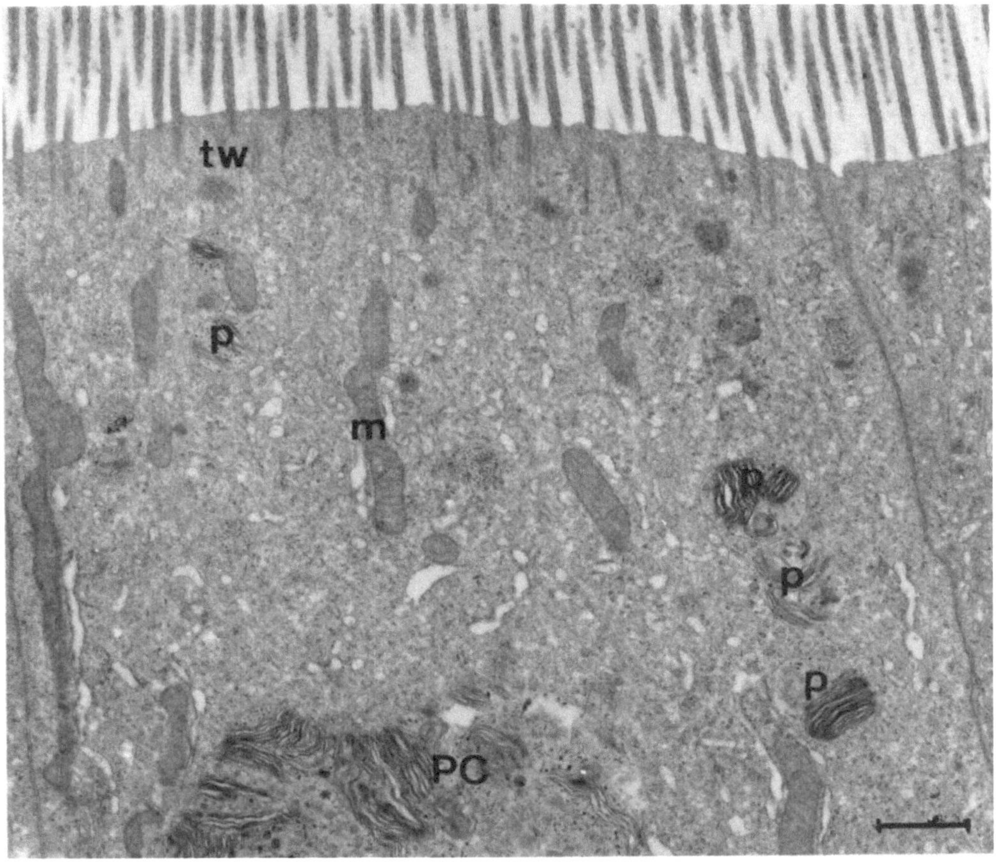

FIGURE 6. Electron micrograph of the apical region of a midgut epithelial cell. Under the terminal web (tw) long mitochondria (m) are oriented in the microvillar axis, while small phagolysosomes (p) coalesce to a large phagolysosomal cytoplasmic body (PC). (Bar = 1 μm.)

glycogen is predominant in the posterior region.[8] Occasionally extrusions of the cytoplasm resembling apocrine secretion occur on the luminal surface in the anterior region of the gut, although degeneration is not excluded.[39] Schrehardt[8] believes this is a fixation artifact and presumes that the observed degeneration and extrusion of whole cells into the gut lumen are a holocrine secretion. This degeneration is more frequent in the gastric caeca than in the posterior portion of the gut. Using histochemical techniques Reeve[40] has shown the chitinous nature of the peritrophic membrane. Snyder and Wolfe[41] studied its formation, but unfortunately we were not able to study their paper. The peritrophic membrane is thought to be derived from the glycocalyx covering the gut microvilli.[42] Kikuchi[39] did not find a peritrophic membrane but a mucous layer covering the food inside the gut. Schrehardt[8] does not even mention its existence.

The basement membrane of the midgut is composed of two layers.[8,39] A thick, often-striated, inner layer adjacent to the epithelial cells is densely packed with rough granules intermixed with amorphous material. An outer layer facing the hemocoel is thin, loose, and finely granular. It does not penetrate into the basal infoldings of the cell membranes.

In the striated circular muscle, nine filaments surround the thick filament[39] instead of six as in limb muscles[28] (Figure 4b). Using the scanning electron microscope, Schrehardt[8] was unable to see the innervation of the anterior part, although the posterior part is innervated by multipolar neurons. Tyson and Lyon[43] also studied the gut with the scanning electron microscope.

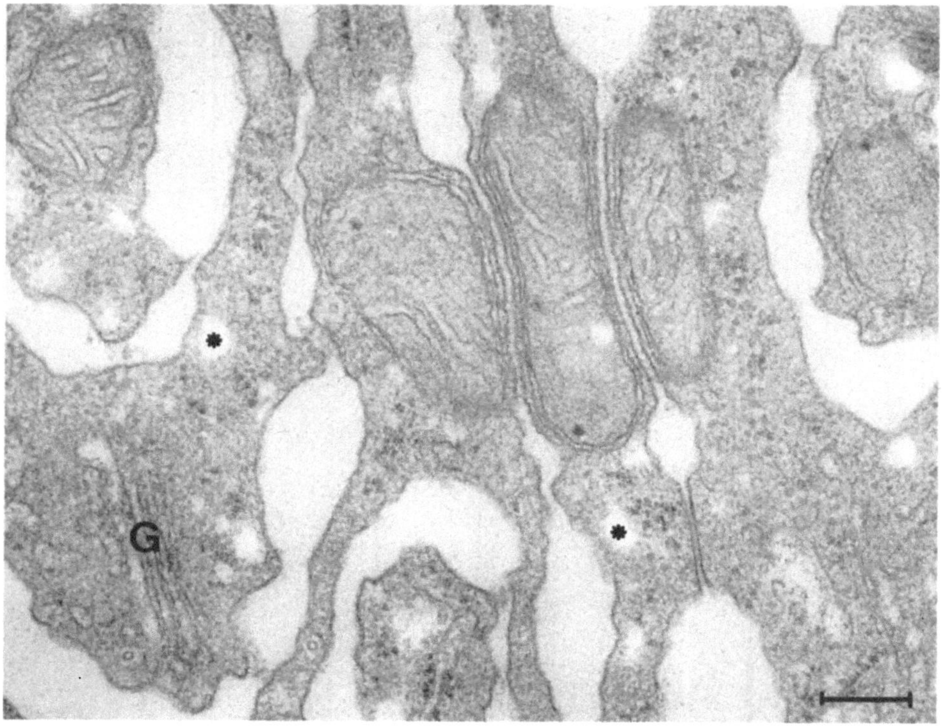

FIGURE 7. Electron micrograph of the basal region of an epithelial cell in the distal part of the midgut showing the special association between mitochondria and infoldings of the plasma membrane. G, Golgi apparatus; *, clear spaces due to elution of glycogen during fixation; bar = 200 nm.

Hootman and Conte[42] studied the naupliar midgut and found it broadly similar to the adult midgut. They noted the absence of a glycocalyx on the microvilli but nonetheless observed a peritrophic membrane. Here it possibly differentiates from numerous clear vesicles found between the microvilli. These vesicles, however, might also represent digestive enzymes. No phagocytolysosomes are found. Basal infoldings associated with mitochondria are present. As they have showed the naupliar gut to be rich in Na+K-activated ATPase, they suggest the midgut is active in Na+ exchanges.

Schrehardt's[8] is the only ultrastructural study of the esophagus and hindgut; however, he does not mention the many tendonal cells for the attachment of the dilatator muscles[25] (Figure 8). He finds both epithelia to be cuboidal rather than cylindrical, and covered by a thin cuticle. They only store small amounts of glycogen and lipid. In the hindgut many myelin figures are found. The basement membrane is monolayered.

At the junction of esophagus and gastric caeca a fine, granular matrix is secreted between the cell apices and the cuticle by the esophagus cells, whereas in this region the cells of the gastric caeca contain electron dense granules of unknown function.[8]

VII. NERVOUS SYSTEM

A. BRAIN AND VENTRAL NERVE CORD

The nervous system of *Artemia* is of the typical primitive Branchiopod type[44] consisting of a dorsal brain, or supraesophageal ganglion, circumesophageal connectives, and a double ventral row of segmental paired ganglia united longitudinally by connectives and laterally by a large anterior and a smaller posterior commissure.

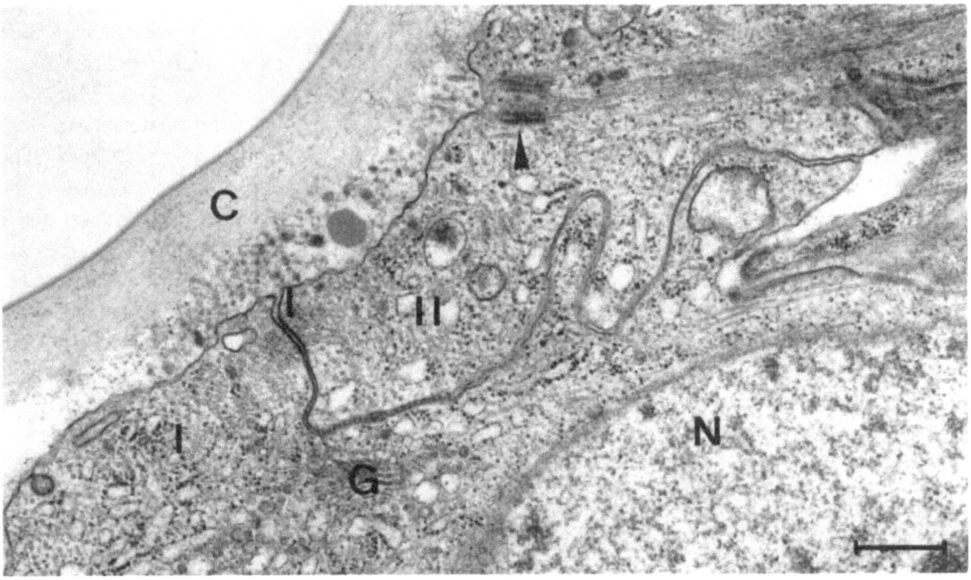

FIGURE 8. Electron micrograph of the hindgut epithelium. A gut epidermal cell (I) and a tendonal cell (II) are shown. C, cuticle; G, Golgi apparatus; N, nucleus; arrow, dense apical rod characteristic for tendonal cells; bar = 500 nm.

Benesch[16] mentions two types of neurons: the larger "ganglia cells" having a well-developed cytoplasm and a nucleus with little chromatin, and the small "globuli cells" which are defined by Hanström[44] as unipolar associative neurons with little cytoplasm and a relatively large nucleus full of chromatin. Their highly branched dendritic trees form a typical associative region in the neuropile called glomeruli. These are thought to have a special function as areas of higher association.[45]

The nervous system has been fully described, but unfortunately only scarcely illustrated.[16,17,25,30,46,47]

The description and nomenclature used by Benesch[16] can be fitted into the common pattern of brain structure of Arthropods proposed by Horridge.[45] It can be correlated with that of Claus.[17] The descriptions of Warren[47] and Cassel[25] are slightly different; those of Spencer[46] are rather schematic. We will try to collate the findings of Benesch[16] with the other descriptions.

The protocerebrum receives nerves from the complex eyes, the nauplius eyes, and the frontal organs. It consists of neuropile masses covered by an interrupted layer of cell bodies. Globuli cells occur in defined regions of the neuropile. Two larger frontal lobes, and two smaller posterior-lateral lobes can be discerned, both linked by a commissure. A transversally oval "central body" built by nerve fibers running in all directions, lies in between both commissures[17] or ventrally from them.[16] Dorsally the frontal lobes are covered by the "Dorsallappen"[17] which are probably homologous to the protocerebral bridge[16,45] which shows two lateral swells separated by a median groove. Probably, this is the part thought by Elofsson[48] to represent the *medulla terminalis*. Frontally the protocerebral bridge is connected to the frontal organs; laterally, to the optic lobes.[17,46] According to Spencer[46] the frontal lobes innervate the nauplius eyes and other sensory organs of the cephalic region. Benesch[16] postulates that the central body is connected to the nauplius eyes, and Elofsson[48] finds the nauplius eye center clearly separated from the rest of the protocerebrum.

The brain as described by Warren[47] consists of three pairs of lobes: an anterior dorsal pair, an anterior ventral, and a posterior ventral pair. The optic nerve arises from the anterior ventral lobes.

The deutocerebrum is the region where the ganglia of the first antenna are found. Right and left antenna ganglia are linked by a commissure.[44] The deutocerebrum is clearly separated from the protocerebrum. The course of the antennal nerve is extensively described;[25] it contains sensory and motor neurons.[16]

The so-called circumesophageal connectives pass backward from the posterior surfaces of the brain on either side of the gut and connect the brain to the ventral nerve cord. The cell bodies which constitute the tritocerebrum are found on these connectives. Close to the brain the second antenna ganglion is rather diffuse.[47] Three ventral nerves and a dorsal nerve arise;[16] the cranial one contains motor and sensory neurons, the medial and caudal ones are motor nerves. Cassel[25] extensively describes the nerve of the second antenna. It innervates sensory receptors both in males, where it innervates the clasper, and in females. More caudally, Warren[47] and Cassel[25] describe the so-called postesophageal ganglia on the connectives. Nonetheless, Claus,[17] Hanström[44] and Benesch[16] do not mention them. The postesophageal ganglia consist of a large dorsal lobe and a smaller ventral mass[25] both right and left lobes being linked with a commissure. Warren[47] finds the commissure of the ventral lobes double. The dorsal one is the largest and gives off the connectives to the mandibular ganglia. The ventral commissure, however, is more important because it is the origin of the stomatogastric visceral nervous system: it gives off a large ventral branch which meets its homolog to form the circumoral ring which passes ventrally and laterally from the esophagus into the labrum where it connects to the labral ganglion[47] or esophageal ganglion.[16] Three nerves arise from this ganglion[16] or from three visceral ganglia of the circumoral ring:[25] a nerve to the anterior part of the esophagus which innervates the circular muscles of the esophagus and probably also its dilatator muscles,[25] and two labral nerves[16] which innervate the muscle of the tip of the labrum and supply the sensory areas of the mouth.[25]

The description of the ventral nerve cord by Claus[17] is brief. Warren[47] (Figure 9), Cassel[25] and Benesch[16] describe it more extensively. The latter, moreover, names all the innervated muscles. The mandibular ganglia are linked by connectives to the dorsal lobe of the postesophageal ganglia, and they are connected to each other by a large anterior and a small posterior commissure[25] which are almost linked.[47] The mandibular nerves arise ventrally but then bend dorsally.[16] A large anterior nerve leads to the mandibular muscles[25,47] which contains two motor and one sensible branch and may thus be compared to the nerves of the thoracic segments;[16] a smaller posterior nerve is a motor nerve[16] and leads to the longitudinal body muscles.[25] In the metanaupliar region a dorsal nerve arises from each ganglion which innervates the muscles of the next segment. The total number of nerves varies up to the first thoracic segment. The first pair of ganglia of the metanaupliar region are the first maxillary ganglia. They deliver a large branched nerve to the muscles of the region and a smaller nerve to the muscles of the first maxilla and possibly a nerve to the maxillary glands.[47] Cassel[25] only mentions the nerves to the first maxilla, whereas Benesch[16] refers to a ventral nerve, containing motor and sensory neurons, and a dorsal motor nerve. Warren[47] observes remnants of a maxillulary ganglion, whereas Cassel[25] describes a slight enlargement in the connectives in this region and a small commissure. Benesch[16] finds two nerves arising from the maxillulary ganglion, a ventral mixed nerve, and a dorsal motor nerve. Next come the eleven pair of thoracic ganglia. The two ganglia of each segment are linked by a larger anterior and a smaller posterior commissure. The nerves arising from the ganglia are similar in structure all over the nerve cord. Cassel[25] provides a more detailed description than Warren.[47] Benesch[16] in 1969, in agreement with Leydig,[30] provides a different description; i.e., he finds three nerves arising from each thoracic ganglion: a cranial motor nerve, a medial sensory nerve, and a caudal motor nerve splitting up into two branches, the dorsal one of these innervating the muscles of the following segment. The first and second pairs of genital ganglia are almost fused to two large masses of nervous tissue. The innervation of the male genital tract is described in great detail by Cassel.[25] Claus[17] finds a single anterior

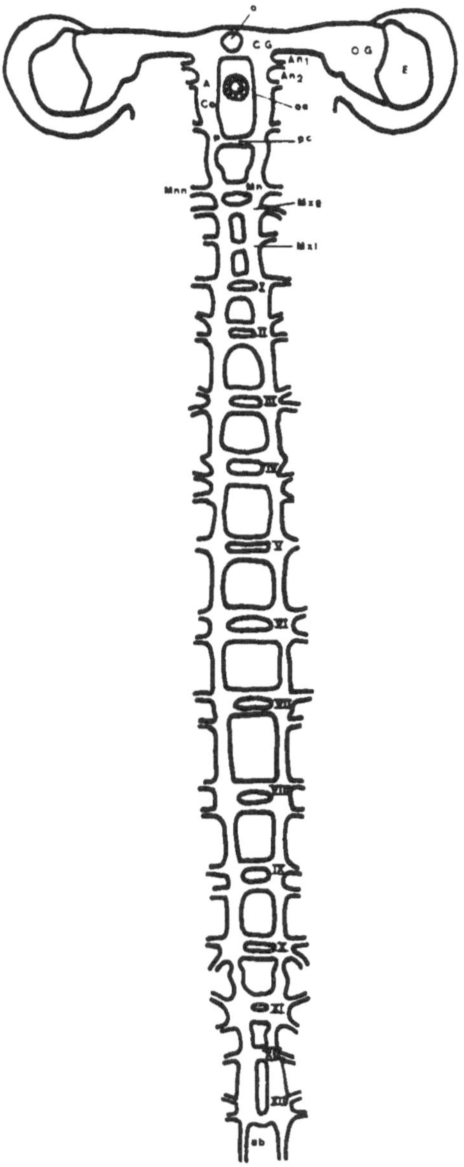

FIGURE 9. Reconstruction of the dorsal projection of the entire central nervous system. A, antennal ganglion; An_1 and An_2, antennular and antennal nerves; C.G., cerebral ganglion; Co, circumesophageal connective; E, compound eye; I to XI, ganglia of the respective thoracic segments, each with its commissures and nerves; XII and XIII, the genital ganglia; Mn, mandibular ganglion; Mnn, mandibular nerves; Mxg, maxillary ganglion; Mxl, maxillular ganglion; o, ocellus; O.G., optic ganglion. (Modified after Warren, H. S., *Trans. Am. Microscop. Soc.*, 49, 198, 1930. With permission.)

and a double posterior commissure. Warren[47] finds both masses united by three commissures. From the side of each mass he finds three nerves leaving. Benesch[16] only finds two nerves leaving from these ganglia: a dorsal motor and a mixed ventral nerve. No ganglia are found in the abdomen. From the posterior end of the genital ganglia one[47] or several[25] nerves pass posteriorly. One of them gives off branches in each segment which innervate the longitudinal body muscle of the following segment. Nerves pass to the heart, midgut, and to the muscles of the hindgut. Benesch[16] describes two nerve cords in the abdomen which give off one dorsal branch in each segment. Hanström[44] reports a net of caudal visceral nerves innervating the hindgut muscles.

Amarant and Elofsson[49] studied the distribution of monoaminergic nerves in the nervous system of *Artemia*. The medullae terminales have an evenly distributed net of fluorescent fibers which are connected with the frontal and dorsal areas. In the protocerebrum three distinct areas of monoaminergic neuropile are found. In the deutocerebrum one fluorescent cell body is distinguished while fluorescent fibers run in the antennula nerve. In the tritocerebrum, monoaminergic neurons enter the esophageal connectives, the labral commissure, and the labral ganglia. In the ventral nerve cord a limited number of beaded axons is concentrated in the middle of each connective.

B. ONTOGENY OF THE NERVOUS SYSTEM[16]

The nervous system originates from the inward proliferation of numerous cells from the ectoderm. The first differentiation of the nervous system is seen in nauplius stage Na-1: the presumptive nervous cells proliferate and migrate from places where the ectoderm has become multilayered and further differentiate into ganglia cells, globuli cells, neurosecretory and neurohemal cells, and sensory cells. The protocerebral ganglia are developed from the protocerebral ectoderm; the antennular and antennal ganglia from the lateral presumptive ectoderm of these segments; and a paired mid-ventral ganglion is formed from the ventral ectoderm of each postnaupliar segment.[36] Benesch[16] traces back some presumptive brain regions to the nauplius stage Na-3. Further development consists in an increase in number, a differentiation of the cells, and formation and expansion of the neuropile. The longitudinal nerve fibers differentiate first (nauplius stage Na-3). The first transverse nerve fibers arise in nauplius stage Na-6 in the proto- and tritocerebrum. At the emergence stage-0, different parts of the protocerebrum can be discerned among which are the protocerebral bridge and also the first and second antenna nerves which arise from the deuto- and tritocerebrum. The protocerebral neuropile further develops in the postlarval stages: the central body differentiates in larval stage-II (first thoracic segment at the end of proliferation from the budding zone), the optical lobe is differentiated in larval stage-V (seventh thoracic segment proliferating from the budding zone). The commissures of the deutocerebrum are built at the end of the larval stages. From the visceral nervous system, the esophageal ganglion is migrating from the labrum ectoderm as early as nauplius stage Na-3. Later it makes contact with the tritocerebrum.

VIII. SENSORY SYSTEM

A. COMPOUND EYES

Artemia has two widely separated compound eyes mounted on unjointed flexible stalks (Figure 10). They are of the apposition type and consist of a distal layer of about 300 eucone ommatidia separated by a basement membrane from the optic lobe which is linked to the lateral protocerebral lobe by an optic peduncle.[45,50,51] An unspecialized cuticle covers the compound eye.

1. Structure of the Ommatidia

Earlier studies on the ommatidial structure[17,25,30,47,52] were surpassed by the compre-

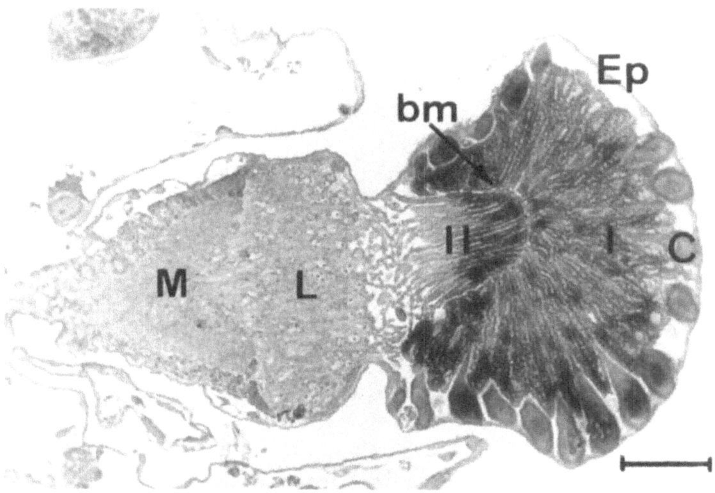

FIGURE 10. Light micrograph of a 2 μm section through an epon-embedded eyestalk showing the eye and the optic-lobe. Basement membrane (bm); C, cristal cone; Ep, epidermis and cuticle; I, zone of the cone-cell roots and the retinula cells; II, zone of the photoreceptor axons; L, lamina ganglionaris; M, medulla of the optic lobe; bar = 50 μm.

hensive and well-illustrated study of Debaisieux[53] and supplemented by electron microscopic studies.[54-56] There is quite a difference of views between Elofsson and Odselius[54] and Hertel,[55] but as Nilsson and Odselius[56] gave Hertel[55] a plausible reply, we prefer to give the view of Elofsson and Odselius[54] (Figure 11 and 12).

The ommatidia are built up by crystalline cone cells and retinula cells. There is a pronounced regional difference in the morphology of the crystalline cones; i.e., short and slender cones are found ventrally, short and thick cones, dorsally, and long and thick cones, in the middle of the eye. These regional differences are important for the light-dark adaptation mechanism.[56] The top of each cone is enveloped by two juxta-crystalline epidermal cells and this complex is surrounded by a clear intercellular space which often contains blood cells. The crystalline cone is composed of four tightly apposed cone cells (sometimes three, five, or six[53]). They consist of the main body of the cone, the crystalline tract which abuts on the rhabdom, and the "cone cell root", which extends from each cell between the retinula cells to the basement membrane. Each cone cell contains a lens-like structure suspended within it, i.e., the highly refractive glycogen body.[54] The light-dark adapting mechanism involves a change in focal length of the glycogen body and a corresponding change in shape of the cone.[56] The cone-cell roots are mostly filled with microtubuli. Their proximal ends widen and flatten to terminal feet containing a conspicuous endoplasmic reticulum. The terminal feet interlock with each other to form the basal plate. The basement membrane of the compound eye is composed of the basal plate, a cellular structure, and an acellular basal lamina which is probably secreted by the cone cells.[54] Moreover each ommatidium contains six retinula cells which contribute to a rhabdom of the fused type. The retinula cells are the only pigmented cells of the eye. Debaisieux[53] as well as Elofsson and Odselius[54] demonstrate a constancy in arrangement of the retinula cells, the rhabdom, and the cone-cell roots. The rhabdoms have a smaller diameter in the ventral ommatidia and a larger diameter in the dorsal ommatidia. The rhabdom is formed by the rhabdomeres in a characteristic way; i.e., its main body is built by five retinula cells, the sixth retinula cell which is often referred to as accessory cell contributes to the rhabdom only two portions for very short distances. Around the rhabdom a palisade is formed consisting of intracellular cisternae. The rhab-

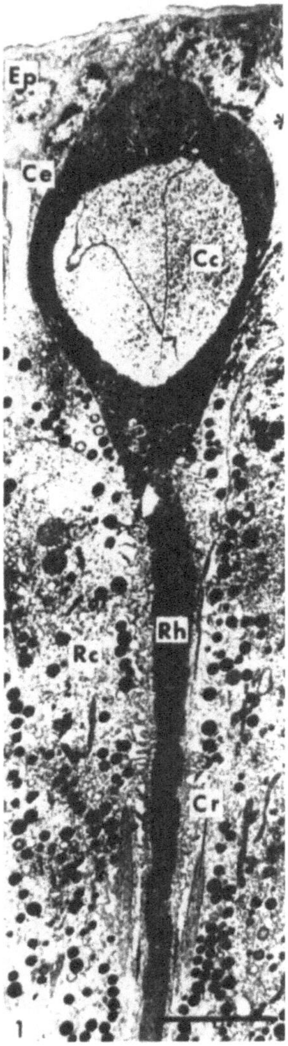

FIGURE 11. Electron micrograph of a paracentral longitudinal section through an ommatidium. Cc, cristalline cone; Ce, juxta-cristalline epidermal cell; Cr, cone-cell root; Ep, epidermis; Rc, retinula cell; Rh, rhabdome; bar = 5 μm. (From Elofsson, R. and Odselius, R., *Acta Zoologica*, 56, 141, 1975. With permission.)

domere space is sealed off by zonulae adhaerentes extending along the whole rhabdom. The cone-cell roots run just outside these zonulae. The pigment granules of the retinula cells are found all over the cell body. The axons penetrate the basement membrane and during this passage become thinner and lose their pigment. Below the basement membrane the axons from each ommatidium unite in bundles of six.[57] Debaisieux[53] found that with dark adaptation the crystalline tract shortens and that the rhabdom simultaneously becomes longer. Electron microscopy[56] moreover shows that the most distal part of the rhabdome becomes thicker and that the palisade moves distally and surrounds the proximal tip of the cone; no adaptational movement of pigment occurs in the retinula cells. Multivesicular bodies are numerous in the retinula cells of light-adapted *Artemia*.

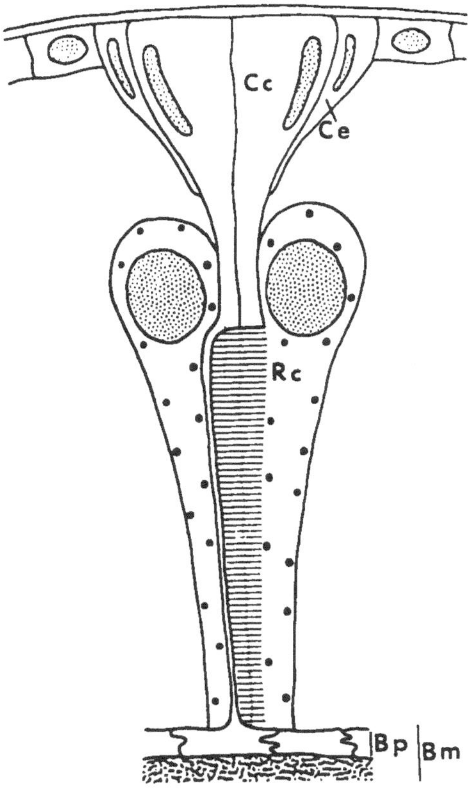

FIGURE 12. Schematic drawing of an ommatidium in the compound eye of *Artemia*. Bm, basement membrane; Bp, basal plate; Cc, cristalline cone; Ce, juxta-cristalline epidermal cell; Rc, retinula cell. (From Elofsson, R. and Odselius, R., *Acta Zool.*, 56, 141, 1975. With permission.)

2. Structure of the Optic Lobe

Two optic neuropiles are found in the eyestalk: a distal lamina ganglionaris and a proximal medulla. As in other Branchiopoda no chiasma is found between both neuropiles.[51,58] Other optic neuropilar regions are retracted into the protocerebrum.

Neuronal cell bodies occur on the distal body of the lamina and on the periphery of the lateral medulla. Monoaminergic perikarya of the globuli type are present in the cell layer covering the medulla medially.[59] The knowledge of the neural connections which were already described by Hanström[44] was further elucidated by the study of Nässel et al.[57] (Figure 13). The dilated terminals of the photoreceptor axons are distributed in two layers of the lamina. They form three types of specialized contacts one of which involves a dyad. One axon of each ommatidium bypasses the lamina and terminates in the medulla. In addition to these photoreceptor axons four other classes of neurons are found in the lamina. Monopolar neurons connect the lamina and the medulla, while tangential T-neurons connect the lamina with the protocerebrum. In their search for monoaminergic neurons Amarant and Elofsson[49] found a specific bilayered pattern of monoaminergic neurons in the lamina ganglionaris, whereas the fluorophore they used in the determination was spread all over the neuropile in the medulla.

3. Muscles of the Eyestalks

One,[53] two,[47] or three[17] muscles are found to be responsible for the movements of the eyestalks.

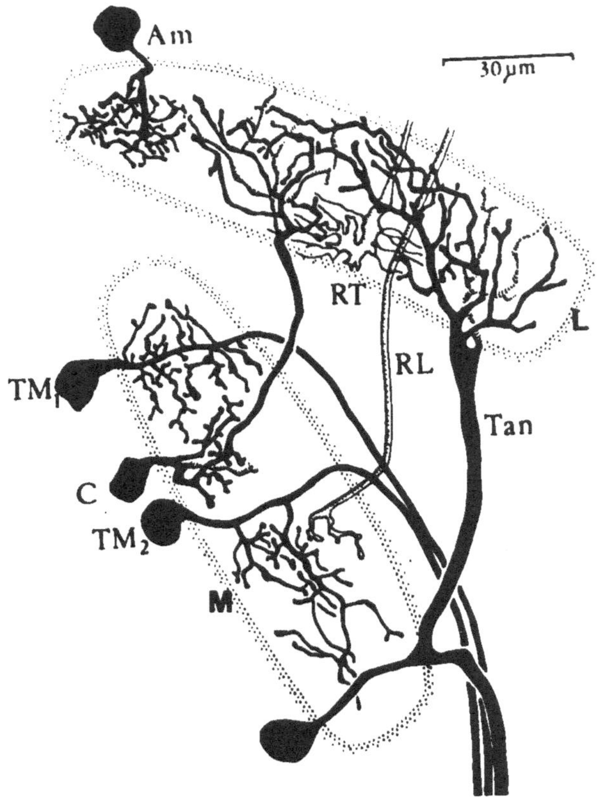

FIGURE 13. Schematic drawing of laminar (L) and medullar (M) neurons in the eyestalk. Am, amacrine neuron; C, centrifugal neuron; RT and RL, photoreceptor axons; RT, receptor terminal; RL, long photoreceptor axon; Tan, tangential T-neuron connecting the lamina and the protocerebrum; TM_1 and TM_2, transmedullary neurons. (Modified after Nässel, D. R., Elofsson, R., and Odselius, R., *Cell Tissue Res.*, 190, 435, 1978. With permission.)

4. Ontogeny of the Compound Eye

Benesch[16] claims to trace back the cells from which the optic lobes originate to nauplius stage Na-3. Debaisieux[53,60] describes the formation of the ommatidia from a proliferation zone; i.e., ommatidial buds of ectodermal origin and neuroblasts both contribute to the formation of the compound eye. However, Debaisieux doubts about the origin of the neuroblasts and suggests they might induce the differentiation of ectodermal cells into ommatidia.

Elofsson and Dahl[58] described the formation of the whole distal part of the eye (ommatidia, lamina ganglionaris, and medulla) from a common proliferation zone. They explain the absence of a chiasma between both optic neuropiles but do not trace back the origin of this proliferation zone. Claus[17] presumes the proliferation zone to be completely derived from ectodermal cells. Dahl[61] states that even if the growth of the distal part of the compound eye and of the optic ganglia is accomplished by cells from the ectodermal proliferation zone, brain cells too seem to have a part in the differentiation although this was not proven. He finds it tempting to assume that the cells which provide the local connections within the lobus are those derived from the proliferation zone, while those which establish the connection with the brain are those derived from a brain rudiment.

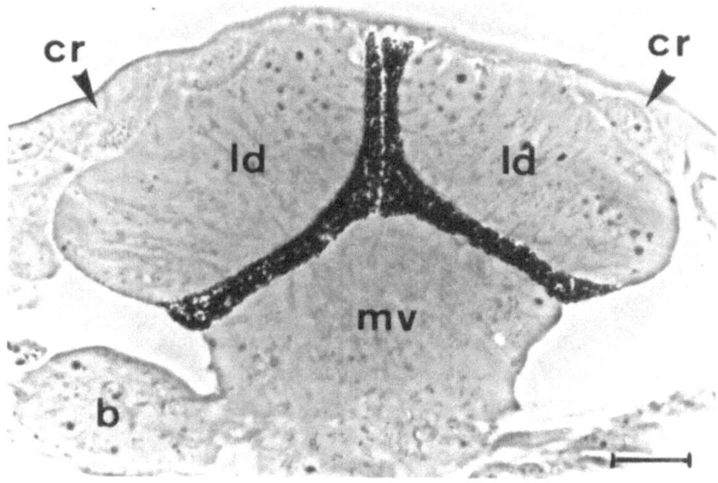

FIGURE 14. Light micrograph of a 2-μm thick epon-section through the nauplius eye and cavity receptor organ. Two pigment cells forming an inverted Y separate the two latero-dorsal (ld) and the medial ventral (mv) cup of sensory cells. b, brain; cr, cavity receptor organ; bar = 20 μm.

B. THE MEDIAN EYE OR NAUPLIUS EYE AND VENTRAL FRONTAL ORGANS

1. Structure

The median eye is the only functional, optical, sense organ in the nauplius. It persists through the various instars, becomes slightly larger, but maintains its general structure until the adult stage.[47]

It is situated in the center of the anterior surface of the head, covered by an unmodified cuticle and epidermis. Posteriorly and dorsally it is separated from the gut diverticles by a basement membrane. Ventrally it lies close to the protocerebrum[62] which may also extend in front of the nauplius eye.[63] It is surrounded by a wide hemocoel except where it is adjacent to the epidermis, the cavity receptor organ (see later), and the cerebrum.[63] It is pear shaped with its thinner part pressed against the epidermis, its thicker part facing the brain, or, in larvae, even protruding into it.[64] Two pigment cells which form an inverted Y are its most obvious feature (Figure 14). Its arms form two latero-dorsal and one medial-ventral cup containing the sensory retinula cells. Claus[65] already defined the nauplius eye as an inverse cup-shaped structure. The lateral cups are deep and depending on the specimen contain 25 to 75 cells each; the ventral cup comprises between 25 and 75 cells.[48] Anadon and Anadon[63] reported fewer cells. Both Elofsson[48] and Anadon and Anadon[63] paid special attention to the innervation of the nauplius eye. The axons of the sensory cells of each cup gather to a nerve running to the nauplius eye center which Elofsson[48] finds clearly separated from the rest of the brain. It is composed of one[48] or three[16,61] neuropiles surrounded with ganglion cells. Elofsson[48] situates this center posteriorly to the nauplius eye, whereas Anadon and Anadon[63] situate it cranially.

Early morphological studies[47,53,64-68] have been replaced by more recent optical[48] and ultrastructural studies.[62-63] Although Elofsson[48] suspected two kinds of sensory cells, electron microscopy only revealed one type. Retinula cells and pigment cells are extensively described by Rasmussen.[62] The facing cell membranes of retinula cells form rhabdoms of the closed type. Rasmussen also finds perirhabdomeric vacuoles and pinocytotic processes at the base of the rhabdom but doubts whether their extent is related to light-dark adaptation. Debaisieux[53] claims that the rhabdoms are partly resorbed in light-adapted eyes and fully developed in the dark. Anadon and Anadon[63] developed a better insight into the three-dimensional structure of the nauplius eye with the use of serial sections. They concluded that its organization does not permit the vision of forms. The three groups of retinula cells have a different orientation and receive luminous rays proceeding from different points, though some rays may reach more than one ocellus.

The ventral frontal organs are also thought to have a photosensory function. Hanström[50] describes them as one group, Elofsson[48] and Rasmussen[6] as two groups, whereas Anadon and Anadon[63] find several groups of sensory cells with an irregular distribution. Although Rasmussen[62] finds them separated from their surroundings by a basement membrane, Elofsson[48] and Anadon and Anadon[63] stress the contiguity of their cells with the epidermis. The retinula cells are similar to those of the nauplius eye. No pigment cells are found nor do they contain pigment granules.[62] Elofsson[48] points out that although there are similarities, they are not homologous with the frontal organs of the Malacostraca, but together with the three naupliar ocelli they may represent five ancestral frontal eyes.

2. Ontogeny

Benesch[16] traces back to the nauplius stage Na-1 the differentiation of the cells which will form the ganglia of the nauplius eye center, the pigment cells, and the retinula cells of both the dorsal and the ventral nauplius eye. In stage Na-3 these cells, as well as the ventral frontal organs can easily be recognized. The nauplius eyes are fully developed from larval stage-III onward. Vaissière[67] describes the larval development of the nauplius eye and traces back its origin to a trilobated bud of the protocerebrum. At emergence it is not yet fully separated from the protocerebrum. The pigment cells mainly contain carotenoids and few melanin granules. During the metanauplius stages the center of the nauplius eye separates from the protocerebrum. The nerves between the nauplius eye and the nauplius eye center become apparent and elongate. Melanin replaces the carotenoid in the pigment cells. After the fourth molt the nauplius eye is complete. Dahl[61] studied the nauplius eye at the time of hatching and found it to be topographically separated from the brain. He presumed that, later in the larval development, two proliferation zones form in the dorsolateral lobes of the protocerebrum which provide additional material for both medial and lateral nauplius eyes and are only separated by a notch in the integument from the more lateral proliferation zone of the compound eye. Benesch[16] has doubts about that late proliferation zone and proposes that all the material proliferates from the optical lobes at once during early nauplius stages, and further expands by mitoses. Moroff[64] considers the nauplius eye to be of purely epidermal origin.

C. DORSAL FRONTAL ORGANS OR CAVITY RECEPTOR ORGANS

Various descriptions and divergent speculations were reported on the function of the dorsal frontal organs which terminate in the epidermis laterally and slightly dorsally to the lateral cups of the nauplius eye[17,46-48,50,62,66] (Figure 14). However, Elofsson and Lake[69] proposed the most plausible hypothesis based on an electron microscopic study. They abandoned their earlier presumptions (X-organ, Elofsson[48]) and called it the cavity receptor organ (Figure 15). The organ consists of ciliated neurons whose cilia protrude into a cavity beneath the cuticle. The neuronal dendrites penetrate a giant accompanying cell, as well as epidermal cells before entering the cavity. Axons of these bipolar nerve cells form nerves which run on both sides of the nauplius eye straight to the medulla terminalis. They do not enter the nauplius eye center. Previously the organ was considered neurosecretory but in the light of present knowledge it is regarded as sensory, although a double function cannot be excluded. Amarant and Elofsson[49] found monoaminergic bodies in the cavity receptor organ. Provided these cells are indeed sensory, this is the first example of monoaminergic sensory cells in Arthropods. Benesch[16] traces back this organ to nauplius stage Na-3. From stage Na-6 a

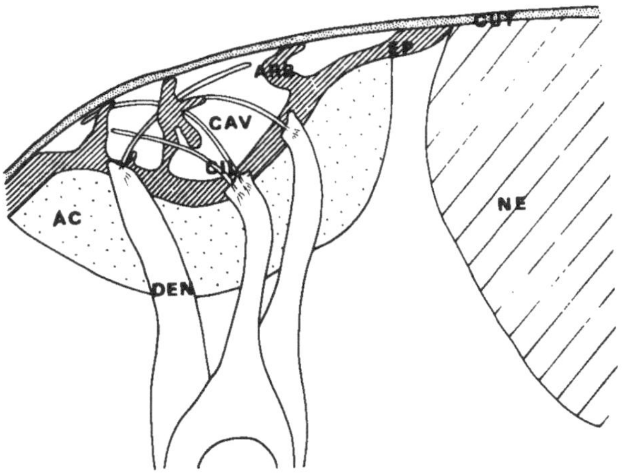

FIGURE 15. Schematic drawing of the distal portion of the cavity receptor organ. AC, accompanying cell; ARB, arborizations from the epidermal cells; CAV, cavity; CIL, cilia; CUT, cuticle; DEN, dendrites from the receptor cells; EP, epidermis; NE, nauplius eye. (From Elofsson, R. and Lake, P. S., *Z. Zellforsch.*, 121, 319, 1971. With permission.)

small elevation on the surface of the embryo marks the presumed location of the organ. In his drawings of later stages it comprises more cells than found in the adult stage.

D. OTHER SENSE ORGANS

The sense organs of the cuticle have been discussed in Section I of this chapter dealing with the external form. They have not yet been described in detail although already in 1886 Claus[17] considered these to be associated with subcuticular ganglion cells. Mandibular and maxillar sense organs are mentioned by Warren.[47]

IX. ENDOCRINE SYSTEM

Little remains to be added to the theories of van den Bosch de Aguilar[70] on the entomostracan endocrinology in general and *Artemia* in particular. Neurosecretion is the only endocrine system presently known to exist in the Entomostraca since no endocrine glands have been identified in any of the species studied. The neurosecretory system seems to be primitive and is characterized by the absence of a storage organ clearly defined as such. Investigations of the neurosecretory system in *Artemia* yielded conflicting results. Van den Bosch de Aguilar[70] thought this may be due to the diversity in histological techniques and to the lack of physiological data on the animals examined.

Lochhead and Resner[71] were the first to study the neurosecretory system. They found "what appears" to be 35 to 40 neurosecretory cells in the anterior part of the brain and three cells on each ventro-posterior side. In the eyestalks no neurosecretory structures were found nor did their removal cause any physiological disturbance. Baid and Ramaswami[72] classified the neurosecretory cells in three groups based on the nature of their secretion; i.e., two cell types occur in clusters in the supraesophageal ganglion, and one type is found in what they presume to be the X-organ of the eyestalk. Ventrally in the middle of the brain they identify a region receiving axons from the above-cited cells which they presume to be the Y-organ. Hentschel[73-74] compares the neurosecretory system of *Chirocephalus* and *Artemia*. Fewer neurosecretory cells are found in *Artemia*, and due to their small size, they are difficult to trace. He localizes the sinus gland in the same region where Baid and Ramaswami[72] localize the X-organ. Hentschel[74] furthermore finds several neurosecretory cells in the proto, deutero-, tritocerebrum, and on the periphery of each ganglion of the ventral nerve cord. The statement of van den Bosch de Aguilar[70] that the presumed neurohemal organ of the eyestalk could be confused with the zone of proliferation and growth of the ommatidia[53,58,60] seems to be confirmed by Nässel et al.[57] who in an electron microscopic study did not find any neurosecretory fibers. Van den Bosch de Aguilar[70] only found four protocerebral neurosecretory cells. He writes "...before labeling a structure as an endocrine organ it is essential to define its function in the physiology of the animal." That is what he did[75] with the two median protocerebral neurosecretory cells. He showed them to be stimulated in hypotonic media and as they are related to the cavity receptor organ,[69] this suggested an osmoregulatory function. The activity of the two lateral protocerebral neurosecretory cells is maximal at the time of reproduction. He does not go into the function of the neurosecretory cells he found separately or in small groups in the deuto- and tritocerebrum, often associated with the roots of the antennula and antenna nerves, and in the ventral nerve cord.

Recently Van Beek[76] tried to demonstrate neurosecretory cells by an immunocytochemical technique tracing neuropeptide-like material. Both FMRF-amide and neuropeptide-Y were found in few median protocerebral cells and in two cell bodies of the subesophageal ganglion. Some small cells were visualized with the neuropeptide-Y antiserum in the protocerebrum. Leuk- and met-enkephalins were localized in the same median protocerebral cells. Separate protocerebral cells were found to be immunoreactive to one of the enkephalins. No immunoreactivity was discovered using LHRH or FSH antisera. Weak immunoreactivity to LH antiserum marked two or three small cells in the lateral parts of male and female protocerebrum. Interpretation of these findings is speculative as no function was demonstrated for any of these peptides in *Artemia*.

Benesch[16] traces back an X-organ and neurosecretory cells to nauplius stage Na-3.

X. EXCRETORY SYSTEM

A. STRUCTURE

Excretion is probably performed by two pairs of glands: the maxillary glands which remain functional in adults and the antennal glands which are functional only in early developmental stages.[77,78] Both have a similar structure composed of a blind-end sac and efferent duct (Figure 16). Lochhead[78] describes the maxillary glands briefly. Gross morphology and fine structure are given in detail by Tyson.[79-82] The maxillary glands form rounded prominences on the sides of the body just behind the mandibles. Each consists of a blind, centrally located, end sac around which an efferent duct makes three loops before continuing as a terminal duct which opens via a small aperture on the second maxilla. The end sac is anchored in place by numerous connective tissue strands which attach it to the efferent duct. Slender strands connect the coils of the efferent duct to one another, and to neighboring structures. The organ is surrounded by hemocoelic spaces. The epithelium of the end sac bears a striking resemblance to the podocytes of Bowman's capsule of the vertebrate nephron; i.e., the cells possess regularly arranged foot processes and junctional elements similar to the "filtration slit membranes". On the basis of these ultrastructural findings it is postulated that the formation of primary urine involves ultrafiltration. The efferent duct is specialized into two morphologically distinct regions: an efferent tubule and a terminal duct (excretory duct and terminal duct, Warren[77]). An apical microvillous border and basal membranous configurations associated with mitochondria here suggest a role in modifying the luminal contents. Adjacent coils of the efferent tubule are linked by structures designated as "intercoil connections".[82] The terminal duct is lined with a cuticle and therefore probably merely functions as a duct for the final urine.

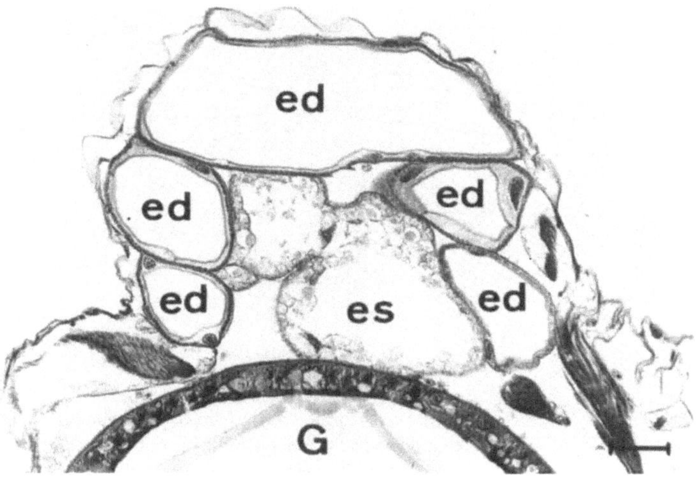

FIGURE 16. Light micrograph of a section through the maxillary gland. The centrally located end sac (es) is surrounded by coils of the efferent duct (ed); G, gut; bar = 50 μm.

In the adults, only rudiments of the antennal glands are found.[17,25,77,78] In larvae the antennal glands are located laterally in the head on either side of the digestive tract in the region from which the second antenna arises.[25] They open on the lateral side of the posterior portion of the protopodite.[77] The end sac consists of sixteen[77] or eleven to fourteen cells.[25] The efferent duct consists of a syncytium of four cells and an intracellular duct,[77] or of three cells,[16,25] the first two surrounding an intercellular proximal part of the duct and a distal third cell with an intracellular duct.[25]

B. ONTOGENY

Warren[77] describes the evolution of the antennal and maxillary gland throughout the larval stages. Although the antennal gland is already developed at hatching, it grows further and reaches its greatest development in the sixth instar; thereafter it degenerates to remnants only in the ninth and tenth instars. The maxillary gland develops in the nauplius and has become completely developed in the sixth instar when it may be functional. During the sixth and seventh instar, both excretory glands are thus fully developed. According to Benesch[16] the antennal gland is completely mesodermal, whereas Warren[77] is convinced that at least the terminal portion of the terminal duct is derived from the ectoderm. Both authors also disagree on the ontogeny of the maxillary gland. Warren[77] is confused about the first and second maxilla, and Benesch[16] is not clear in his description of their development. According to the latter, both the end sac and the excretory duct develop from the coelom of the second maxilla. He furthermore admits that perhaps a few cells near the opening are from ectodermal origin. Warren[77] sees the excretory duct arising as a solid rod from the first maxilla mesoderm coming into contact with the end sac derived from the schizocoel of the first maxilla and with the terminal duct derived from the ectoderm of the second maxilla. The finding of a cuticular lining in the terminal duct[81] corroborates Warren's assumption[77] that this part of the duct has an ectodermal origin.

XI. REPRODUCTIVE SYSTEM

A. MALE GENITAL TRACT

The male reproductive system was described by Cassel.[25] Wolfe[83] added histological

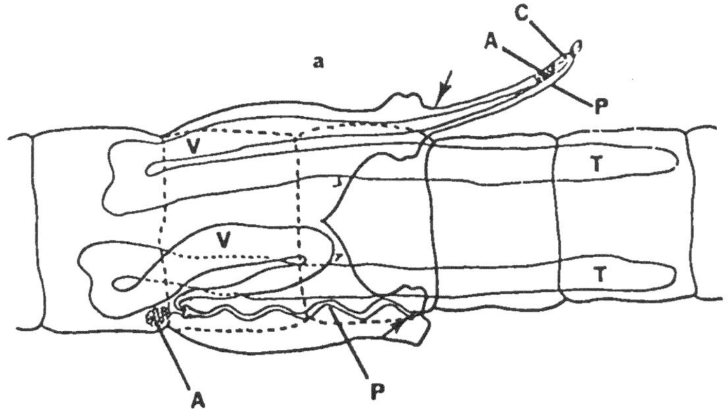

FIGURE 17. Schematic drawing of ventral view of the male reproductive system, showing an everted penis on one side while the other penis is in normal retracted condition. A, accessory gland; C, collecting duct; P, eversible penis; T, testis; V, vas deferens. (Modified after Wolfe, A. F., *J. Morphol.*, 135, 117, 1971. With permission.)

and histochemical details. The reproductive tract extends from the first genital segment to the second or third abdominal segment. It consists of paired testes, vasa deferentia, accessory glands, and penes (Figures 17, 18).

The testes are simple tubular structures surrounded by a thin hyaline membrane. Two cell types are found: somatic supporting cells and germ cells. The supporting cells are located immediately beneath the testicular membrane and form a continuous sheet along the entire length of the testes. Germ cells are located just beneath the supporting cells, usually near the junction of two of those cells. They are arranged in clusters. During the various stages of spermatogenesis all of the germ cells within a given cluster are at approximately the same stage of differentiation. Wolfe[83] does not go into the description of spermatogenesis. Brown[84-85] and Wingstrand[86] studied the ultrastructure of *Artemia* sperm and spermatogenesis. Spermatogonia in various stages of mitosis are located along the periphery of the testes. Spermatogonia and dividing spermatocytes are often linked with intercellular bridges and thus form a syncytium. Synaptonemal complexes are common in the primary spermatocytes. Presumably each spermatocyte eventually divides to form two spermatids. Originally these are closely associated with the supporting cells but then become free and move into the testicular lumen where they are found in various stages of spermiogenesis. Few have become mature spermatozoa. This view of Brown[85] is not completely in agreement with Wingstrand[86] who found that spermatids most often ripen in cyst-like dilatations of the intercellular spaces of the testis wall. Nonetheless he admitted that most sperm found in the testicular lumen were immature. Sperm presumed to be mature were observed in the vas deferens. The general outline of sperm is ovoid or spherical with densely set small papillae.[86] The nucleus is irregular in shape and contains dense material, probably chromatin. In the rather abundant cytoplasm, different kinds of vesicles are found as well as mitochondria, smooth endoplasmic reticulum, and ribosomal type particles. These are the general features. On closer view of individual sperm cells Brown[85] as well as Wingstrand[86] find such a variation in details that they cannot conclude which ones are mature. In sperm released from the dissected vas deferens Brown[84] observed the appearance of numerous "arms" protruding from the cell body. It was noted that similar protrusions can be seen in sperm stored in the female tract during the time lapse between copulation and fertilization[87] (Figure 19 a, b). This sperm equally shows this hitherto unexplained variation in ultrastructural detail (personal observation). The vasa deferentia join the testes at their ventral surface near the anterior end, form a loop, and terminate at the top of both penes. Their somewhat distended distal part

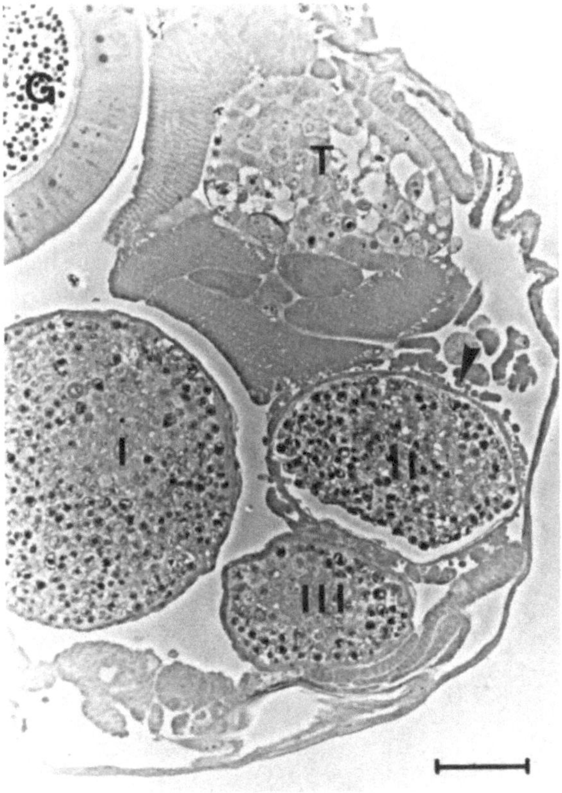

FIGURE 18. Light micrograph of a semi-thin transverse epon-section through the male genital system. G, gut; I, II, and III, section through the proximal, medial, and distal part of the vas deferens, respectively, all filled with sperm; T, testis; arrow, well-developed muscle layer of the intermediate part of the vas deferens, bar = 50 μm.

may be called the seminal vesicle.[25] Their histology is similar throughout the entire length, i.e., squamous epithelial cells surrounded by a thin layer of longitudinal muscle and a thick layer of circular muscle. The epithelial cells contain a neutral polysaccharide which is secreted into the lumen and constitutes the major portion of the seminal fluid. Sperm cells are stored along the entire length of the vas. Rhythmic contractions of the vas muscles keep the sperm well mixed. Each vas deferens leads into a penis.

The penis consists of two parts: a noneversible and an eversible portion. The noneversible portion is single but bifurcated. Its epithelium is covered by a cuticle similar to the general body surface. The eversible part is a tortuous tube that connects the vas deferens with the outside. In the retracted state it is lined with a cuticle and an epithelium surrounded by several bands of muscles. In the everted state the lining of the lumen becomes the outer covering of the penes and the vas deferens becomes the distal part of the reproductive tract. Three to five spines are located near the junction of the eversible and the noneversible penis.

Accessory glands composed of approximately 20 pairs of gland cells are located near the junction of the vas deferens and the eversible penis and have a separate opening on its top. Their function is unknown.

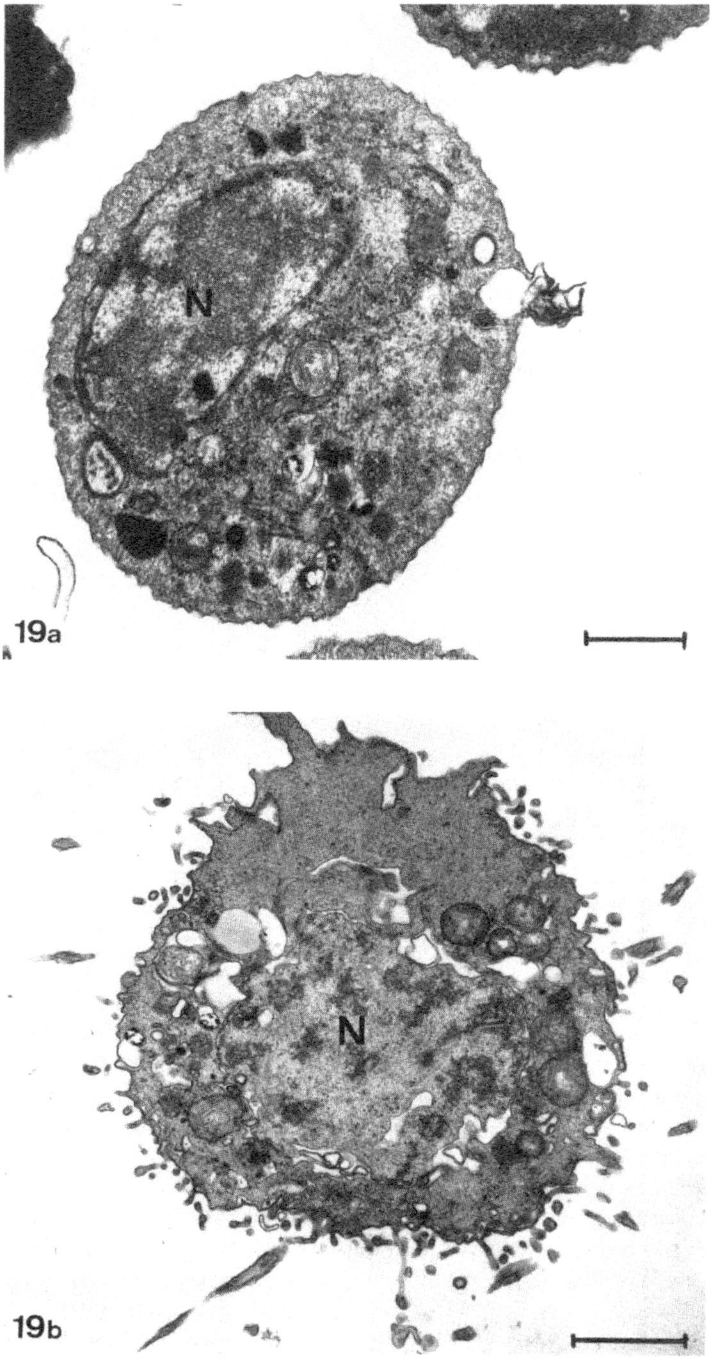

FIGURE 19. Electron micrographs through sperm cells; 19a, sperm from the female receptacle immediately after copulation; 19b, sperm from the female receptacle at the moment of the opening of the shutter between the lateral pouches and the uterus; N, nucleus; note the appearance of pseudopods and the widening of the space between both layers of the nuclear envelope; bar = 1 μm.

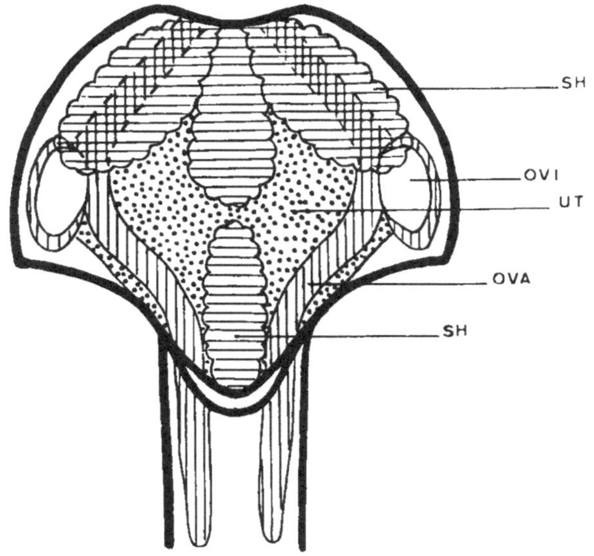

FIGURE 20. Schematic drawing of a ventral view of the female reproductive system. The uterus (UT) whose volume is determined by the presence or absence of developing cysts or nauplii is a pouch within the ovisac. Two antero-lateral and a lobed medial group of shell glands (SH) open into it. The lateral ovaries (OVA) stretch more caudally along the ovisac, and open into the oviducts (OVI) which are found in the lateral horns of the ovisac. The oviducts which temporarily store the eggs between ovulation and fertilization in lateral pouches are separated from the uterus by a shutter.

B. FEMALE REPRODUCTIVE TRACT

The female reproductive tract consists of paired ovaries and oviducts leading into a single, median uterus, wherein several clusters of shell glands open (Figure 20).

The ovaries are paired tubular structures which extend from the eleventh thoracic segment, pass through the genital segments, and end in the abdominal segments.[25] Macroscopically they show cyclic changes which reflect the events of vitellogenesis. A review of current knowledge on their structure and pertaining literature[87] was recently supplemented with more ultrastructural information.[88] As in the testes, two different cell types are found: germ cells and somatic cells (Figure 21). The germ cells consist of oogonia and oocytes which later differentiate into oocytes and nurse cells. Premeiotic oocytes form clusters linked to remnants of the mitotic spindle. At the onset of meiosis synaptonemal complexes form simultaneously in all oocytes of the cluster; the spindle remnants disappear and the oocytes which remain linked to each other by cytoplasmic bridges align into ribbons. Oocyte-nurse cell differentiation becomes obvious at the end of previtellogenesis by the increase in cytoplasm of one oocyte of the ribbon. Its nucleus transforms into an excentric germinal vesicle, accessory nuclei spread over the oocyte surface, and a yolk nucleus appears in the center. Four yolk products are formed, i.e., protein yolk platelets, lipid droplets, clear vesicular yolk bodies, and small intracisternal granules. Protein yolk platelets arise both by endogenous yolk formation and by exogenous protein uptake. Engulfing of nurse cells as described at the light microscopic level[89] was not confirmed by electron microscopy. Nurse cells become polyploid, degenerate, and are phagocytosed by the somatic cells. The somatic cells form a continuous layer under the basement membrane and surround the oocyte-nurse cell complexes. A well-developed network of indented plasma membranes leads to the assumption that in addition to their phagocytosing function, they play a role in transport.

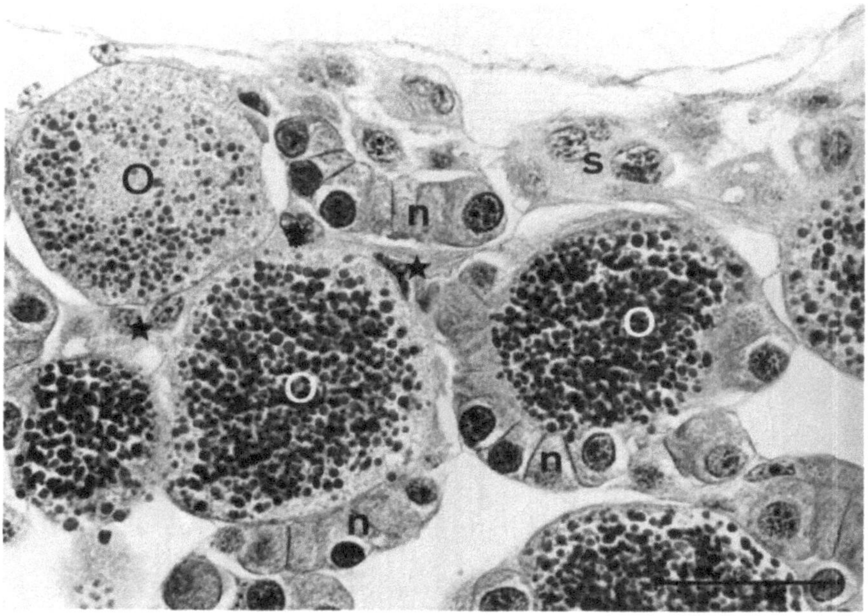

FIGURE 21. Light micrograph of the ovary during vitellogenesis. The maturing oocytes (O) are surrounded by nurse cells (n). Somatic cells separate the groups of oocytes and nurse cells (*) and form a lateral more or less continuous layer (s); bar = 50 μm.

The oviducts arise from the ovaries at the level of the anterior part of the third abdominal segment.[25] Each leads down, outward, and forward through the lateral areas of the ovisac to empty into the antero-lateral border of the uterus. Their secretory epithelium is surrounded with longitudinal and circular muscle fibers. After ovulation the eggs enter the oviducts where they are temporarily stored, bathing in a secretion product of the oviduct. The expanded oviducts or lateral pouches are separated by a shutter from the median and frontal parts of the uterus.[87] These function as a seminal receptacle during the time lapse between the copulation and fertilization; the stored sperm cells further mature and acquire peripheral arms or disintegrate.[87] The oviduct cells show a secretory cycle which has been studied at the light[87] and electron-microscopic level.[90]

The shell glands consist of several clusters of gland cells. Their color varies in different reproductive cycles: all intermediate colors between dark brown and white or colorless glands may occur. The exact role of the shell glands in the determination of ovoviviparity and oviparity is not yet elucidated. The brown shell glands (Figure 22) were studied extensively at the light and electron microscopic level.[91-93] In the white shell glands, two types are distinguished based on the appearance of the secretory granules[87,93] (Figure 23 a,b). One type is thought to be related to the development of nauplii, the other type is postulated to form the thin shell of embryos hatching shortly after oviposition.

C. ONTOGENY OF THE REPRODUCTIVE SYSTEM

The development of the gonads starts with the differentiation of primordial germ cells.[16] At the emergence stage twelve to sixteen such cells lie between the mesodermal somites of the second maxillar and first thoracic segment. They move along to the first genital segment while they are multiplying.[16,36,94] Benesch[16] considers the gonoducts as the segmental organs of the first genital segment. In the female they develop into the oviducts which attach first to the differentiating ovary, then later to the uterus which has invaginated from the ectoderm

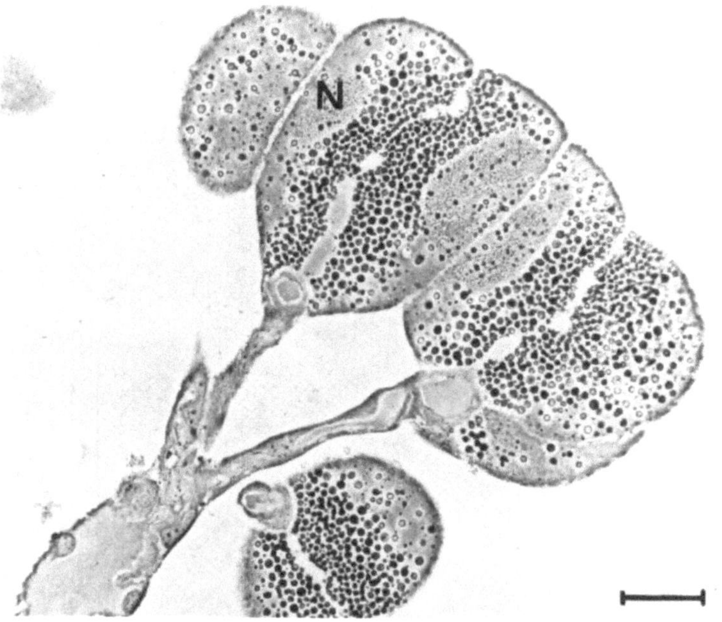

FIGURE 22. Light micrograph of a 2-μm epon-section of a brown shell gland. Note the cytoplasm filled with homogeneous dark granules. N, nucleus; bar = 25 μm.

together with the shell glands. In males they develop into vasa deferentia and seminal vesicles which attach cranially to the developing testis and caudally to the ejaculatory duct (or the eversible part of the penis[83]) which has arisen as an ectodermal invagination together with the accessory glands. Claus[17] considers the whole genital tract as mesodermal.

XII. SALT-SECRETING CELLS OF THE GILLS

Copeland[13,95] performed an ultrastructural study of the first ten pairs of gills which were shown by Croghan[96] to be the site of active NaCl excretion in hypertonic media and probably of active uptake from hypotonic media. They are shown to become functional when the neck organ degenerates. The branchial cuticle is lined with a single layer of cells which can be classified into dark cells and light cells (Figure 24). Both occur in about equal numbers and are arranged in an alternate interdigitating manner. Both cell types show marked differences in morphology. The so-called "mitochondrial pumps" presumed to be concerned with salt transport only occur in the dark cells. The light cells have a regular pattern of rough endoplasmic reticulum that could serve a transfer function. The complicated and characteristic interdigitations of light and dark cells suggest that a functional relationship is involved. The light cells are separated from the hemolymph by at least a thin overlay of dark cell cytoplasm. Which cell type serves as the final pathway for the release of salt to the external environment is not certain, although it is likely that the dark cells are responsible. The morphology of the branchiae changes with a change in physiological activity: animals raised in dilute sea water isotonic with the body fluid have gills of normal size, but the epithelial cell lining has degenerated, is thin, and possesses few mitochondria. In contrast, in increased external salinity the epithelium thickens with many mitochondria and greater cellular complexity.

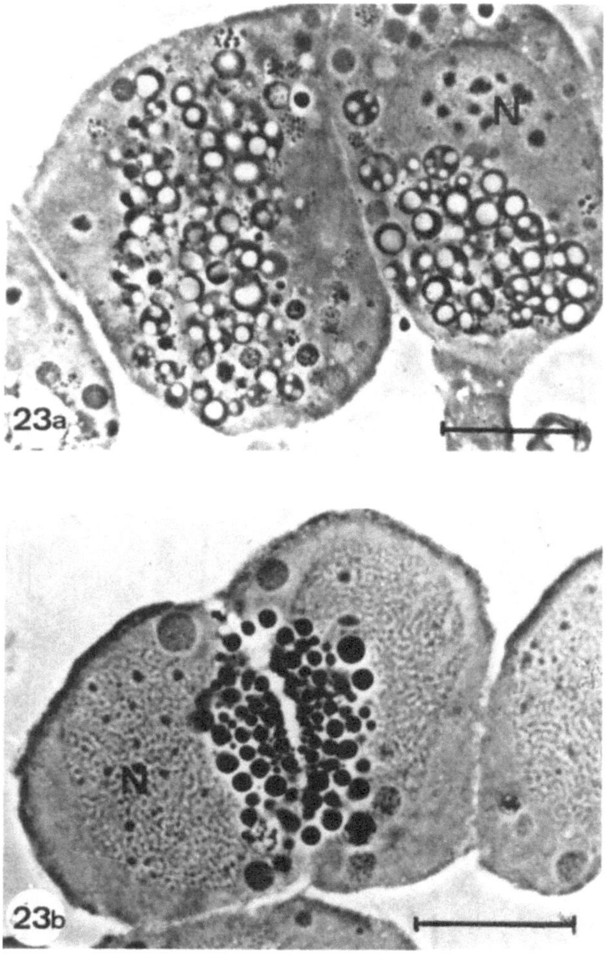

FIGURE 23. Light micrographs of 2-μm epon-sections through transparent or slightly colored shell glands. (23a) Cells of a gland presumed to provide material for the nauplius developing in utero; note the cytoplasm filled with large granules of irregular density. (23b) Cells of a gland presumed to provide nondiapausing cysts with a thin shell; note the cytoplasm only partly filled with homogene dense granules, while some heterogeneous larger granules are also seen in the periphery. N, nucleus; bar = 25 μm.

XIII. INFECTIOUS INCLUSIONS IN *ARTEMIA*

Codreanu[97] and Codreanu-Balescu[98] found four microsporidian species and two species of yeast in a parthenogenetic *Artemia* from Roumania, i.e., *Nosema exigua* develops in the gut cells; *Glugea artemiae* in the muscles and phagocytic storage cells, as well as in the maxillary glands, the nervous system, and the epidermis; *Gurleya dispersa* is found in the phagocytic storage cells, thoracic, and limb hemocoel; and *Plistophora myotropha* gradually infects the trunk muscles. Two *Metschnikova* yeast species are found in the hemocoel. The microsporidians never infest the ovaries. All these infections cause premature death.

Tyson[24,99-101] describes a spirochete infection. Spirochetes were found in the cytoplasm of cells of the maxillary gland, epidermal cells, phagocytic storage cells, and muscle cells.

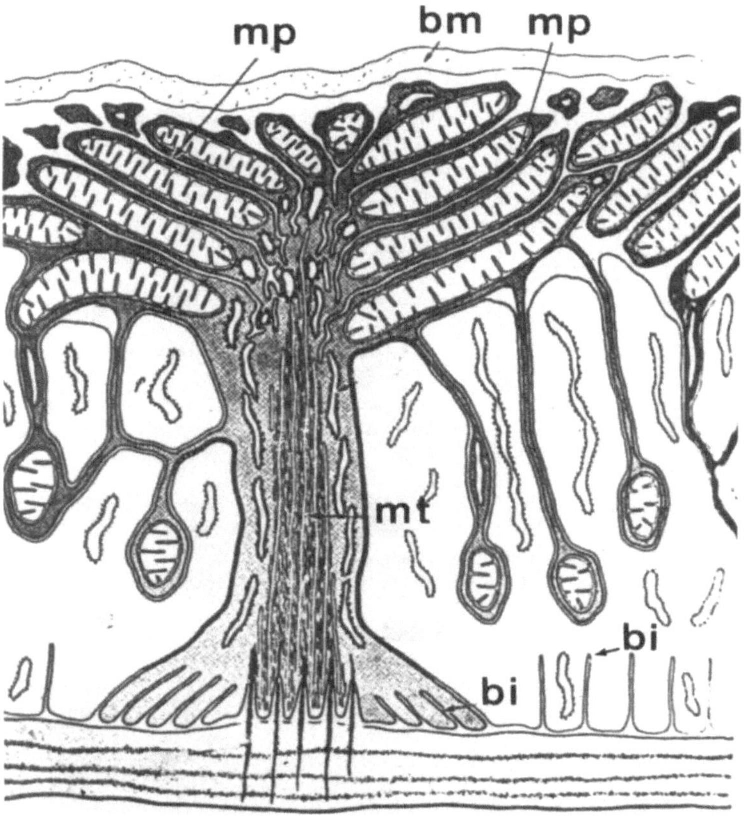

FIGURE 24. Salt-secreting cells of the gills: schematic drawing depicting the interdigitations of the light and dark cell cytoplasm and the general arrangement of the mitochondrial pump systems in the dark cells. The cuticle is below. Hemolymph spaces and basement membrane (bm) are at the top. The dark cell is located centrally; bi, basal indentations for dark cells and light cells; mp, mitochondrial pumps; mt, microtubules and microfibrils in dark cell. (Modified after Copeland, D. E., *Protoplasma*, 63, 363, 1967. With permission.)

Extracellular spirochetes occurred in the lumen of the end sac of the maxillary gland and in the hemocoel.

Artemia of the Camargue region in France are shown to be the intermediate host of some hymenolepid and cyclophyllid cestodes. The cysticercoids are found in the hemocoel.[102]

A procaryotic symbiont is found in the gut cells of *Artemia* of Great Salt Lake North Arm.[103]

Intensive cultures of *Artemia* favor the external infection with a *Leucothrix* species. It attaches to all external surfaces of adults and late instars and causes premature death.[104]

Overton and Bland[105] experimentally infected *Artemia* larvae with the fungus *Haliphtoros milfordensis* and described the evolution of the infection both with scanning and transmission electron microscopy.

A presumed viral infection of the shell glands and ovaries has also been described.[87,93] Unfortunately the phagocytic storage cells were not included in this study. Viral masses, different from those cited above may occur in the phagocytic storage cells (personal observation).

REFERENCES

1. **Tyson, G. E. and Sullivan, M. L.,** Frontal knobs of the male brine shrimp: scanning electron microscopy, *Am. J. Zool.*, 19, 891, 1979.
2. **Tyson, G. E. and Sullivan, M. L.,** Scanning electron microscopy of the frontal knob of the male brine shrimp, *Trans. Am. Microscop. Soc.*, 99, 167, 1980.
3. **Wolfe, A. F.,** A light and electron microscopic study of the frontal knob of *Artemia* (Crustacea, Branchiopoda), in *The Brine Shrimp Artemia*, Vol. 1, *Morphology, Genetics, Toxicology*, Persoone, G., Sorgeloos, P., Roels, O., and Jaspers, E., Eds., Universa Press, Wetteren, Belgium, 1980, 117.
4. **Tyson, G. E. and Sullivan, M. L.,** Scanning electron microscopy of the antennular sensilla of the brine shrimp, *Am. Zool.*, 18, 632, 1978.
5. **Tyson, G. E. and Sullivan, M. L.,** Antennular sensilla of the brine shrimp, *Artemia salina, Biol. Bull.*, 156, 382, 1979.
6. **Tyson, G. E.,** Fine structure of the type 2 antennular sensillum of the brine shrimp, *Am. Zool.*, 20, 816, 1980.
7. **Tyson, G. E. and Sullivan, M. L.,** Scanning electron microscopy of cuticular sensilla of *Artemia*: setae of the adult trunk segments, in *The Brine Shrimp Artemia*, Vol. 1, *Morphology, Genetics, Radiobiology, Toxicology*, Persoone, G., Sorgeloos, P., Roels, O., and Jaspers, E., Eds., Universa Press, Wetteren, Belgium, 1980, 99.
8. **Schrehardt, A.,** Ultrastructural investigations of the filter-feeding apparatus and the alimentary canal of *Artemia*, in *Artemia Research and Its Applications*, Sorgeloos, P., Bengtson, D. A., Decleir, W., and Jaspers, E., Eds., Universa Press, Wetteren, Belgium, 1987, 33.
9. **Tyson, G. E. and Sullivan, M. L.,** Scanning electron microscopy of the molar surfaces of the mandibles of the brine shrimp, *Am. Zool.*, 21, 957, 1981.
10. **Tyson, G. E. and Sullivan, M. L.,** A scanning electron microscopic study of the molar surfaces of the mandibles of the brine shrimp (Cl.Branchiopoda: O.Anostraca), *J. Morphol.*, 170, 239, 1981.
11. **Freeman, J. A.,** The integument of *Artemia* during early development, in *Biochemistry and Cell Biology of Artemia*, MacRae, T. H., Bagshaw, J. C., and Warner, A. H., Eds., CRC Press, Boca Raton, FL, 1989, 233.
12. **Criel, G. R. J. and Walgraeve, H. R. M. A.,** Molt staging in *Artemia* adapted to Drach's system, *J. Morphol.*, 199, 41, 1989.
13. **Copeland, D. E.,** A study of salt secreting cells in the brine shrimp *(Artemia salina), Protoplasma*, 63, 363, 1967.
14. **Horst, M. N.,** Molecular and cellular aspects of chitin synthesis in larval *Artemia*, in *Cell and Molecular Biology of Artemia Development*, Warner, A., Mac.Rae, T., and Bagshaw, J. C., Eds., Plenum Press, New York, 1989, 59.
15. **Dornesco, G. T. and Steopoe, J.,** Les glandes tégumentaires des phyllopodes anostracés, *Ann. Sci. Nat. Zool.*, 20, 29, 1958.
16. **Benesch, R.,** Zur ontogenie und morphologie von *Artemia salina* L., *Zool. Jordbrukstek., Anat.*, 86, 307, 1969.
17. **Claus, C.,** Untersuchungen über die Organisation und Entwicklung von Branchipus und *Artemia* nebst vergleichenden Bemerkungen über anderen Phyllopoden, *Arb. Zool. Inst. Wien*, 6, 267, 1886.
18. **Spangenberg, F.,** Zur Kenntniss van *Branchipus stagnalis, Z. Wiss. Zool.*, Suppl. 25, 1, 1875.
19. **Lochhead, J. H. and Lochhead, M. S.,** Studies on the blood and related tissues in *Artemia* (Crustacea, Anostraca), *J. Morphol.*, 68, 593, 1941.
20. **Bruntz, L.,** Etude physiologique sur les phyllopodes branchiopodes, *Arch. Zool. Exp. Gén.*, 4, 183, 1905.
21. **Kollmann, M.,** Notes histologiques sur les phyllopodes branchiopodes, *Bull. Soc. Zool. Fr.*, 49, 495, 1924—1925.
22. **Barigozzi, C.,** I fenomeni cromosomici nelle cellule somatiche di Artemia salina Leach, *Z. Zellforsch. B. Chromosoma*, 2, 251, 1941.
23. **Van Beek, E., Van Brussel, M., Criel, G., and De Loof, A.,** A possible extra-ovarian site for synthesis of lipovitellin during vitellogenesis in *Artemia* sp. (Crustacea; Anostraca), *Int. J. Invertebr. Reprod. Dev.*, 12, 227, 1987.
24. **Tyson, G. E.,** Phagocytosis and digestion of spirochetes by amebocytes of infected brine shrimp, J. Invert. Pathol., 26, 105, 1975.
25. **Cassel, J. D.,** The morphology of Artemia salina (Linnaeus). M.A. thesis, Leland Stanford Junior University, CA, 1937.
26. **Cannon, H. G.,** On the postembryonic development of the fairy shrimp *(Chirocephalus diaphanus), J. Linn. Soc.*, 36, 401, 1926.
27. **Hsü, F.,** Studies on the anatomy and development of a freshwater phyllopod *Chirocephalus nankinensis* Shen., *Contrib. Biol. Lab. Sci. Soc. China*, (Shangai), 9, 119, 1933.

28. **Reger, J. F.**, The fine structure of limb muscle fibres from the crustacean *Artemia salenus*, *Anat. Rec.*, 142, 323, 1962.
29. **Joly, M.**, Histoire d'un petit crustacé (*Artemia salina* Leach), auquel on a faussement attribué la coloration en rouge des marais salans méditerrenéens, suivie de recherches sur la cause réelle de cette coloration, *Ann. Sci. Nat.* (Paris) *Zool. Biol. Anim.*, 13, 225, 1840.
30. **Leydig, F.**, Über *Artemia salina* und *Branchipus stagnalis*, *Z. Wiss. Zool.*, 3, 280, 1851.
31. **Greene, W.**, The circulatory system of the brine shrimp, *Science*, 60, 411, 1924.
32. **Vehstedt, R.**, Über Bau, Tätigkeit und Entwicklung des Rückengefässes und des lacunären System von Artemia salina var. arieta, *Z. Wiss. Zool.*, 154, 1, 1940.
33. **Okland, S., Tjonnenland, A., Larsen, L. N., and Nylund, A.**, Heart ultrastructure in *Branchinecta paludosa*, *Artemia salina*, *Branchipus schaefferi*, and *Streptocephalus* sp. (Crustacea, Anostraca), *Zoomorphology*, 101, 71, 1982.
34. **Weisz, P. B.**, The histological pattern of metameric development in *Artemia salina*, *J. Morphol.*, 81, 45, 1947.
35. **Steopoe, J. and Dornesco, G. T.**, La cytologie et les propriétés des éléments sanguins des phyllopodes anostracés, *Acad. Roumaine Bull. Sect. Sci.*, 28, 220, 1945.
36. **Anderson, D. T.**, Embryology and Phylogeny in Annelids and Arthropods, International Series of Monographs in Pure and Applied Biology, Zoology Division, Pergamon Press, 50, 263, 1973.
37. **Frenzel, J.**, The mid-gut of *Artemia salina*, *J. Roy. Microsc. Soc.*, *London*, 11, 37, 1893.
38. **Kuenen, J. D.**, Systematical and physiological notes on the brine shrimp, *Artemia*, *Arch. Néerl. Zool.*, 3, 365, 1939.
39. **Kikuchi, S.**, The fine structure of the alimentary canal of the brine shrimp, *Artemia salina*: the midgut, *Annu. Rep. Iwate Med. Univ. Sch. Lib. Arts* Sci., 7, 15, 1972.
40. **Reeve, M. R.**, The filter-feeding of *Artemia*, III. Fecal pellets and their associated membranes, *J. Exp. Biol.*, 40, 215, 1963.
41. **Snyder, S. D. and Wolfe, A. F.**, A histological study of the digestive system of *Artemia* with reference to the production of its peritrophic membrane, *Proc. Pa. Acad. Sci.*, 54, 123, 1988.
42. **Hootman, R. and Conte, F. P.**, Fine structure and function of the alimentary canal in *Artemia salina* nauplii, *Cell Tissue Res.*, 155, 423, 1974.
43. **Tyson, G. E. and Lyon, D. S.**, Intrinsic musculature of the digestive system of the brine shrimp: scanning microscopy of the intestine, *Amer. Zool.*, 25, 114A, 1985.
44. **Hanström, B.**, Vergleichende Anatomie des Nervensystems des wirbellosen Tiere, Springer-Verlag, Berlin, 1928.
45. **Horridge, G. A.**, Arthropoda: general anatomy in Structure and Function in the Nervous System of Invertebrates, Bullock, T. H. and Horridge, G. A., Eds., W. H. Freeman, San Francisco, 1965, 810.
46. **Spencer, W. K.**, Zur Morphologie des Centralnervensystems des Phyllopoden, nebst Bemerkungen über deren Frontalorgane, *Z. Wiss. Zool.*, 71, 508, 1902.
47. **Warren, H. S.**, The central nervous system of the adult *Artemia*, *Trans. Amer. Micrscop. Soc.*, 49, 198, 1930.
48. **Elofsson, R.**, The nauplius eye and frontal organs of the non-malacostraca (Crustacea), *Sarsia*, 25, 1, 1965.
49. **Amarant, R. and Elofsson, R.**, Distribution of monoaminergic neurons in the nervous system of non-malacostracan crustaceans, *Cell Tissue Res.*, 166, 1, 1976.
50. **Hanström, B.**, Neue Untersuchungen über Sinnesorgane und Nervensystem der Crustaceen, I, *Z. Morphol. Ökol. Tiere*, 23, 80, 1931.
51. **Horridge, G. A.**, Arthropoda: receptors for light and optic lobe, in *Structure and Function in the Nervous System of Invertebrates*, Bullock, T. H. and Horridge, G. A., Eds., W. H. Freeman, San Francisco, 1965, 1063.
52. **Nowikoff, M.**, Über die Augen und die Frontalorgane der Branchiopoden, *Z. Wiss. Zool.*, 79, 432, 1905.
53. **Debaisieux, P.**, Les yeux des crustacés. Structure, développement, réactions à l'éclairement, *La Cellule*, 54, 5, 1944.
54. **Elofsson, R. and Odselius, R.**, The anostrancan rhabdom and the basement membrane. An ultrastructural study of the *Artemia* compound eye (Crustacea), *Acta Zoologica*, 56, 141, 1975.
55. **Hertel, H.**, The compound eye of *Artemia salina* (Crustacea). I. Fine structure when light and dark adapted, *Zool. Jahrb. Physiol.*, 84, 1, 1980.
56. **Nilsson, D. E. and Odselius, R.**, A new mechanism for light-dark adaptation in the *Artemia* compound eye (Anostraca, Crustacea), *J. Comp. Physiol. A*, 143, 389, 1981.
57. **Nässel, D. R., Elofsson, R., and Odselius, R.**, Neuronal connectivity patterns in the compound eyes of *Artemia salina* and *Daphnia magna*, *Cell Tissue Res.*, 190, 435, 1978.
58. **Elofsson, R. and Dahl, E.**, The optic neuropiles of Crustacea, *Z. Zellforsch.*, 107, 343, 1970.
59. **Elofsson, R. and Klemm, N.**, Monoamine-containing neurons in the optic ganglia of crustaceans and insects, *Z. Zellforsch.*, 133, 475, 1972.

60. **Debaisieux, P.,** Histologie et histogénèse chez Chirocephalus diaphanus Prév., *Cellule,* 54, 253, 1952.
61. **Dahl, E.,** The ontogeny and comparative anatomy of some protocerebral sense organs in notostracan phyllopods, *Quart. J. Microsc. Sci.,* 100, 445, 1959.
62. **Rasmussen, S.,** Die Feinstruktur des Mittelauges und des ventralen Frontalorgans von *Artemia salina* L. (Crustacea: Anostraca), *Z. Zellforsch.,* 117, 576, 1971.
63. **Anadon, A. and Anadon, E.,** Nauplius eye and adjacent organs of adult *Artemia,* in The Brine Shrimp *Artemia,* Vol. I, *Morphology, Genetics, Radiobiology, Toxicology,* Persoone, G., Sorgeloos, P., Roels, O., and Jaspers, E., Eds., Universa Press, Wetteren, Belgium, 1980, 41.
64. **Moroff, T.,** Entwicklung und phylogenetische Bedeutung des Medianauges bei Crustaceen, *Zool. Anz.,* 40, 11, 1912.
65. **Claus, C.,** Das Medianauge des Crustaceen, *Arb. Zool. Inst. Univ. Wien,* 9, 225, 1891.
66. **Nowikoff, M.,** Einige Bemerkungen über das Medianauge und die Frontalorgane von *Artemia salina, Z. Wiss. Zool.,* 81, 691, 1906.
67. **Vaissière, R.,** Evolution de l'oeil d'*Artemia salina* Leach (Crustacé, Branchiopode, Phyllopode) au cours de ses stades post-embryonnaires, *C. R. Acad. Sci.,* 242, 2051, 1956.
68. **Vaissière, R.,** Morphologie et histologie comparées des Crustacés Copépodes, *Arch. Zool. Exp. Gén.,* 100, 102, 1961.
69. **Elofsson, R. and Lake, P. S.,** On the cavity receptor organ (X-organ or organ of Bellonci) of *Artemia salina* (Crustacea: Anostraca), *Z. Zellforsch.,* 121, 319, 1971.
70. **Van den Bosch de Aguilar, P.,** Neurosecretion in the entomostraca crustaceans, *La Cellule,* 73, 1, 1979.
71. **Lochhead, J. H. and Resner, R.,** Functions of the eye and neurosecretion in Crustacea Anostraca, *15th Int. Congress Zool., London,* 1958, 397.
72. **Baid, I. C. and Ramaswami, L. S.,** Neurosecretory cells in *Artemia salina* L., *Experientia,* 21, 528, 1965.
73. **Hentschel, E.,** Zum neurosekretorischen System der Anostraca, Crustacea (*Artemia salina* Leach und *Chirocephalus grubei* Dybowsci), *Zool. Anz.,* 170, 187, 1963.
74. **Hentschel, E.,** Neurosekretion und Neurohämalorgane bei *Chirocephalus grubei* Dybowci und *Artemia salina* Leach (Anostraca, Crustacea), *Z. Wiss. Zool.,* 171, 44, 1965.
75. **Van den Bosch der Aguilar, P.,** Neurosécrétion et régulation hydroelectrolytique chez *Artemia salina, Experientia,* 32, 228, 1976.
76. **Van Beek, E.,** Endocrinological aspects of vitellogenesis in the brine shrimp *Artemia* sp. (Crustacea: Anostraca), Ph.D. thesis, Catholic University of Leuven, Belgium, 1988.
77. **Warren, H. S.,** The segmental excretory glands of *Artemia salina,* Lin., var. Principalis Simon, the brine shrimp, *J. Morphol.,* 62, 263, 1938.
78. **Lochhead, J. H.,** *Artemia,* in *Selected Invertebrate Types,* Brown, F. A., Ed., John Wiley & Sons, New York, 1950, 394.
79. **Tyson, G. E.,** Gross morphology and fine structure of the maxillary gland of the brine shrimp, *Artemia salina, Am. Zool.,* 6, 555, 1966.
80. **Tyson, G. E.,** The fine structure of the maxillary gland of the brine shrimp, *Artemia salina*: the end sac, *Z. Zellforsch.,* 86, 129, 1968.
81. **Tyson, G. E.,** The fine structure of the maxillary gland of the brine shrimp, *Artemia salina*: the efferent duct, *Z. Zellforsch.,* 93, 151, 1969.
82. **Tyson, G. E.,** Intercoil connections of the kidney of the brine shrimp, *Artemia salina, Z. Zellforsch.,* 100, 54, 1969.
83. **Wolfe, A. F.,** A histological and histochemical study of the male reproductive system of *Artemia* (Crustacea, Branchiopoda), *J. Morphol.,* 135, 117, 1971.
84. **Brown, G. G.,** Ultrastructural studies on crustacean spermatozoa and fertilization, Ph.D. thesis, Miami University, FL, 1966.
85. **Brown, G. G.,** Some ultrastructural aspects of spermatogenesis and sperm morphology in the brine shrimp *Artemia salina* Leach (Crustacea: Branchiopoda), *Proc. Iowa Acad. Sci.,* 76, 473, 1970.
86. **Wingstrand, K. G.,** Comparative spermatology of the Crustacea Entomostraca 1. Subclass Branchiopoda, *Biol. Skrifter,* 22, 1, 1978.
87. **Criel, G.,** Morphology of the genital apparatus of *Artemia*: a review, in The Brine Shrimp Artemia, Vol. 1, *Morphology, Genetics, Radiobiology, Toxicology,* Persoone, G., Sorgeloos, P., Roels, O., and Jaspers, E., Eds., Universa Press, Wetteren, Belgium, 1980, 75.
88. **Criel, G. R. J.,** Morphological study of the ovary of *Artemia,* in *Cell and Molecular Biology of Artemia Development,* Warner, A., MacRae, T., and Bagshaw, J. C., Eds., Plenum Press, 1989, 99.
89. **Fautrez-Firlefyn, N.,** Etude cytochimique des acides nucléiques au cours de la gamétogénèse et des premiers stades du développement embryonnaire chez *Artemia salina* L., *Arch. Biol.,* 62, 291, 1951.
90. **Criel, G.,** Ultrastructural observations on the oviduct of *Artemia,* in *The Brine Shrimp Artemia,* Vol. 1, *Morphology, Genetics, Radiobiology, Toxicology,* Persoone, G., Sorgeloos, P., Roels, O., and Jaspers, E., Eds., Universa Press, Wetteren, Belgium, 1980, 87.

91. **Anderson, E., Lochhead, J. H., Lochhead, M. S., and Huebner, E.,** The origin and structure of the tertiary envelope in the thick-shelled eggs of the brine shrimp *Artemia, J. Ultrastruct. Res.*, 32, 497, 1970.
92. **Fautrez, J. and Fautrez-Firlefyn, N.,** Contribution à l'étude des glandes coquilières et des coques de l'oeuf d' *Artemia salina, Arch. Biol.* (Liège), 82, 41, 1971.
93. **De Maeyer-Criel, G.,** Elektronenmicroscopisch onderzoek naar de ultrastructuur tijdens de secretiecyclus van de schaalklier bij *Artemia salina.* Verhandelingen van de Koninklijke Academie voor Wetenschappen, Letteren en Schone Kunsten van België, Klasse der Wetenschappen (40), 145, 1978.
94. **Fränsemeier, L.,** Zur Frage der Herkunft des metanaupliales Mesoderms und die Segmentbildung bei *Artemia salina* Leach, *Z. Wiss. Zool.*, 152, 439, 1939.
95. **Copeland, E.,** Salt transport organelle in *Artemia salenis* (brine shrimp), *Science*, 151, 470, 1966.
96. **Croghan, P. C.,** The mechanism of osmotic regulation in *Artemia salina* (L.): the physiology of the branchiae, *J. Exp. Biol.*, 35, 234, 1958.
97. **Codreanu, R.,** Sur quatre espèces nouvelles de microsporidies parasites de l'*Artemia salina* (L.) de Roumanie, *Ann. Sci. Nat. Zool.* (11 série), 19, 561, 1957.
98. **Codreanu, R. and Codreanu-Balescu, D.,** On two *Metschnikowia* yeast species producing hemocoelic infections in *Daphnia magna* and *Artemia salina* (Crustacea, Phyllopoda) from Roumania, *J. Invert. Pathol.*, 37, 22, 1981.
99. **Tyson, G. E.,** The occurrence of a spirochete-like organism in tissues of the brine shrimp *Artemia salina, J. Invert. Pathol.*, 5, 145, 1970.
100. **Tyson, G. E.,** Ultrastructure of a spirochete found in tissues of the brine shrimp, *Artemia salina, Arch. Microbiol.*, 99, 281, 1974.
101. **Tyson, G. E.,** Tubular structures associated with intracytoplasmic spirochetes, *Cell Tissue Res.*, 158, 333, 1975.
102. **Gabrion, C. and MacDonald, G.,** *Artemia sp.* (Crustacé, Anostracé), hôte intermédiaire d'*Eurycestus avoceti*, Clark, 1954 (Cestode, Cyclophyllide) parasite de l'avocette en Camargue, *Ann. Parasitol.* (Paris), 55, 327, 1980.
103. **Post, F. J. and Youssef, N. N.,** A procaryotic intracellular symbiont of the Great Salt Lake brine shrimp, *Artemia salina* (L.), *Can. J. Microbiol.*, 23, 1232, 1977.
104. **Solangi, M. and Overstreet, R. M.,** A filamentous bacterium on the brine shrimp and its control, *Gulf Res. Rep.*, 6, 275, 1979.
105. **Overton, S. V. and Bland, C. E.,** Infection of *Artemia salina* by *Haliphtoros milfordensis*: a scanning and transmission electron microscope study, J. Invertebr. Pathol., 37, 249, 1981.

Chapter 7

ONTOGENY OF *ARTEMIA*

Godelieve R. J. Criel

TABLE OF CONTENTS

I. RESUMPTION OF MEIOSIS, FERTILIZATION, AND FIRST CLEAVAGE

As long as the oocytes are in the ovary, meiosis is arrested in diakinesis. In diploid bisexual[1] and in some parthenogenetic strains,[2] chromosomes are seen in tetrads in the germinal vesicle. In other parthenogenetic stains diads seem to be formed.[3-4] A centrosome is never seen in the ovary. An intranuclear meiotic metaphasic spindle is formed immediately after ovulation,[5] when the eggs are moving slowly to the expanded oviducts, or lateral pouches, where they remain between for a period of 30 min to some hours, depending on the strain, the eventual copulation,[6-8] and the rearing conditions. As long as the eggs are in the oviducts, the spindle is oriented tangentially. Once released into the uterus through the opening of a shutter,[8] the spindle rotates to a radial position. In bisexual strains, normal maturation and cleavage of the eggs can only ensue upon the stimulus of fertilization. In parthenogenetic eggs this inhibition is lacking and the eggs proceed with maturation independently of external influences.

In all strains a first polar body is formed. Reduction of the chromosomes, however, only occurs in diploid bisexual and some parthenogenetic races[1,9] where previous synapsis has occurred in the ovary.[4] In tetraploid and polyploid parthenogenetic strains there is no synapsis and no reduction occurs.[4]

In all strains the vitelline membrane is formed after the emission of the first polar body.[1-2] A description of its ultrastructural formation has been given by De Maeyer-Criel et al.[10] Meiotic divisions probably proceed without preformation of centrospheres.[11-12] Fries[13] is the only one to describe a centrosome near the first metaphasic spindle in the parthenogenetic Odessa (U.S.S.R.) strain; Petrunkewitsch[3] and Artom[1] find a structure resembling a centrosphere in the center of the egg at the moment of the emission of the first polar body.

Formation of a second polar body is evident in bisexual strains. In parthenogenetic strains it still is a matter of debate. In some diploid parthenogenetic strains where reduction has taken place with the formation of the first polar body, the rudiment of a second polar body, which is never expelled, fuses with the female pronucleus; this process restores the initial chromosome number.[9,14] A similar rudiment and fate of a second polar body has also been described in other cases where no previous reduction has occurred. In these cases this procedure gives rise to polyploidization, be it an abnormal process or not.[2,15-16]

After maturation the female nucleus enlarges to a pronucleus, rounds up with the chromatin dispersed in small blocks,[1] and migrates to the center of the egg — with or without the company of the rudiment of the second polar body. In bisexual eggs the male nucleus undergoes changes similar to those just described in the female pronucleus. Both pronuclei meet in the center. In parthenogenetic as well as in bisexual strains one centrosome and an aster arise while the pronucleus is moving to the center of the egg. It soon divides, and both halves move to the poles of the future mitotic spindle. Fautrez-Firlefyn[12] is sure it is the male spermmaster, but its presence even in parthenogenetic eggs gives rise to doubt.[1,7,11] The space between the pronuclei is filled with parallel microtubules joining the nuclear membranes.[17] Male and female chromosomes remain separated during prophase and meet during metaphase. Although Fautrez-Firlefyn[12] describes a basophilic crescent, no polarity is found in the egg.[7]

II. FORMATION OF THE EGG ENVELOPES

A. THE EGG SHELL

Generally two reproductive modes are considered in *Artemia*: an oviparous and an ovoviviparous. However, not all the encysted embryos produced by the oviparous animal

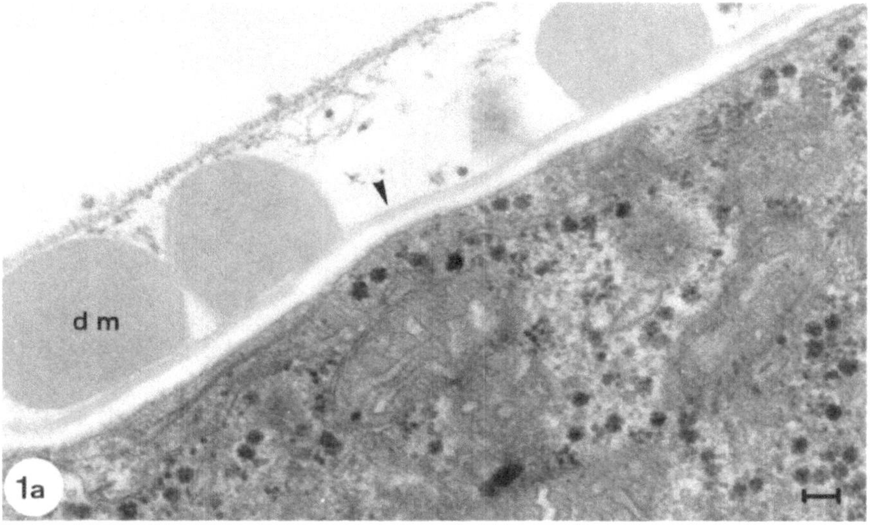

FIGURE 1a. Transmission electron micrograph of the thin shell of a blastula presumed to develop into a nondiapausing cyst. Irregular patches of dense material superficially covered by a fibrous mat (dm); trilaminar fertilization membrane (arrow); bar = 100 nm.

are in diapause. Some animals indeed produce cysts which emerge without a previous dehydration or any other treatment.[7,18-24] In contrast to diapause cysts they are only surrounded by a thin shell[20] (Figure 1a). It only consists of a trilaminar fertilization membrane and a thin chorion with irregular patches of dense material. It is superficially covered by a fibrous mat. An embryonic cuticle is never formed (personal observations).

The detailed description of the shell of the cryptobiotic cyst has been given by Morris and Afzelius.[23] In addition to the hypochlorite, soluble, double-layered, outer chorion secreted by the shell glands (Figure 1b) and the hypochlorite, resistant, embryonic cuticle formed by the blastodermal cells, their studies revealed the presence of two important layers not previously described; i.e., the outer cuticular membrane which is resistant to hypochlorite, impermeable to lead and separates the chorion from the embryonic cuticle, and the inner cuticular membrane which delineates the embryo from the fibrous layer of the embryonic cuticle. The SEM study of Mazzini[25] gives an idea of the three-dimensional structure of three main layers of the cyst envelope. Dutrieu,[21-22] Anderson et al.,[24] Fautrez and Fautrez-Firlefyn,[26] and Mazzini[25] studied the chemical composition of the cryptobiotic cyst wall.

A partial description of the formation of the shell of the cryptobiotic cysts is given by Anderson et al.[24] and Fautrez and Fautrez-Firlefyn.[26] Although its formation in *Artemia* in general terms is similar to that of *Tanymastix lacunae* which was described in detail by Garreau de Loubresse,[27] some differences are found (personal observations): the fertilization membrane of *Artemia* does not split with a diffusion of the shell gland secretion between both layers, but transforms into a trilaminar membrane which becomes covered with a fibrillar matrix in which dense, secretion products accumulate. In *Tanymastix* the outer cuticular membrane is the first layer observed after the completion of the chorion. In *Artemia* this layer gets its final structure after the formation of the fibrous layer of the embryonic cuticle by the blastomeres. De Chaffoy et al.[28] have shown that its structure is not achieved in embryos excised and cultured *ex utero*. The embryo evidently responds to some maternal factors in developing its final structure. As in *Tanymastix*, the inner cuticular membrane is formed immediately after the release of the cysts from the uterus.

In ovoviviparous animals the early embryos are surrounded only by a trilaminar fertilization membrane and a thin fibrous matrix (Figure 1c).

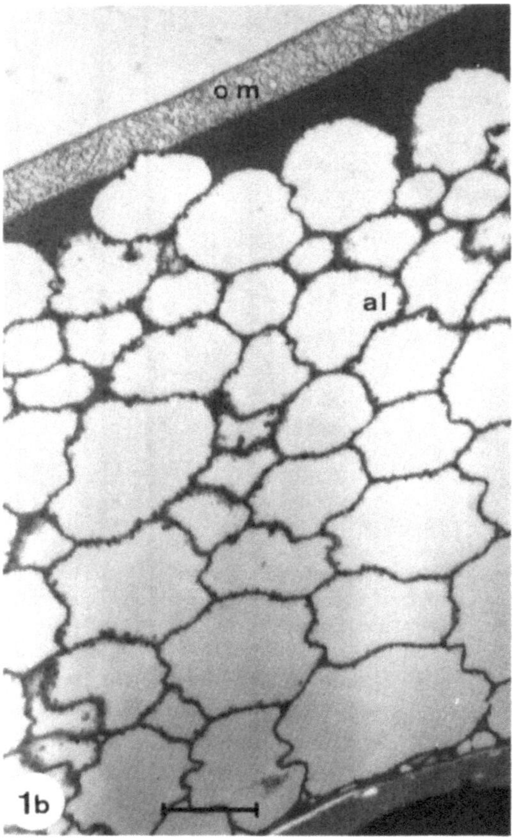

FIGURE 1b. Transmission electron micrograph of the thick shell of an 8-cell-stage embryo presumed to develop into a diapausing cyst. Undifferentiated alveolar layer (al); outer mat layer found in young developing cysts (om);[24] bar = 1 μm.

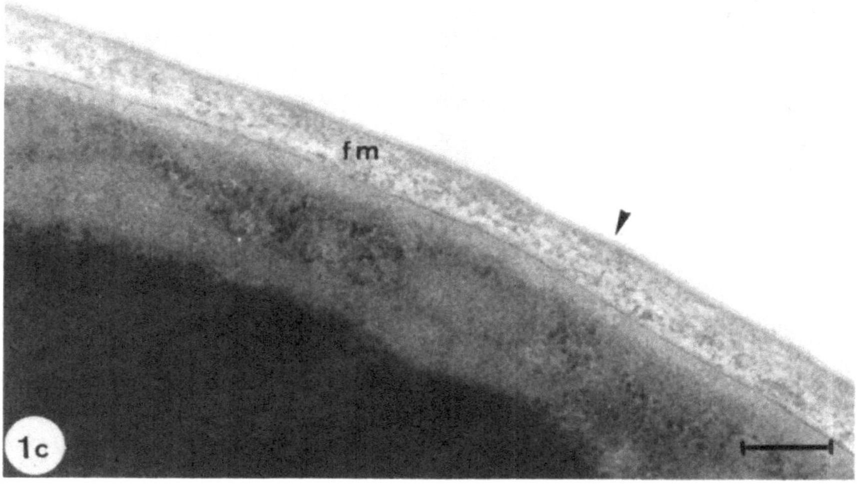

FIGURE 1c. Transmission electron micrograph of the cuticle of a nauplius developing *in utero* Thin fibrous matrix (fm); trilaminar fertilization membrane (arrow); bar = 100 nm.

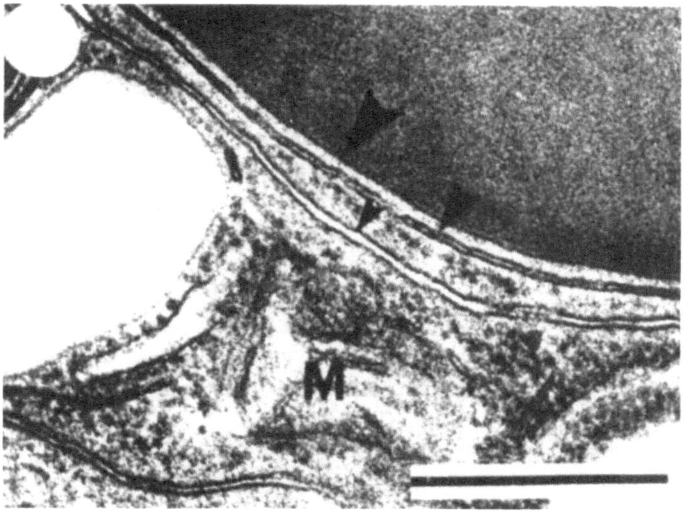

FIGURE 2. Transmission electron micrograph at a high magnification of the cytoplasm near a yolk platelet in dehydrating cyst. Yolk (large arrow), PERM (medium arrow), and plasmamembranes (small arrow) remain clearly visible; M, disrupted mitochondrion; bar = 0.5 μm. (From Jardel, J. L., Ph.D. thesis, Monash University, Australia, 1986. With permission.)

B. EMBRYONIC EXUVIATIONS AND HATCHING MEMBRANE

Oviparous development again has to be distinguished from ovoviviparous development. Personal observations support the description by Morris and Afzelius[23] of the hatching membrane being formed as a molted larval cuticle. If Lochhead, Linder and Nourisson, cited by Benesch[7] on page 348 are right and the embryonic cuticle is the first molt, then we would agree with Belk[29] that *Artemia* has two embryonic molts. However, Benesch[7] on page 349 considers the hatching membrane as the first embryonic cuticle. Moreover Benesch was unaware of the work of Morris and Afzelius[23] and of the existence of an inner cuticular membrane which forms after the release of the encysted embryos. If one considers the fibrous layer of the embryonic cuticle as a molt, what about the inner cuticular membrane? On a purely morphological basis neither can be considered as a molt, even if some of the arched patterns found in the fibrous layer could be similar to some part of an adult crustacean cuticle. In Belk's words,[29] "obviously there is more to learn about embryonic cuticles in these crustaceans."

III. CHANGES IN THE RELEASED CRYPTOBIOTIC CYST

Jardel[30] describes changes in the cysts after their release. His results suggest that initially the released embryos are not metabolically dormant but that metabolic dormancy is gradually established after so-called preparatory measures; the latter can be divided into changes preparative to dehydration and changes related to dehydration itself.

Within 96 h of release, ultrastructural modifications are described and a peak in cytochrome-C-oxidase suggests that these transformations are energy dependent. The most interesting morphological change concerns the rough endoplasmic reticulum which remodels to narrow ribosome free profiles called PERM (paired endoplasmic reticulum membranes). These PERM gradually surround the cell organelles, the yolk platelets, and the lipid yolk droplets and are thought to protect them against dehydration damage (Figure 2). They could be the membranes described around the yolk platelets by Vallejo et al.[31] Golgi complexes are transformed to what must be considered as a nonfunctional cryptic state. The number of gap junctions is reduced and septate junctions are formed.

The finding that the dehydration causes neither severe disruption of cellular organization, nor a loss of any cell components, nor a loss of the typical trilaminar structure of the cellular membranes is noteworthy. Dehydration of cell components is unequal, however, with some components dehydrating to a greater extent than others.

Jardel[30] also describes the rehydration of the dried cyst. In this phase the PERM lamellae surrounding the cell organelles are restored to rough endoplasmic reticulum cisternae. He does not mention any resumption of embryonic development.

Morris[32] investigated the relationship between hydration, the onset of metabolism, and the onset of development in encysted *Artemia*. Although metabolic activity is resumed less than 1 h after immersion in 0.5 *M* NaCl, the newly hydrated cyst enters a latent period of about 12 to 18 h during which time dehydration is reversible and no differentiation occurs. Only after this latent period is the embryo irreversibly committed to differentiation. At the light microscopical level no obvious changes occur in the hydrated cyst during the first 8 h (personal observations).

A point must be made about Benesch's[7] work. Although he specifies the use of embryos developing within ovoviviparous females, in several illustrations of the embryologic events the use of hydrated cysts is suggested by the presence of shells. He does not take any latent period in account when he juxtaposes embryonic development of hydrating cysts and of nauplii developing *in utero*. This lack of accuracy may lead to misinterpretation.

IV. EMBRYONIC DEVELOPMENT

A. CLEAVAGE AND BLASTULA

Unless cited otherwise the data presented here originate from Benesch.[7]

In contrast to earlier descriptions[2,12] cleavage in *Artemia* is total with an equal distribution of yolk among the blastomeres, at least at the light microscopical level. Recent electron microscopic studies might however necessitate a redefinition of what is meant by "total" cleavage: Jardel[30] observes what he calls "fusomes" between some blastomeres, but after personal observations we feel they are better compared with the cytoplasmic bridges observed in the ovary of *Artemia* (Figure 3). The "zones nécrotiques" described by Bes et al.[33] may represent the same cytoplasmic bridges after less successful preservation.

It is not possible to determine whether there is a relationship between the orientation of the first mitotic spindle and the place of elimination of the first polar body.

The first and second cleavage are meridional. As far as the cleavage is concerned the yolky eggs of *Artemia* are most easily compared to these in Cladocera.[34]

Total equal and radial cleavages proceed until the blastula stage is reached at the 512-cell stage. The blastocoel enlarges from the 4 to the 256-cell stage, the length of the blastomeres decreases and the embryo enlarges. The nuclei lie in the center of the blastomeres until stage 64; from stage 128 on they move to the periphery. Between stages 256 and 512 the blastomeres grow and diminish the blastocoel.

It was impossible to follow individual cell lineages although in some regions mitoses seem to occur preferentially. A hollow spherical blastula is formed.

B. GASTRULA

Compared with other non-Malacostraca, gastrulation in *Artemia* starts late. As in the cladoceran *Cyprideis*,[34] it proceeds in two main phases: G-I and G-II.

During G-I, the primordial germ cells and the mesoderm reach the inside. Its onset is marked by the immigration of 1 or 2 nuclei, caused by an increase in peripheral cytoplasm, not by an actual cell movement. Probably there is no determination yet, as this immigration starts from more than one site, of which one soon predominates. Once 3 to 5 nuclei are present, whole mesoderm cells immigrate and protrude in the blastocoel while their mitotic

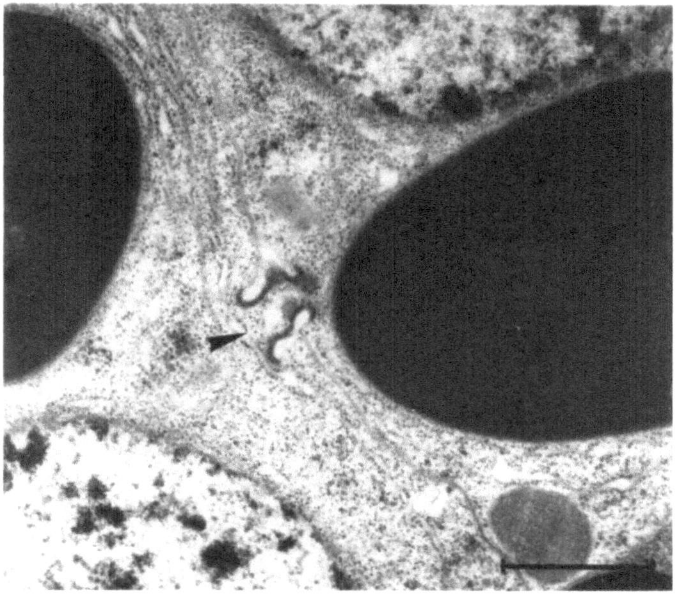

FIGURE 3. Electron micrograph of hydrated cyst. A narrow intercellular bridge (arrow) lined by an electrondense rim is found between two presumed ectodermal cells; bar = 1 μm.

rate increases. The immigration cone grows, moves aside, and reduces the blastocoel. A dent in the immigration cone is the first indication of the blastoporus. At the end of G-I the embryo is clearly symmetrical. Possibly this symmetry is organized by the mesoderm. The mesoderm has immigrated almost completely and only a small remnant of the blastocoel is left. By the structure of their nucleus 12 to 16 primordial germ cells can be distinguished from the other mesoderm cells.

At the onset of G-II a second invagination pole arises where entoderm cells, characterized by large nuclei, enter at an angle of approximately 120° from the caudal region where the mesoderm is invaginating. Presumptive mesoderm and entoderm are thus separated on the blastosphere by the presumptive ectoderm. Meanwhile, a mass of mesodermal cells originating from a ring surrounding the primordial germ cells organize into naupliar segments and remain separated from the encroaching ectodermal cells. In the first immigration pole the metanaupliar mesoderm immigrates last, caudally from the primordial germ cells. This interpretation of Benesch[7] seems better founded than the earlier interpretation of Fränsemeier[34] and Weisz[35] that the metanaupliar mesoderm arises from the ectoderm. Anderson[36] postulates that the metanaupliar mesoderm arises from mesoteloblasts as in many non-Malacostraca.[37] However, as Weisz[35] and Anderson[36] only studied postemergence stages, their interpretation may be insufficiently established.

At the time when naupliar segmentation becomes apparent in the mesoderm, ectodermal cells protrude in those regions where the subepidermal tendonal cells will appear. It is possible that those two phenomena are interdependent. Whereas many mitoses occur in the mesodermal cells, the endodermal cells immigrate without cell divisions and separate the regions of naupliar mesoderm. The somites are not completely separated however, but stay close and form cross-arches. Thus, the mandible mesoderm lays as an arch over the metanaupliar mesoderm, the mesoderm of the second antenna lays over the primordial germ cells and the entodermal cone, and the mesoderm of the first antenna lays over the preantennal mesoderm. A coelom is apparent in mandible and antenna II mesoderm.

Naupliar and metanaupliar mesoderms differ, although both originate from the same invagination pole: the naupliar segments arise and differentiate simultaneously whereas the metanaupliar mesoderm forms an unpair mass of cells at the caudal end of the embryo. In this mass, mitoses occur after hatching when the metanaupliar proliferation starts and somites are formed one after another. The ectoderm gradually differentiates from the blastoderm. Nuclei from the appendage-ectoderm (antenna I and II and mandible) migrate first. Intersegmental ectodermal cells are elongated, i.e., they are the presumptive tendonal cells. In the caudal region the first immigration pole fades away. The hindmost mesodermal cells are the presumptive gut mesoderm; the ectoderm of the same region is the presumptive ectodermal proliferation zone of metanaupliar segments.

The "immigration" of entoderm gradually evolves into an "invagination" when the immigrating cells reach the dorsal ectodermal cells and push the mesoderm aside. No cell divisions occur yet in the entoderm; mitotic activity decreases in the mesodermal cells and increases in the ectodermal cells. The invaginating entoderm pushes the mesoderm whose invagination has come to an end in the following way: the metanaupliar mesoderm, proctodeum, and mandible mesoderm are pushed caudally; the second antenna mesoderm, laterally; the first antenna mesoderm, cranially and laterally; the preantennal and stomodeum mesoderm, cranially; and the primordial germ cells are to be found in the mandible segment.

At the end of gastrulation, ectoderm, mesoderm, and entoderm are clearly differentiated. Proctodeum and stomodeum still lay outside the embryo. Segmental ectoderm and mesoderm are found in the naupliar segments and in the preantennal mesoderm.

Clegg[38] found important differences in chemical composition in intrauterine embryos depending on whether they were developing into cysts or nauplii. An effort should be made to find morphological evidence supporting these findings.

C. NAUPLIUS DIFFERENTIATION

Benesch[7] describes seven stages in naupliar development before emergence: Na-0 to Na-6. He is the only one to have studied this development extensively. Since he was studying ovoviviparously developing nauplii, this can explain the discrepancy between his drawings of mitoses throughout naupliar development and the findings of Fränsemeier,[34] Nakanishi,[39] and Olson and Clegg[40] that no mitoses occur during the hydration of cryptobiotic cysts.

1. Na-0

The lumen of the entoderm is still nonexistent. Invagination of the stomodeum ectoderm follows that of the entoderm. Mitoses are scarce in the mesoderm where some muscle cells differentiate. At the caudal end of the embryo the presumptive proctodeum has not invaginated; its ectodermal cells are smaller than those of the metanaupliar ectoderm. The mesodermal cells of the proctodeum lie in between the metanaupliar mesoderm.

2. Na-1

At the end of the invagination of the stomodeum the cleft between the entodermal cells broadens and the cells order into an epithelium. This leads Benesch[7] to conclude that there has never been a barrier between the stomodeum and the gut cells. The stomodeum, however, remains closed to the outside. Stomodeum and midgut have a dorsoventral orientation. Multiple mitoses cause a proliferation of the entoderm to form the midgut. Not all the mitoses have stopped yet in the mesoderm. Most important changes occur in the ectoderm where ganglion cells move in and arrange into presumptive ganglia, i.e., mandible-, trito- and deutocerebral ganglia, which are separated by tendonal cells, proto- and deutocerebrum. The latter are separated by large, characteristic wedge-shaped cells with a large oval peripheral nucleus. The optic ganglia and the ganglia of the nauplius eye move in, as well as the two pigment cells of the nauplius eye, and ganglion cells and retinula cells for the dorsal and ventral nauplius eye.

3. Na-2

As the midgut proliferates, its junction with the stomodeum moves cranially. Ectodermal and entodermal cells can easily be distinguished as the nuclei of the stomodeal cells lie close to the lumen whereas those of the midgut cells lie closer to the periphery. The future opening of the stomodeum moves caudally as it is pushed by the invaginated cerebral ganglia. Invagination of the cerebral ganglia from one side and proliferation of the midgut cells from the other side again cause movements of the mesoderm. The preantennal mesoderm moves ventrally and cranially from the stomodeum. The dorsal and cranial mesoderm of the first antenna is pushed caudally where it contacts the mesoderm of the second antenna. Primitive blood cells arise in the medial dorsal cells of the first antenna mesoderm and lie between the muscle cells of the second antenna. The mesoderm of the second antenna and mandible remain lateral to the gut. Metanaupliar mesoderm and primordial germ cells move caudally. Constrictor and dilatator muscles arise as elongated spindle-shaped cells in the preantennal mesoderm. The cells of the constrictors lie transversely to the stomodeum, while the dilator cells are pushed laterally from them and reach caudally into the segment of the second antenna. These mesodermal cell movements make it impossible to decide whether the constrictor and dilatator muscles of the esophagus and the labrum muscles arise from the second antennal or from the preantennal mesoderm. The dorsal ectodermal cells enlarge to form what Benesch[7] named the "Atemplatte", correlating with the function Dejdar[41] ascribed to this organ which will later form the naupliar salt gland.[42] The "Atemplatte" shows a threepartite pattern caused by three inside projections which are the prospective attachment places for the antennal and mandibular muscles respectively.

4. Na-3

When the proliferation of the midgut has reached the invaginating proctodeum caudally, it touches the ectoderm cranially in the region dorsally to the presumptive nauplius eye, where the formation of the lateral gut diverticles start. These evaginations will again influence the position of the mesoderm. The esophagus narrows and moves cranially followed by its muscle cells. At the caudal end, the proctodeum invaginates and pushes its own muscle cells and those of the caudal part of the gut laterally and dorsally. In the mesoderm, muscle cells and connective tissue cells can be discerned as well as the second antenna gland. The preantennal mesoderm lies between the first part of the midgut, the esophagus and the brain. The mesoderm of the first antenna lies laterally to the preantennal mesoderm, dorsally and laterally from the midgut. Both mandible mesoderms touch dorsally, cranially, and laterally and are well separated from the metanaupliar mesoderm which lies laterally to the caudal part of the gut and the proctodeum invagination. The primordial germ cells lie ventrally and laterally in between the mandible and metanaupliar mesoderm. Prospective gut muscle cells surround the midgut. Ectoderm is further differentiated, i.e., the seta-forming cells appear in the presumptive appendages, caudally from them a notch starts forming where the appendage will appear. All parts of the nervous system and of the neurosecretory system can be traced, as can the larval salt gland.

5. Na-4

The proctodeum terminates its invagination but has not yet broken into the midgut. The stomodeum has found its final position. The entodermal cells are the biggest cells while the ectodermal cells are smaller and epithelial-like. A notch marks the appendages. In the naupliar mesoderm the first fibrils appear in the muscles. Only the preantennal mesoderm has not yet reached its final position.

6. Na-5

The embryo starts growing but lies rounded within its surrounding membrane; the organs are compressed, especially the midgut. The mesoderm shows muscles, the antennal gland, blood cells, fat-storing cells, and muscle cells of the stomodeum and the proctodeum. On either side of the proctodeum at the transition of midgut and hindgut, two giant cells appear between the muscle cells. According to Benesch[7] they are unrelated to the future innervation of the end-gut region. Although he cannot ascribe a specific function to them, he does not agree with Fränsemeier[34] who named them ganglion cells, nor does he think they can represent mesoteloblasts, as they do not divide. Naupliar mesoderm is delineated from the metanaupliar mesoderm. The labrum grows caudally. The appendages grow and lie around the head; the first three pairs of head appendages are present in rudimentary form. The antennules form lateral knob-like protuberances near the front end of the embryo. Behind them the rudimentary antennae project somewhat beyond the general body surface. A vertical split appears and setae develop. The mandibles occur on a level just caudal to the tips of the antennal rami.[35] The ectodermal cells of the telson make an inward fold. At this stage there is some variation in the outer aspect of the embryos: some are more stretched than others; their inner structures, however, are all alike. Blood cells and fat-storing cells are filled with yolk platelets. The blood cells lie caudally and dorsally from the first antenna mesoderm and form the primary hemocytes. Benesch[7] reports an enlargement of the hemocoel at this stage but the moment of its first appearance is not mentioned.

7. Na-6

The thoraco-abdomen elongates without mitoses. The gut is opened cranially but the hindgut is still closed. Differentiation of naupliar muscles is complete. The antennal gland is functional. The metanaupliar mesoderm consists of a compact mass of cells. The primordial germ cells are found in the maxillar region. In the nervous system the nerve fibers are visible. The periesophagal nerve ring is formed. Nerve fibers are seen to leave the protocerebrum to the ganglion of the nauplius eye. Benesch[7] even discerns the nerves innervating the X-organ. Rudiments of compound eyes are present as slight lateral proliferations just behind the bases of the antennules.[35] The nauplius eyes have differentiated.

8. Stage 0: Hatching

Just before and after hatching, *Artemia* is a true nauplius with an undifferentiated preantennal and an unsegmented metanaupliar mesoderm, but with a function neuromuscular system. It is pear shaped with rudiments of the three naupliar segments, i.e., antennular, antennal, and mandibular, each with its rudimentary appendages. The telson fold is more or less smooth. The gut has stretched and at its cranial end, opens into the stomodeum. Caudally, a double layer of epithelial cells still separates the gut from the proctodeum (Figure 4). Over the whole length, the gut muscles can be distinguished from the somatic muscles. The fat-storing cells from the labrum still lie dorsally between the nauplius eye, brain, and stomodeum, and ventrally before the gut diverticles. The primary blood cells are rounding off, eventually becoming rounded while losing their yolk. The primordial germ cells lie in the region of the maxillary and first thoracic segments. In the ectoderm, multilayered cell groups form the rudiments of the protocerebral ganglion and the ganglia, up to the mandibulary ganglion. The ectoderm is unilayered with elongated cylindrical cells in the neck organ and in the zone of ectoderm proliferation. Subepidermal tendonal cells protrude between the naupliar segments. Ectodermal cells forming the integument are flattened. In the brain the presumptive parts can be recognized. The hemocoel is relatively large. Yolk platelets are present in all the cells.

V. LARVAL DEVELOPMENT

A. EXTERNAL DEVELOPMENT

1. Description of the Larval Stages

The development of the external form in larval *Artemia* can be summarized in the

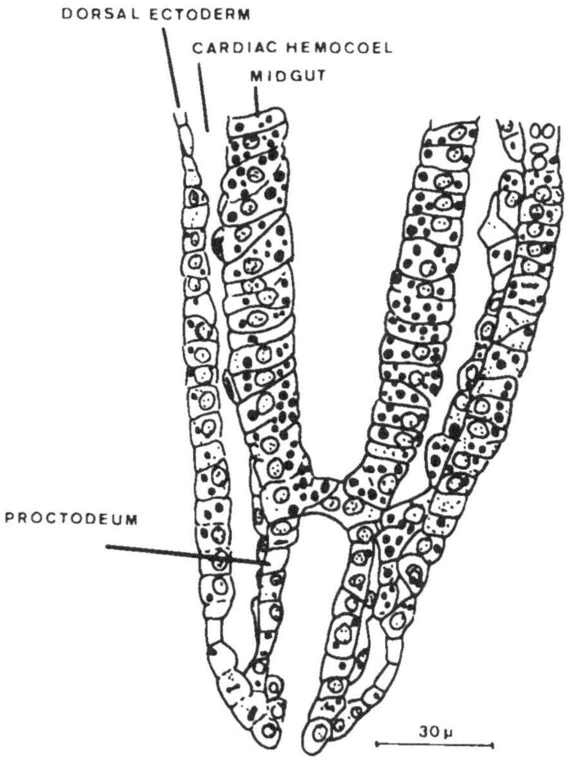

FIGURE 4. Sagittal section of postmandibular region of young nauplius. A double cell layer separates the lumen of the midgut from that of the proctodeum. (From Anderson, D. T., *Austr. J. Zool.*, 15, 47, 1967. With permission.)

successive appearance of the maxillar, thoracic, and genital segments with their appendages, and of the abdominal segments. Authors studying larval development of *Artemia* have proposed a subdivision in stages, based mainly on changes in external form, although often a histological survey is also given of the internal events. Several authors relate these changes to the larval molts,[43-49] but as there still is no agreement on the number of larval molts (nine,[36] eleven,[43] thirteen,[44] seventeen,[45,47] and nineteen[48-49]), it is not surprising that there is no agreement in the staging systems either. Weisz[35] found a variability in the number of larval molts with the rearing conditions, and Benesch[7] reports that formation and differentiation of the segments proceeds individually, unrelated to moltings and that more than one developmental step may occur during an intermolt period. Therefore the system Weisz[35] proposes appears to be the most logical; i.e., nineteen stages whereby each stage number is defined by the number of delineated segments present. Freeman[48-49] also discerns nineteen stages, but as he links the appearance of each segment to a molt, he discerns 19 instars. Benesch[7] arbitrarily describes twenty-four larval stages without clear evidence. With scanning electron microscopy, Schrehardt[47] established irrefutably the most significant details of the sequence of events leading to the adult form. The following description summarizes his study. Although he based his description on the molts, as did Hentschel,[45] he has also attempted to fit *Artemia*'s development into a more general scheme proposed by Kaestner[50] for Euanostracans. One naupliar (Figure 5), 4 metanaupliar (Figures 6—9), 7 postmetanaupliar (Figures 10—16), and 5 postlarval stages (Figure 17 shows the first postlarval stage) are discerned corresponding to one of the stages described by Hentschel,[45] and numbered L-1 to L-17. Differences in the development or in the reduction of antenullae, antennae,

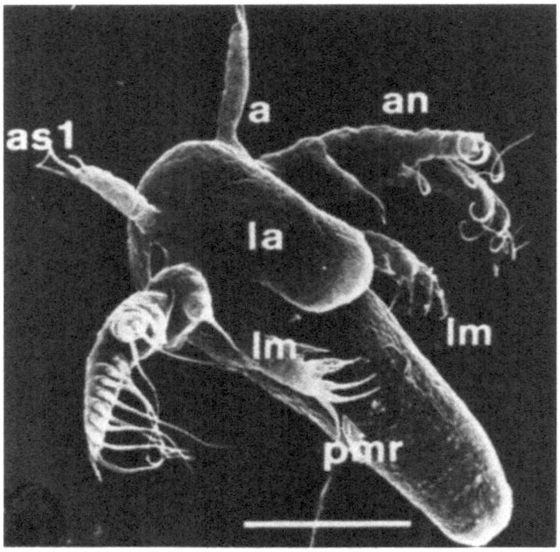

FIGURE 5. Ventral view of a Nauplius-I (L-1) just before molting. (a) Antennule; (an) antenne; (as1) antennula-seta-type 1; (1a) labrum; (lm) larval mandible; (pmr) postmandibular region; bar = 100 μm. (From Schrehardt, A., *Artemia Research and Its Applications,* Vol. I, Sorgeloos, P., Bengtson, D. A., Decleir, W., and Jaspers, E., Eds., Universa Press, Wetteren, Belgium, 1987, 5. With permission.)

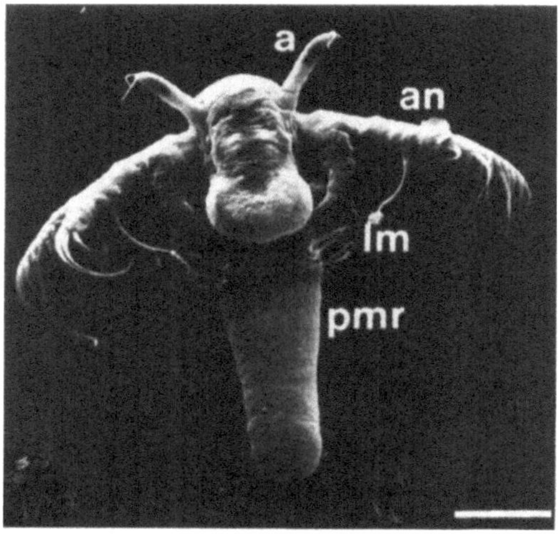

FIGURE 6. Ventral view of a Metanauplius-I (L-2). Antennule (a); antenne (an); larval mandible (lm); postmandibular region (pmr); bar = 100 μm. (From Schrehardt, A., *Artemia Research and Its Applications,* Vol. I, Sorgeloos, P., Bengtson, D. A., Decleir, W., and Jaspers, E., Eds., Universa Press, Wetteren, Belgium, 1987, 5. With permission.)

and larval mandibles can be used to distinguish the naupliar and metanaupliar stages. They are compiled in Table 1. Meta- and postmetanaupliar stages are characterized by the segmentation and development of the maxillules, maxilles, thoracomeres (thoracic segments), and thoracopods. Benesch,[7] who describes the formation of the first maxilla more extensively than Schrehardt,[47] concludes from its musculature that it is a modified telopodite or exopodite of which the bristles are the modified endites.

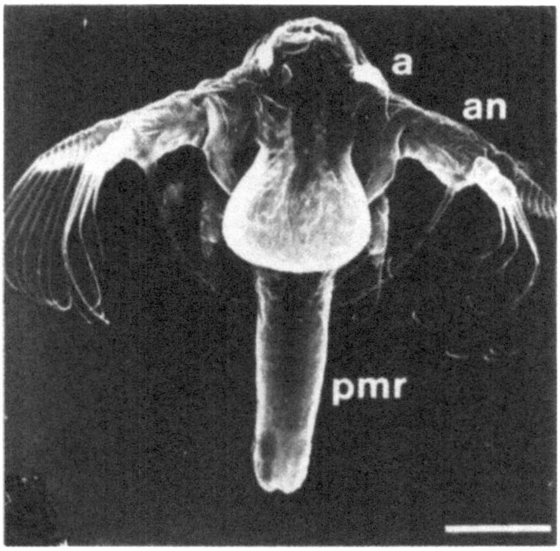

FIGURE 7. Ventral view of a late Metanauplius-II (L-3). Antennule (a); antenne (an); postmandibular region (pmr); bar = 100 μm. (From Schrehardt, A., *Artemia Research and Its Applications*, Vol. I, Sorgeloos, P., Bengtson, D. A., Decleir, W., and Jaspers, E., Eds., Universa Press, Wetteren, Belgium, 1987, 5. With permission.)

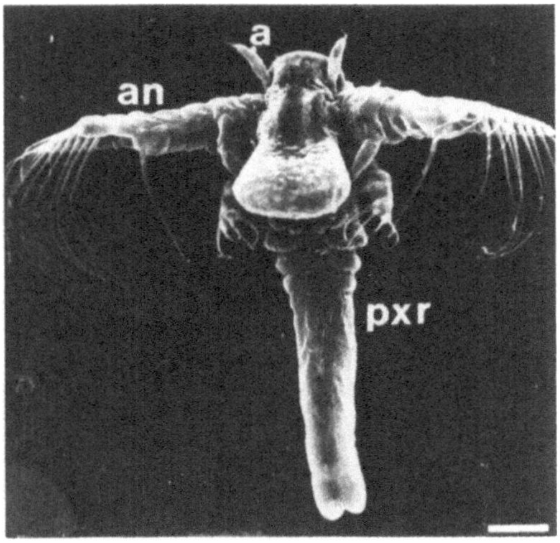

FIGURE 8. Ventral view of a late Metanauplius-III (L-4). Antennule (a); antenne (an); postmaxillar region (pxr); bar = 100 μm. (From Schrehardt, A., *Artemia Research and Its Applications*, Vol. I, Sorgeloos, P., Bengtson, D. A., Decleir, W., and Jaspers, E., Eds., Universa Press, Wetteren, Belgium, 1987, 5. With permission.)

Table 2 demonstrates how segmentation as well as different steps in appendage differentiation proceed two by two during each intermolt period. The work of Schrehardt[47] invalidates Anderson's[36] presumption that the first six thoracic segments become simultaneously functional coincident with the loss of the specialized naupliar feeding apparatus at the ninth molt. The postlarval stages are characterized by the transformation of the antennae toward

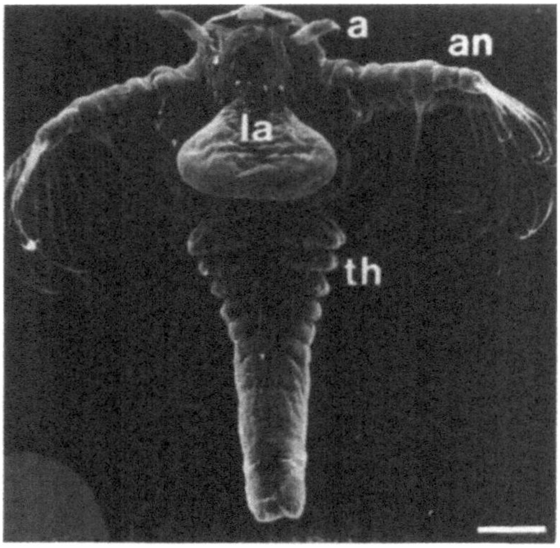

FIGURE 9. Ventral view of a late Metanauplius-IV (L-5). Antennule (a); antenne (an); labrum (la); thoracopods (th); bar = 100 μm. (From Schrehardt, A., *Artemia Research and Its Applications*, Vol. I, Sorgeloos, P., Bengtson, D. A., Decleir, W., and Jaspers, E., Eds., Universa Press, Wetteren, Belgium, 1987, 5. With permission.)

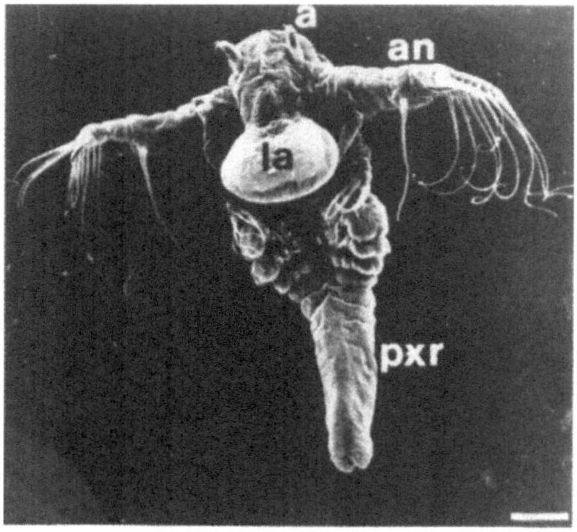

FIGURE 10. Ventral view of an early Postmetanauplius-I (L-6). Antennule (a); antenne (an); labrum (la); postmaxillar region (pxr); bar = 100 μm. (From Schrehardt, A., *Artemia Research and Its Applications*, Vol. I, Sorgeloos, P., Bengtson, D. A., Decleir, W., and Jaspers, E., Eds., Universa Press, Wetteren, Belgium, 1987, 5. With permission.)

their definitive shape and function (Figures 18—20), and by the development of the genital structures. The seventeenth molt concludes the postembryonic development and both sexes attain the adult organization.

In addition to the two tables it may be interesting to follow some developments in particular.

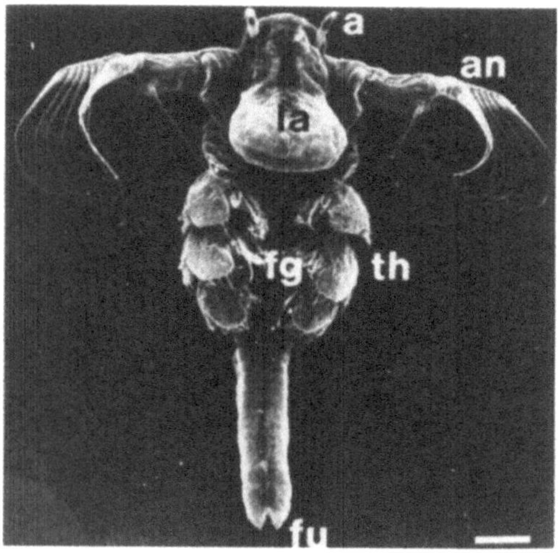

FIGURE 11. Ventral view of Postmetanauplius-II (L-7). Antennule (a); antenne (an); food groove (fg); furca (fu); labrum (la); thoracopod (th); bar = 100 μm. (From Schrehardt, A., *Artemia Research and Its Applications*, Vol. I, Sorgeloos, P., Bengtson, D. A., Decleir, W., and Jaspers, E., Eds., Universa Press, Wetteren, Belgium, 1987, 5. With permission.)

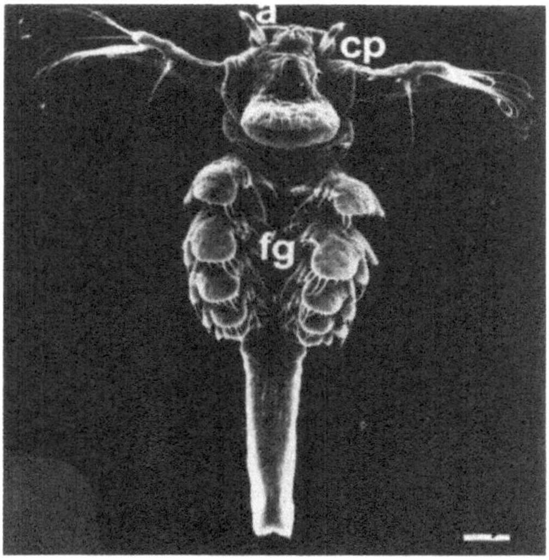

FIGURE 12. Ventral view of an early Postmetanauplius-III (L-8). Antennule (a); compound eye (cp); food groove (fg); bar = 100 μ. (From Schrehardt, A., *Artemia Research and Its Applications*, Vol. I, Sorgeloos, P., Bengtson, D. A., Decleir, W., and Jaspers, E., Eds., Universa Press, Wetteren, Belgium, 1987, 5. With permission.)

2. Development of the Filter-Feeding Apparatus

The development of the filter-feeding apparatus and filter-feeding itself start in the Metanauplius-I (L-2) together with the formation of setulae on the terminal setae of the antennae and of the mandibular exopodites. The filter-feeding apparatus becomes more complicated in Metanauplius-II (L-3) where the buds of the adult mandibles appear. Haired

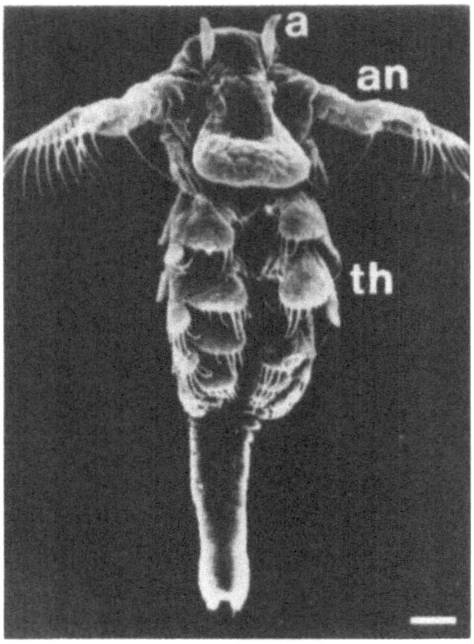

FIGURE 13. Ventral view of a Postmetanauplius-IV (L-9). Antennule (a); antenne (an); thoracopod (th); bar = 100 μm. (From Schrehardt, A., *Artemia Research and Its Applications*, Vol. I, Sorgeloos, P., Bengtson, D. A., Decleir, W., and Jaspers, E., Eds., Universa Press, Wetteren, Belgium, 1987, 5. With permission.)

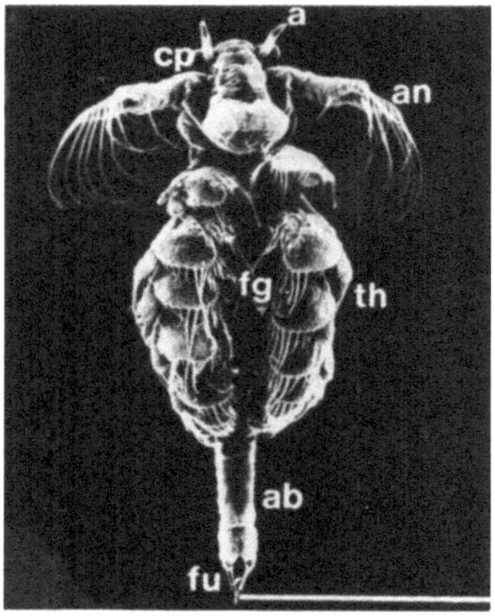

FIGURE 14. Ventral view of a Postmetanauplius-V (L-10). Antennule (a); abdomen (ab); antenne (an); compound eye (cp); food groove (fg); furca (fu); thoracopod (th); bar = 1 mm. (From Schrehardt, A., *Artemia Research and Its Applications*, Vol. I, Sorgeloos, P., Bengtson, D. A., Decleir, W., and Jaspers, E., Eds., Universa Press, Wetteren, Belgium, 1987, 5. With permission.)

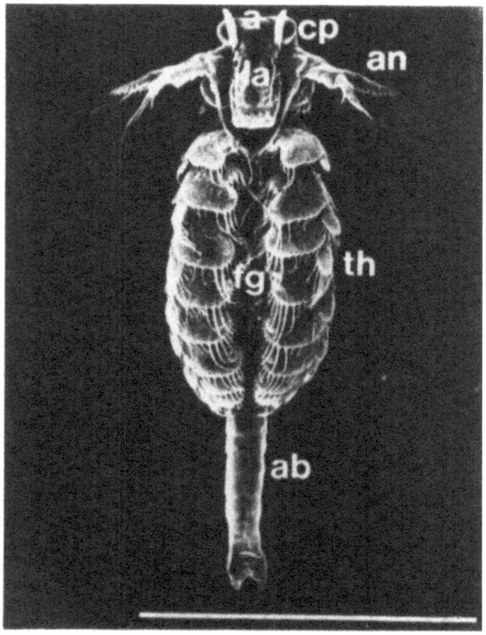

FIGURE 15. Ventral view of a Postmetanauplius-VI (L-11). Antennule (a); abdomen (ab); antenne (an); compound eye (cp); food groove (fg); labrum (la); thoracopod (th); bar = 1 mm. (From Schrehardt, A., *Artemia Research and Its Applications*, Vol. I, Sorgeloos, P., Bengtson, D. A., Decleir, W., and Jaspers, E., Eds., Universa Press, Wetteren, Belgium, 1987, 5. With permission.)

paragnaths develop during Metanauplius-III (L-4), and the rudiments of the adult mandibula expand without further differentiation. The stoma shifts from a para-antennal into a paramandibular position in Metanauplius-IV (L-5), and the rudiments of the adult mandibles bend towards the sagittal axis of the larva. Maxillules and maxilles become active in food transport in Postmetanauplius-I (L-6). In this stage the mandibles form their cutting edges. In Postmetanauplius-II (L-7) the mandibles develop cuticular teeth for the mechanical treatment of the filtered food. The upper lip of the larva starts changing during Postmetanauplius-III (L-8), by forming a trapezoid border. During Postmetanauplius-IV (L-9) it develops a nose-shaped portion inside the trapezoid border which already begins to resorb in Postmetanauplius-V (L-10), collapses during Postmetanauplius-VI (L-11), and is completely resorbed in Postmetanauplius-VII (L-12). From then on, the labrum gradually gets in definitive tongue shape as seen in the adult. This aspect is reached in Postlarval Stage-II (L-14). During Postlarval Stage-IV (L-16) the adult form is achieved: a part of the dorsal surface of the cephalon deepens to form a transverse cervical groove joining the basal portions of the adult mandibles. Its morphological significance is the strengthening of the attachment site for the dorsal suspension of tendons and muscles.

3. Development of the Compound Eyes

This development occurs in the metanaupliar and postlarval periods. Rudiments of the compound eyes arise as rounded elevations, dorsal to the antennules during Metanauplius-IV (L-5). They begin their enlargement during Postmetanauplius-III (L-8): ommatidia are not yet identifiable with the SEM although they are already visible with the light microscope. The rudiments of the compound eyes protrude and a single ommatidium can be distinguished

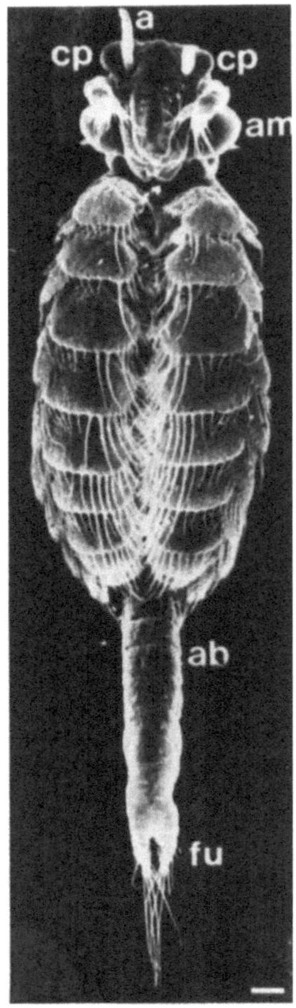

FIGURE 16. Ventral view of a Postmetanauplius-VII (L-12). Antennule (a); abdomen (ab); adult mandible (am); furca (fu); bar = 100 μm. (From Schrehardt, A., *Artemia Research and Its Applications*, Vol. I, Sorgeloos, P., Bengtson, D. A., Decleir, W., and Jaspers, E., Eds., Universa Press, Wetteren, Belgium, 1987, 5. With permission.)

in Postmetanauplius-IV (L-9). The ommatidia multiply during Postmetanauplius-V (L-10) until Postlarval Stage-I (L-13) where initial extension of the eyestalks can be noticed. The extension of the eyestalks advances during Postlarval Stage-II (L-14).

4. Differentiation of Male and Female Antennae

This differentiation can be used to distinguish stages of the postlarval period. The differentiation between male and female antennae starts after they have shifted into a ventrolateral position in Postlarval Stage-I (L-13). During Postlarval Stage-II (L-14) the protopodites are no longer visible and from the endopodites only stumps remain at the bases of the exopodites. Setae develop on the basal portions of the exopodite and in males a rudimental frontal knob develops. During Postlarval Stage-III (L-15) the antennae are tied together both in males and in females by a basal tissue bridge covering the basal portion of the labrum; the remnants of the endopodite decrease in size; the exopodites become sickle shaped, and in males, initial formation of cuticular spines can be noticed on the frontal knobs (Figure 18). In Postlarval Stage-IV (L-16) the endopodites are completely reduced while the setae

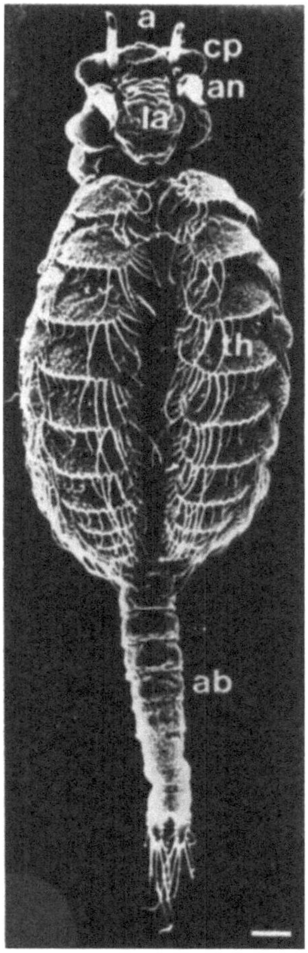

FIGURE 17. Ventral view of a Postlarval Stage-I (L-13). Antennule (a); abdomen (ab); antenne (an); compound eye (cp); labrum (la); thoracopod (th); bar = 100 μm. (From Schrehardt, A., *Artemia Research and Its Applications,* Vol. I, Sorgeloos, P., Bengtson, D. A., Decleir, W., and Jaspers, E., Eds., Universa Press, Wetteren, Belgium, 1987, 5. With permission.)

increase on the exopodite. In the male antenna a secondary joint develops which is absent in females (Figure 19). The frontal knobs enlarge and cuticular spines and rudiments of mechanoreceptors[51] appear. The frontal knobs further enlarge during Postlarval Stage-V (L-17) (Figure 20) and develop ramified and nonramified spines, which occur singly or in pairs; occasionally three spines arise from the same region.

5. Development of the Genital Structures

The differentiation of the genital structures can also be used in the determination of the postlarval stages. During Postmetanauplius-VII (L-12) the buds of the penes appear as globular swellings on the ventral portions of the genital segments and the ovisac is built by fusion of ventral tissue lobes of the two genital segments. In the Postlarval Stage-I (L-3) the buds of the penes and the rudiments of the ovisac enlarge. The orifices of the vasa deferentia are first seen in Postlarval Stage-II (L-4). At this stage the ventral portion of the intersegmental furrow separating the genital segments is reduced so that a ventro-lateral segment fusion can be distinguished. Genitals of males and females further expand through

TABLE 1
Evolution of the Larval Appendages

		Antenna			Mandible		
	Antennule	**Protopodite**	**Endopodite**	**Exopodite**	**Protopodite**	**Endopodite**	**Exopodite**
Nauplius-I (L-1)	Three setae type I	Two endites each with two setae: gnatobasenseta hook bristle	One short and one long seta	One terminal seta	Two endites + each a bristle	Two terminal setae	Three terminal setae
Metanauplius-I. (L-2)	Two buds of seta type II	—	Short bristle elongates + four setae develop	Terminal setae get setulae	—	—	Terminal setae get setulae
Metanauplius-II. (L-3)	Buds develop two setae type II	Gnatobasenseta bifurcates	Second order setulae are formed			—	—
Metanauplius-III. (L-4)	—	—	—	—	—	Decrease in size	—
Metanauplius-IV. (L-5)	Bud of third seta type II	—	—	—	—	—	—
Postmetanauplius-I (L-6)	Third seta type II differentiates	Reduction of second order setulae			—	Decrease in size	—
Postmetanauplius-II (L-7)	Gains adult form three setae type I four setae type II	—	—	—	—	—	—
Postmetanauplius-II (L-8)	Reduction of the setulae	—	—	—	Decrease in size + reduction of the setulae		
Postmetanauplius-IV (L-9)	Seta type II gets distal pore	—	—	—	—	—	—
Postmetanauplius-V (L-10)	Distal pore	Reduction of setulae of endites	—	—	—	—	—
Postmetanauplius-VI (L-11)		Vestigial endites, gnatobasenseta + hook bristle deteriorate	Cease locomotory and filtering functions		Reduced to rudimentary extremities on adult structures		
Postmetanauplius-VII (L-12)		Gnatobasenseta rudimentary, hook bristle disappeared	Terminal setae deteriorate		Button-shaped tissue in adults		

TABLE 2
Development of the Thoracopods

	Appearance of segmentation furrow	Budding of appendages	Formation of protopodite + endopodite form	Formation of exopodite and epipodite	Develop filter setae on prodopodite and endite	Get final shape
Metanauplius I. (L-2)	Maxillular + maxillar segment	—	—	—	—	—
Metanauplius II. (L-3)	Thoracomers I and II	Maxillules maxillae thoracopod I	—	—	—	—
Metanauplius III. (L-4)	Thoracomers III and IV	Thoracopod II	Thoracopod I	—	—	—
Metanauplius IV. (L-5)	Thoracomers V and VI	Thoracopods III + IV	Thoracopods II sometimes thoracopod III	Thoracopod I	—	—
Postmetanauplius-I. (L-6)	Thoracomers VII and VIII	Thoracopods V and VI	Thoracopods III and IV	Thoracopod II	Thoracopods I and II	—
Postmetanauplius-II. (L-7)	Thoracomers IX and X indentation marks formation of the furca + short seta	Thoracopods VII and VIII	Thoracopods V and VI	Thoracopod III and IV	Thoracopods III and IV	Maxillules maxillae
Postmetanauplius-III. (L-8)	Thoracomer XI + first genital segment	Thoracopods VIII, IX, and X	Thoracopod VII	Thoracopods V, VI, VII	—	—
Postmetanauplius-IV. (L-9)	2nd genital + 3rd abdominal segment	Thoracopod XI	Thoracopods VIII and IX	Thoracopods VI and VII	Thoracopod V	—
Postmetanauplius-V. (L-10)	Abdominal segments IV + V, each furca ramus + 3 to 4 feathered setae	—	Thoracopods X and XI	Thoracopods VIII + IX	Thoracopods VI and VII	—
Postmetanauplius-VI (L-11)	Abdominal segment VI + VII, furca rami elongate	—	—	Thoracopods X and XI	Thoracopods VIII + IX	—
Postmetanauplius-VII (L-12)	Abdominal segment VII telson	—	—	—	Thoracopods X and XI	—

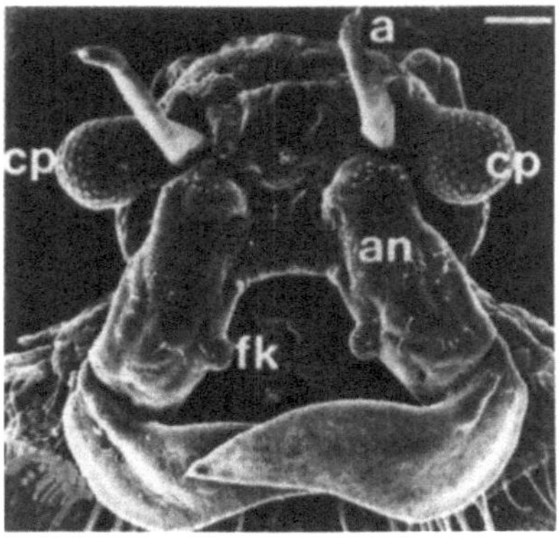

FIGURE 18. Cephalon of a Postlarval Stage-III (L-15). Antennule (a); adult mandible (am); compound eye (cp); endopodite (en); exopodite (ex); labrum (la); bar = 100 μm. (From Schrehardt, A., *Artemia Research and Its Applications*, Vol. I, Sorgeloos, P., Bengtson, D. A., Decleir, W., and Jaspers, E., Eds., Universa Press, Wetteren, Belgium, 1987, 5. With permission.)

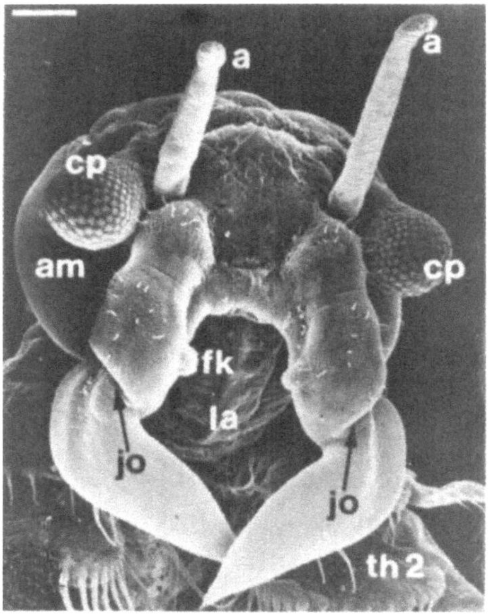

FIGURE 19. Cephalon of a Postlarval Stage-IV (L-16). Antennule (a); adult mandible (am); compound eye (cp); frontal knob (fk); secondary joint at the antenne (jo); labrum (la); thoracopod 2 (th 2); bar = 100 μm. (From Schrehardt, A., *Artemia Research and Its Applications*, Vol. I, Sorgeloos, P., Bengtson, D. A., Decleir, W., and Jaspers, E., Eds., Universa Press, Wetteren, Belgium, 1987, 5. With permission.)

the Postlarval Stage-III (L-15), -IV (L-16), and -V (L-17) when the ovisacs form a spine and ovisac button on each lateral side.

B. INTERNAL DEVELOPMENT[7]

1. Endoderm

The midgut elongates by mitoses occurring over its whole length, with a slight concen-

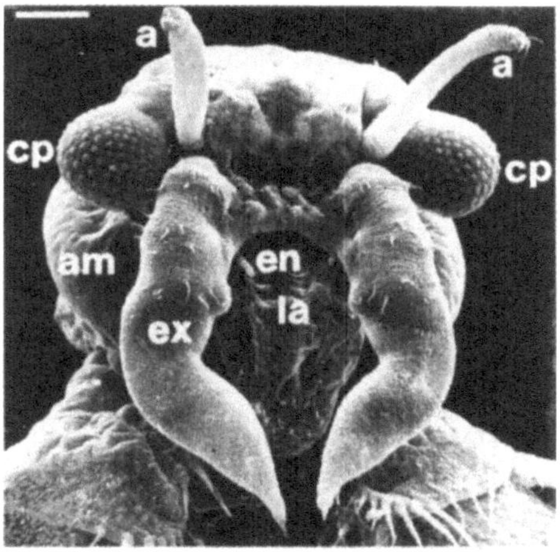

FIGURE 20. Cephalon of a Postlarval Stage-V (L-17). Antennule (a); antenne (an); compound eye (cp); frontal knob (fk); bar = 100 μm. (From Schrehardt, A., *Artemia Research and Its Applications*, Vol. I, Sorgeloos, P., Bengtson, D. A., Decleir, W., and Jaspers, E., Eds., Universa Press, Wetteren, Belgium, 1987, 5.)

tration in its distal part. The partition between the midgut and the hindgut breaks so that the gut is now open to the outside (Stage IV) and feeding can start. Most of the yolk has disappeared. The gut diverticles differentiate further; i.e., the gut wall bulges as it attaches to the ectoderm by tendons. The midgut tube is uniform over its entire length. As the yolk content of the midgut cells disappears their cytoplasm clears and an apocrine secretion can be observed together with the formation of a peritrophic membrane.

2. Mesoderm

Visceral mesoderm develops and differentiates independently from the somatic mesoderm. Elongated mesodermal cells discernible from the somatic mesodermal cells surround the whole alimentary canal. The circular and longitudinal muscles of the esophagus are derived from the preantennal mesoderm; its dilatator originates perhaps partly from the preantennal mesoderm, and partly from the second antennal mesoderm. The circular midgut musculature arises from antenna 1 and 2 mesoderm, mandible mesoderm, and metanaupliar mesoderm. The hindgut constrictors, longitudinal muscles, and dilatators originate from the metanaupliar mesoderm. From the moment the yolk is resorbed, fibrils start forming in the cytoplasm of the muscle cells. The gut muscles are striated but unlike trunk muscles they are mononucleated. The two giant mesodermal cells still lie on both sides of the endgut.

Somatic mesoderm is found in the naupliar as well as in the metanaupliar region. In the naupliar region it forms multilayered arcs; in the metanaupliar region it differentiates by cell division from the proliferation site as broad-paired lateral bands beneath the naupliar embryonic ectoderm. From these bands, cells group as somites which become separated. Formation of mesodermal somites goes together with the external appearance of segments.

In the naupliar region segmentation of the mesoderm starts in the Gastrulation Stage-II and occurs simultaneously over the whole region by a close conjunction of mesodermal and tendonal cells at a moment when many mitoses still occur. It ends when the mesodermal cell movements have ceased, the organs can be distinguished, and the ganglia of the ectoderm are obvious.

Benesch[7] describes the segmentation events of the metanaupliar region. His view is given here although it should be mentioned that his description of the formation of the maxillar segment is unclear: first it is described to develop as a single structure; later it seems divided into Maxilla I and Maxilla II. Moreover his timing differs from the timing Anderson[36] proposes, but Anderson's timing is at least partly outdated by the work of Schrehardt.[47]

The mesodermal somites which are individually formed during segmentation are thought to arise from the metanaupliar mesoderm[7] which is found as a U-shaped double layer of densely packed fusiform cells between ectoderm and entoderm on both sides of the hindgut and separated by a small gap from the mandible mesoderm. Anderson[36] limits the mesodermal budding zone to 6—7 mesoteloblasts on each side of the gut. He postulates that they bud off a band of mesodermal cells which grow ventrally by cell divisions to form the typically U-shaped sheet of the somite.

After hatching, the pear-shaped nauplius becomes conical due to stretching by cell transformation and division mitoses. Mitoses occur sporadically in the ectoderm and in the somatic mesoderm, but frequently in the gut muscles and the midgut region. The long cylindrical ectodermal cells become cubical; the smaller ones between the mandibular and the budding zone and the telson ectoderm become even smaller; the telson fold straightens. The hemocoel between mesoderm and gut, and between the budding zone and endgut, broadens. Visceral mesoderm of the gut separates from the budding zone of metanaupliar mesoderm.

During the embryonal period the metanaupliar mesoderm has accumulated sufficient material to form maxillar and first thoracic somites and to initiate the formation of the second thoracic somite. The third somite has to be formed entirely after hatching. Maxillar and first thoracic somites are the first to separate from the mesodermal budding zone. Later the somites are formed one by one. They are fused ventrally, open dorsally, and comprise 2 to 3 cell rows. Before the somites separate, prospective tendonal cells emerge laterally and ventrolaterally in the intersegmental region. Medially these tendonal cells do not appear. Mitoses in the mesoderm cause progressive enlargement of the previously formed somite, development of the somite fused to the budding mesoderm, and finally regeneration of the budding mesoderm. Mitoses in the ectoderm cause elongation and set the following somite free. In each developing segment, as well as in the budding mesoderm, mitotic waves proceed in a cranio-caudal direction. It is impossible to state whether ectoderm or mesoderm initiates the segmentation; in any case, initially, more mitoses occur in the ectoderm. This may be due to the fact that the ectodermal cells are smaller than the mesodermal cells; i.e., they do thus need more mitoses to obtain the same elongation. Later, during the growth and differentiation phase of the somites, the mesodermal cells become smaller, and the mitotic proportion changes. The segmentation furrow becomes externally visible when the dorsoventral muscle cells, which have first differentiated cranially in the somite, attach to tendonal cells (Figure 21). Caudally, this somite is still attached to the proliferation mesoderm. Its loosening is preceded by the formation of epidermal tendonal cells at its caudal end together with a straightening of the ectoderm by mitoses. After the insertion of the dorsoventral muscle cells, mitoses cause a bulge of the segmental ectoderm, while the mesoderm becomes multilayered and fills the gap before the ectoderm also becomes multilayered and forms ganglia.

In the naupliar region the preantennal mesoderm forms the esophagus muscle, the labrum levator muscle, the eyestalk adductors, the transverse eye muscle, the connective tissue of the brain, and the fat-storing cells of the labrum. The antennula mesoderm forms the antennula muscles, the connective tissue cells of the brain, and the primary blood cells. The antenna mesoderm forms the muscles, the connective tissue cells for the mandibular ganglion, and the mandibular gland (secondary blood-forming organ). The preantennal segments differentiate somewhat slower than the other naupliar segments, i.e., the levator muscle of the labrum and the muscles of the esophagus differentiate simultaneously with other naupliar

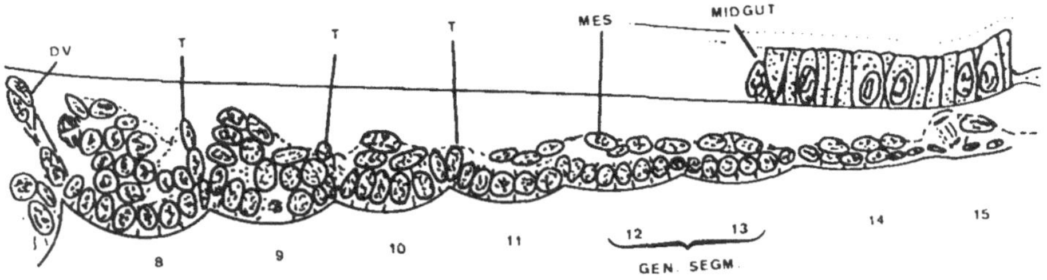

FIGURE 21. Sagittal section through the ventral side of a late larval stage. The sequence of the segmentation events is shown in the thoracic segments 8 to 11, the genital segments (GEN.SEGM) 12 and 13, and the abdominal segments 14 and 15. The dorsoventral muscle (DV) is differentiated in front of the 8th segment. Tendonal cells (T) delimit the segments 8 to 10. MES, mesoderm. (From Fränsemeier, L., *Z. Wiss. Zool.*, 152, 439, 1939.)

muscles but prior to the eye muscles, the fat-storing cells of the labrum, and the connective tissue of the brain. As the muscles of the naupliar region are segmental, their disposition can be compared to that of the maxillae and the thorax.

In the metanaupliar region the segmental somites start differentiation upon evagination of the thoracopod. They differentiate into heart and pericardial septum, connective tissue and fat-storing cells, muscular tissue, as well as coelom and segmental organs. After segmentation, the somites separate first ventrally, then divide into dorsal and ventral mesoderm. The dorsal mesoderm forms a fold which attaches to the gut muscles and which will form the heart and the pericardial septum. In the cranial part of the somite the dorsoventral muscles are first to differentiate and to attach dorsally and ventrally. Initially the somite consist of one layer of cells, later becoming multilayered by mitoses. The somite then divides to form three dorsoventral strings which are tied together dorsally by the heart, the pericardial septum, and the developing dorsal longitudinal muscles and ventrally, by the developing ventral longitudinal muscles and the thoraco-brachial limb muscles. The anterior string forms the cranial, dorsal, thoraco-brachial, limb muscles; the middle string forms connective tissue and the coelom; and the posterior string forms the caudal dorsal thoraco-brachial limb muscles and connective tissue.

3. Ectoderm

Just before emergence, the ectoderm forms a high cylindrical epithelium in the naupliar region, lowers immediately after the mandibular ectoderm, and becomes longer again in the proliferation zone. At the end of the proliferation zone a fold delimits the zone where lower cells form the telson ectoderm and reach the ectoderm of the proctodeum. After emergence the length of all ectodermal cells decreases. They differentiate into the integument and the exoskeleton; the nervous system; the esophagus and the hindgut, which are easily discerned from the midgut by the presence of longitudinal muscles; stronger circular muscles and a chitinous inner lining; the paired ductus ejaculatorius in the male; the unpaired uterus in the female; as well as the integumentary glands — i.e., the ectodermal glands of the labrum, the thoracopodal glands in the thoracic segments, the shell glands in the genital segments of the female, and the accessory glands in the genital segments of the male.

The larval integument was extensively studied by Freeman.[48-49] Characteristic features of the integument are the secretion of a cuticle to the outside and a basement membrane to the inside. Unlike most crustaceans the larval cuticle of the *Artemia* nauplius is much thinner than the underlying epidermal cells. It is divided into an outer epicuticle and a fibrous procuticle of 0.3 to 1.0 μm thickness. The cuticle has the same appearance in all the regions of the integument. In nauplii the epidermis is a monolayer of cells variable in thickness; i.e., in the head and antennal region the cells are thinner than in the developing thoracic and abdominal regions where they are cuboidal with nuclei that fill most of the volume of the cell. The tendonal cells, the seta-forming cells, and the salt gland or neck organ are differentiations of the integument. The salt gland or neck organ is discussed subsequently. The tendonal cells[49] are epidermal cells which are modified for the attachment of muscles or which act as an endoskeleton. It is not yet clear whether they stem from arthrodial membrane cells or from thoracopod bud cells. Benesch[7] finds them from the Gastrula Stage-II onward and ascribes an important role to them in the visual distinction of the segmentation. Their function is probably to form a permanent connection of the arthrodial membrane to the dorsal surface of the segment. The seta-forming cells[49] are epidermal cells that have reached the differentiated stage before emergence and hatching. Their fine structure has not yet been examined but light microscopic observations have shown that the nuclei of these cells are tetraploid. Benesch[7] does not describe the setal cells but shows them for the first time in his drawings of the Na-4 stage.

Freeman[48-49] brought a new insight in the segmentation phenomena in *Artemia*. He studied the segmentation events in the ectodermal cells during the first two instars and considered growth and differentiation as two distinct but related processes. His results indicate that segment morphogenesis in the thorax involves a spatio-temporal pattern of cell replication in the ventral epidermis. The division between the segments can be seen as a result of differential replication activities: regions of lower cell density due to minor replication activity will become intersegmental arthrodial membranes (places of articulation for the thoracopods) whereas regions with high mitotic activity will achieve a great cell density and evaginate. In addition to differences in amount of growth, he also stated a difference in the pattern of the orientation of the division axes in the segmenting and nonsegmenting regions. Although the epidermal cells are arranged in a hexagonal manner their nuclei appear to be arranged in rows or files running in an anterior-posterior direction. The orientation of the division axes influences the direction of the rows, which is shown to be a determining factor in the differentiation by the inhibitor bromodeoxyuridine. He also studied cell differentiation during thoracopod morphogenesis, an aspect which is scarcely dealt with by Benesch.[7] The cells on the ventro-lateral surface of the segment (the budding regions of the future appendages) are the most active during early development. The regions which will become the arthrodial membrane initially show no mitotic activity and consist of two transverse files of cells (later other cells are added). The nuclei of these cells are displaced to the side (probably by action of microtubules) in such a way that the nucleus-free zones of the cells in each file are contiguous and a region of pure cytoplasm separates the segments. The cells of the arthrodial membrane change from cuboidal to squamous, spindle shape while adjacent cells of the thoracopodal bud remain cuboidal. Cells of the thoracopodal bud replicate to a cell density 60% greater than that of the arthrodial membrane. This causes invagination of the arthrodial membrane and evagination of the thoracopod epithelium. Freeman presumes that molting cycle events could also play a role in the evagination, i.e., in larval *Artemia* apolysis proceeds in a cranio-caudal direction, so a loss of contact with the restrictive exoskeleton in the anterior regions may permit the region with greater density to curve slightly out and adjacent thinner regions to deform inwardly. The more posterior regions of the thorax would not be susceptible to deformation since the density of epidermal cells remains roughly equivalent in all areas of this region. Microtubules and microfilaments may play a role in the evagination process.

As the thoracopod develops, the endites, exites, and epipodites are differentiated from the central body of the appendage. Arthrodial membranes and tendonal cells also will form in each of these structures.

C. NECK ORGAN OR SALT GLAND

Ever since this "enigmatic organ" was described by Leydig[52] in *Branchipus*, controversy

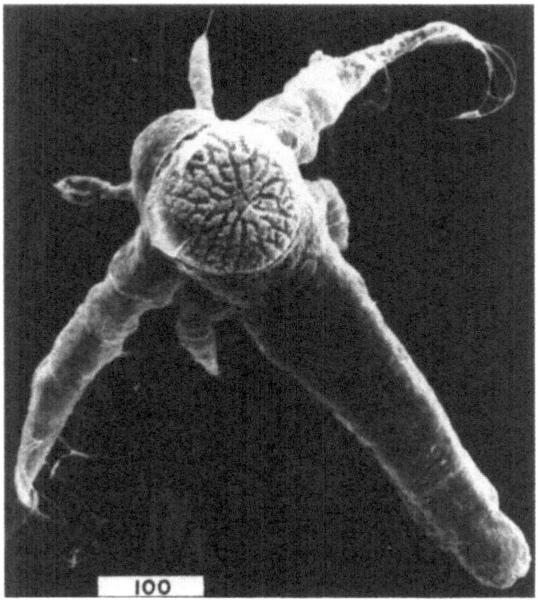

FIGURE 22. Dorsal view of an early stage nauplius showing the salt gland (arrow). The cuticular layer has been removed during preparation; bar = 100 μm. (From Hootman, S. R., Harris, P. J., and Conte, F. P., *J. Comp. Physiol.*, 79, 97, 1972. With permission.)

has risen over its function and various names have been offered to designate this organ, which appears early in embryonic life and regresses before adulthood. Spangenberg[53] and Claus[54] both see it as a shield, the first with a glueing, and the second with an undefined glandular function. Weisz[35] and Benesch[7] both merely consider it as an anchoring site for antennal and mandibular muscles. Using selective staining in living animals Dejdar[41] demonstrated a reducing activity in the neck organ of various phyllopods and concluded that it had a respiratory function. Since more recent investigators considered this reducing activity to be indicative of salt or chloride movements,[55-56] the neck organ is now regarded as a salt-secreting organ.[42,57-60] Evidence suggesting that the neck organ may be a salt-regulatory organ was obtained by micropuncture studies.[61] The finding of Na^+K-activated ATPase[62] has reinforced this view.

Conte et al.[42] and Hootman and Conte[60] define the neck organ as a dome-like gland situated atop the cephalothorax of the larva (Figure 22), composed of 50 to 60 cuboidal epithelial cells, much larger than the surrounding epidermal cells, having a large nucleus with two to five prominent nucleoli, a diffuse membranous cytoplasm, yolk platelets, and other storage products. Some ultrastructural aspects were determined by Kikuchi,[58] and Conte et al.[42] Hootman and Conte[60] studied the gland more thoroughly. Electron micrographs (Figure 23) show no indication of the typical tendonal cell morphology.[49] The ultrastructure is similar to that of the branchial gills of the adult described by Copeland.[63-64] As in the gills, dark and light cells were found initially.[42,58] Later, improved techniques only revealed one cell type[60,65] in which an apical, a central, and a basal zone can be discerned. The apical plasmalemma is infolded into irregular loops which are in contact with the cuticle. A major ultrastructural characteristic which suggests a salt secretory role is the smooth tubular network which fills the central cytoplasmic zone. Sinusoids from this labyrinth open directly into the hemocoel on the basal and lateral cell surfaces. Kikuchi[58] has shown that the extent of

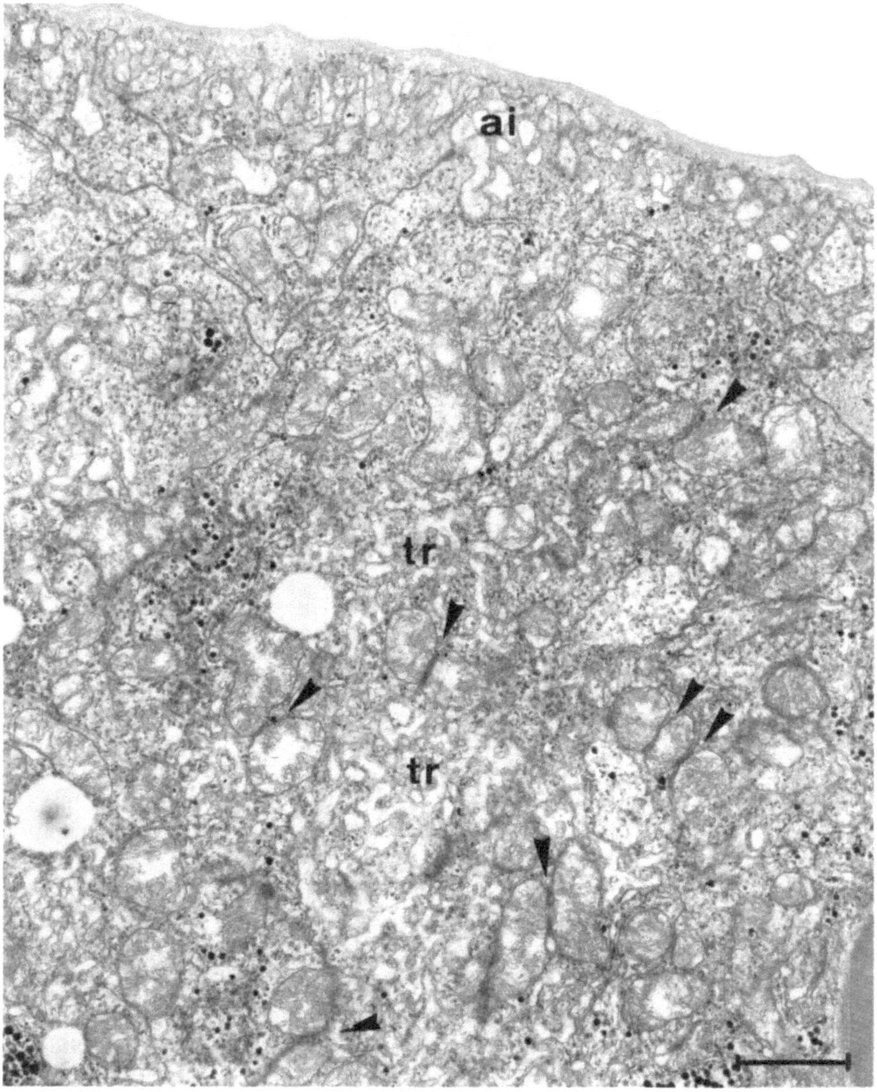

FIGURE 23. Electron micrograph of the apical zone of a transversally sectioned neck organ. Apical infoldings (ai); tubular reticulum (tr); arrows point to the presumed mitochondrial pumps; bar = 1 μm.

this network, both in the neck organ and in the gills, is related to the salinity of the culture medium. The most abundant organelles are the mitochondria. They often form aggregations in intimate contact with the tubular membranes reminiscent of mitochondrial pumps.[63-64] Narrow channels are provided for hemolymph penetration close to the mitochondria. $Na^{+}K$-ATPase is found on the apical and baso-lateral cell boundaries. An unusual feature of the neck organ epithelium is the absence of a basal lamina. Scanning electron microscopy of the organ in young nauplii[59] shows continued development of the gland by growth and enlargement of the epithelium from a flat cap-like structure into a hemispherical dome having deep channels through the epithelium and covered by a thin cuticular film. Zograf[66] reported a marked difference between the cells bordering the neck organ and the central cells. From differential staining he presumed the border cells to be nerve cells. The preemergence polyploidization of the salt-gland nuclei poses a problem to the prevailing opinions on DNA synthesis in encysted embryos.[49] Fränsemeier[34] and Weisz[35] indeed noted its appearance in hydrating cysts while Benesch[7] draws it for the first time in a late Stage G-II and then describes it in the Na-3 stage of the developing nauplius.

The salt glands are resorbed before the adult stage is attained.

REFERENCES

1. **Artom, A.**, La maturazione, la fecondazione e i primi stadii di sviluppo dell' *Artemia salina* Lin. di Cagliari, *Biologica* (Torino), 1, 24, 1908.
2. **Brauer, A.**, Zur Kenntniss der Reifung des parthenogenetisch sich entwicklendes Eies von *Artemia salina, Arch. Mikrosk. Anat.*, 43, 162, 1984.
3. **Petrunkewitsch, A.**, Die Reifung der parthenogenetischen Eier von *Artemia salina, Anat. Anz.*, 21, 256, 1902.
4. **Barigozzi, C.**, I fenomeni cromosomici delle cellule germinali in *Artemia salina* Leach, *Z. Zellforsch. B. Chromosoma*, 2, 550, 1943.
5. **Gross, F.**, Intranuclear spindle formation and mitosis in *Artemia salina, Nature*, 1, 985, 1934.
6. **Bowen, S. T.**, The genetics of *Artemia salina* I. The reproductive cycle, *Biol. Bull.*, 122, 25, 1962.
7. **Benesch, R.**, Zur Ontogenie und Morphologie von *Artemia salina* L., *Zool. Jahrb. Anat.*, 86, 307, 1969.
8. **Criel, G.**, Morphology of the genital apparatus of *Artemia*: a review, in *The Brine Shrimp Artemia*, Vol. 1, *Morphology, Genetics, Radiobiology, Toxicology*, Persoone, G., Sorgeloos, P., Roels, O., and Jaspers, E., Eds., Universa Press, Wetteren, Belgium, 1980, 75.
9. **Artom, C.**, Osservazioni preliminari sulla maturazione dell' *Artemia salina* di Cette, *Boll. Soc. Med. Chir.*, 43, 571, 1929.
10. **De Maeyer-Criel, G., Fautrez-Firlefyn, N., and Fautrez, J.**, Formation de la membrane de fécondation de l'oeuf d' *Artemia salina, W. Roux's Arch.*, 183, 223, 1977.
11. **Gross, F.**, Die Reifungs- und Furchungsteilungen von *Artemia salina* im Zusamenhang mit dem Problem des Kernteilungsmechanismus, *Z. Zellforsch.*, 23, 522, 1935.
12. **Fautrez-Firlefyn, N.**, Etude cytochimique des acides nucléiques au cours de la gamétogénèse et des premiers stades due développement embryonnaire chez *Artemia salina* L., *Arch. Biol.*, 62, 291, 1951.
13. **Fries, W.**, Die Entwicklung der Chromosomen im Ei von *Branchipus* Grub., und der parthenogenetischen Generationen von *Artemia salina, Arch. Zellforsch.*, 4, 44, 1909.
14. **Stefani, R.**, Differenze nel ciclo annuale tra biotipo amfigonico e biotipo partenogenetico nell' *Artemia salina* di Cagliari, *Riv. Biol.* (Perugia), 54, 457, 1961.
15. **Gross, F.**, Untersuchungen über die Polyploidie und die variabilität bei *Artemia salina, Naturwissenschaften*, 20, 962, 1932.
16. **Goldschmitt, E.**, Fluctuation in chromosome number in *Artemia salina, J. Morphol.*, 91, 111, 1952.
17. **Anteunis, A., Fautrez-Firlefyn, N., and Fautrez, J.**, L'accolement des pronuclei de l'oeuf d' *Artemia salina, J. Ultrastruct. Res.*, 20, 206, 1967.
18. **Jensen, A. C.**, Some observations on *Artemia gracilis*, the brine shrimp of Great Salt Lake, *Biol. Bull.*, 34, 18, 1918.
19. **Mathias, P.**, Biologie des crustacés phyllopodes, *Actual. Sci. Ind.*, 447, 1, 1937.
20. **Lochhead, J. H. and Lochhead, M. S.**, The egg shells of the brine shrimp *Artemia, Anat. Rec.*, Suppl. 75, 78, 1940.
21. **Dutrieu, J.**, Quelques observations biochimiques et physiologiques sur le développement d' *Artemia salina* Leach, Rendiconti degli Istituti Scientifici della Universita di Camerino, 9, 196, 1960.
22. **Dutrieu, J.**, Observations biochimiques et physiologiques sur le développement d'*Artemia salina* Leach, *Arch. Zool. Exp. Gén.*, 99, 1, 1960.
23. **Morris, J. E. and Afzelius, B. A.**, The structure of the shell and outer membranes in encysted *Artemia salina* embryos during cryptobiosis and development, *J. Ultrastruct. Res.*, 20, 244, 1967.
24. **Anderson, E., Lochhead, J. H., Lochhead, M. S., and Huebner, E.**, The origin and structure of the tertiary envelope in the thick-shelled eggs of the brine shrimp *Artemia, J. Ultrastruct. Res.*, 32, 497, 1970.
25. **Mazzini, M.**, Scanning electron microscope morphology and amino-acid analysis of the egg-shell of encysted brine shrimp *Artemia salina* Leach (Crustacea, Anostraca), *Monitore Zool. Ital.* (N.S.), 12, 243, 1978.
26. **Fautrez, J. and Fautrez-Firlefyn, N.**, Contribution à l'étude des glandes coquillières et des coques de l'oeuf d' *Artemia salina, Arch. Biol.* (Liège), 82, 41, 1971.

27. **Garreau de Loubresse, N.,** Etude chronologique de la mise en place des enveloppes de l'oeuf d'un crustacé phyllopode *Tanymastix lacunae, J. Microscop.*, 20, 21, 1974.
28. **De Chaffoy, D., De Maeyer-Criel, G., and Kondo, M.,** On the permeability and formation of the embryonic cuticle during development in vivo and in vitro of *Artemia salina* embryos, *Differentiation*, 12, 99, 1978.
29. **Belk, D.,** Embryonic cuticles of *Artemia* during diapause and hatching: insights from comparison with other Branchiopoda, *J. Crustacean Biol.*, 7, 691, 1987.
30. **Jardel, J.-L.,** Ultrastructural modifications associated with anhydrobiosis in the brine shrimp *Artemia* (L.), Ph.D. thesis, Monash University, Australia, 1986.
31. **Vallejo, G. C., Günter Sillero, M. A., and Marco, R.,** Mitochondrial maturation during *Artemia salina* embryogenesis. General description of the process, *Cell. Mol. Biol.*, 25, 113, 1979.
32. **Morris, J. E.,** Hydration, its reversibility, and the beginning of development in the brine shrimp, *Artemia salina, Comp. Biochem. Physiol.*, 39A, 843, 1971.
33. **Bes, J.-C., Caratero, C., Caratero, A., Gaubin, Y., and Planel, H.,** Aspects ultrastructuraux de l'embryon dormant d'*Artemia, Bull. Assoc. Anat.*, 67, 255, 1983.
34. **Fränsemeier, L.,** Zur Frage der Herkunft des metanaupliales Mesoderms und die Segmentbildung bei *Artemia salina* Leach, *Z. Wiss. Zool.*, 152, 439, 1939.
35. **Weisz, P. B.,** The histological pattern of metameric development in *Artemia salina, J. Morphol.*, 81, 45, 1947.
36. **Anderson, D. T.,** Larval development and segment formation in the branchiopod crustaceans *Limnadia stanleyana* King, (Conchostraca) and *Artemia salina* L. (Anostraca), *Austr. J. Zool.*, 15, 47, 1967.
37. **Anderson, D. T.,** Embryology and Phylogeny in Annelids and Arthropods, International Series of Monographs in pure and applied Biology, Zoology Division, Pergamon Press, 50, 263, 1973.
38. **Clegg, J. S.,** The origin of trehalose and its significance during the formation of encysted dormant embryos of *Artemia salina, Comp. Biochem. Physiol.*, 14, 135, 165.
39. **Nakanishi, J. H., Okigaki, T., Kato, H., and Iwasaki, T.,** Cytological studies of *Artemia salina*, II. Desoxyribonucleic acid (DNA) content and the chromosomes in encysted dry eggs and nauplii, *Proc. Jpn. Acad.*, 139, 306, 1963.
40. **Olson, C. S. and Clegg, J. S.,** Cell division during the development of *Artemia salina, W. Roux's Arch.*, 184, 1, 1978.
41. **Dejdar, E.,** Die Korrelationen zwischen Kiemsäckchen und Näckenschild bei Phyllopoden, *Z. Wiss. Zool.*, 136, 422, 1930.
42. **Conte, F. P., Hootman, S. R., and Harris, P. J.,** Neckorgan of *Artemia salina* nauplii. A larval salt gland, *J. Comp. Physiol.*, 80, 239, 1972.
43. **Heath, H.,** The external development of certain phyllopods, *J. Morphol.*, 38, 453, 1924.
44. **Reynier, M.,** Recherches sur le développement et la reproduction d' *Artemia salina, Bull. Soc. Sci. Nancy*, 18, 155, 1959.
45. **Hentschel, E.,** Die postembryonalen Entwicklungsstadien von *Artemia salina* Leach bei verschiedenen Temperaturen (Anostraca, Crustacea), *Zool. Anz.*, 180, 372, 1968.
46. **Blake, R. W.,** The development of the brine shrimp *Artemia salina* (L.): a morphometric approach, *Zool. J. Linn. Soc.*, 65, 255, 1979.
47. **Schrehardt, A.,** Scanning electron microscopic study of the post-embryonic development of *Artemia*, in *Artemia Research and Its Applications*, Vol. I., Sorgeloos, P., Bengtson, D. A., Decleir, W., and Jaspers, E., Eds., Universa Press, Wetteren, Belgium, 1987, 5.
48. **Freeman, J. A.,** Epidermal cell proliferation during thoracic development in larvae of *Artemia, J. Crust. Biol.*, 6, 37, 1986.
49. **Freeman, J. A.,** The integument of *Artemia* during early development, in *Biochemistry and Cell Biology of Artemia*, MacRae, T. H., Bagshaw, J. C., and Warner, A. H., Eds., CRC Press, Boca Raton, FL, 1989, 233.
50. **Kaestner, A.,** *Lehrbuch der speciellen Zoologie*, Vol. 1/3 Arthropoda, Jena, East Germany, 1967.
51. **Wolfe, A. F.,** A light and electron microscopic study of the frontal knob of *Artemia* (Crustacea, Branchiopoda), in *The Brine Shrimp Artemia*, Vol. 1, *Morphology, Genetics, Toxicology*, Persoone, G., Sorgeloos, P., Roels, O., and Jaspers, E., Eds., Universa Press, Wetteren, Belgium, 1980, 117.
52. **Leydig, F.,** Über *Artemia salina* und *Branchipus stagnalis, Z. Wiss. Zool.*, 3, 280, 1851.
53. **Spangenberg, F.,** Zur Kenntniss van *Branchipus stagnalis, Z. Wiss. Zool.*, Suppl. 25, 1, 1875.
54. **Claus, C.,** Untersuchungen über die Organisation und Entwicklung von *Branchipus* und *Artemia* nebst vergleichenden Bemerkungen über anderen phyllopoden, *Arb. Zool. Inst.*, Wein, 6, 267, 1886.
55. **Koch, H. J.,** The absorption of chloride ions by the anal papillae of diptera larvae, *J. Exp. Biol.*, 15, 152, 1938.
56. **Wigglesworth, V. B.,** The regulation of osmotic pressure and chloride concentration in the hemolymph of mosquito larvae, *J. Exp. Biol.*, 15, 235, 1938.

57. **Conte, F. P., Peterson, G. L., and Ewing, R. D.,** Larval salt gland of *Artemia salina* nauplii. Regulation of protein synthesis by environmental salinity, *J. Comp. Physiol.*, 82, 277, 1973.
58. **Kikuchi, S.,** Three-dimensional network of a tubular system in the salt-transporting cells of the gill and the neck organ of *Artemia salina* (brine shrimp), *Ann. Rep. Iwate Med. Univ. Sch. Liberal Arts Sci.*, 7, 15, 1972.
59. **Hootman, S. R., Harris, P. J., and Conte, F. P.,** Surface specialization of the larval salt gland in *Artemia salina* nauplii, *J. Comp. Physiol.*, 79, 97, 1972.
60. **Hootman, R. and Conte, F. P.,** Functional morphology of the neck organ in *Artemia salina* nauplii, *J. Morphol.*, 145, 371, 1975.
61. **Russler, D. and Mangos, J.,** Micropuncture studies of the osmoregulation in the nauplius of *Artemia salina, Am. J. Physiol.*, 234, R216, 1978.
62. **Conte, F. P., Droukas, P. C., and Ewing, R. D.,** Development of sodium regulation and de novo synthesis of Na^+K- activated ATPase in larval brine shrimp, *Artemia salina, J. Exp. Zool.*, 202, 3, 1977.
63. **Copeland, E.,** Salt transport organelle in *Artemia salenis* (brine shrimp), *Science*, 151, 470, 1966.
64. **Copeland, D. E.,** A study of salt secreting cells in the brine shrimp *(Artemia salina), Protoplasma*, 63, 363, 1967.
65. **Lowry, R. J. and Conte, F. P.,** Morphology of isolated crustacean larval salt glands, *Am. J. Physiol.*, 248, R709, 1985.
66. **Zograf, N.,** La calotte cervicale chez les nauplius de l' *Artemia salina, C. R. Acad. Sci.*, 141, 903, 1905.

Chapter 8

THE EXTRACELLULAR HEMOGLOBINS OF *ARTEMIA*: STRUCTURE OF THE OXYGEN CARRIER AND RESPIRATION PHYSIOLOGY

L. Moens, G. Wolf, M. L. Van Hauwaert, I. De Baere, J. Van Beeumen, S. Wodak, and C. N. A. Trotman

TABLE OF CONTENTS

I. INTRODUCTION: THE EXTRACELLULAR HEMOGLOBINS OF THE INVERTEBRATES

In contrast with the hemoglobins of the vertebrates, which are almost invariably intracellular and tetrameric, the intra- and extracellular hemoglobins of the invertebrates show a wide variety in their molecular size (M_r ~16,000 up to ~1.7 × 10^6) and architecture.[1-5] With the exception of the hemoglobins of the chironomid larvae,[6] practically all of the extracellular hemoglobins have a high M_r which is probably necessary to avoid elimination from the hemolymph by excretory processes and an excessive osmotic pressure.[4] Despite the heterogeneity of invertebrate hemoglobins, Svedberg and Hedenius[7] suggested that all these pigments are composed of myoglobin-like polypeptide chains of M_r ~16,000 containing one heme group and able to bind oxygen reversibly. Polypeptide chains or fragments of much longer chains having these characteristics (M_r ~16,000; one heme) were defined by Vinogradov[5] as "heme-binding domains". Based in the number of domains and subunits in the native molecule, the invertebrate extracellular hemoglobins can be classified into four groups:

1. Single domain, single-subunit hemoglobins consisting of a single polypeptide chain, containing one heme group and having an M_r ~16,000 (*Chironomus*)
2. Single-domain, multisubunit hemoglobins consisting of aggregates of monomeric subunits, some of which are disulfide bonded (*Annelida*)
3. Two-domain, multisubunit hemoglobins consisting of aggregates of dimeric polypeptide chains (M_r 30,000 to 40,000), each containing two heme-binding domains (*Arthropoda, Nematoda*)
4. Multidomain, multisubunit hemoglobins consisting of two or more polypeptide chains each comprising from eight to twenty heme-binding domains (*Arthropoda, Mollusca*)

Although the majority of Crustacaea have hemocyanin as an oxygen-carrier, extracellular hemoglobins are mainly found in the hemolymph of the Branchiopoda. The carapaced branchiopods, such as *Lepidurus, Daphnia, Cyzicus*, and *Caenestheria*, all have hemoglobins with molecular masses between 250,000 and 800,000, consisting of 10 to 24 two domain (M_r 30,000 to 40,000) subunits.[8,11] The carapaceless branchiopods, including *Artemia, Parartemia*, and *Streptocephalus*, have hemoglobins of M_r 240,000 to 300,000 and consist of two multidomain (six to eight) subunits. Similar hemoglobins, with M_r of 1.5 to 1.75 × 10^6 and containing several multidomain (10 to 20) subunits, are also found in gastropod snails and clams.[12-15] The *Artemia* hemoglobins thus represent a vast group of invertebrate extracellular hemoglobins on which very little information is available. In this review we discuss their structure, ontogenetically programmed synthesis, and function.

II. THE STRUCTURE OF THE OXYGEN CARRIER

A. PHYSICOCHEMICAL CHARACTERIZATION AND SUBUNIT STRUCTURE

The isolation of intact hemoglobins from *Artemia* is greatly hampered by the presence of a variety of proteases present in crude extracts of the total organism.[16-24] The use of protease inhibitors, low temperature, and speed of preparation cannot completely inhibit their activity.[25] This results in the presence of variable amounts of nicked molecules that partially copurify with the native hemoglobins.[25,26] However, starting from fresh *Artemia*, carefully purified and stored (−70°C), intact hemoglobins can be isolated by a procedure which has been applied successfully to several other invertebrate hemoglobins.[27,28]

Briefly, hemoglobin is precipitated from a crude extract by polyethyleneglycol or ammonium sulphate and freed of major contaminants by DEAE-Sephadex A, 50 chromatog-

TABLE 1
Physicochemical Characteristics of the *Artemia* Hemoglobins

	Native hemoglobin			
	Hb I	**Hb II**	**Hb III**	**Ref.**
$S_{20,w}$	11.3—11.7	11.3—11.7	11.3—11.7	37
$D_{20,w}$ (cm^2/sec)	—	(3.7 ± 0.1) 10^{-7}	—	37
v (cm^3/g)	—	0.737	—	37
Form factor (f/f°)	—	1.03	—	37
Axial ratio	—	1.7	—	37
EM dimension (Å)	—	120 × 70	—	38
M_r (× 10^{-3})	230—270	240—270	230—270	17,19,37,39
pI	5.6	5.7	5.9	18,26
Heme content (%)	—	3.6	—	17,26,42
Iron content (%)	—	0.29	—	17,26,42
Genotype	α_2	$\alpha\beta$	β_2	41
		Globin chains		
M_r	α = 131,800		β = 130,300	41
$S_{20,w}$		α,β = 6,3		38
Min M_r/Min AA composition		17,500		17,19

raphy using a stepwise elution. Glycoproteins are separated from the hemoglobins by affinity chromatography on a Concanavalin A Sepharose 4B column.[16-18,29,30] The major glycoproteins in hemoglobin fractions are the artemocyanins, blue-green pigments carrying a biliverdin chromaphore.[31-35] The majority of artemocyanin 1, which is a zymogen (M_r = 1.2 10^6; polypeptide M_r = 104,000/87,000) is strongly bound on to the affinity column whereas the hemoglobins and the smallest polypeptide of artemocyanin 2 (M_r = 67,000) which is a latent protease are not retained.

The latter is separated from the hemoglobins by preparative polyacrylamide gel electrophoresis under native conditions.[29,30] The three phenotypes, Hb, 1, 2, and 3 can be separated by gradient elution from a DEAE-column using low or high pressure techniques.[16-18,36] The three hemoglobin phenotypes and the smallest polypeptide from artemocyanin 2 are separated by a single preparative polyacrylamide gel electrophoresis run. However, this step is a limiting factor for large-scale preparations which are necessary for structural work. Some physicochemical parameters of the *Artemia* hemoglobins are summarized in Table 1.

The apparent M_r and sedimentation coefficients of the native hemoglobins have been estimated to be 230,000 to 270,000 and 11.3 to 11.7 S, respectively.[17,19,37,38] Iron and heme content indicate one protoporphyrin IX for every M_r 18,000, suggesting the presence of functional units of that size.[26] This value is confirmed by the minimum M_r of 17,000 calculated from amino acid composition.[17,19,26]

Electron microscopic observations of the hemoglobin molecules reveal the presence of two morphologically distinct structures: a circular lobed one with a diameter of 120 ± 10 Å and a rectangular one with dimensions 120 ± 10 Å × 70 Å. A possible interpretation is that the former is a top view of the hemoglobin molecule and the latter a side view of two, stacked, lobed discs.[38] The axial ratio and form factor are in agreement with this observation.[37]

The amino acid composition of the hemoglobins is comparable with those of other invertebrate hemoglobins[1,4,27] showing a high Asp, Glu, Leu, Val and Ala content but a low Cys content. The latter is in agreement with the amino acid composition of all other Mollusc

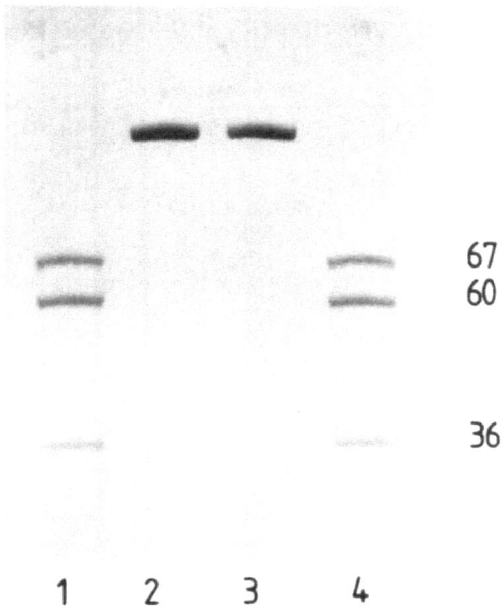

FIGURE 1A. Subunit structure of the *Artemia* hemoglobin. SDS-polyacrylamide gel electrophoresis of Hb 2 purified in the presence of protease inhibitors. A 12.5% gel according to Laemmli[43] was used; lanes 1 and 4: M_r markers; lanes 2 and 3: Hb 2.

and Arthropod extracellular hemoglobins suggesting that a high M_r is obtained without disulfide bridges between low M_r chains.

Due to the activity of proteases, especially artemocyanin 2 (peptide M_r = 67,000), the M_r of the globin chains has long been controversial. When carefully purified hemoglobin is denatured in 6 *M* guanidine-HCl, SDS or at extreme pH values (pH $\leqslant$ 4; pH $\geqslant$ 11), a M_r for the globin chains of 118,000 to 131,000 and a sedimentation coefficient of $S_{20,w}$ = 6.17—6.30 has been determined by gel filtration, electrophoresis, and analytical ultracentrifugation[19,25,26,30,39] (Figure 1A). However, when partially nicked hemoglobin molecules are analyzed, together with the authentic chains, a series of polypeptides which are simple multiples of M_r 15,000 to 17,000 are revealed, and a limit degradation product of that size accumulates. The latter is in accordance with the minimum M_r calculated from amino acid composition, heme, and iron content (Table 1).[19,26]

As total inhibition of proteolytic activity is extremely difficult, variable amounts of this degradation product are commonly observed in routinely purified hemoglobin samples.

The high M_r of the globin chains was confirmed by the translation of naupliar poly A^+ RNA *in vivo* (*Xenopus* oocytes) and *in vitro* (rabbit reticular lysate) followed by immunoprecipitation, showing authentic globin chains of M_r 130,000.[39,40]

Analysis of the Hb 2 with high resolution electrophoresis techniques (SDS-PAGE and Triton X-100 − urea PAGE) and HPLC reveals the presence of two different globin chains, α and β.[41] The α chain is slightly but consistently larger and more acidic than the β chain, but elutes later from a C_4 column (Figure 1B and C). Comparison of the globin composition of Hb 1 and Hb 3 clearly shows that they are the homo-dimers of the α and β chain respectively whereas Hb 2 is the $\alpha\beta$ heterodimer. This subunit structure is confirmed by *in vitro* reconstruction experiments and by the immunological cross-reactivity of Hb 1 and Hb 3 with an affinity purified anti-Hb 2 antibody.[41] In addition the presence of two different

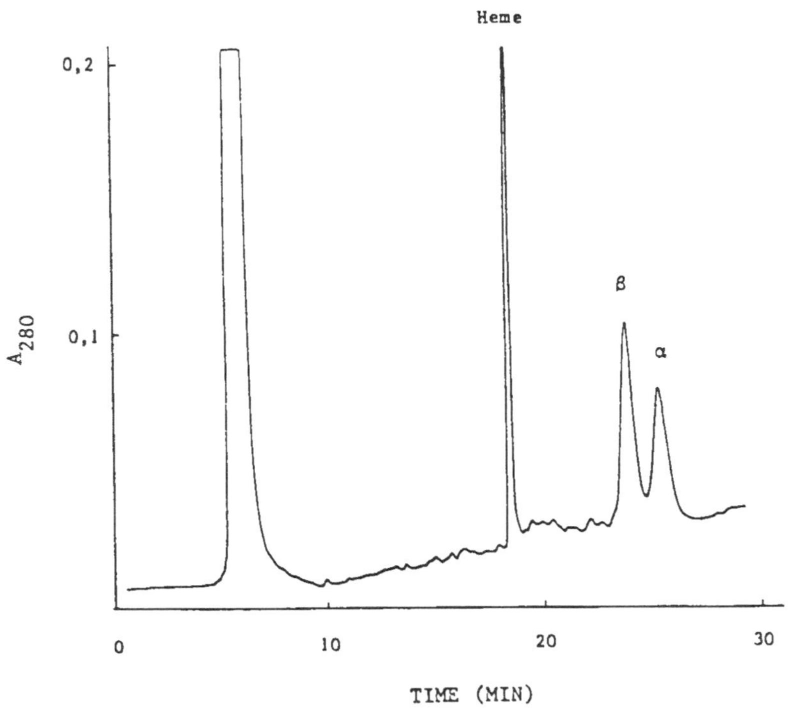

FIGURE 1B. Separation of the globin chains by reversed phase chromatography. Chromatography of guanidinium-HCl denatured Hb 2 was performed on a 300 Å C_4 column in 0.1% trifluoro-acetic acid using an acetonitrile gradient.

genetic loci and the different reaction of Hb 1 and Hb 3 at low oxygen pressure is consistent with the presence of two different globin types. Hemoglobin preparations of *Artemia* of different geographical and ecological origin (Tientsin, China; Great Salt Lake, U.S.; Chaplin Lake, Canada; Macau, Brazil; and Galera Zamba, Columbia) were compared by different electrophoretic techniques in order to detect possible structural variants. Separation by charge, using polyacrylamide gel electrophoresis and cellulose acetate electrophoresis[16] shows that, although in different ratios (Part 2), Hb 1 and Hb 2 are detectable in all samples and have a similar relative mobility with exception of Hb Tientsin and Hb Macau. The latter hemoglobins are more negatively charged, suggesting possible chain differences (Figure 1D). Analysis of the globin chains by SDS-PAGE confirms this observation. Under conditions where both chains can be separated, only a single band was detected for Hb in Tientsin, Macau, and Galera Zamba confirming the existence of structural variants in *Artemia* globins.

B. GLOBIN STRUCTURE

Analysis of the hemoglobins under native and denaturing conditions clearly shows that the molecule is a dimer of two very similar globin chains of M_r 130,000 (Figure 1). Heme and iron analysis, as well as minimum M_r based on amino acid analysis indicates the presence of eight structural and functional units (domains) of about M_r 16,000 (Table 1). Globin chains with similar characteristics are found in mollusc and arthropod extracellular hemoglobins.[1,2,5]

Accepting the presence of eight myoglobin-like domains in the M_r 130,000 globin chains we can expect that the will form conformationally compact structures separated by stretches of more unfolded "linkers" which will be more accessible for proteolytic enzymes.[45] Treatment of native hemoglobins with different proteases (trypsin, chymotrypsin, and subtilisin)

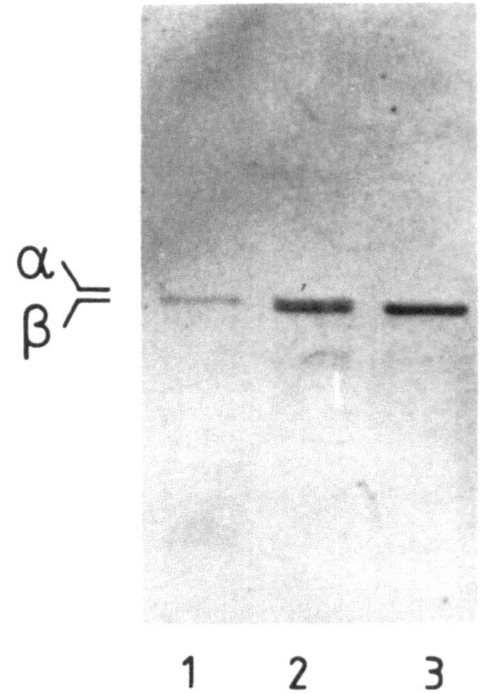

FIGURE 1C. Separation of the globin chains by SDS polyacrylamide gel electrophoresis. A 7.5% gel containing 0.05% *N-N'*-bis-methylene-acrylamide according to Laemmli[43] was used. Gels were run at 12.5 mA; lane 1: Hb 1; lane 2: Hb 2; lane 3: Hb 3.

creates a series of stable fragments which are more or less integral multiples of M_r 16,000 (Table 2).[44,45] The relative amount of each fragment depends mainly on the enzyme/substrate ratio being used. However, increasing proteolytic activity results in a decrease in the M_r of the fragments and the partial accumulation of a temporally stable set of fragments of M_r 15,000 to 17,000 representing single domains.[45] At a low enzyme/substrate ratio (1/150; 120 min), trypsin, in contrast to subtilisin, preferentially cleaves the globin chains into fragments of 80,000 and 50,000. This indicates the presence of a region in the structure of the intact hemoglobin that is very susceptible to this enzyme.[44] This pattern, which is also observed after cleavage of other high M_r globin chains,[1,2,5] is comparable with the random scission of a polymer into its monomers.[46]

The majority of globins sequenced show a conservative Trp residue at position A_{12}.[47] Cleavage of the *Artemia* hemoglobins with *N*-bromosuccinimide or BNPS-skatole [2-(2-nitrophenylsulfenyl)-3-methyl-3-bromo-indolenine] creates a series of fragments similar to that obtained by enzymatic cleavage (Table 2).[41] This result not only confirms the presence of M_r 16,000 domains in the *Artemia* globin chains, but also confirms the presence of a conserved Trp residue in them.

Limited proteolysis of the native hemoglobins was optimized for subtilisin in terms of a maximal generation of single domains. This was obtained by cleavage at 25°C for 60 min at an enzyme to substrate ratio of 1/150.[45] The cleaved molecules were dissociated in alkaline conditions (pH 10.5) and one of the fractions, (fraction E) containing single domains (M_r = 16,000) was selected from the resulting fragment mixture by gel filtration on a Sephacryl S200 column. The properties of fraction A are similar to those of the intact hemoglobin

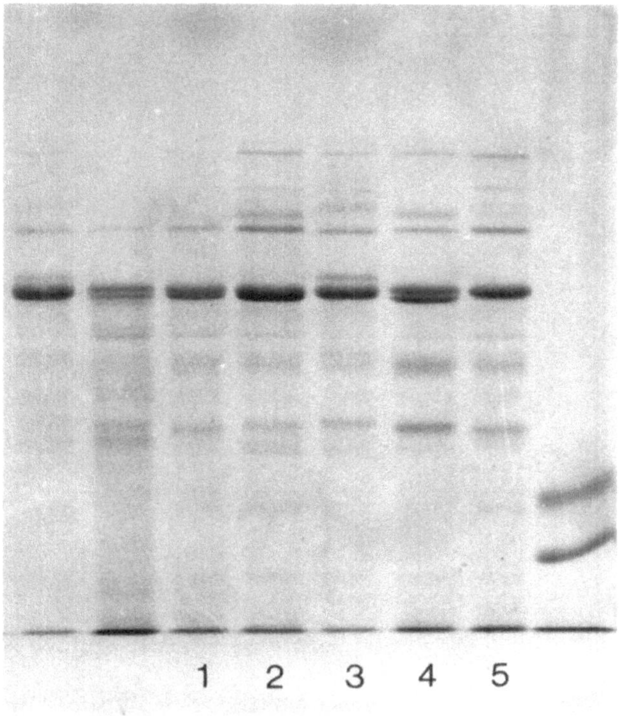

FIGURE 1D. Separation of the globin chains from *Artemia* from different geographical and ecological origins. The same gel system as in Figure 1C was used; lane 1: Chaplin Lake, Canada; lane 2: Macau, Brazil; lane 3: Tientsin, China; lane 4: Great Salt Lake, U.S.; lane 5: Calera-Zamba, Columbia.

TABLE 2
M_r of the Fragments Resulting from Enzymatic and Chemical Cleavage of *Artemia* Hemoglobin

	Relative Concentration of Fragments			Theoretical Size	
Experimentally determined $M_r \times 10^{-3}$ (SDS-PAGE)[43]	Enzymatic cleavage: Tryptic[44]	Enzymatic cleavage: Subtilisin[45]	Chemical cleavage *N*-Bromo Succinimide[41]	Multiple of M_r 16,000	$M_r \times 10^{-3}$
130	+	—	—	8	128
~112	+	Trace amt.	Trace amt.	7	112
~97	+	Trace amt.	Trace amt.	6	96
75—84	+ + + +	+	—	5	80
62—67	+ +	+ +	+	4	64
50—55	+ + + +	—	+	—	—
46—49	—	+ +	+ + +	3	48
32—35	—	+ + +	+ + +	2	32
22—25	+ + +	+ +	+	—	—
18—19	+ +	—	+	—	—
16—17	+ + +	+ + + +	+ + + +	1	16
13—15	+	+ +	+	—	—
≤10	+ +	Trace amt.	Trace amt.	—	—

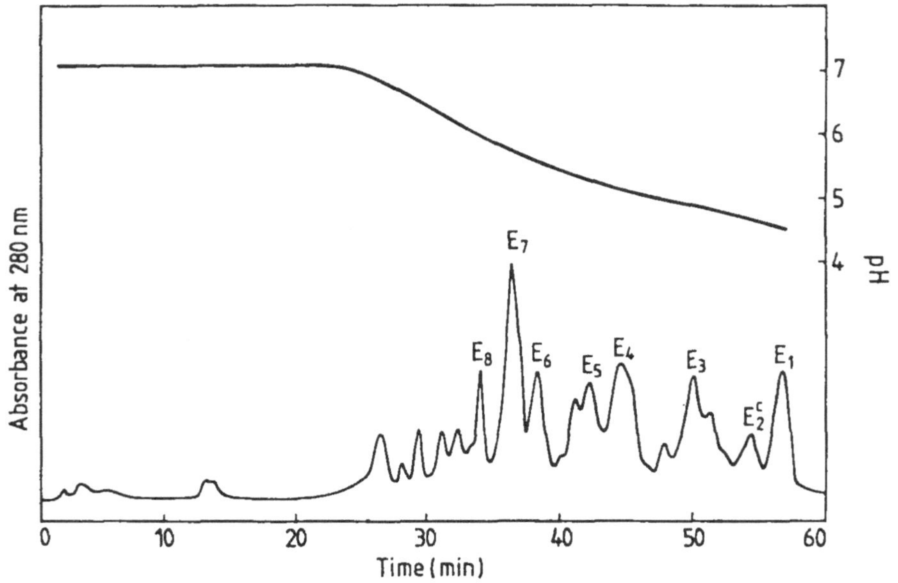

FIGURE 2. Separation of a monodomain fraction (Fraction E) by chromatofocusing. Chromatofocusing was performed on a Mono P HR 5/20 column using 0.025 *M* bis-tris-iminodiacetic acid at pH 7.1 and 9% polybuffer 74 iminodiacetic acid at pH 4.0 respectively, as starting and elution buffer.

(Table 1). It is recognized by a polyclonal antibody raised against Hb 2 indicating the presence of common immunological determinants. It contains one heme group for every M_r 16,000 and its amino acid composition is not significantly different from that of the intact molecules. Finally, it binds oxygen reversibly and its absorption maxima are indistinguishable from any other hemoglobin. This strongly suggests that fraction E represents the structural and functional units from which the globin chains are constructed.[46]

Analysis of fraction E under native and denaturing conditions indicates the presence of a population of single domains, each being slightly different in M_r and charge. This heterogeneity might well be artificially generated by the subtilisin digestion or may reflect a heterogeneity within the domains or linker regions within the authentic globin chains. Purification of the domains to homogeneity was dependent on the preceding elimination of the contaminating and copurifying subtilisin activity. This was accomplished by ion exchange chromatography of fraction E on a DE52 column at pH 7.6.[48] Preparative separation of the domains was realized by isoelectrofocusing on granulated gel or by chromatofocusing on a mono P column.[48] (Figure 2) Chromatofocusing reveals at least eight fractions with an apparent pH of elution between 4.6 to 5.98 (Figure 2, Table 3). Analysis of these fractions by two dimensional electrophoresis[49] reveals a single spot for domains E1 and E7, thus illustrating their relative purity.

C. AMINO ACID SEQUENCE OF THE GLOBIN CHAINS AND THEIR DOMAINS

The expected chain length of the *Artemia* globin chains is about 1,200 amino acid residues. Their sequence determination is correspondingly more complex than that of vertebrate globins and has therefore been approached by sequencing both at the protein and at the cDNA level.

1. Sequencing at the Protein Level

The amino acid sequence of domains E1 and E7, purified as described in Section 2,[48] was determined by automated Edman degradation of the aminoterminal segment[50] and by

TABLE 3
Apparent Isoelectric Points and Molecular Weights of the *Artemia* Hemoglobin Domains Separated by Chromatofocusing

Fraction	pH of elution (Fig. 2)	$M_r \times 10^{-3}$ (Ref. 43)
E1	4.60	16
E2	4.70	16
E3	4.92	16
E4	5.11	17
E5	5.38	16
E6	5.60	17
E7	5.79	16
E8	5.98	16/17

```
                      A1   5    10   15      B1   5    10   15   C1   5     CD1 5     D1   5    E1   5    10   15   20 EF1 5
                      :    :    :    :       :    :    :    :    :  :      :   :     :    :    :    :    :    :    :  :   :
MYWHP              VL SEGEWQLVLHVWAKVE A  DVAGHGQDILIRLFKS HPETLEK FDRFKHLK TEAEMKA SEDLKKHGVTVLTALGAILK K   K
HBHU              VHL TPEEKSAVTALWGKV     NVDEVGGEALGRLLVV YPWTQRF FESFGDLS TPDAVMG NPKVKAHGKKVLGAFSDGLA HL  D
E1        SLERVDPITGL SGLEKNAILDTWGKVR G  NLQEVGKATFGKLFAA HPEYQQM FRFFQGVQ LAFLVQ  SPKFAAHTQRVVSALDQTLL ALNRP
E7          GLVAP KAL TALEKQSIQDIWTILK AV GLEFLQVKMFGKLFAD HPEYKAH FDNFLTAI FSVAEDL VPKLRAHLHRVIDAFDLVIF AL  G
ALFA AR       AE VSGI LVSDKATIKRTWSIVN     DLPSFGRNVFLSNFAA KPIYDXX
BETA AR     ` AE IRGI LXDSKMTIXLTXAQVT     DLPSFGNNVDASVRAA KXEYDIG F
E6            VDVITGL SLXEKNAILHTWGKVK     PNLXXGGNKFEK

                 10   F1   5    10 FG1  G1   5    10   15     GH1 5 H1   5    10   15   20   25   HC1
                 :    :    :    :  :    :    :    :    :      :   : :    :    :    :    :    :    :
MYWHP      GHHEAE LKPLAQSHAT KHKI PIKYLEFISEAIIHVLHSR HPGDF GADAQGAMNKALELFRKDIAAKYKEL GYQG
HBHU        NLKGT FATLSELHCD KLHV DPENFRLLGNVLVCVLAHH FGKEF TPPVQAAYQKVVAGVANALAHKYH
E1        SDQFVYM IKELGLDHIN RGTD RSFVEYLKESLGDSVDEFT TVQSF GAVIVNFLNEGLRQA
E7        RESLRGS LKDLGIFHTG RDIV DPVESLTGFKLMVAVIEEG LDT F   RAVPEYSKGLEGRFGNVDNINENA PFR
ALFA AR
BETA AR
E6
```

FIGURE 3. Alignment of the amino acid sequences of domains E1 and E7, and of the aminoterminal sequence of domain E6 and the intact α and β chains, with sperm whale myoglobin and the human β chain. The standard numbering system based on myoglobin is used.

the manual sequencing[51] of peptides obtained by cleavage with trypsin (E1 and E7), chymotrypsin (E1 and E7), *Staphylococcus aureus* V8 protease (E1) and thermolysin (E1).[52-54] In addition the amino-terminal sequence of another domain E6 and of the intact chains purified from Hb 2 by reversed phase chromatography has also been obtained.[53] Details of the methods have been published.[52] Comparison with all sequences in the NBRF data bank showed the sequences obtained to be most closely related to myoglobin and β chain hemoglobins, thus belonging to the globin family. As homologous proteins have regions that retain the same general fold and regions where the folds differ, the sequences were interpreted in terms of the myoglobin fold.[55] An alignment was thus constructed, using existing programs[56,57] and taking into account the conserved residues known from the Bashford templates and from X-ray crystallo-graphic data.[47,55] The consensus alignment is shown in Figure 3. Helices A to H could be identified unequivocally in domains E1 and E7. The greatest discrepancy in length and in similarity occurred at the loops between the helices and in the H region.

The intact globin chains both have accessible amino terminals and clearly show the motifs of the A, B, and C helices.[53] Compared to the vertebrate globins both chains have an aminoterminal extension of four residues. Similar extensions, up to ten residues[47] are commonly observed in many other invertebrate globins. Although the α and β chains show extensive similarity, their specific differences suggest the existence of two structurally different globin chains. This confirms earlier physicochemical, genetic, and physiological observations.[58] The aminoterminal sequences of the domains E1, E6, and E7 are different from those of the α and β chains. Consequently, they are not aminoterminally located in the intact globin chains.

During sequencing of the domains E1 and E7 several PTH amino acids were released at each cycle, indicating the presence of more than one peptide. Careful analysis of the quantitative data clearly shows that in both cases a contaminating polypeptide chain, with the same sequence but starting some residues earlier, was present (Figure 3). This proves that the cleavage of the hemoglobin molecule with subtilisin is not uniform and that it creates domains with ragged ends hindering their purification.

Comparison of the sequences of E1, E6, and E7 illustrates their differences. This strongly suggests that the globin chains are heteropolymers of eight structurally similar, but consistently different domains. As the domains are produced by limited proteolysis of heterodimeric Hb 2, they cannot be assigned to the α or β chain. However, the similarity of the sequences of domains E1 and E6 and their relative difference from E7 may suggest the possible presence of two different sets of domains; one characteristic for the α chain, the other for the β chain (Figure 3). Homology with the E1 and E7 sequence can be traced in globins from diverse phyla and an alignment is provided in Figure 4. This alignment is based on protein structural rules expressed by the Bashford templates[47] and differs from the alignment constructed by Goodman et al.[59] The major difference is that in our alignment, no insertions and deletions within conserved patterns are allowed. Discrepancies thus occur at the beginning or end of the patterns, or between them. Based on this alignment all invertebrate globin sequences show an identity with sperm whale myoglobin of between 13 and 24% (Figure 4), whereas about 8% is statistically random of the similarity. A hypothetical ancestral globin sequence was constructed by applying the maximum parsimony method to the given alignment using the PROTPARS program of Felsenstein.[60] The majority of residues known from vertebrate globins to determine the globin fold, are recognized in the ancestral sequence (Figure 4).

2. Sequencing at the cDNA Level: the Nature of the Domain Linkage

Based on the sequence of domain E1 an oligonucleotide (17—mer) corresponding to the putative C-helix was synthesized.[61] cDNA was synthesized from poly A^+ RNA enriched in globin mRNA using this oligonucleotide as primer. The resulting cDNA was cloned in the SmaI restriction site of the cloning vector pUC9. One clone pAGp27 (280 bp) was identified by sequencing[61] to be globin specific. The cDNA from clone pAGp27 was then used to screen a cDNA library constructed in λgt10 from 44 h poly A^+ RNA.

The positive clones contained cDNA greater than 4 to 5 kb and fell into two classes by restriction nuclease analysis, possibly representing the two globin gene families. A restriction fragment of the cDNA coding for a region towards the carboxyterminus of the polypeptide was identified and designated as AG1E2.[61] The amino acid sequences derived from both sequences is shown in Figure 5. Both sequences extend across a putative linker region connecting the H region of the preceding domain into the A region of the next domain and thus confirm the covalent nature of the domain linkage. The amino acid sequence derived from pAGp27 extends from the C region of E1 domain to the F region of the preceding domain. With the exception of positions A10 (D/S) and B15 (A/T) (not shown) the cDNA sequence confirms the aminoterminal sequence of domain E1. These discrepancies may reflect either species differences or polymorphism at the DNA level affecting the amino acid sequence. The amino acid sequence derived from the cDNA of AG1.E2 can be aligned with a GH to A region. The alignment shown in Figure 5 strongly suggests that the "pre A helix" region represents the linker region between two domains as has been suggested.[52] The consensus sequence -VDPITGL- is not only present at the amino terminals of E1, E6, and E7 but can also be recognized in the H helix motif of pAGp27, AG1.E2 and E7 (Figure 5). This strongly suggests that the end of the H-helix region is adapted to form a linker

FIGURE 4. Alignment of some selected globin sequences. Sequences were used as listed in the NBRF data bank. Standard numbering system, based on myoglobin, is used. HAHU: human hemoglobin; α chain, HBHU: human hemoglobin; β chain, MYWHP: sperm whale myoglobin, GGICE: *Chironomus thumni thumni* globin 1 to 10, GGNW1B: *Glycera dibranchiata*, GGGAA: *Aplysia limacina*, GGGA2A: *Aplysia kurodai*, GGNKIB: *Anadara broughtonii*, GGNKIT: *Anadara trapezia*, GGGAB: *Busicon canaticulatum*, GGACR: *Cerithidea rhizophorarum*, GGWNS: *Tylorrhynchus heterochaetus* I, GGWNSA: *Tylorrhynchus heterochaetus* IIA, GGWNSB: *Tylorrhynchus heterochaetus* IIB, GGWNSC: *Tylorrhynchus heterochaetus* IIC, E1AR: domain E1 *Artemia sp.*, E7AR: domain E7 *Artemia sp.*, GGEWA1: *Lumbricus terrestris* 1, GGEWA2: *Lumbricus terrestris* 2.

between the two domains rather than the insertion during evolution of a specialized linker sequence.[61] The highly conserved nature of the linker sequence suggests the tight structural constraint on the amino acid sequence of this region. The presence of a Pro residue in this region is possibly indicative for a bend between the H-region of the preceding domain and the A-helix of the next domain.

D. COMPUTER MODELING OF DOMAINS E1 AND E7

Sequence alignment of domain E1 and E7 (Figure 3) clearly shows their similarity with the classic globin family and raises the question of whether they are also compatible with a tertiary structure of the myoglobin type. This was investigated by modeling, using the BRUGEL computer graphics software.[63] A three dimensional model of domains E1 and E7 was built using the crystallographic structures of sperm whale myoglobin for E1 and human

```
                    H1  5   10   15   20      A1  5   10   15
                    :   :    :    :    :      :   :    :    :
ALFA AR                               AE VSGI LVSDKATIKRTWSIVN...
BETA AR                               AE IRGI LXSDKMTIKLTXAQVT...
E6AR NTERM                            VDVITGL SLXEKNAILHTWGKVK...
E7AR CTERM         ...RAVPEYSKGLEGRFGNVDNINEN APFR
E7AR NTERM                          GLVAP KAL TALEKQSIQDIWTLK...
E1AR CTERM       ...GEVIVNFLNEGLR QA
E1AR NTERM                        SLERVDPITGL SGLEKNAILDTWGKVR...
P27               ...DAFEKYISMGLS SLKRVDPITGL SGLEKNAILSTWGKVR...
AG1.E2            ...EVIISFLNEGLR QADIVDPVTHL TGRHKE...
 Consensus >                          VDPITGL-SGLEKNAILDTWGKV-
```

FIGURE 5. Alignment of the amino acid sequence of the linker regions connecting two globin domains. Sequence information was obtained by sequencing on the protein level (α and β chain; domain E1, E6, E7) and cDNA level (clone pAGp27 and AG1.E2) as described in the text. NTERM: aminoterminal region; CTERM: carboxyterminal region; standard numbering system based on myoglobin is used.

β chain for E7 as a template and following the sequence alignment shown in Figure 3. The side chains of the template were replaced by those corresponding to the E1 and E7 sequences respectively, and the compatibility of the new side chains was studied in terms of bond distances, energies, and physicochemical character. In regions of length discrepancy, fragments were found in the Brookhaven data base with similar coordinates in main-chain flanking regions and similar in length and sequence.[64] The fragments were then substituted into the globin template using coordinate superposition[65] and their sequences were changed to match the *Artemia* sequences. This approach clearly shows that both domains E1 and E7 are, overall, highly compatible with a myoglobin folded structure as illustrated in Figure 6. Only a few areas gave difficulties, where functionally new side chains were introduced or length discrepancy occurred. However, in each case a sterically plausible solution could be obtained. Some of these discrepant regions will be discussed as examples.

1. CD-D and GF-G Region of E1

After the very conserved -PheXXPhe-, the E1 sequence is one residue shorter than myoglobin, leading to the question of whether a D-helix is present or not (Figure 7). A search in the data base provided an equivalent region of *Chironomus* globin 3A with the same length.[66] This was fitted into the myoglobin chain and mutated to E1 side chains.[64,65] The fit was excellent with the one residue deficiency being taken up in the vicinity of D3. Side chains could be well accommodated. Only Phe at position D4 had to be rotated away from Gln at CD8.

An interesting structural exchange of two side chains was predicted. In domain E1, Leu at position D5 replaces functionally the Leu at position CD7 in myoglobin. Similarly in domain E1, Leu at position D1 replaces Met at position D5.[52] The FG-turn is also one residue shorter than myoglobin and again *Chironomus* globin 3A[66] provided the best fitting fragment (Figure 8). The one residue deletion is solved by a tighter winding of the end of the F-helix. In myoglobin, the last turn of the F-helix is in a π-conformation whereas in domain E1 it is substituted by a tighter 3_{10} conformation. This brings the much longer side chain of Arg in FG_1 of domain E1 into the same position as the His side chain in FG_2 of myoglobin where it performs the role of hydrogen binding to the heme propionate.[52]

2. GH and BC-C Region of Domain E7

In the GH corner of domain E7, three deletions occur when compared to the human β chain; one residue before the conserved Phe at GH_5 and two after it. The best fitting fragment was again from *Chironomus* globin 3A. However, this fragment shows two deletions before GH_5 and one after it. Nevertheless, the fit on the anchoring regions is excellent and brings

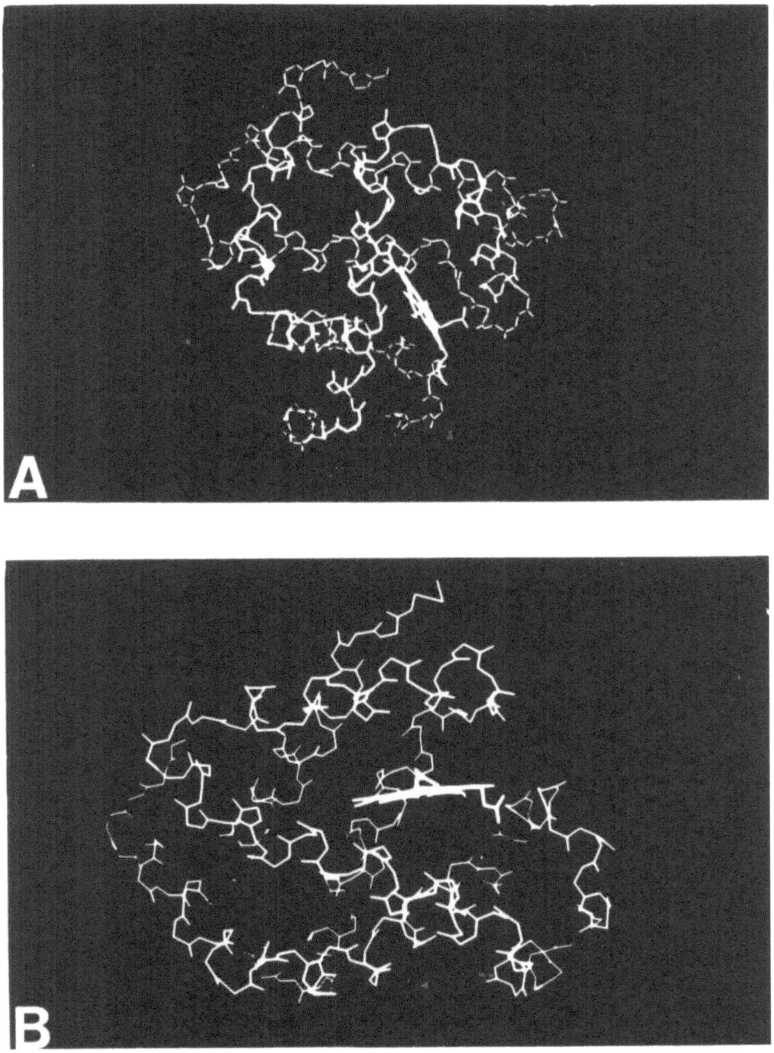

FIGURE 6. A computer model for the tertiary structure of domains E1 (A) and E7 (B). Only the polypeptide backbone and heme ring are shown.

the Phe side chain of GH_5 into the same place and orientation as in the β chain. It therefore shows that the GH corner of domain E7, despite the deletions, is comparable with that of the human β chain (Figure 9).[53,54]

In domains E1 and E7 as well as in the amino terminals of the intact chains the rather conservative Thr at position C_4 is substituted by the much more bulky Tyr (Figure 10). In myoglobin the C and G helices cross rather near, thus creating a problem of accommodating at C_4 in domain E1. This could only be solved by a concomitant adjustment of the neighboring residues. In the β chain however the C and G helices are more spaced such that the Tyr residue at C_4 of domain E7 has enough space to turn the phenolytic group to the surface whereas the hydrophobic C-G contact is realized by the aromatic ring.[53,54]

These examples illustrate that discrepant regions of domain E1 and E7 can be interpreted in terms of the globin fold. This implies that globin folded structures not only occur free as in myoglobin or in molecular aggregates as found in the vertebrate tetramer and many

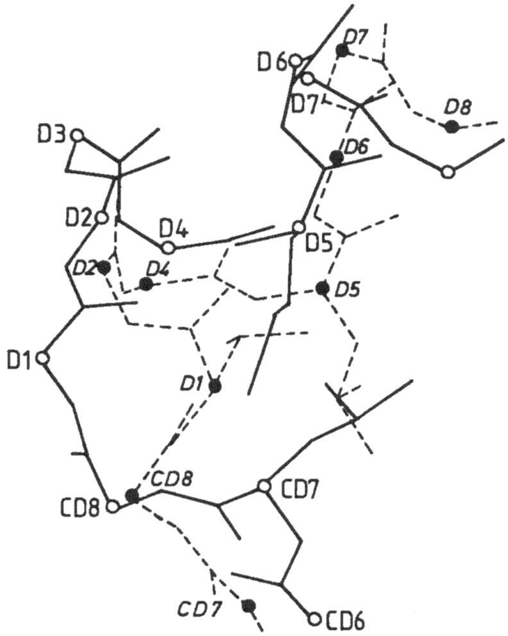


```
        CD1 5     D1  5    E1  5
        :   :     :   :    :   :
MYWHP   FDRFKHLK  TEAEMKA  SEDLKKH
E1AR    FRFFQGVQ  LAFLVQ-  -SPKFAAH
GGICE3  FTQFAG-F  DLESIKG  TAPFETH
```

FIGURE 7. The CD and D region of domain E1. The CD and D region of domain E1 was modeled with BRUGEL as described in text. MYWHP: sperm whale myoglobin (—); E1AR: *Artemia* domain E1 (---); GGICE3: *Chironomus* globin 3A.

invertebrate hemoglobin, but also in covalently linked polymeric globin chains. This illustrates the adaptability of the globin fold towards the quaternary structure in which it occurs.

E. HYPOTHETICAL MODEL OF THE NATIVE HEMOGLOBIN

Electron-microscopical observation of the hemoglobin molecule suggests the presence of two stacked discs with dimensions 120 × 70 Å (Table 1).[38] Each disc (120 × 35 Å) represents a globin chain, constructed of eight myoglobin folded domains with dimensions of an ellipsoid of 4.5 × 4.5 × 2.5 nm (Figure 11). Sequence analysis suggests that no extensive linker sequence is inserted between two domains but rather, the equivalent of the HC and NA regions are adapted to form a linker sequence (see Section C.2). Consequently two neighboring domains must be oriented in such a way that the respective H and A regions are close (Figure 5). This may be achieved by a ring-shaped orientation of the eight domains with the E and F helices coplanar, as shown in Figure 11B. Heme is thus orientated towards the hollow center and the EF turn toward the periphery. Combination of two such octomers at their E-F faces will provide a stable dimer (Table 1). The increased hydrophobicity of the EF region of both domains E1 and E7 compared to other globins is compatible with this hypothesis.[53,54]

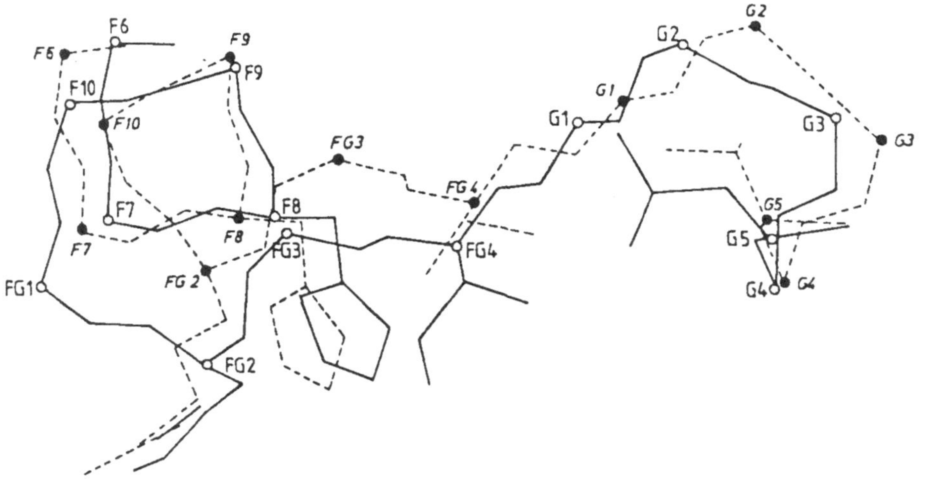


```
         F1  5    10  FG1   G1  5    10
         :   :    :   :     :   :    :
MYWHP    LKPLAQSHAT   KHKI  PIKYLEFYSE
E1AR     IKELGLDHIN   RGT-  DRSFVEYLKE
GGICE3   VNTFVASHKP   RG-V  THDQLNNFRA
```

FIGURE 8. The FG and G region of domain E1. The FG and G region of domain E1 was modeled with BRUGEL as described in text. MYWHP: sperm whale myoglobin (—); E1AR: *Artemia* domain E1 (---); GGICE3: *Chironomus* globin 3A.

F. THE EXPRESSION OF THE HEMOGLOBINS

The identification of hemoglobin as a major protein (20% of soluble protein) induced during post-gastrular development has been confirmed by many authors.[58] No globin could be detected by immunological methods until 20 to 22 h of development and appearance of the first nauplii, after which the hemoglobin concentration rises rapidly.[61] The absolute concentration of the hemoglobins and the ratio of the three phenotypes depend on external factors as will be discussed in Section III.

As *Artemia* are rich in stored mRNA,[67] the globin mRNA might be present in a latent form and its activation for translation during the emergence of nauplii may be an ontogenetic or externally regulated process. To test this hypothesis, globin mRNA was assayed by different methods. Poly A^+ was extracted at different stages of development and translated in an *in vitro* system, (rabbit reticulocyte) or in an *in vivo* system (*Xenopus* oocytes).[39,40,61] Although the translation of the globin mRNA *in vitro* is strongly inhibited by the secondary structure of the RNA, authentic globin chains could be detected by immuno-precipitation of the translation products beginning at 20 h of development and later. This was confirmed by translation *in vivo*[68] showing not only that the globin mRNA is not stored in a latent form but also confirming the high M_r (130,000) of the authentic globin chains (Table 1).[19]

The detection of the globin mRNA by its translated products is an indirect method that takes no account of possible translational control. Therefore, globin mRNA was also detected using several oligonucleotide probes constructed from available protein sequences (see Section II.C.2).[61] It was possible to identify a single large globin mRNA species of 4 to 5 kb consistent with the globin size. This mRNA is not stored in a latent form and was only detectable from 20 h of development, confirming the results from translation experiments.

It can be concluded that the appearance of the hemoglobin mRNA and the protein it

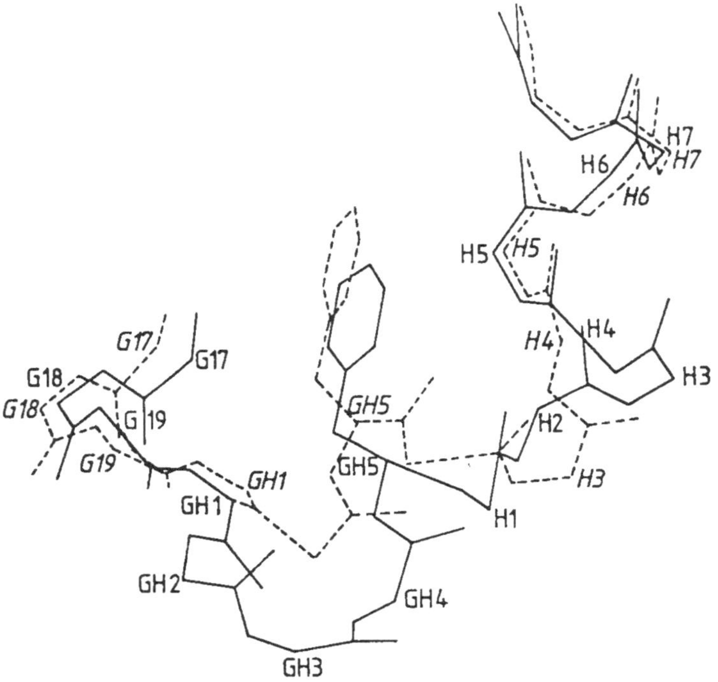

	G15	GH1 5	H1 5 10
	:	: :	: : :
HBHU	VCVLAHH	FGKEF	TPPVQAAYQK
E7AR	VAVIEEG	LDT-F	--RAVPEYSK
GGICE3	VSYMKAH	T--DF	A-GAEAAWGA

FIGURE 9. The GH region of domain E7. The GH region of domain E7 was modeled with BRUGEL as described in text. HBHU: human β chain: (—); E7AR: *Artemia* domain E7; GGICE3: *Chironomus* globin 3A.

encodes are closely related. Consequently the initiation of the transcription of the globin mRNA and its subsequent translation to a functional protein is correlated with the metabolic switch at the emergence of the nauplii. The mechanism by which this physiological switch is mediated, remains unknown.

III. RESPIRATORY PHYSIOLOGY OF THE OXYGEN CARRIER

A. PHYSIOLOGY OF THE OXYGEN CARRIER

1. The Oxygen Carrier

The oxygen binding sites of the respiratory protein are iron-porphyrins, also called heme groups. Each globin chain (M_r 130,000) consists of eight domains and a heme content of one per M_r 17,000, suggesting the existence of eight oxygen binding sites.[16,19,26,38,68] For more details of the molecular structure see Section II of this chapter.

2. The Oxygen Equilibrium

The heme iron in each globin chain domain acts as a site for reversible oxygen binding.

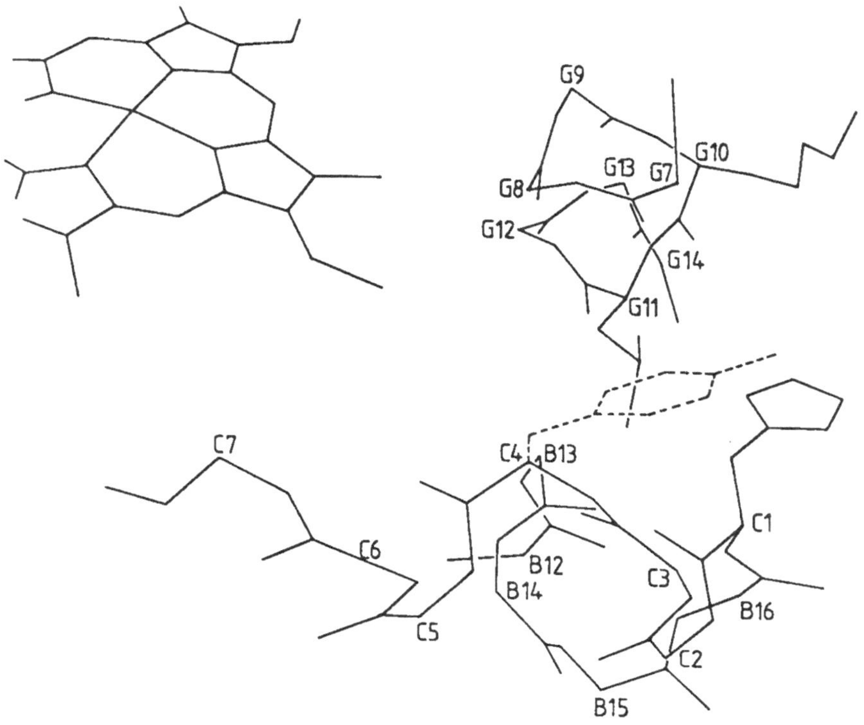

	B10 15	C1 5	CD1 5
	: :	: :	: :
MYWHP	LIRLFKS	HPETLEK	FDRFKHLK
HBHU	LGRLLVV	YPWTQRF	FESFGDLS
E1AR	FGKLFAA	HPEYQQM	FRFFQGVQ
E7AR	FGKLFAD	HPEYKAH	FDNFLTAI

FIGURE 10. The B-C region of domain E7. The B-C region of domain E7 was modeled with BRUGEL as described in text. MYWHP: sperm whale myoglobin; HBHU: human β chain; E1AR: *Artemia* domain E1 (—); side chain of C_4: (---); E7AR: *Artemia* domain E7.

Oxygenation is a function of the environmental oxygen partial pressure with which the pigment is in equilibrium, high pO_2 driving the equilibrium to the oxygenated state and low pO_2 driving it to the deoxygenated state.

$$Hb + O_2 \leftrightarrow HbO_2$$

The proportion of the total amount of hemoglobin molecules in the oxygenated state (% saturation) is related to the pO_2 in a relatively complex manner, compared with the relation between pO_2 and dissolved oxygen, which is linear. The oxygen equilibrium curve closely fits the Hill equation:

$$y = \frac{kx^n}{1 + kx^n}$$

where y is the proportion of pigment in the oxygenated form (% saturation), x is the equilibrium pO_2, k is the binding constant defining the oxygen affinity of the pigment and n is the Hill coefficient which expresses cooperativity.

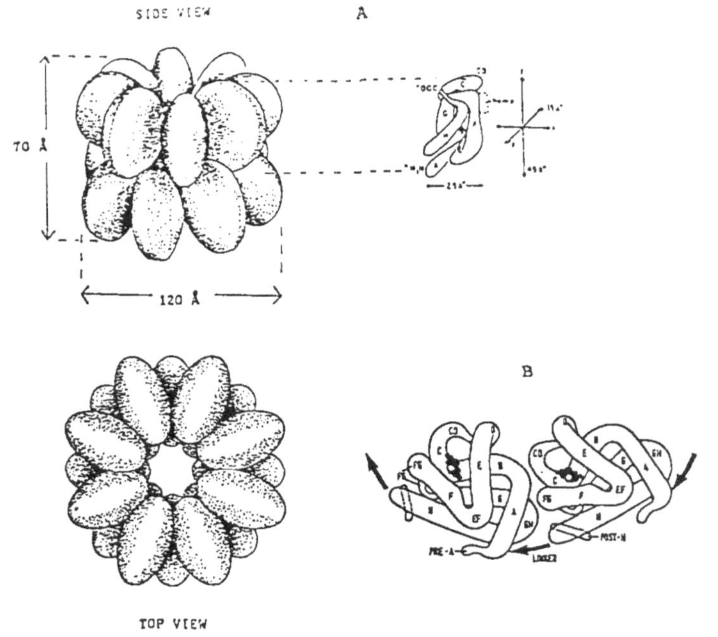

FIGURE 11. A hypothetical model of *Artemia* hemoglobin molecule. A: total molecule; B: detail of two linked domains.

The oxygen binding characteristics of the three Hb-types of *Artemia* can be summarized as follows: Hb I has a low oxygen affinity and a high cooperativity; Hb II, an intermediate oxygen affinity while cooperativity is variable but always exceeds 1; and Hb III, a high oxygen affinity and the lowest cooperativity.[69-71] The purified extracts (containing Hb I, II, and III) show a moderate oxygen affinity and a rather low cooperativity as reported by Wolf et al.[72]

Cooperativity gives a measure of the site-site interactions between the different oxygen binding sites in the molecules; i.e., binding of successive oxygen atoms to the hemoprotein is facilitated by earlier ones. n-Values of approximately 1 indicate that no heme-heme interactions take place in the molecule. Cooperativity of purified hemoglobin extracts shows n-values between 1 and 2. Cooperativity in fresh hemolymph can reach n-values as high as 4.20, while for the three hemoglobin-types it is found between ≈1.4 and 3.2.[72,73]

Oxygen affinity is expressed by the p_{30}-value (the partial pressure of oxygen at half saturation of the hemoprotein). The oxygen affinity of an oxygen-binding pigment is species specific. It is determined by the influence of groups in the polypeptide chain upon the equilibrium constants of the reactions at individual binding sites and is represented collectively by k in Hill's equation. Oxygen affinity is sensitive to changes in pH (Bohr effect). In the pH range from 7.25 to 8.60, p_{50}-values of highly purified *Artemia* hemoglobin range between ≈2 and 9 mm Hg at 22°C. The Bohr factor ϕ ($d \log p_{50}/d$ pH) appears to range between −0.19 and −0.45.[72]

Oxygen affinity is also sensitive to changes in temperature. As the reaction between oxygen and hemoglobin is exothermic, the oxygen affinity decreases with rising temperature. At 10°C the p_{50}-values of highly purified hemoglobins measured for different *Artemia* strains range between 1.55 and 2.80 mm Hg. These values increase to 10.39 and 16.33 mm Hg at 34°C. The free oxygen-binding energy, dH, ranges between −34.7 and −56.2 kJ/mole.[74]

The oxygen-binding characteristics of the purified intact pigment containing the two globin chains (260,000) show a p_{50} of 6.20 mm Hg and a Hill coefficient n of 1.67 at 18°C

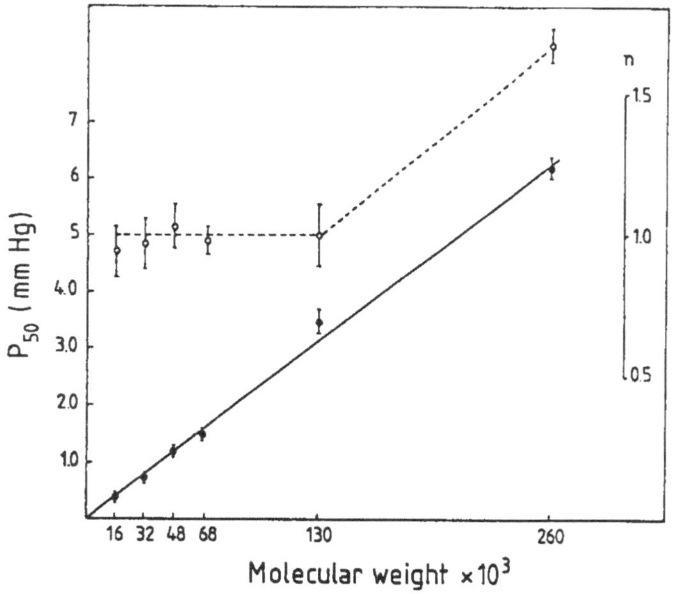

FIGURE 12. The oxygen affinity decreases linearly with increasing molecular weight of the *Artemia* hemoglobin fragments. Cooperativity (n > 1) exists only for the dimers (M_r 260,000) and not for the monomers and the smaller fragments (n = 1).

and pH 7.5.[42] After the cleavage of the molecule by limited digestion with subtilisin,[45] between one and four domain fragments are obtained. The determination of the oxygen-binding properties of these fragments shows that the oxygen affinity increases linearly with decreasing molecular weight. Cooperativity (n = >>1) is only expressed in the intact dimeric molecule and does not exist at the subunit and fragment level. The oxygen-binding properties of the whole molecule do not reflect the sum of the behavior of the different subunits or domains. Indeed, the association of polypeptide chains into stable aggregates and these into complex subunits (or, covalent linkage of O_2-binding sites into multidomain polypeptide chains) results in the lowering of the overall affinity for oxygen of the whole molecule compared to the affinity of the free polypeptide chains, subunits, or domains. This strongly suggests that the site-site interactions in the intact molecule only occur between sites from one polypeptide chain to the other[42] (Figure 12).

3. The Shape of the Oxygen-Binding Curve

For both the highly purified total hemoglobin extracts and for the individual Hb-types I, II, and III, the oxygen-binding curves show cooperativity which is reflected by the sigmoid shape of the curves and by n-values exceeding one. Cooperativity was determined for different *Artemia* strains and the values ranged between 1.07 and 2.14 for highly purified extracts and between 1.41 and 3.15 for the individual hemoglobin types.[71,72]

The oxygen-binding curves of fresh hemolymph, however, have a particular shape. Next to sigmoidality, a biphasic course of the curves is observed. The oxygen equilibrium shows a kink between 30 and 40% O_2-Hb (Figure 13). These observations were made on two *Artemia* strains, i.e., Great Salt Lake (U.S.) and Kazakhstan (U.S.S.R.).[73] Some literature on vertebrate hemoglobin also mentions biphasic curves.[75] The authors explain the biphasic character of those curves by processes of aggregation, gelation, and crystallization — effects that may affect the functional properties of hemoglobin. We do not known why this phenomenon is observed in *Artemia* hemolymph and not in purified total Hb extracts or in the

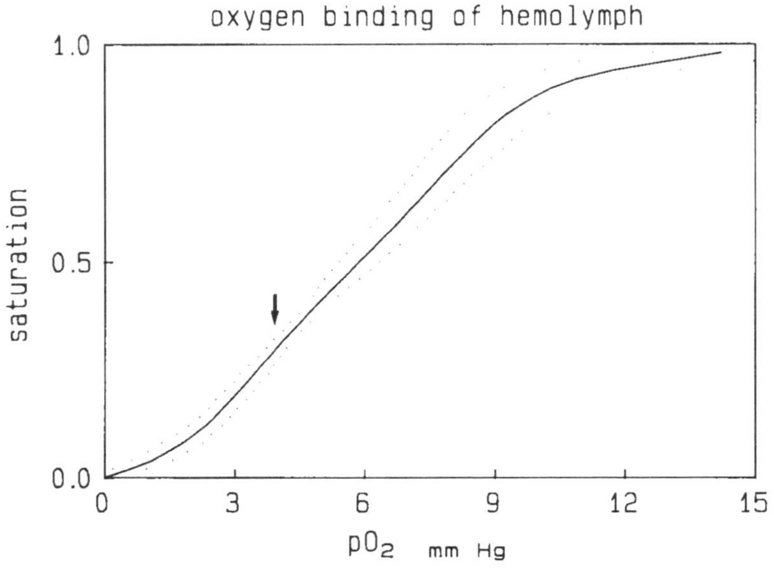

FIGURE 13. Biphasic oxygen-binding curve of hemolymph (Kazakhstan, U.S.S.R. strain) with 95% probability limits (N = 10 measurements).

individual hemoglobin I, II, and III. Nor it is clear whether the same explanations given above apply to *Artemia* hemoglobin dissolved in hemolymph. Nevertheless, the occurrence of biphasic oxygen-binding curves seems to be linked to factors present in the hemolymph.

Cooperativity in fresh hemolymph can reach relatively high values (n = 4.20). No cooperative oxygen binding exists at the subunit and multidomain level ($M_r \leq 130,000$). The oxygen-binding curves of hemoglobin fragments are hyperbolic, with n-values approximating one.[42]

4. Physiological Aspects of the Genetic Expression of Hemoglobins

Hemoglobins are present in the hemolymph of *Artemia* throughout their life. *Artemia* hemoglobins begin to be synthesized after the larvae have hatched into nauplii.[76,77] The induction of hemoglobin in freshly hatched nauplii should be accompanied by a *de novo* synthesis of globin polypeptides. The hemoglobins are synthesized in detectable amounts 4 h after the start of larval life.[78,79] Hb II is the first to be detected in the freshly hatched nauplii, followed immediately by Hb III. The point and level of induction of *Artemia* hemoglobin during development have been found to be controlled ontogenetically under a wide range of salt and oxygen concentrations as long as the oxygen concentration is above 0.09 m*M*.[77] Hb I synthesis decreases and Hb III increases with decreasing environmental oxygen concentrations.[68,80] Females have slightly more Hb I and less Hb III than males. The presence of Hb III in females decreases during larval development. As they mature either Hb III synthesis is decreased or its rate of degradation is increased.

In older cultures the hemolymph of adult *Artemia* frequently contains a blue-green pigment called artemocyanin which shows no oxygen-binding activities.[71] It contains carbohydrate, lipid, and copper, and the chromophore responsible for the blue-green color which has been identified as a bile pigment. This bile pigment may be derived from turnover of the hemoglobin. Artemocyanin is also found to exhibit a latent protease activity.[33,81]

In adult *Artemia*, Bowen et al.[68,82] describe four different hemoglobins (Hb I, II, X, and III) while Moens and Kondo[19] describe only three hemoglobins (Hb I, II, and III). Van den Branden et al.[69] and D'Hondt et al.[70] have shown that the three pigments show a moderate oxygen affinity but differ with Hb III ≥ Hb II ≥ Hb I. As a response to changing environmental parameters (temperature, partial oxygen pressure, and salinity) *Artemia* has developed adaptational mechanisms. One adaptation to hypoxia is the increasing concentration of respiratory pigment. The three hemoglobins are each adapted to cope with different environmental pO_2-values; Hb I being more efficient in oxygen-rich water, while Hb III is most efficient in poorly oxygenated water. Vos et al[80] consider the biosynthesis of Hb III as a long term adaptation to hypoxia.

Comparative work on the oxygen-binding characteristics of *Artemia* hemoglobins has been done with solutions containing the three highly purified hemoglobins in known proportions,[72] or with each separately.[69-71] In these studies, several *Artemia* strains were used.

Complex techniques were used to extract hemoglobins from different *Artemia* strains, reared in mass cultures.[19] As the different strains had characteristic hypoxic tolerances in mass culture, it was not possible to subject them to identical hypoxic stress conditions.[83] During the extraction of the hemoglobin, the use of proteolytic enzyme inhibitors was necessary[19] and possibly they were less than fully effective. This was reflected in the deterioration of the oxygen-binding properties of the extracts and the formation of considerable amounts of methemoglobin.[84] Consequently, the amounts and proportions of the three hemoglobin types in these extracts may have reflected nonidentical experimental conditions. In spite of these preservations, the physiological properties of these whole extracts showed oxygen affinity to be lower when Hb I was predominant and higher when Hb III was predominant.

An interesting question emerges from this comparative study. Do the differences in the functional characteristics of the extracted hemoglobins of the different *Artemia* strains depend solely on the proportions of the three Hbs present in the hemolymph, or are the differences among the Hb I, II, and III of one *Artemia* strain compared to another strain responsible for the differences between these physiological properties?

It has already been mentioned that the three hemoglobins within an *Artemia* strain differ in their oxygen-binding characteristics.[69,70] Recently, Decleir et al.[71] showed that the oxygen-binding properties of the Hb I, II, and III of the Great Salt Lake (U.S.) population (*A. franciscana*) differ significantly from the Kazakhstan (U.S.S.R.) population (*A. parthenogenetica*) under identical experimental conditions. Genetical distance between these two *Artemia* strains is as such that they are considered as belonging to different species.[85] These findings are reinforced by the observed physiological differences of the three hemoglobin-types, which reflect adaptation to different environmental conditions. Their response to hypoxia differs. The Kazakhstan population synthesizes up to 23 mg ml^{-1} hemoglobin with predominantly Hb I and Hb II; the oxygen-carrying capacity of the hemolymph is increased by making more respiratory pigment. The Great Salt Lake population synthesizes about one third as much hemoglobin but also synthesizes Hb III (16.2%), which is much better adapted to low oxygen pressures (Figure 14). It increases the oxygen carrying capacity of the hemolymph by making a pigment that is more specialized to bind oxygen at low oxygen pressures (Table 4).

The cooperativity of Great Salt Lake (U.S.) strain is about 1.75 times higher than that of Kazakhstan (U.S.S.R.). However, the p_{50} values of the hemolymph of both populations are similar (≈5.8 mm Hg at 25°C). Nevertheless, these similar p_{50} values of the hemolymph are the result of differing physiological characteristics of the composing Hb-types (Table 4). In both populations Hb III shows the highest and Hb I the lowest affinity, but the Hb III of Great Salt Lake is adapted to lower environmental oxygen concentrations that the Hb III of Kazakhstan. The affinity divergence between Hb I and Hb III is 4.63 and 2.45 mm Hg, respectively for Great Salt Lake and Kazakhstan, indicating that the pO_2 range over which the three Hbs of Great Salt Lake are active is greater (Figure 15).

Addition of GTP to the separated hemoglobins of the Kazakhstan population results in

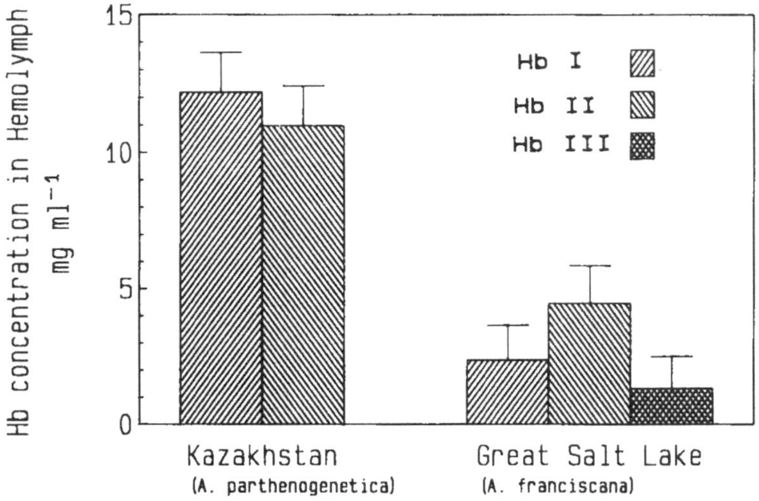

FIGURE 14. Distribution of the three hemoglobin types in preadult females of the *Artemia* strains Kazakhstan (U.S.S.R.) and Great Salt Lake (U.S.), culturing conditions: 25°C, 3.5 g/l salinity and 1 ml O_2/l.

TABLE 4
Oxygen-Binding Characteristics of Hemolymph and of Three Hemoglobin Types in the *Artemia* Populations Great Salt Lake, U.S. and Kazakhstan, U.S.S.R. (25°C)

	N	Measurement conditions	p_{50} (mm Hg)	n_{50} (Hill coeff.)
Great Salt Lake				
hemolymph[a]	10	fresh	5.81 (±0.17)	4.27 (±0.57)
Hb I[a]	4	pH 8.58	7.26 (±0.52)	3.15 (±0.38)
Hb II[a]	12	pH 8.58	5.07 (±0.32)	3.27 (±0.52)
Hb III[a]	6	pH 8.58	2.63 (±0.20)	1.41 (±0.25)
Hb-extract[b]	8	pH 8.60, 26°C	6.83 (±0.24)	2.18 (±0.14)
Kazakhstan				
hemolymph[a]	10	fresh	5.86 (±0.24)	2.44 (±0.17)
	10	pH 8.58	5.61 (±0.56)	2.88 (±0.53)
	10	(GTP) pH 8.58	6.48 (±1.22)	2.61 (±0.52)
Hb I[a]	7	pH 8.58	6.27 (±0.40)	2.39 (±0.39)
	6	(GTP) pH 8.58	7.80 (±0.49)	2.73 (±0.33)
Hb II[a]	10	pH 8.58	4.40 (±0.43)	1.56 (±0.21)
	7	(GTP) pH 8.58	5.40 (±0.52)	1.59 (±0.21)
Hb III[a]	7	pH 8.58	3.82 (±0.22)	1.59 (±0.23)

Note: GTP concentration: 3.1 mg/ml (N): number of analyses performed, using 2 to 3 animals/assay.

[a] Reference: Decleir et al.[71]
[b] Reference: Wolf et al.[72]

a significant decrease of oxygen affinity. Although it is unlikely that organic phosphates play an important regulatory role in the control of oxygen binding to high molecular weight hemoproteins in free solution, the binding of such phosphates and their action on the molecule seems to be a general property of most hemoglobins.[71]

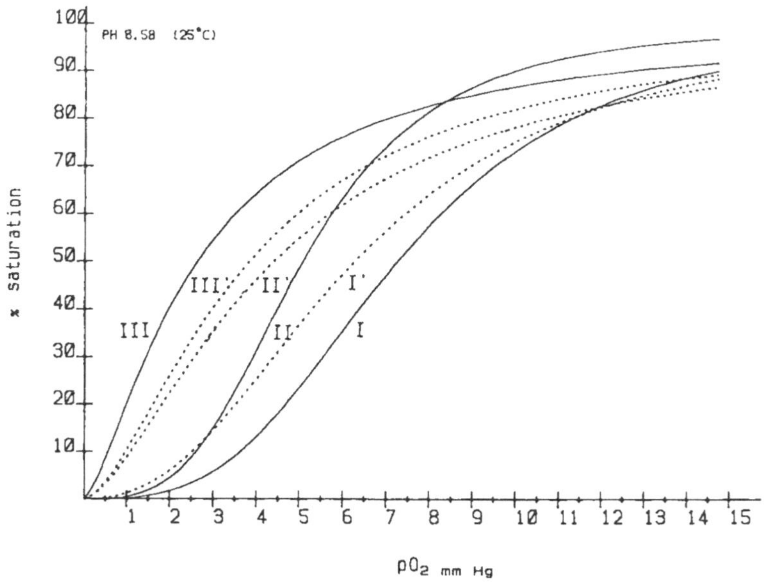

FIGURE 15. Oxygen-binding curves of electrophoretically separated hemoglobin from the two *Artemia* strains: Great Salt Lake, U.S. (Hb I, II, and III: —) and Kazakhstan, U.S.S.R. (Hb I', II', and III': . . .).

B. PHYSIOLOGICAL ASPECTS OF RESPIRATION DURING ONTOGENY

1. Introduction

Respiratory metabolism is influenced both by endogenous factors (activity, ontogeny, sex, etc.) and by exogenous factors (temperature, salinity, pH, oxygen availability, etc.). Waterbreathing animals are confronted with low concentrations of oxygen (6.3 ml/l at 20°C and 1 atm, for fresh water), compared to air (210 ml/l). Atmospheric gases including oxygen dissolve in water by diffusion, which is a slow process. This process is enhanced by natural streaming and by mixing of the water. However, fluctuations of physical and chemical environmental factors affect the oxygen solubility. Seawater with a salinity of 3.5 g/l contains only 5.12 ml O_2/l (20°C, 1 atm). The oxygen content of aquatic systems containing living organisms depends on a number of complex circumstances. In shallow standing waters the oxygen content can quickly become critical when salinity and temperature increase in such a way that the slow oxygen diffusion cannot compensate consumption. The demand for oxygen will exceed supply and aerobic animals will only survive in these environments when the available oxygen is used efficiently.

Artemia inhabits salt lakes, salt ponds and lagoons which show a great diversity in environmental parameters.[91] Occurrence of *Artemia* populations in different geographical areas is mainly controlled by temperature, salinity, and oxygen availability. Brine shrimp have been found alive in supersaturated brine at salinities as high as 34.0 g/l.[92] Temperature tolerance ranges from 6 to 40°C with an optimum range between 25 and 30°C. In most habitats the *Artemia* biotopes show important diurnal and seasonal temperature fluctuations. This implies that *Artemia* have to cope with low oxygen concentrations. *Artemia* show adaptations to hypoxic conditions, that include changes in reproductive behavior,[87,88] a modified energy balance,[89] and synthesis of respiratory pigments within a few hours after hatching.[90] Adaptational mechanisms which increase the oxygen-carrying capacity of the hemolymph seem to be specific for each individual *Artemia* strain.[71] Hatching yield decreases when oxygen concentrations drop to ≈ 1.5 to 2.0 m/l.[93] Preadult and adult *Artemia* cope

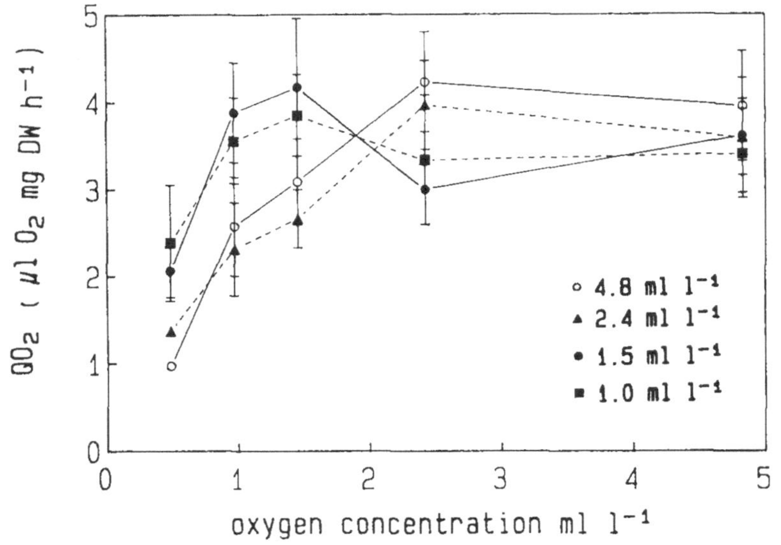

FIGURE 16. Relation between the respiration rate and the environmental oxygen concentration for four groups of adult *Artemia* which were acclimatized to respectively 4.8, 2.4, 1.5, and 1.0 ml O_2/l. The critical partial oxygen pressure (Pc) is situated between 1.5 and 2 ml O_2/l.

with decreasing pO_2 until the critical partial oxygen pressure is reached ($\approx$ 1.5 to 2 ml/l)[94] (Figure 16). Nauplii are more sensitive to low oxygen levels and survive only at concentrations of 3 ml/l.

2. Respiratory Organs and Pigments

Adult *Artemia* has 11 pairs of thoracopods, where the epithelial structure of the exopodital surfaces (metepipodites) is apparently so thin that they function as respiratory surfaces (branchial elements), while the endopodites are concerned with locomotion and filter feeding. The metepipodite of *Artemia* is a flattened leaf-like structure with a finely dimpled surface (Figure 17). The internum of the cuticle of the metepipodite is lined with a single layer of cells which can be classified into two types (referred to as dark and light cells) showing marked differences in morphology, e.g., the dark cells containing many mitochondria and the light containing only a few.[95]

The cuticular exoskeleton of the brine shrimp is permeable to chemical substances such as K, Na, and Mg.[96] The rate of oxygen diffusion through the body surface primarily depends on the thickness of the cuticle and of the number of cell layers to be crossed before reaching the hemolymph. Oxygen has only to cross the cuticle of the metepipodites; once in the intercellular space of the monolayer it reaches the hemolymph and can bind to the respiratory pigments. The hemolymph is part of an open circulatory system and as such, it enables the oxygen-rich hemoglobin to be transported to those tissues where it is required. As *Artemia* is an obligatory filter feeder, its thoracopods are constantly in motion, and a stream of water constantly irrigates the metepipodites so that the oxygen supply is continuous.

There seems to exist a relation between salinity and the morphology of the epithelial structure. At low salinities there are relatively few mitochondria and thin epithelia, while at high salinities there are many mitochondria and thick epithelia. Copeland[95] links these morphological features to osmoregulation and mitochondrial pumps.

3. Respiration of Cysts

When discussing cyst respiration it is important to consider separately dry (dormant)

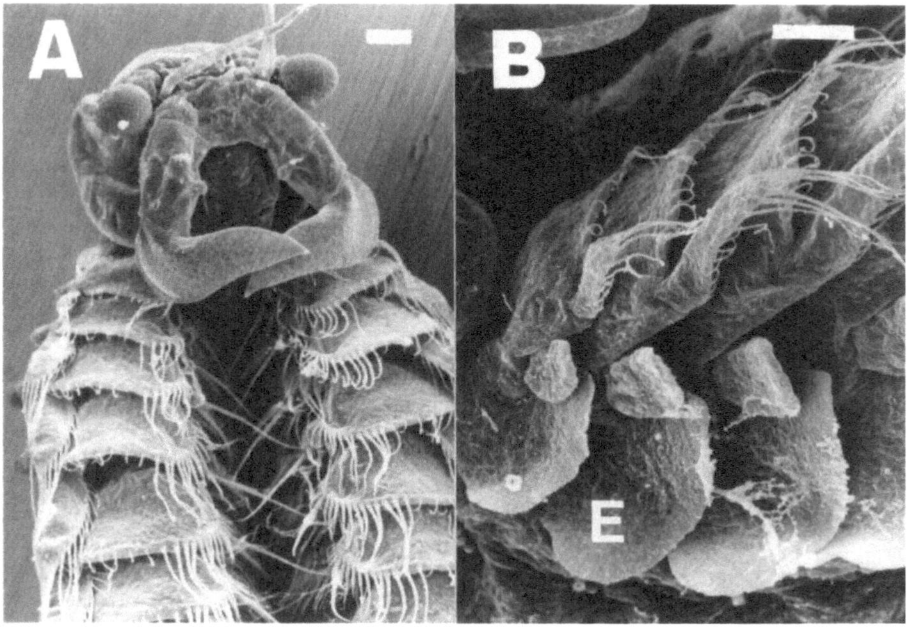

FIGURE 17. Ventral view of the cephalon and thoracopods of an adult female (A); scale: 100 μm. Lateral view of the gills (B); E: epipodite; scale: 100 μm. (With the authorization of Drs. L. Andries and J. Van Landuyt who were responsible for the preparation and photography of the SE specimen.)

and hydrated cysts. Diapause deactivation and developmental processes involved with dormancy in cysts have been extensively reviewed.[97,98] Dormant cysts are dehydrated encysted gastrulae. They show an impressive resistance to various external conditions such as extreme temperatures,[87,99-101] low pressure,[102] extreme dehydration,[103] and irradiation.[99] Cysts with a water content below 0.1/g are certainly ametabolic, and are considered as cryptobiotic.[101,104-108] Diapause is characterized by absent or extreme low respiratory metabolism.[109] Although Sundess et al.,[110] who incubated dried cysts (containing about 0.1 g H_2O/g) for 780 days found a very small increasing oxygen consumption with increasing temperatures, Clegg and Conte[97] believe that the observation of oxygen consumption in dry biological systems need not reflect metabolic activity.

Metabolism of cysts is reinitiated by rehydration.[111,112] At a minimal rehydration of 0.65 g/g metabolism begins but at a low rate. The preemergence development occurs without participation of DNA synthesis[113,114] or cell division.[115-118] The rise of the internal pH (from 6.2 up to 6.8 and more) can lead to breaking of dormancy.[119] Busa and Nuticelli[120] showed that nondormant cysts do not respire in an aerobic environment when internal pH is kept depressed by CO_2.

Oxygen consumption during hydration has been examined extensively during the first three hours after the onset of a cyst development. Initially the respiration increases very fast (0 to 1.2 μl O_2/mg dry weight/h), then at the ≈ 6th h it decreases for a short time to 1.0 μl O_2, whereafter it increases to 1.5 μl O_2 at 13 h. Thereafter, respiration remains constant until emergence (18 to 24 h) whereafter oxygen consumption increases up to 1.7 μl O_2. A third plateau is reached with an increase up to 2 μl O_2 during hatching.[121]

The initial steps of increased respiration are related to important changes in the structure and function of free mitochondria.[122,123] These observations are in accordance with the findings of Van der Linden et al.[124,125] who studied the influence of light on the hatching of hydrated cysts. They found that large numbers of commercial cysts when hydrated and incubated in aerobic conditions remain developmentally arrested as long as they are deprived of light. Upon transient illumination, metabolism resumes, as measured by a rapid increase in respiration. At the same time mitochondria lose their circular outline and become elongated. Evidently, the initial increase in respiration is related to an increase of the inner membrane surface of the mitochondrial cristae rather than an activation of specific mitochondrial sites of oxidative phosphorylation. The second increase in respiration is explained by the enlargement of the surface during preemergence embryonic development.[87,126,127] The respiratory quotient (RQ: CO_2/O_2) of the preemergent embryo is situated between 0.94 and 1.0,[87,109,128] suggesting that carbohydrate is the main substrate for respiration. Trehalose, which is degraded during development, is thought to be the substrate for the synthesis of glycerol.[129] After hatching the RQ decreases to ≈ 0.7 to 0.8 indicating a switch to other substrates.

The increase of oxygen uptake is related to the initiation or increase of metabolic activity since there is increased lipase activity,[130] DNA synthesis and cell division,[114,131,132] and the possible use of GP4G for ATP formation.[103,133] RNA synthesis increases rapidly upon resumption of development in encysted embryos,[78] and metabolization of about 50% of the yolk platelets occurs.[134,135] Trehalose, a prominent yolk material, starts to decrease.[129]

4. Respiration after Hatching

Bernaerts et al.[136] followed the oxygen consumption of *Artemia* (San Francisco Bay, at 25°C and 35 g/l salinity) during ontogeny. At the breaking of the cyst shell (after 20 h of hydration) oxygen consumption reached 5 μl/mg DW/h. After 35 h of hydration hatching takes place with oxygen uptake of 11 μl mg/DW/h. During the molting from Instar I to Instar II the uptake increases further reaching a maximum value of 20 μl/mg DW/h at the molting from Instar II to Instar III. Less marked oxygen consumption peaks are found for molting stages IV to VI. Finally, respiration slowly decreases until a typical rate of 5 μl O_2/mg DW/h is reached. The increase of respiration at cyst-break age is explained by the enlargement of the embryonic surface, facilitating oxygen diffusion. The respiratory values found by Bernaerts et al.[136] at hatching are not confirmed by Emerson[121] who studied the respiration during 38 h after hatching and found 2 μl O_2/mg DW/h, in the latter case a different population (Great Salt Lake, Utah) was used. Further, the study of Bernaerts et al.[136] revealed that molting stages until Stage VI were correlated with peaks in oxygen consumption (Figure 18).

As the synchronism of growth in a mass culture of *Artemia* only lasts until Instar V or VI, the synchronous, oxygen-consumption peaks of each molting, can only be detected as long as synchronism exists. Therefore a brine shrimp's oxygen consumption may be higher when it molts. Surface enlargement is probably the most important factor governing increased oxygen consumption.

The relation between respiratory metabolism and increasing weight during ontogeny is illustrated in Figure 19. The b-factor (log respiration-log weight ratio) is 0.819 which is significantly more than the expected value of 2/3 or 3/4. Eliasse[137] determined three values for the b-factor (0.75, 1.00, and 0.60) with increasing age. von Bertalanaffy et al.[126] concluded that for *Artemia*, the surface rule is valuable (b = 3/4) while Khmeleva et al.[138] determined b = 0.70. Bernaerts et al.[136] determined the highest b-value but this parameter can be influenced by temperature. Figure 19 illustrates that embryos and nauplii with yolk reserve have b-values which are much lower and do not fit into the equation.

Oxygen consumption of adult *Artemia* is not affected by mild hypoxia until a critical value (Pc) is reached, corresponding to 1.5 to 2 ml O_2/l.[80,94] Below this Pc, *Artemia* can no longer regulate its respiration. Nevertheless, animals which have been acclimatized during a prolonged period to lower pO_2 show a decrease in their Pc.[80]

The influence of temperature on respiratory metabolism (between 5 and 45°C), was

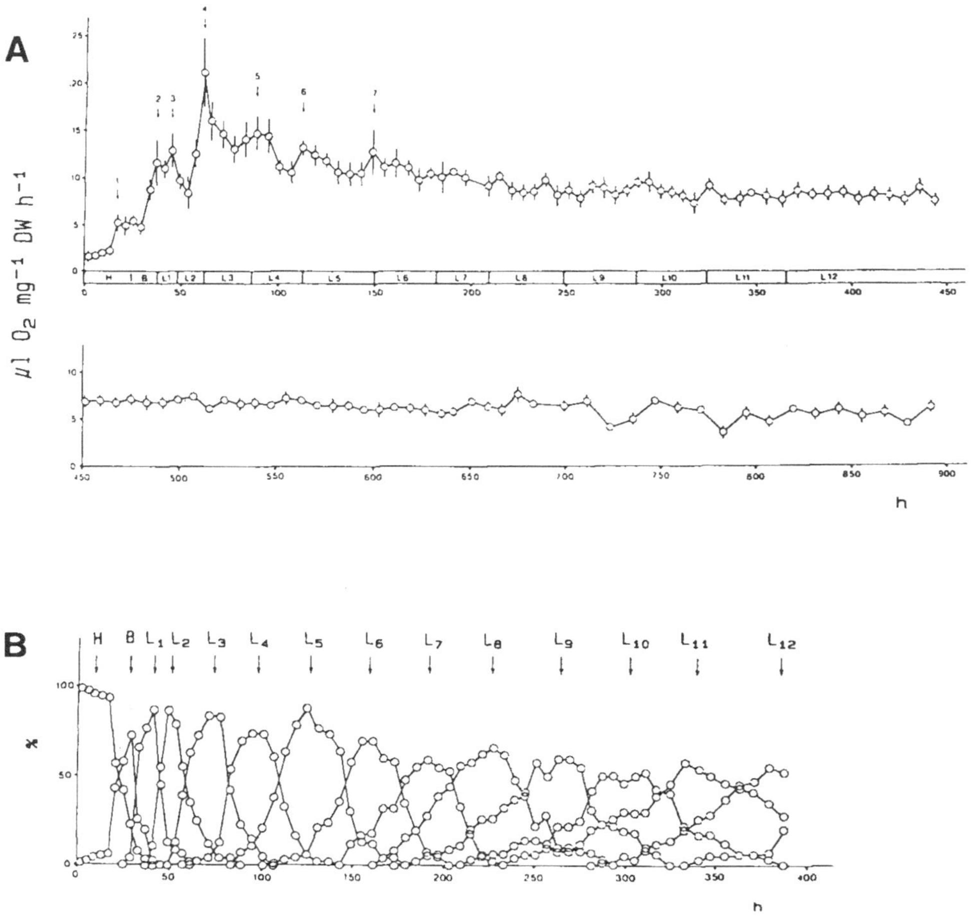

FIGURE 18. Oxygen uptake during ontogeny of *Artemia* (A). Arrows 1 and 2 correspond with increasing oxygen consumption during hydration and breaking of the shell; arrows 3—7 correspond with respiratory peaks coinciding with molting peaks. Percentage of the succeeding ontogenetic stages in the culture (B); sample points at which respiration was measured are indicated. H, hydration; B, breaking of the shell; and L, larval stages.

studied for males and females of different strains.[139] Both sexes show an equal lethal response to temperatures as low as 5°C, with an oxygen consumption of about 1 μl/mg DW/h. The lethal response to high temperatures for males and females is respectively at 45 and 40°C, except for San Francisco Bay males which tolerate only 35°C. For several *Artemia* strains, oxygen consumption at 35°C is between 7 and 10 μl/mg DW/h for both sexes.

The effect of salinity on respiration seems less important. Oxygen consumption in newly hatched nauplii appears to be independent of the NaCl concentration in the incubation medium.[140] For adult *Artemia* the observations are somewhat confusing. According to Dutrieu,[87] Herbst et al.,[141] and Gilchrist,[89,142] males are sensitive and females nonsensitive to changing salinities, although Khmeleva and Yurkevitch[138] mention that salinities between 4.5 and 16 g/l have no effect on respiratory rate.

C. CONCLUDING REMARKS

Artemia begins to synthesize hemoglobin in detectable amounts with a few hours after hatching. Apparently the three Hb-types are induced at specific times during postlarval development.

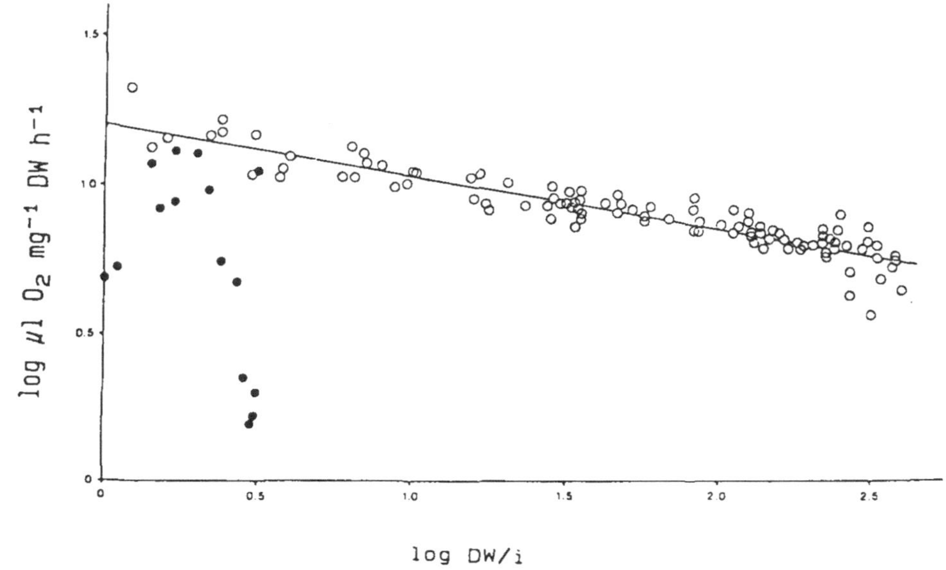

FIGURE 19. Relation between the respiratory metabolism and the body length (DW/i: dry weight/individual). Embryos and larvae with (●) and without (○) yolk sac.

Subjection of adult *Artemia* to low partial pressure of oxygen results in an adaptive response expressed by increased hemoglobin biosynthesis and a change in the proportions of the three hemoglobin types. Each hemoglobin type is adapted to cope with a different environmental pO_2-value. Significant differences between the oxygen-binding characteristics of the three hemoglobin types were observed between *A. franciscana* and *A. parthenogenetica*. The plasticity of each species to adapt to the applied low environmental pO_2 by synthesizing the appropriate hemoglobin is genetically determined.

Respiration of *Artemia* is influenced by the oxygen, partial pressure, temperature, and possibly by salinity. Oxygen consumption is different for males and females. The relation between weight and oxygen consumption (b-factor) is high (0.819).

Respiration during ontogeny shows a pattern (especially at the beginning of development, from Instar I to Instar VI larvae) with peaks of oxygen consumption which can be correlated with the molting stages which in turn are correlated with a marked increase in larval surface after each molt.

REFERENCES

1. **Chung, M. C. M. and Ellerton, M. D.,** The physico-chemical and functional properties of extracellular respiratory hemoglobins and chlorocruorins, *Prog. Biophys. Molec. Biol.*, 35, 53, 1979.
2. **Vinogradov, S. N., Shlom, J. M., Kapp, O. N., and Frossard, P.,** The dissociation of annelid extracellular hemoglobins and their quanternary structure, *Comp. Biochem. Physiol.*, 67B, 1, 1980.
3. **Terwilliger, R. C.,** Structure of invertebrate hemoglobins, *Ann. Zool.*, 20, 53, 1980.
4. **Wood, E. J.,** The oxygen transport and storage proteins of invertebrates, *Essays in Biochemistry*, 16, 1, 1980.
5. **Vinogradov, S. N.,** The structure of invertebrate extracellular hemoglobins (Erythrocruorins and chlorocruorins), *Comp. Biochem. Physiol.*, 82B, 1, 1985.
6. **Goodman, M., Braunitzer, G., Kleinschmidt, T., and Aschauer, H.,** Evolution of monomeric and homodimeric hemoglobins of *Chironomus thummi thummi*, *Hoppe-Seyler's Z. Physiol. Chem.*, 364, 205, 1983.

7. **Svedberg, T. and Hedenius, A.**, The sedimentation constants of the respiratory pigments, *Biol. Bull.*, 66, 191, 1934.
8. **Ilan, E. and Daniel, E.**, Structural diversity of arthropod extracellular hemoglobins, *Comp. Biochem. Physiol.*, 63B, 303, 1979.
9. **Ilan, E. and Daniel, E.**, Erythrocruorin from the water flea *Daphnia magna*. Quaternary structure and arrangements of subunits, *Biochem. J.*, 207, 197, 1982.
10. **Dangott, L. J. and Terwilliger, R. C.**, Structural studies of a branchiopod crustacean (*Lepidurus bilobatus*) extracellular hemoglobin, *Biochim. Biophys. Acta*, 579, 452, 1979.
11. **Daniel, E.**, Subunit structure of arthropod erythrocruorins, *Life Chem. Rep.*, Suppl., 1, 157, 1983.
12. **Wood, E. J. and Mosby, L. J.**, Physicochemical properties of *Planorbis corneus* erythrocruorin, *Biochem. J.*, 149, 437, 1975.
13. **Terwilliger, N. B., Terwilliger, R. C., and Schabtach, E.**, The quaternary structure of a mollusc (*Helisoma trivolvis*), *Biochim. Biophys. Acta*, 453, 101, 1976.
14. **Terwilliger, N. B. and Terwilliger, R. C.**, Oxygen binding domains of a clam (*Cardita borealis*) extracellular hemoglobin, *Biochim. Biophys. Acta*, 537, 77, 1978.
15. **Terwilliger, R. C. and Terwilliger, N. B.**, Oxygen binding domains in invertebrate hemoglobins, *Life Chem. Rep.*, Suppl. 1, 227, 1983.
16. **Waring, G., Poon, M. C., and Bowen, S. T.**, The hemoglobins of *Artemia salina*. II. Isolation of three hemoglobins, *Int. J. Biochem.*, 1, 537, 1970.
17. **Bowen, S. T., Moise, H., Waring, G., and Poon, M. C.**, The hemoglobins of *Artemia salina*. III. Characterization, *Comp. Biochem. Physiol.*, 55B, 99, 1976.
18. **Moens, L. and Kondo, M.**, The structure of *Artemia salina* hemoglobins, *Eur. J. Biochem.*, 67, 397, 1976.
19. **Moens, L. and Kondo, M.**, Evidence for a dimeric form of *Artemia salina* extracellular hemoglobins with high molecular weight subunits, *Eur. J. Biochem.*, 82, 65, 1978.
20. **Osuna, C., Olalla, A., Sillero, A., Sillero, M. G., and Sebastian, J.**, Induction of multiple proteases during the early larval development of *Artemia salina*, *Develop. Biol.*, 61, 94, 1977.
21. **Olalla, A., Carmen, O., Sebastian, J., Sillero, A., and Sillero, M. G.**, Purification and properties of three proteases from the larvae of the brine shrimp *Artemia salina*, *Biochim. Biophys. Acta*, 523, 181, 1978.
22. **Nagainis, P. A. and Warner, A. M.**, Evidence for the presence of an acid protease and protease inhibitors in dormant embryos of *Artemia salina*, *Dev. Biol.*, 68, 259, 1979.
23. **Garesse, R., Perona, R., Marco, R., and Vallejo, C. G.**, The unmasking of proteolytic activity during the early development of *Artemia salina*, *Eur. J. Biochem.*, 106, 225, 1980.
24. **Warner, A. H. and Shridda, V.**, Purification and characterization of a cytosol protease from dormant cysts of the brine shrimp *Artemia*, *J. Biol. Chem.*, 260, 7008, 1985.
25. **Mansfield, B. C., Krissansen, G. W., Smith, M. G., and Tate, W. P.**, Isolation of intact larval hemoglobin from the brine shrimp *Artemia salina*, *Biochem. Biophys. Acta*, 623, 163, 1980.
26. **Moens, L. and Kondo, M.**, Characterization of the extracellular hemoglobins of *Artemia salina*, *Biochem. J.*, 165, 111, 1977.
27. **Peeters, K., Mertens, J., Van Hauwaert, M.-L., and Moens, L.**, The structure of *Daphnia pulex* hemoglobin; a comparison with other invertebrate hemoglobins, *Comp. Biochem. Physiol.*, in preparation.
28. **De Baere, I. and Moens, L.**, The structure of *Ascaris lumbricoides* hemoglobin, Symposium on "Invertebrate oxygen carriers", Leuven, Belgium, 1989.
29. **Krissansen, G. W., Trotman, C. N. A., and Tate, W. P.**, A novel protease may explain widely differing models for the structure of *Artemia salina* hemoglobin, *Biochim. Biophys. Acta*, 671, 104, 1981.
30. **Krissansen, G. W., Trotman, C. N. A., and Tate, W. P.**, Purification of stable hemoglobin from the adult brine shrimp *Artemia*, *Biochem. Int.*, 7, 519, 1983.
31. **Krissansen, G. W., Trotman, C. N. A., and Tate, W. P.**, Two forms of the glycoprotein artemocyanin from the brine shrimp *Artemia*. A relationship between latent proteolytic activity and structural modifications to the protein, *Biochim. Biophys. Acta*, 747, 159, 1983.
32. **Krissansen, G. W., Trotman, C. N. A., and Tate, W. P.**, Purification of artemocyanin, a hemolymph protein from the brine shrimp *Artemia*, *Biochem. Int.*, 7, 369, 1983.
33. **Krissansen, G. W., Trotman, C. N. A., and Tate, W. P.**, Physical and biochemical characterization of artemocyanin, an abundant glycoprotein complex with latent proteolytic activity from the brine shrimp *Artemia*, *Biochim. Biophys. Acta*, 747, 151, 1983.
34. **Krissansen, G. W., Trotman, C. N. A., and Tate, W. P.**, Identification of the blue-green chromophore of an abundant biliprotein from the haemolymph of *Artemia*, *Comp. Biochem. Physiol.*, 77B, 249, 1984.
35. **Krissansen, G. W., Trotman, G. N. A., and Tate, W. P.**, Artemocyanin: A proteinase zymogen rather than a proteinase inhibitor, *Comp. Biochem. Physiol.*, 80B, 883, 1985.
36. **Marshall, C. J., Cutfield, J. F., Trotman, C. N. A., and Tate, W. A.**, Purification of the hemoglobin 1 and 3 from the brine shrimp, *Artemia*, *Biochem. Int.*, 12, 693, 1986.

37. **De Voeght, F. and Clauwaert, J.**, Structural characterization by biophysical methods of extracellular hemoglobin of *Artemia salina*, in *Structure, Active Site and Function of Invertebrate Oxygen Carriers*, Lamy, J., Ed., Marcel Dekker, New York, 1980.
38. **Wood, E. J., Barker, C., Moens, L., Jacob, W., Heip, J., and Kondo, M.**, Biophysical characterization of *Artemia salina* extracellular hemoglobin, *Biochem. J.*, 193, 353, 1981.
39. **Manning, A. M., Ting, G. S., Trotman, C. N. A., and Tate, W. P.**, Faithful translation of the mRNA for the haemoglobin polypeptide of *Artemia* nauplii in *Xenopus* oocytes, *Proc. Univ. Otago Med. School*, 63, 24, 1985.
40. **Manning, A. M., Ting, G. S., Mansfield, B. C., Trotman, C. N. A., and Tate, W. P.**, The isolation and faithful translation of *Artemia* naupliar hemoglobin mRNA, *Biochem. Int.*, 12, 715, 1986.
41. **Moens, L.**, The extracellular hemoglobins of *Artemia* sp. A biochemical and ontogenetical study, *Academiae analecta*, 44, 3, 1982.
42. **Wolf, G., Van Pachtenbeke, M., Moens, L., and Van Hauwaert, M. L.**, Oxygen binding characteristics of *Artemia* hemoglobin domains, *Comp. Biochem. Physiol.*, 76B, 731, 1983.
43. **Laemmli, U. K.**, Cleavage of structural proteins during the assembly of the head of bacteriophage T_4, *Nature*, 227, 680, 1970.
44. **Geelen, D., Moens, L., Heip, J., Hertsens, R., Donceel, K., and Clauwaert, J.**, The structure of *Artemia* sp. hemoglobins. I. Isolation and characterization of oxygen binding domains obtained by limited tryptic digestion, *Int. J. Biochem.*, 14, 991, 1982.
45. **Moens, L., Geelen, D., Van Hauwaert, M. L., Wolf, G., Blust, R., Witters, R., and Lontie, R.**, The structure of *Artemia* sp. hemoglobin. Cleavage of the native molecules into functional units by limited subtilisin digestion, *Biochem. J.*, 223, 861, 1984.
46. **Tanford, C.**, *Physical Chemistry and Macromolecules*, John Wiley & Sons, New York.
47. **Bashford, D., Chothia, C., and Lesk, A.**, Determinants of a protein fold. Unique features of the globin amino acid sequence, *J. Mol. Biol.*, 196, 199, 1987.
48. **Moens, L., Van Hauwaert, M.-L., and Wolf, G.**, The structure of *Artemia* sp. hemoglobins. III. Purification of a structural unit to homogeneity, *Biochem. J.*, 227, 917, 1985.
49. **O'Farell, P. H.**, High resolution two-dimensional electrophoresis of proteins, *J. Biol. Chem.*, 250, 4007, 1975.
50. **Hewick, R. M., Hunkapiller, M. W., Hood, L. E., and Dreyer, W. J.**, A gas-liquid solid phase peptide and protein sequenator, *J. Biol. Chem.*, 256, 7990, 1981.
51. **Chang, J. Y., Creaser, E. H., and Bentley, K. W.**, 4-N,*N'*-dimethylaminoazoben-zene-4'-isothiocyanate. A new chromophoric reagent for protein sequence analysis, *Biochem. J.*, 153, 607, 1976.
52. **Moens, L., Van Hauwaert, M.-L., De Smet, K., Geelen, D., Verpooten, G., Van Beeumen, J., Wodak, S., Alard, P., and Trotman, C. N. A.**, A structural domain of the covalent polymer globin chains of *Artemia*, *J. Biol. Chem.*, 263, 4679, 1988.
53. **Moens, L., Ver Donck, K., De Smet, K., Van Hauwaert, M.-L., Van Beeumen, J., Alard, P., Wodak, S., and Trotman, C. N. A.**, The structure of *Artemia* hemoglobin and hemoglobin domains, in *Cell and Molecular Biology of Artemia Development*, NATO Advanced Study Institute Series, Plenum Press, New York, 1989, in press.
54. **Moens, L., Van Hauwaert, M.-L., De Smet, K., Ver Donck, K., Trotman, C. N. A., Van Beeumen, J., Alard, P., and Wodak, S.**, Structural interpretation of the amino acid sequence of a second domain from the *Artemia* covalent polymeric globin chains, *J. Biol. Chem.*, in preparation.
55. **Lesk, A. M. and Chothia, C.**, How different amino acid sequences determine similar protein structures; the structural and evolutionary dynamics of the globins, *J. Mol. Biol.*, 136, 225, 1980.
56. **Lipman, D. J. and Pearson, W. R.**, Rapid and sensitive protein similarity searches, *Science*, 227, 1435, 1985.
57. **Stockwell, P. A. and Peterson, G. B.**, HOMED; a homologous sequence editor, *CABIOS*, 3, 37, 1987.
58. **Moens, L., De Smet, K., Van Hauwaert, M.-L., Geelen, D., Verpooten, G., and Van Beeumen, J.**, The hemoglobins of *Artemia*; oxygen carriers with multidomain globin chains, in *Biochemistry and Cell Biology of Artemia*, MacRae, T. H., Bagshaw, J. C., and Warner, A. H., Eds., CRC Press, Boca Raton, FL, 1989, 133.
59. **Goodman, M., Pedwaydon, J., Czelusniak, J., Suzuki, T., Gotoh, T., Moens, L., Shishikura, F., Walz, D., and Vinogradov, S.**, An evolutionary tree for invertebrate globin sequences, *J. Mol. Evol.*, 27, 236, 1988.
60. **Felsenstein, J.**, Confidence limits on phylogenies; an approach using the bootstrap, *Evolution*, 39, 783, 1985.
61. **Manning, A., Marshall, C., Powell, R., Trotman, C., and Tate, W.**, The molecular biology of *Artemia* hemoglobins, in *Cell and Molecular Biology of Artemia Development*, Nato Advanced Study Institute Series, Plenum Press, New York, 1989, in press.

62. **De Smet, K., Van Hauwaert, M.-L., Moens, L., and Van Beeumen, J.**, The structure of the *Artemia* hemoglobins. A comparison of the structural units composing the *Artemia* globin chains, in *Artemia Research and Its Applications*, Vol. 2, Decleir, W., et al., Eds., Universa Press, Wetteren, Belgium, 1987, 41.
63. **Delhaize, P., Van Belle, D., Bardiaux, M., Hamers, P., Van Cutsem, E., and Wodak, S.**, Analysis of data from computer simulation on macromolecules using the CERAM packet, *J. Mol. Graphics*, 3, 116, 1986.
64. **Jones, A. T. and Thirup, S.**, Using known substructures in protein model building and crystallography, *EMBO J.*, 5, 819, 1986.
65. **McLachlan, A. D.**, Gene duplications in the structural evolution of chymotrypsin, *J. Mol. Biol.*, 128, 49, 1979.
66. **Huber, R., Epp, O., Steigemann, W., and Formanek, M.**, The atomic structure of erythrocruorin in the light of the chemical sequence and its comparison with myoglobin, *Eur. J. Biochem.*, 19, 42, 1971.
67. **Slegers, H.**, Enzyme activities through development: A synthesis of the activity and control of the various enzymes as the embryo matures, in *Artemia Biology*, Browne, R. A., Sorgeloos, P., and Trotman, C. N. A., Eds., CRC Press, Boca Raton, FL, 1990.
68. **Bowen, S. T., Lebherz, H. G., Poon, M., Chow, V. H. S., and Grigliatti, T. A.**, The hemoglobin of *Artemia salina*. I. Determination of phenotype by genotype and environment, *Comp. Biochem. Physiol.*, 31, 733, 1969.
69. **Van den Branden, C., D'Hondt, J., Moens, L., and Decleir, W.**, Functional properties of the haemoglobin of *Artemia salina* L., *Comp. Biochem. Physiol.*, 60 A, 185, 1979.
70. **D'Hondt, J., Moens, L., Heip, J., D'Hondt, A., and Kondo, M.**, Oxygen binding characteristics of three extracellular haemoglobins of *Artemia* salina, *Biochem. J.*, 171, 705, 1978.
71. **Decleir, W., Wolf, G., and De Wachter, B.**, Adaptation to hypoxia in *Artemia*, in *Cell and Molecular Biology of Artemia Development*, Plenum Press, New York, 1989.
72. **Wolf, G., Vanpachtebeke, M., Hens, F., Decleir, W., Moens, L., and Van Hauwaert, M. L.**, Study of the oxygen binding properties of *Artemia* hemoglobin, in *Artemia Research and Its Applications*, Vol. 2, Decleir, W., Moens, L., Slegers, H., Sorgeloos, P., and Jaspers, E., Eds., Universa Press, Wetteren, Belgium, 1987, 221.
73. **Wolf, G., De Wachter, B., and Hens, F.**, Hemolymph of *Artemia*. Oxygen binding characteristics, *Comp. Biochem. Physiol.*, in press, 1989.
74. **Wolf, G., Hens, F., Moens, L., and Decleir, W.**, Comparative study of hemoglobin from different *Artemia* populations. The influence of temperature on oxygen equilibria, *Comp. Biochem. Physiol.*, 88(B), 285, 1987.
75. **Gill, S., Skold, J., Fall, L., Shaeffer, T., Spokane, R., and Wyman, J.**, Aggregation effects on oxygen binding of sickle cell hemoglobin, *Science*, 201, 362, 1978.
76. **Heip, J., Moens, L., and Kondo, M.**, Effects of concentrations of salt and oxygen on the synthesis of extracellular hemoglobin during development of *Artemia*, *Develop. Biol.*, 63, 247, 1978.
77. **Heip, J., Moens, L., Hertsens, R., Wood, E. J., Heyligen, H., Van Broekhoven, A., Vrints, R., Dechaffoy, D., and Kondo, M.**, *Artemia* extracellular hemoglobin: ontogeny, structure and *in vivo* radiolabeling, in *The Brine Shrimp Artemia*, Vol. 2, Persoone, G., Sorgeloos, P., Roels, O., and Jaspers, E., Eds., Universa Press, Wetteren, Belgium, 1980, 427.
78. **Bagshaw, J.**, Biochemistry of *Artemia* development. Report on a symposium held in Toronto (Canada) in July 1979, in *The Brine Shrimp Artemia*, Vol. 2, Persoone, G., Sorgeloos, P., Roels, O., and Jaspers, E., Eds., Universa Press, Wetteren, Belgium, 1980, 3.
79. **Bagshaw, J. and Acey, R.**, Stage specific gene expression in *Artemia*, in *Biochemistry of Artemia Development*, Bagshaw, J. and Warner, A., Eds., University Microfilms International, Ann Arbor, Michigan, 1979.
80. **Vos, J., Bernaerts, F., Gabriels, I., and Decleir, W.**, Aerobic and anaerobic respiration of adult *Artemia salina* L., acclimated to different oxygen concentrations, *Comp. Biochem. Physiol.*, 62(A), 545, 1979.
81. **Trotman, C., Krissansen, G., Kastner, B., and Tate, W.**, Occurrence and properties of the glycoprotein artemocyanin, in *Artemia Research and Its Applications*, Vol. 2, Decleir, W., Moens, L., Slegers, H., Sorgeloos, P., and Jaspers, E., Eds., Universa Press, Wetteren, Belgium, 1987, 159.
82. **Bowen, S. T., Moise, H. W., Waring, G., and Poon, M. C.**, The hemoglobin of *Artemia salina*. III. Characterization, *Comp. Biochem. Physiol.*, 55(B), 99, 1976.
83. **Lavens, P. and Sorgeloos, P.**, The cryptobiotic state of *Artemia* cysts, its diapause deactivation and hatching: a review, in *Artemia Research and Its Applications*, Vol. 3, Sorgeloos, P., Bengtson, D. A., Decleir, W., and Jaspers, E., Eds., Universa Press, Wetteren, Belgium, 1987, 20.
84. **Wolf, G., Blanckaert, R., Moens, L., and Vos, J.**, *Artemia* haemoglobin: influence of isolation and conservation procedures on the physiological properties of the molecule, *Arch. Int. Physiol. Bioch.*, 91, B 47, 1983.

85. **Abreu-Grobois, F. and Beardmore, J.,** International study on *Artemia*. II. Genetic characterization of *Artemia* populations, an electrophoretic approach, in *The Brine Shrimp Artemia,* Vol. 1, Persoone, G., Sorgeloos, P., Roels, O., and Jaspers, E., Eds., Universa Press, Wetteren, Belgium, 1980, 133.
86. **Abreu-Grobois, F.,** A review on the genetics of *Artemia*, in *Artemia Research and Its Applications,* Vol. 1, Decleir, W., Sorgeloos, P., Bengtson, D. A., Decleir, W., and Jaspers, E., Eds., Universa Press, Wetteren, Belgium, 1987, 61.
87. **Dutrieu, J.,** Observations biochimiques et physiologiques sur le developpement d'*Artemia salina* Leach, *Arch. Zool. Exp. Gen.,* 99, 1, 1960.
88. **Sorgeloos, P.,** De invloed van abiotische en biotische faktoren op de levenscyclus van het pekelkreeftje *Artemia salina* L., Ph.D. thesis, Rijksuniversiteit, Gent, Belgium, 1975.
89. **Gilchrist, B. M.,** The oxygen consumption of *Artemia salina* (L.), *Hydrobiologia,* 8, 54, 1956.
90. **Gilchrist, B. M.,** Haemoglobin in *Artemia, Proc. R. Soc. London,* 143, 136, 1954.
91. **Vanhaecke, P., Tackaert, W., and Sorgeloos, P.,** The biogeography of *Artemia*, an updated review, in *Artemia Research and its Applications,* Vol. 1, Sorgeloos, P., Bengtson, D. A., Decleir, W., and Jaspers, E., Eds., Universa Press, Wetteren, Belgium, 1987, 129.
92. **Post, E. J. and Youssof, N. N.,** A procaryotic intracellular symbiont of the Great Salt Lake brine shrimp *Artemia salina* (L.), *Can. J. Microbiol.,* 23, 1232, 1977.
93. **Nimura, Y.,** Note on hatching the cyst of *Artemia, Aquaculture,* 16, 105, 1968.
94. **Decleir, W., Vos, J., Bernaerts, F., and Van den Branden, C.,** The respiratory physiology of *Artemia* sp., in *The Brine Shrimp Artemia,* Vol. 2, Persoone, G., Sorgeloos, P., Roels, O., and Jaspers, E., Eds., Universa Press, Wetteren, Belgium, 1980, 137.
95. **Copeland, D. E.,** A study of salt secreting cells in the brine shrimp *Artemia, Protoplasma,* 63, 363, 1967.
96. **Croghan, P. C.,** The survival of *Artemia salina* (L.), in various media, *J. Exp. Biol.,* 35, 213, 1958.
97. **Clegg, J. S. and Conte, F.,** A review of the cellular and developmental biology of *Artemia*, in *The Brine Shrimp Artemia,* Vol. 2, Persoone, G., Sorgeloos, P., Roels, O., and Jaspers, E., Eds., Universa Press, Wetteren, Belgium, 1980, 11.
98. **Lavens, P. and Sorgeloos, P.,** Controlled production of *Artemia* cysts under standard conditions in a recirculating culture system, *Aquacult. Eng.,* 3, 221, 1987.
99. **Iwasaki, T.,** Tolerance of *Artemia* dry eggs for temperature, vacuum and radiation, *Bull. Inst. Intern. Froid. Annexe (Suppl.),* 5, 79, 1973.
100. **Skoultchi, N. and Morowitz, H.,** Information storage and survival of biological systems at temperatures near absolute zero, *J. Biol. Med.,* 37, 158, 1964.
101. **Hinton, H.,** Reversible suspension of metabolism and the origin of life, *Proc. R. Soc. London,* 171, 43, 1968.
102. **Gaubin-Blanquet, Y., Pianezzi, B., and Massue, J. P.,** Influence de differents facteurs physiques sur les capacites de developpement de l'oeuf d' *Artemia salina, C. R. Seanc. Soc. Biol.,* 170, 1305, 1976.
103. **Clegg, J. S., Zettlemoyer, A. C., and Hsing, H. H.,** On the residual content of dried but viable cells, *Experientia,* 34, 734, 1978.
104. **Keilin, D.,** The problem of anaerobiosis or latent life: history and current concept, *Proc. R. Soc. London,* 150, 149, 1959.
105. **Clegg, J. S.,** Metabolic studies of cryptobiosis in encysted embryos of *Artemia salina, Comp. Biochem. Physiol.,* 20, 801, 1967.
106. **Clegg, J. S.,** Do dried cryptobiotes have a metabolism?, in *Anhydrobiosis,* Crowe, A. and Clegg, J., Eds., Dowden, Hutchinson & Ross, Stroudsbourg, PA, 1973, 141.
107. **Crowe, J. H.,** Anhydrobiosis: an unsolved problem, *Amer. Nat.,* 105, 563, 1971.
108. **Crowe, J. and Clegg, J.,** *Dry Biological Systems,* Crowe, J. and Clegg, J., Eds., Academic Press, New York, 1978.
109. **Muramatso, S.,** Studies on the physiology of *Artemia* embryos. I. Respiration and its main substrate during early development of the encysted embryo, *Embryologia,* 5, 95, 1960.
110. **Sundess, G. and Valen, E.,** Respiration of dried cysts of *Artemia salina*, *J. Cons. Perm. Int. Explor. Mer.,* 32, 413, 1969.
111. **Clegg, J. S.,** Interrelationship between water and cellular metabolism in *Artemia* cysts. VIII. Sorption isotherms and derived thermodynamic quantities, *J. Cell. Physiol.,* 94, 123, 1978.
112. **Clegg, J. S.,** The hydration dependence of metabolism in *Artemia* cysts, in *The Biochemistry of Artemia Development,* Warner, A. and Bagshaw, J., Eds., University Microfilm International, Ann Arbor, MI, 1979, 223.
113. **Bellini, L.,** Osservazioni sugli acidi nucleici nello sviluppo di *Artemia salina* Leach, *Ricera Sci.,* 30, 816, 1960.
114. **Nakanishi, Y. H., Okigaki, T., Kato, H., and Iwasaki, T.,** Cytological studies of *Artemia salina* deoxyribonucleic acid (DNA) content and the chromosomes in encysted dry eggs and nauplii, *Proc. Jpn. Acad.,* 39, 306, 1963.

115. **Nakanishi, Y. H., Iwasaki, T., Okigaki, T., and Kato, H.,** Cytological studies of *Artemia salina*. I. Embryonic development without cell multiplication after the blastula stage in encysted dry eggs, *Annot. Zool. Jpn.*, 35, 223, 1962.
116. **Finamore, F. J. and Clegg, J. S.,** Biochemical aspects of morphogenesis in the brine shrimp *Artemia salina*, in *The Cell Cycle Gene-Enzyme Interactions*, Padilla, G. M., Whitson, G. L., and Cameron, I. L., Eds., Academic Press, New York, 1969, 249.
117. **MacLean, D. K. and Warner, A. H.,** Aspects of nucleic acid metabolism during development of the brine shrimp *Artemia salina, Dev. Biol.*, 1971.
118. **Olson, C. S. and Clegg, J. S.,** Cell division during the development of *Artemia salina, Wilhelm Roux's Arch.*, 184, 1, 1978.
119. **Crowe, J. H., Crowe, L. M., Drinkwater, L., and Busa, W. B.,** Intracellular pH and anhydrobiosis in *Artemia* cysts, in *Artemia Research and Its Applications*, Vol. 2, Decleir, W., Moens, L., Slegers, H., Sorgeloos, P., and Jaspers, E., Eds., Universa Press, Wetteren, Belgium, 1987, 19.
120. **Busa, W. B. and Nuticelli, R.,** Metabolic regulation via intracellular pH, *Am. J. Physiol.*, 245, R 409, 1984.
121. **Emerson, D. N.,** Surface area respiration during the hatching of encysted embryos of the brine shrimp *Artemia salina, Biol. Bull.*, 132, 156, 1967.
122. **Smitt, H., Grossfeld, H., and Littauer, V. Z.,** Mitochondrial biogenesis during differentiation of *Artemia salina* cysts, *J. Cell. Biol.*, 58, 643, 1973.
123. **Jardel, J. P. and Nayudu, P. R. V.,** The maturation of *Artemia salina, Micron*, 13, 365, 1982.
124. **Van der Linden, A., Blust, R., Van Laere, A. J., and Decleir, W.,** Light induced release of *Artemia* dried embryos from diapause. Analysis of metabolic status, *J. Exp. Zool.*, 247, 131, 1988.
125. **Van der Linden, A., Domisse, R., Criel, L., and Decleir, W.,** Light induced development of *Artemia* cysts, in *Cell and Molecular Biology of Artemia Development*, Plenum Press, New York, in press, 1989.
126. **von Bertalanaffy, L. and Krywienczyk, J.,** Surface rule in crustaceans, *Am. Nat.*, 87, 107, 1953.
127. **Zeuthen, E.,** Oxygen uptake as related to body size in organisms, *Quart. Rev. Biol.*, 28, 1, 1953.
128. **Emerson, D. N.,** The metabolism of hatching embryos of the brine shrimp *Artemia* salina, *Proc. S. D. Acad. Sci.*, 42, 131, 1963.
129. **Clegg, J. S.,** The control of emergence and metabolism by external osmotic pressure and the role of free glycerol in developing cysts of *Artemia salina, J. Exp. Biol.*, 41, 879, 1964.
130. **Bellini, L.,** Studio delle dipeptidase e proteinasi nello sviluppo di *Artemia* salina Leach, *Atti Accad. naz. Lincei Rc.*, 22, 340, 1957.
131. **Emerson, D. N.,** Incorporation of tritiated thymidine by *Artemia salina, Proc. S. D. Acad. Sci.*, 43, 90, 1964.
132. **Cervera, M., Martin, E., Domingo, A., Vallejo, C. G., and Marco, R.,** DNase activity during early development in *Artemia*, in *The Brine Shrimp Artemia*, Vol. 2, Persoone, G., Sorgeloos, P., Roels, O., and Jaspers, E., Eds., Universa Press, Wetteren, Belgium, 1980, 293.
133. **Warner, A. H.,** The biosynthesis, metabolism and function of dinucleoside phosphates in Artemia embryos: a compendium, in *The Brine Shrimp Artemia*, Vol. 2, Persoone, G., Sorgeloos, P., Roels, O., and Jaspers, E., Eds., Universa Press, Wetteren, Belgium, 1980, 105.
134. **Warner, A. H., Poudziukas, J. G., and Finamore, F. J.,** Yolk platelets in brine shrimp embryos. Site of biosynthesis and storage of the diguanosinenucleotides, *Exp. Cell. Res.*, 70, 365, 1972.
135. **Olalla, A., Renart, J., Sillero, M. A. G., and Sillero, A.,** Metabolismo de las plaquetas de vitelo durante el desarollo de la *Artemia salina*, in *Avances de la bioquimica*, Cornudella, L., Heredia, C. F., Oro, J., and Sols, A., Eds., Editorial Salvat, Barcelona, Spain, 1977, 111.
136. **Bernaerts, F., Doumen, C., Sebrechts, J., Kupers, L., Van der Linden, A., Van den Branden, C., and Decleir, W.,** The aerobic metabolism during ontogeny of *Artemia, Biol. Jordbrukstek. Dodonaea*, 49, 49, 1981.
137. **Eliasse, N.,** The energy metabolism of *Artemia salina* in relation to body size, seasonal rhythms, and different salinities, *Univ. Bergen. Arb. Naturv.*, 11, 1, 1952.
138. **Khmeleva, N. N. and Yurkevich, G. N.,** Energy metabolism in *Artemia salina* (L) and some other Crustacea, *Dokl. Akad. Nauk.*, SSSR, 183, 978, 1968.
139. **Bernaerts, F., Blust, R., Thoeye, C., and Van der Linden, A.,** Comparison of the aerobic metabolism for four strains of *Artemia*, in *Artemia Research and Its Application*, Vol. 2, Decleir, W., Moens, L., Slegers, H., Sorgeloos, P., and Jaspers, E., Eds., Universa Press, Wetteren, Belgium, 1987, 3.
140. **Conte, F., Lowry, J., Carpenter, J., Edwards, A., Smith, R., and Ewing, R.,** Aerobic and anaerobic metabolism of *Artemia* nauplii as a function of salinity, in *The Brine Shrimp Artemia*, Vol. 2, Persoone, G., Sorgeloos, P., Roels, O., and Jaspers, E., Eds., Universa Press, Wetteren, Belgium, 1980, 125.
141. **Herbst, D. B. and Dana, G. L.,** Environmental physiology of salt tolerance in an alkaline salt lake population of *Artemia* from Mono Lake, California, USA, in *The Brine Shrimp Artemia*, Vol. 2, Persoone, G., Sorgeloos, P., Roels, O., and Jaspers, E., Eds., Universa Press, Wetteren, Belgium, 1980, 157.
142. **Gilchrist, B. M.,** The metabolism of *Artemia salina*, Ph.D. thesis, University London, UK, 1959.

Chapter 9

TAXONOMY AND POPULATION GENETICS OF *ARTEMIA*

Robert A. Browne and Sarane T. Bowen

TABLE OF CONTENTS

I. INTRODUCTION

The genus *Artemia* Leach 1819[1] is an exceptionally favorable system for the study of evolutionary processes. The ease with which the environmental parameters may be quantified and the convenience with which cysts may be stored and nauplii reared in the laboratory make this material ideal for the study of evolutionary adaptation. With regard to the study of evolutionary genetics, sexual populations may be compared with asexual forms (both diploid and polyploid). There are sibling species and superspecies (clusters of incipient species in *statu nascendi*) and the sequential array of evolutionary stages allows us to see evolution in progress. Because many populations have undergone Holocene radiations, we might anticipate minimal morphological differences, a range in the degree of reproductive isolation, and populations intermediate between races and species. We may anticipate some taxonomic problems with examples of partial barriers to gene flow. Since the phylogenetic systematics of *Artemia* has recently been reviewed,[2] we will limit our discussion to "alpha taxonomy".

II. TAXONOMY

The genus *Artemia* is a complex of sibling species and superspecies defined by the criterion of reproductive isolation. Endemic to the Old World are *A. parthenogenetica* Barigozzi 1974,[3] Bowen and Sterling 1978[4] (in Europe, Asia, and Australia), *A. tunisiana* Bowen and Sterling 1978[4] (in the Mediterranean region), and *A. urmiana* Gunther 1900[5] (from Iran). *A. sinica* Yaneng 1989[5] (from China) is probably distinct from *A. urmiana*. Endemic in the New World are *A. persimilis* Piccinelli and Prosdocini 1968[7] (in Argentina) and the *A. franciscana* superspecies (in the Americas and the Caribbean). Within the superspecies are populations which are reproductively isolated in nature: *A. (franciscana) franciscana* Kellogg 1906,[8] *A. (franciscana) monica* Verrill 1869,[9] and *A. franciscana sp.*, the cluster of populations in Nebraska, U.S. (details in section III).

Because many populations are now being characterized, scientists are urged to use the application *Artemia* sp. unless they have carefully studied biochemical, cytogenetic, or morphological traits that are diagnostic of the sibling species. All papers should cite the location and date of collection of the adults or cysts. These recommendations are the consensus of scientists attending the two International *Artemia* Symposia.

Early taxonomists assigned species names to populations with different morphologies although they were collected at different temperatures and salinities. In the period from 1910 to 1968, the profusion of names was abandoned and most workers referred to all brine shrimps as *A. salina* Linnaeus 1758[10] (see Barigozzi[3]). Some authors continue the practice today. In this review, we assign different names to reproductively isolated populations of clusters of populations. The binomen *Artemia salina* is assigned to the extinct population at Lymington, England first described by Schlosser in 1755 (reprinted in Kuenen and Baas-Becking[11]) and named by Linnaeus.[10] This has several advantages. First, it discontinues the use of *A. salina* for any living species, reducing the confusion between the entire complex and one species. Second, this greatly facilitates computer searches of the literature; a request for papers on *A. tunisiana* accesses only research on the Mediterranean biparental species, whereas a search for papers on *A. salina* accesses references on many species. This is in accord with Bowen and Sterling,[4] Bowen et al.,[12] and Abreu-Grobois.[2] The alternative would be an attempt to demonstrate an affinity between the shrimp described by Linnaeus and one of the three living Old World bisexual species or to *A. persimilis* which has been reported in the Mediterranean.[13] In our opinion, this last alternative would be unwise at this time because it might restore the confusion in terminology.

There are many problems in recognizing the species designation of a newly discovered shrimp population. There may be two populations in the same saline habitat. For example, mixtures of parthenogenetic and biparental populations have been reported in Mediterranean salterns.[14] Commercial aquaculture ventures have seeded salterns with imported cysts on many occasions. *A. franciscana* has been introduced throughout Asia, Australia, and South America. The Second International Symposium on *Artemia* adopted the resolution contained in the last paragraph of the paper by Geddes and Williams[15] recommending that measures be taken to conserve the genetic resources of natural populations.

The greatest problem in recognizing the species designation of an unknown population is the lack of reliable and convenient diagnostic characteristics. Abreu-Grobois[2] tabulated the known species differences in chromosome and chromocenter numbers and in morphology of caudal furca, ovisac, penes, and clasper knob. However, the morphological studies are usually reported from only a few races and may not be identical for all populations within one species. Although allozymes have been studied in many populations within each species, often only the dendrograms have been published. Thus, the tables of electrophoretic mobilities are not available to enable other workers to assign an unknown population to a particular species.

The electrophoretic patterns and relative mobilities have been presented for two esterases, three hemolymph proteins (two hemoglobins and artemocyanin-I), and Alp-1.[4,12,16] These characters were selected as being of value in diagnosing species or identifying mixtures of populations. They distinguish between endemic and introduced populations, and provide a key for identification of the species.[16] The control hemolymphs were from the San Francisco Bay population; for reference, the mobility of unity was assigned to the most prevalent San Francisco Bay allozyme of hemoglobin. Among publications of lists of mobilities, the proteins of greatest value as diagnostic characters (low variation within species, great variation between species) are the MDH, esterase, and ALP enzymes, and the two fastest hemoglobins.[4,12,16] In addition, Abreu-Grobois[17] has suggested recognition of sibling species based on polymorphism in these enzymes: esterases, GOT, SOD, CAT, MDH, MPI, and PGI.

The importance of thorough documentation of cyst collections and painstaking attention to morphologies and biochemical markers is illustrated by the problems encountered with *A. urmiana*.[2,5,18] The population in Lake Urmia, Iran is the only known example of the biparental *A. urmiana*.[3,11,13] However, a collection of cysts reported to be from Lake Urmia was found to be parthenogenetic.[19] Recent evidence supports the co-occurrence of bisexual and parthenogenetic in Lake Urmia.[20] The biparental *A. urmiana* sibling species is defined by four biochemical markers unique to biparental populations and is the only population known with Est-1 and Est-2 relative mobilities of 108 and 104, respectively.[4,16] It is also the only known biparental population with Hemoglobin-2 mobility greater than unity, a trait shared only by the parthenogenetic populations.[12] For each of these three markers, at least 17 individuals were scored and all had the same electrophoretic pattern. The presence of a unique GOT band was noted by Abreu-Grobois.[17] It will be interesting to see if *A. urmiana* is present at additional Asian sites and if it is the senior synonym for *A. sinica*.

The question arises as to what is the appropriate criterion for species or semi-species status? We recognize that cross-fertility is a plesiomorphic trait which is inadequate for assessing the evolutionary history of a group. However, we choose the criterion of reproductive isolation as the most appropriate in predicting the future evolution of the genus *Artemia*; the recent increase in rate of dissemination of cysts into new environments suggests that the gene pool of many endemic populations may change if there is no barrier to gene introgression.

We will now review the cross-fertility tests which were the basis of the diagnosis of sibling species in *Artemia*. Among the bisexual populations, the first demonstrations of reproductive isolation were between wild-type Mediterranean saltern shrimps and either

"California shrimps"[13,21-22] or *A. persimilis.*[13,18] Reproductive isolation was later shown for *A. urmiana,*[16,18] *A. monica,*[22,23] and for *A. sp.*, the cluster of populations from Nebraska.[24]

In an ideal experimental design, each of the known *Artemia* populations would be crossed in a matrix. Because there are more than 300 *Artemia* habitats,[25] this is impractical. One solution is to mate many populations to one well-characterized wild-type population[12,18,22,23] or to a stock which carries a genetic marker.[12,16,18,22,26] The marker mutation offers the advantage of excluding the possibility of both parthenogenesis and gynogenesis (the parthenogenetic development of an egg after stimulation by a sperm). To rule out mixtures of cysts, the last mentioned studies of reproductive isolation were performed with cyst collections which were characterized by electrophoretic assays. In all cases, the genetic diversity in PGI allozymes and hemoglobin-1 bands revealed gene frequencies predicted by the Hardy-Weinberg law for single populations.[4,15,16] In this way, reproductive isolation was noted between all groups that are now designated sibling species with one exception: to our knowledge, no report has been made of an attempt to mate *A. urmiana* with either *A. persimilis* or *A. sinica*.

The presence of partial reproductive isolation within the *A. franciscana* superspecies is discussed subsequently under the headings "Ecological Isolation" and "Hybridization".

III. CYTOGENETICS AND CLASSICAL GENETICS

A. CYTOGENETICS

The bisexual species of *Artemia* have a chromosome number of 42 with the exception of *A. persimilis* which has a diploid number of 44. Parthenogenetic populations show diploid, triploid, tetraploid, and pentaploid forms. Within one saltern, the parthenogenetic *Artemia* may have several ploidy numbers as well as aneuploidy. The literature on chromosome counts, the documentation of several types of aneuploidy (in polyploid and diploid populations), and the proposed mechanisms for generation of polyploid and aneuploid series have been reviewed extensively.[2,3]

Artemia populations can be characterized by the number of chromocenters (heavily staining heterochromatic areas in the interphase nuclei which are rich in repeats of adenine and thymine residues). *A. tunisiana, A. urmiana*, and *A. parthenogenetica* have no chromocenters whereas the *A. persimilis* population at Buenos Aires has an average of 3 chromocenters and populations within the *A. franciscana* superspecies have an average of 4 to 17 chromocenters (reviewed by Abreu-Grobois[2]). The Alu I family of repeats is positively correlated with chromocenter number and intensity of C-banding.[2,19,27]

Abatzopoulos et al.[28] have reviewed the earlier literature on karyotypes and chromosome measurements, publishing karyotypes of *A. franciscana* (Great Salt Lake) and *A. parthenogenetica* (three Greek tetraploid populations). Heteromorphic chromosomes were not evident. In two Italian populations, the total chromosome length in the tetraploid was just twice that in diploid cells.[29] Endoreduplication of chromosomes is common in Crustacea[30] and has been noted in *Artemia*. Although the chromosome count is 2n, the proliferating epidermal cells will have 2C, the gut cells 4C—8C, and the salt gland cells >8C units of DNA.[31]

B. CLASSICAL GENETICS

Studies of the inheritance of mutations affecting morphology or eye color revealed that the female is the heterogametic sex[22,26] and that the female does not store sperm from one reproductive cycle to the next, as is the case in *Drosophila*.[32] Examples of epistasis and gene interaction were found among the mutations governing the ommochrome eye pigments.[33,34] Photographs of the mutations of *A. franciscana*[33] and a comparison of these with the mutations of *A. tunisiana*[35] have been published.

The white eye mutation has been observed in *A. persimilis*, *A. tunisiana*, and *A. franciscana*. The locus is partially sex linked (located on the pairing segments of the sex chromosomes). The amount of crossing over between the white locus and the "sex locus" (junction of pairing and non-pairing regions of the sex chromosome) varies from 20 to <0.1% among the populations. The white-eye mutation facilitated the first genetic analysis of sex determination in the Crustacea.[36] Because the female is heterogametic, a stock characterized by low cross-over rate produces only white-eyed males and black-eyed females in each generation.[26,37]

Mosaics have been described.[33] Eleven were sex mosaics (including a perfect bilateral gynandromorph) and five shrimps were mosaic for eye color as well. In each case, the internal organs were clearly male or female, with no intersexual region. Thus, if a sex hormone is present in *Artemia*, it does not suppress the differentiation of cells with the chromosome constitution of the opposite sex. Three shrimps were mosaic for mixtures of genitalia and phyllopod tissue, illustrating the homology of these structures.

Artemia hemoglobins and artemocyanins have been followed in genetic crosses. Genetic analyses revealed that the hemoglobin with fastest electrophoretic mobility, Hb 1, is an alpha homodimer.[12,39] Hb 2 is an alpha-beta heterodimer, and Hb X (HbIII of Heip et al.[38]) is a beta homodimer.[12,39] The polypeptide composition of these three hemoglobins was confirmed on two-dimensional electrophoresis.[40] The beta polypeptide is differentially induced by low oxygen[15,41,42] and the beta homodimer has the highest affinity for oxygen.[42] Two-dimensional electrophoresis revealed that another heme-containing protein is a complex between the beta homodimer and artemocyanin.[43] Artemocyanin is a glycoprotein bound to a bilatriene blue-green pigment.[44] The hemoglobin-artemocyanin I complex, Hb 3, was observed only in shrimps reared to adulthood under high oxygen. The artemocyanin locus (erroneously termed "Hb-3") in Sterling et al.[45] is not linked to the locus governing the alpha polypeptide of hemoglobin. Although many other biochemical genetic markers have been described, there have been no reports of linkage except for that of the white eye gene and the sex-locus.

C. HETEROZYGOSITY LEVELS

The influence of heterozygosity on female reproductive success has recently been examined.[46] Studies with *A. franciscana* from Great Salt Lake, UT show that higher heterozygosity results in greater overall reproductive success and an increased capacity to switch resources into the oviparous mode of reproduction.

IV. DIVERGENCE OF SEXUAL POPULATIONS

A. ECOLOGICAL ISOLATION

Profound differences among *Artemia* populations in tolerance of the anionic composition of the medium have been observed.[2,23] Shrimps from low-chloride lakes (which are high in sulfate or carbonate) have reduced viability in concentrated sea water. Nauplii from salterns or high chloride lakes die in high carbonate lake waters.[22,24] Thus, many populations within the *A. franciscana* superspecies are ecologically isolated in nature.[24] The tests of hybridization among these groups are discussed subsequently.

Surprisingly, all nauplii within the *A. franciscana* superspecies were readily cultured in simple synthetic media containing only chloride, carbonates, sulfate, sodium, potassium, magnesium, and calcium. Other minerals were supplied by the yeast diet; dietary additions were relatively small and were minimal compared to the ionic differences of the culture media (or the native source waters). Viability of nauplii was observed over a three week period in pairs of simple media which differed in regard to only one parameter. In this way, it was found that the inability of saltern populations to live in high carbonate waters (such as Fallon Pond, NV, or Mono Lake, CA), could be attributed to their inability to tolerate calcium concentrations <20 mg/l. The comparison of viabilities in permissive and selective media revealed population differences in the tolerances of each of the four anions and four cations in the synthetic media.[24] For each of the eight ions, the range of concentrations encountered by each population in the habitat is narrower than the range affording high viability in laboratory media. As expected, the midpoints of the two ranges are sometimes similar. In many cases, however, the narrower range of ionic concentrations reported for lake water is at the end of the range affording high viability in the laboratory. In plots of the eight ions, the realized niche (ionic composition of the lake) often lies at the edge of the potential niche (based on genetic potential, determined by laboratory tests).[24] The populations in inland salt lakes of North America have experienced enormous changes in ion compositions during the Holocene and it appears that certain *A. franciscana* semispecies have not succeeded in closely tracking the environment.

B. CROSS-FERTILITY AMONG NORTH AMERICAN POPULATIONS

Twenty bisexual populations from high-chloride habitats in Mexico, U.S.A., Canada, two Caribbean islands, one Galapagos Island, and Australia were tested for cross fertility.[4,22,23] Although all possible crosses were not made, there were no examples of reproductive isolation among the wild populations. Furthermore, the 20 populations were crossfertile with inbred stocks homozygous for recessive mutations. These mutant stocks were derived from the Utah and/or the San Francisco Bay population(s). In all cases, fertile F_2 progeny, the expected sex ratio, and Mendelian segregations of the marker were obtained. The discovery that the Rockhampton (Australia) population was cross-fertile with the San Francisco Bay population was a surprise. However, the two populations had the same PGI and hemoglobin profiles.[4] Evidently, the San Francisco Bay population was recently introduced into the Rockhampton salterns.

Recent studies show that a number of additional populations can be assigned to *A. franciscana*: Yucatan and Sonora (Mexico), Araya (Venezuela),[19] Loving Salt Lake, NM,[47] and Laysan Island in the Hawaiian archipelago.[23] Vanhaecke et al.[25] have published an extensive table of locations of *Artemia* habitats in North America.

There is evidence of genetic barriers between *A. franciscana* populations that inhabit the same type of ionic habitats but are geographically isolated. Recently, Browne et al.[48] crossed populations from Monterey Bay, CA, (MB) and Zuni Salt Lake, NM, (ZUN), a geographically isolated population that is morphologically distinguishable by one female characteristic from other *A. franciscana* populations.[17] Separate lines originating from reciprocal crosses, along with control populations, were maintained through the F_3. For each generation, 11 life history traits were quantified, including components of female lifespan. The F_1 and F_2 populations exhibited a decline in reproductive output of 20 and 80% respectively compared to P populations, with a high incidence of sterility and semisterility in F_2 populations. Female (but not male) life-span also showed significant decreases, with the length of the female reproductive period decreasing significantly more than pre- or post-reproductive components. Partial recovery was found for the F_3, with offspring production levels approaching those of the F_1. Encystment/ovoviviparity ratios also appear to be influenced by hybridization events. Despite the electrophoretic similarity between ZUN and MB, there appear to be serious barriers to hybridization between the two populations due to maladaptive epistatic interactions and/or the break-up of coadapted gene complexes. Thus, the ZUN population may best be described as a subspecies of the *A. franciscana* complex that is approaching distinct speciation. Supporting evidence has been found,[43] where low fertility was found in F_2 between Zuni and Fallon populations.

After low-chloride synthetic media were devised, it became possible to test for postmating reproductive isolation among the ecologically isolated inland lake populations in the U.S.

Artemia from Kiatuthlanna, AZ, Mono Lake, CA, Fallon Pond, NV, Penley Lake, WA, and Jesse Lake, NB were all cross-fertile with each other or with San Francisco Bay shrimps.[23] The location and composition of the lake waters have been described.[24] The greatest ecological isolation was found for the population from Jesse Lake where chloride makes up $<12\%$ of the molar anionic composition. No synthetic medium could be devised which permitted both Jesse nauplii and San Francisco nauplii to thrive. However, Jesse was successfully mated to Fallon shrimps in solution A and Fallon shrimps were mated to San Francisco Bay shrimps in solution B. The two types of F_1 progeny were mated to produce fertile F_2 progeny. Because most ionic tolerance genes are dominant or codominant,[43] some offspring were obtained in the F_3. However, reduced viability (hybrid breakdown) was observed in the F_2 and F_3 in each culture medium, presumably due to segregation of genes for ionic tolerance. Although some *A. franciscana* populations are reproductively isolated in nature, they can hybridize in carefully contrived laboratory media. All these populations are part of the *A. franciscana* superspecies. That is, they represent a monophyletic group of allopatric populations which includes sibling species, incipient species (semispecies), and subspecies.[2,23,24] The Mono Lake population has a unique egg morphology and distinctive life history traits which make it uniquely suited to this deep cold permanent lake.[49-51] It is correctly called either *Artemia monica* or *A. (franciscana) monica* Verrill 1869.[9]

C. CROSS-FERTILITY AMONG OLD-WORLD BISEXUAL POPULATIONS

A number of Old-World bisexual populations have been hybridized in the laboratory. For the populations from Larnanca (Cyprus), Megrine (Tunisia), and Santa Pola (Spain), crosses were both round robin and reciprocal. Additional crosses, although limited in number, were conducted with populations from San Pedro del Pinatar, Cadiz, Bonmati, and Bras de Port (all from Spain). In general, the number of F_1 offspring produced was intermediate compared with control crosses.[47,52] Recent work[53] has shown *Artemia* from Wadi Natrun, Egypt can also interbreed with Tunisian *Artemia*. Thus, all bisexual populations from the Mediterranean basin that were tested were found to be capable of interbreeding. It is possible that reproductive barriers may occur in post-F_1 generations but given the similar life-history characters of the Mediterranean basin populations,[54] and close genetic similarity,[55] that possibility is not considered likely for most populations. However, partial reproductive isolation has been found between the Tunisian and San Bartolomeo (Cagliari, Sardinia) populations, although some F_2 progeny were obtained.[16]

V. DIVERGENCE OF PARTHENOGENETIC POPULATIONS

A. INTRODUCTION

Diploid parthenogenetic populations may be composed entirely of one genotype (monoclonal) or two or more clones (polyclonal). Abreu-Grobois and Beardmore[55,56] looked at 16 enzyme systems representing 23 loci and found a large degree of clonal variation within the majority of the diploid asexual populations examined. Thus, *Artemia* are similar to the majority of asexual species, which have been found to consist of multiple clones.[57] However, *Artemia* are exceptional with regard to the extent of polyclonality, with 60+ clones reported at one site.[58]

The electrophoretic evidence from numerous parthenogenetic species are generally consistent with the hypothesis that most clonal diversity arose via multiple independent origins from sexual populations.[59-61] For example, in *Daphnia* although some clones reproduce by obligate parthenogenesis, they may be of very recent origin, perhaps as little as a decade or two.[62] However, the situation is quite different in *Artemia*, where there is no continual asexual genetic variation generated via ancestral sexual recombination. Both Bowen et al.[12] and Abreu-Grobois and Beardmore[56] have estimated from electrophoretic data that parthenogenesis in *Artemia* is of monophyletic origin, the latter suggesting that the parthenogenetic lineage branched from an Old World sexual lineage approximately 6×10^6 years ago. This makes the high clonal diversity found in *Artemia* asexual populations (as well as that between populations) a very interesting and important question since clonal diversity may be in relatively long-term equilibrium. However, allozymic data are less than reliable for constructing phylogenies. Further insight should occur when phylogenetic relationships based on mtDNA restriction fragment length polymorphisms are constructed.[58,63]

B. CLONAL VARIATION IN THE SALIN DE GIRAUD POPULATION: A CASE STUDY

The asexual diploid population at Salin de Giraud, France has been extensively studied with regard to the evolutionary aspects of polyclonality.[52,58,64,65] The population occurs in the compact yet extremely variable environments of a commercial salina. Located at the Rhone delta, the salt works consists of approximately 100 ponds covering 11,000 ha. The ponds, however, are not arranged along a geographical cline according to salinity but instead the distribution is patchy.[64] This exceedingly steep and stable salinity gradient offers an ideal opportunity to study the effects of environmental factors on clonal distribution.[52] Salinity levels, along with pond depth, dictate not only abiotic factors such as osmotic pressure, temperature and dissolved oxygen but also such biotic factors as phytoplankton quality and quantity, and the diversity and density of competitors and predators. The Salin de Giraud population consists of at least 60 clones.[58] The clones are relatively closely related in terms of electrophoretic differences[52,56,58] but are widely divergent for numerous life history traits.[58,65] The clones are most likely of local origin since allozymically determined genotypes of other populations of *Artemia* are not present.[55,66]

Five reproductive and five life-span characteristics for nine clonal lineages from the Salin de Giraud population have been quantified.[58,65] The differences between clones are surprisingly large. For example, the total number of offspring produced per female varies seven-fold between clones, while the percent of offspring encysted (as opposed to ovoviviparous production) ranges from 0 to 99.5%. The length of the prereproductive period, a key factor in determining population growth rates, varies 2.9-fold.[58]

Artemia from eight ponds within the Salin de Giraud complex which were collected during 1985, and whose salinity levels varied from 73 to 290 g/l, have been examined for allozymic composition.[52] A significant correlation ($r = 0.78$; $p < 0.05$) exists between percent of loci polymorphic (%P) and the salinity levels of the eight sampled ponds. There also is a positive (but non-significant) correlation between salinity and average individual heterozygosity level ($\overline{H}$). Thus, as salinity increases, %P significantly decreases while $\overline{H}$ increases. Salinity also correlates significantly with allozymic variation at three loci: phosphoglucose isomerase-1 (Pgi-1), leucine alanine-1 (La-1), and indophenol oxidase-1 (Ipo-1).

The significant correlations between allozymic frequencies and salinity levels lend support to the hypothesis that polymorphic loci are maintained by environmental heterogeneity. This may be due to the fact that the salina at Salin de Giraud represents an optimal environmental regime for maintaining genetic variation: a spatial coarse-grained environment with habitat selection.[67] In addition, environmental conditions in each pond are seasonally variable and there is some opportunity for mixing of pond populations.

Similar allozymic correlations were found when genotype analyses were performed on Salin de Giraud pond populations collected in 1987. Genotype selection at different salinities for the Salin de Giraud population has also been conducted in the laboratory.[58] Random aggregations of clones were raised at 100, 150 or 190 g/l salinity. When the populations were subsequently electrophoretically surveyed, very rapid clonal selection had taken place, with clonal survivorship mimicing what had been found in earlier distributions in the salinas.

In the laboratory selection experiments, the number of clones decreased from 46 to 32 to 7 as salinity increased from 110, 150 to 190 g/l, respectively. Population survivorship also dropped from 152 to 132 to 22 individuals, respectively. The change in clonal composition is attributable to differential clonal survival and not differential reproduction since the duration of the experiments was limited to 77 days and involved only a single generation. Browne and Hoopes[58] concluded that salinity effectively acts as a filter, selectively removing clones, until at the highest salinity, one clone dominates the population. Despite extensive differences between the Salin de Giraud population and other asexual populations, mitochondrial restriction fragment length polymorphisms are identical for nine restriction enzymes for all clones from Salin de Giraud,[58] suggesting that the clonal diversity at Salin de Giraud is of very recent origin.

The results from the investigation of the Salin de Giraud population lead to a number of conclusions: (1) a large number of clonal genotypes occur despite obligate asexuality; (2) the genotypes appear to be monophyletic and due to their mtDNA similarity are presumed to be of relatively recent origin; (3) the genotypes differ in fitness and face rapid sorting (selection) according to salinity. This suggests that rapid evolution can occur in an obligately asexual *Artemia* population leading to diverse microspeciation.[58]

C. *ARTEMIA* AS A MODEL FOR EXAMINING THE ADAPTIVENESS OF SEXUAL VS. ASEXUAL REPRODUCTION

The conventional wisdom regarding asexual populations is that they are expected to be low in genetic variability since sexually generated recombination is not possible.[68,69] However, this expectation has not proven to be valid for asexual *Artemia* populations that are diploid. In these populations, genetic recombination can occasionally occur ameiotically during gamete production (for further details[2,3]). Even if a population was founded by a single asexually reproducing female with no subsequent immigration, genetic variation could still be generated via mutation, and different genetic combinations could be obtained in subsequent generations, much as in sexually reproducing populations. The product is an asexual population that is genetically variable, which in turn, could result in variability in lifespan and reproductive traits. Thus, in diploid asexual *Artemia* populations successful genotypes are faithfully perpetuated but occasional recombination allows the production of a few "experimental" genotypes, some of which may be successful in different sets of biotic and/or abiotic conditions. Exact replication is combined with occasional experimentation, but not at the expense of the elimination of previously proven genotypes. Favorable mutations can also potentially be expressed as homozygotes via ameiotic recombination. Ameiotic recombination can also permit a decrease in genetic load. Genetic load (Mueller's rachet) has been hypothesized as one of the long-term detriments to asexual reproduction.[70,71] The impression given by work on other asexual species has been that genetic diversity is considerably greater when asexual lineages are recurrently recruited from sexual populations than when parthenogenesis is obligate.[60] The data on *Artemia* challenges this belief since the evidence points to a high degree of interclonal genetic diversity accompanied by major differences in ecologically important life-history traits between clones.

D. RARE MALES IN PARTHENOGENETIC *ARTEMIA* POPULATIONS

Males occur in low numbers in at least nine different *A. parthenogenetica* populations.[72] Stefani[73] has suggested that the rare males in asexual populations of *Artemia* arise as the result of fusion of two haploid X cells after the second meiotic division. Recent allozymic work[74] has found that rare males carry allozymic markers that are identical to their maternal genotypes. Rare male mtDNA profiles also are identical to the maternal type.[66]

The occurrence and potential reproductive role of rare males produced by obligately asexual brine shrimp were examined by MacDonald and Browne.[72] The study of sex ratios in offspring from 30 different parental lines through three generations indicated the existence of two groups of females (Group A and Group B), which were genetically distinct with regard to their tendency to produce males. Group A females consistently produced fewer males than did the Group B females, and thus had a lower ratio of males to females among their offspring than the average population sex ratio. A significant increase in ratios of males to females in broods of the Group B females in the third generation suggests that selection for females expressing this trait may have occurred. Another possibility is that selection may have occurred between the many clonal lines which are known to occur at Salin de Giraud.

There are at least three possible explanations for the production of males by asexual female *Artemia*.[72] The first is that these males are functionally sterile and represent a disadvantageous trait that will be selected against. This view regards rare males as "recombinational mistakes". The second is that the production of males is directly advantageous to the parental female; that is, they contribute in some unknown way to genome proliferation or survival. The third is that rare males are nonfunctional in their own population, but their production is associated with another genetic trait which is itself adaptive. For example, as Hebert[75] has pointed out, male reproduction in obligately parthenogenetic clones of *Daphnia* may serve as a vector in the displacement of sexual populations by clones reproducing by obligate parthenogenesis. However, since *Artemia* males (unlike *Daphnia*) are XX and females are XY, the situation is not directly analogous. No evidence was found by MacDonald and Browne[72] for the ability of asexually produced males to outbreed or hybridize with sexual females. However, Bowen et al.[12] succeeded in crossing rare males with females of *A. franciscana* and *A. urmiana*. In two cases, fertilization was confirmed by Mendelian segregation patterns of genetic markers in the offspring. Fertilization by rare males might also result in the generation of polyploidy in parthenogenetic females.[56]

VI. EVOLUTIONARY DIVERGENCE BASED ON MITOCHONDRIAL DNA ANALYSIS

Hoopes and Browne[63] have begun a general survey of a number of sexual and asexual populations using mtDNA restriction fragment length polymorphisms (RFLPs). This has been done using 9.6 and 5.7 Kb fragments that represent the entire mitochondrial genome which have been cloned into the BamH I site of pBR322 plasmid.[76]

Mitochondrial DNA approaches an ideal molecular system for phylogenetic analysis due to its simple genetic structure, straightforward mode of transmission, and relatively fast rate of evolution. It has provided the first extensive data available to evolutionists for inferring genealogical relationships at the intraspecific level.[77] Meaningful phylogenetic comparisons using mtDNA surveys should normally be confined to conspecific populations and closely related species whose separations date to within the last 10 million years.[77] The analysis for *Artemia* fits these criteria since the allozymic evidence points to a very approximate date of divergence between Old World sexual and parthenogenetic lineages of 6 million years ago.[56]

The RFLP data for the enzyme EcoR I is presented in Table 1 (see also Browne and Hoopes[58]). Three patterns (A, B, and C) have been found for asexual populations. The five sample populations from Salin de Giraud, France include two lab-reared monoclonal lines (each derived from a single female) and three populations from ponds of different salinity collected in the field. All populations show the same RFLP pattern suggesting monophyletic origin (at least with regard to this RFLP). Pattern C appears to be representative of tetraploid populations since Margerita, Commachio, and Lavalduc are known tetraploids and Huelva is suspected of being mixed ploidy.[2] Some populations have more than one pattern, suggesting that some populations are mixed. Pattern D appears in *A. franciscana* while two patterns (E and F) are found for the two Old World bisexual populations (*A. tunisiana*). Two four-

TABLE 1
Restriction Fragments for EcoR I for the 9.6 kb mtDNA Fragment[a]

Patterns

A	B	C	D	E	F
8.0 kb	6.2	6.0	6.2	9.6	7.2
5.4	5.4	4.2	3.5	4.0	4.8
1.3	1.3	3.1	2.7		3.7
0.6	0.6	1.9	1.4		
0.3	0.3		0.9		

Populations with Above Patterns

A	B	C	D	E	F
Kutch, India	Lavalduc, France	Margherita, Italy	Monterey Bay, CA	Larnaca, Cyprus	Megrine, Tunisia
Madras, India	Sete, France	Commachio, Italy	Zuni, NM		
Izmir, Turkey	Salin de Giraud, France	Huelva, Spain	Loving Lake, NM		
Tientsin, China	Clone SGI (monoclonal line)	Lavalduc, France			
Namibia, South Africa	Clone A (monoclonal line)				
Cabo de Gata, Spain	Mullets (pond population)				
LaRoche, Morocco	Mas du Village (pond population)				
Kazakhistan, USSR	Grand Rascaillon (pond population)				
Huelva, Spain					
Sete, France					

[a] The populations found for each pattern in the upper half of the table are listed in the lower half of the table.

Note: Columns A, B, and C are the *Artemia parthenogenetica*; Column D, *A. franciscana;* columns E and F, *A. tunisiana*.

base cutting enzymes (Dde I and Mbo I) have been used on the five Salin de Giraud samples listed in Table 1. All show the same restriction patterns, again pointing to a monophyletic origin of the Salin de Giraud population.

Preliminary information[63] obtained from a number of restriction enzymes points to *A. tunisiana* as the parental group, with the New World bisexual species (*A. franciscana* and *A. persimilis*) having branched off a relatively long time ago (millions of years). In contrast, *A. parthenogenetica* appear to be a relatively recent branch. Currently, *A. parthenogenetica* cannot be linked to a particular Old World bisexual population (*A. tunisiana* or possibly *A. urmiana*) but that may be possible when additional populations are examined and with the use of additional restriction enzymes. Preliminary work[63] supports, at least in very broad terms, the phylogenetic relationships proposed by Abreu-Grobois and Beardmore.[55,56] However, questions concerning divergence times, the high variability found among *A. tunisiana* populations, and the evolutionary relationships of populations inhabiting athalassic waters, remain to be resolved. Hoopes and Browne[63] have also found that two restriction endonucleases (Hind I and Bam) produce repetition sequence RFLPs in polyploid populations, supporting the occurrence of repetition sequences reported by Batuecas et al.[76]

Marco et al.[78] have also found considerable mtDNA divergence among parthenogenetic and bisexual species found in Spain, as well as divergence between diploid and tetraploid asexuals. This data, as well as that of Hoopes and Browne,[63] reinforce the earlier conclusions based on allozymic analyses,[55] that bisexual species from North America and Eurasia diverged millions of years ago, and that the parthenogenetic lineage is a relatively ancient branch (on the order of millions of years) from the Old World sexual lineage. One discrepancy between allozyme and mtDNA data is that the former suggests a closer association among populations of *A. parthenogenetica* of different ploidy. One intriguing hypothesis is that this may be due to selective pressures on the nuclear genome as opposed to neutral changes of the mtDNA genome.[58] Since the results from both groups are preliminary, further investigation is needed to clarify exact phylogenetic relationships, as well as underlying mechanisms.

As noted in the Chapter on Molecular Genetics by Marco et al., *Artemia* species show differences in organization of at least some nuclear genes, namely satellite I sequences,[79,80] the large ribosomal RNA genes,[81,82] and the 5S-histone genes.[82]

Satellite I DNA is only present in the North American bisexual species. Marco et al.[78] have hypothesized that since current theories hold that *Artemia* originated in the Old World, then satellite I is a relatively recent acquisition, i.e., after the geographical isolation of Old and New World *Artemia*. However, 5S rRNA and histone genes are linked in all species examined. Therefore, if the parthenogenetic lineage branched from a bisexual ancestor, then the link between the 5S and histone genes occurred early in *Artemia* evolution.

ACKNOWLEDGMENTS

We thank Dr. Denton Belk and Dr. F. Alberto Abreu-Grobois for helpful comments throughout our work and for a critical reading of the manuscript. Sarane T. Bowen has received support from the National Science Foundation. Support to Robert A. Browne was provided by Whitehall Foundation Grant 86 M14 and an Archie Faculty Enrichment Grant.

REFERENCES

1. **Leach, W. E.**, Entomostraca, *Dictionaire des Sciences Naturelles*, 14, 524, 1819.
2. **Abreu-Grobois, F. A.**, A review of the genetics of Artemia, in *Artemia Research and Its Applications*, Vol. 1, Sorgeloos, P., Bengtson, D. A., Decleir, W., and Jaspers, E., Eds., Universa Press, Wetteren, Belgium, 1987, 61.
3. **Barigozzi, C.**, *Artemia*: a survey of its significance in genetic problems, *Evol. Biol.*, 7, 221, 1974.
4. **Bowen, S. T. and Sterling, G.**, Esterase and malate dehydrogenase isozyme polymorphisms in 15 *Artemia* populations, *Comp. Biochem. Physiol.*, 61(B), 593, 1978.
5. **Gunther, R. T.**, Contributions to the natural history of Lake Urmi, N.W. Persia, and its neighborhood, *J. Linn. Soc. London Zool.*, 27, 345, 1900.
6. **Yaneng, C.**, New *Artemia* sibling species from PR CHINA, *Artemia Newsl.*, 11, 41, 1989.
7. **Piccinelli, M. and Prosdocimi, T.**, Descrizione tassonomica delle due specie *Artemia salina* L. e. *Artemia persimilis* Piccinelli e Prosdocimi, *Rend. Inst. Lomb. Sci. Lett. Classe Sci.*, 102(B), 113, 1968.
8. **Kellogg, V. L.**, A new *Artemia* and its life conditions, *Science*, 24, 594, 1906.
9. **Verrill, A. E.**, Contributions to zoology from the museum of Yale College. III. Descriptions of some new American Phyllopod Crustacea, *Am. J. Sci. Arts Ser.*, 2, 43, 244, 1869.
10. **Linnaeus, C.**, *Systema naturae, Editio decima, reformata, Vol. I, Regnum animale, Impensis Direct*, Laurentiii Salvii, Holmiae, 1758.
11. **Kuenen, D. J. and Baas-Becking, L. G. M.**, Historical notes on *Artemia salina* (L.), *Rijksmuseum van Natuurlijke Hist. Zool. Meded.*, 20, 222, 1938.
12. **Bowen, S. T., Durkin, J. P., Sterling, G., and Clark, L. S.**, *Artemia* hemoglobins: genetic variation in parthenogenetic and zygogenetic populations, *Biol. Bull.*, 155, 273, 1978.
13. **Halfer-Cervini, A. M., Piccinelli, M., Prosdocimi, T., and Baratelli-Zambruni, L.**, Sibling species in *Artemia* (Crustacea: Branchiopoda), *Evolution*, 22, 373, 1968.
14. **Perez, R.**, Cyst production of *Artemia* in salt ponds in south-eastern Spain, in *Artemia Research and Its Applications*, Vol. 3, Sorgeloos, P., Bengtson, D. A., Decleir, W., and Jaspers, E., Eds., Universa Press, Wetteren, Belgium, 1987, 215.
15. **Geddes, M. C. and Williams, W. D.**, Comments on *Artemia* introductions and the need for conservation, in *Artemia Research and Its Applications*, Vol. 1, Sorgeloos, P., Bengtson, D. A., Decleir, W., and Jaspers, E., Eds., Universa Press, Wetteren, Belgium, 1987, 19.
16. **Bowen, S. T., Davis, M. L., Fenster, S. R., and Lindwall, G. A.**, Sibling species of *Artemia*, in *The Brine Shrimp, Artemia*, Vol. 1, Persoone, G., Sorgeloos, P., Roels, O., and Jaspers, E., Eds., Universa Press, Wetteren, Belgium, 1980, 155.
17. **Abreu-Grobois, F. A.**, personal communication.
18. **Clark, L. and Bowen, S. T.**, The genetics of *Artemia salina*. VII. Reproductive isolation, *J. Heredity*, 67, 385, 1976.
19. **Badaracco, G., Baratelli, L., Ginelli, E., Meneveri, R., Plevani, P., Valsasnini, P., and Barigozzi, C.**, Variations in repetitive DNA and heterochromatin in the genus *Artemia*, *Chromosoma*, 95, 71, 1987.
20. **Sorgeloos, P.**, Two strains of *Artemia* in Urmia Lake (Iran), *Artemia Newsl.*, 13, 5, 1989.
21. **Gilchrist, B. M.**, Growth and form of the brine shrimp *Artemia salina* (L.), *Proc. Zool. Soc. London*, 134, 221, 1962.
22. **Bowen, S. T.**, The genetics of *Artemia salina*. IV. Hybridization of wild populations with mutant stocks, *Biol. Bull.*, 126, 333, 1964.
23. **Bowen, S. T., Fogarino, E. A., Hitchner, K. N., Dana, G. L., Chow, V. H. S., Buoncristiani, M. R., and Carl, J. R.**, Ecological isolation in *Artemia*: population differences in tolerance of anion concentrations, *J. Crustacean Biol.*, 5, 106, 1985.
24. **Bowen, S. T., Buoncristiani, M. R., and Carl, J. R.**, *Artemia* habitats: ion concentrations tolerated by one superspecies, *Hydrobiologia*, 158, 201, 1988.
25. **Vanhaecke, P., Tackaert, W., and Sorgeloos, P.**, The biogeography of *Artemia*: an updated review, in *Artemia Research and Its Applications*, Vol. 1, Sorgeloos, P., Bengtson, D. A., Decleir, W., and Jaspers, E., Eds., Universa Press, Wetteren, Belgium, 1987, 129.
26. **Bowen, S. T.**, The genetics of *Artemia salina*, V. Crossing over between the X and Y chromosomes, *Genetics*, 52, 695, 1965.
27. **Barigozzi, C., Valsasnini, P., Ginelli, E., Badaracco, G., Plevani, P., and Baratelli, L.**, Further data on repetitive DNA and speciation in *Artemia*, in *Artemia Research and Its Applications*, Vol. 1, Sorgeloos, P., Bengtson, D. A., Decleir, W., and Jaspers, E., Eds., Universa Press, Wetteren, Belgium, 1987, 103.
28. **Abatzopoulos, J., Triantaphyllidis, C. D., and Kastritsis, C. D.**, Preliminary studies on some *Artemia* populations from northern Greece, in *Artemia Research and Its Applications*, Vol. 1, Sorgeloos, P., Bengtson, D. A., Decleir, W., and Jaspers, E., Eds., Universa Press, Wetteren, Belgium, 1987, 107.

29. **Baratelli, L.**, First metrical data on the length of the prophasic chromosomes of diploid and tetraploid parthenogenetic *Artemia*, in *Artemia Research and Its Applications*, Vol. 1, Sorgeloos, P., Bengtson, D. A., Decleir, W., and Jaspers, E., Eds., Universa Press, Wetteren, Belgium, 1987, 101.
30. **Hedgecock, D., Tracey, M. L., and Nelson, K.**, Genetics, in *The Biology of Crustacea*, Vol. 2, *Embryology, Morphology, and Genetics*, Abele, L. G., Ed., Academic Press, New York, 1982, 283.
31. **Freeman, J. A. and Chronister, R. B.**, Cell-specific endopolyploidy in developing *Artemia*, *Roux's Arch. Dev. Biol.*, 197, 490, 1988.
32. **Bowen, S. T.**, The genetics of *Artemia salina*, I. The reproductive cycle, *Biol. Bull.*, 122, 25, 1962.
33. **Bowen, S. T., Hanson, J., Dowling, P., and Poon, M. C.**, The genetics of *Artemia salina*, VI. Summary of mutations, *Biol. Bull.*, 131, 230, 1966.
34. **Kiyomoto, R. K., Poon, M. C., and Bowen, S. T.**, Ommochrome pigments of the compound eyes of *Artemia salina*, *Comp. Biochem. Physiol.*, 29, 975, 1969.
35. **Barigozzi, C.**, Problems of speciation in the genus *Artemia*, in *Proceedings of the 5th European Marine Biology Symposium*, Battaglia, B., Ed., Piccin Editore, Padova, Italy, 1972, 61.
36. **Ginsburger-Vogel, T. and Charniaux-Cotton, H.**, Sex determination, in *The Biology of Crustacea*, Vol. 2, *Embryology, Morphology, and Genetics*, Abele, L. G., Ed., Academic Press, New York, 1982, 257.
37. **Bowen, S. T.**, The genetics of *Artemia salina*. II. White eye, a sex-linked mutation, *Biol. Bull.*, 124, 17, 1963.
38. **Heip, J., Moens, L., Hertsens, R., Wood, E. J., Heyligen, H., Van Broekhoven, A., Vrints, R., DeChaffoy, R. D., and Kondo, M.**, *Artemia* extracellular hemoglobins: ontogeny, structure, and *in vivo* radiolabeling, in *The Brine Shrimp Artemia*, Vol. 2, Persoone, G., Sorgeloos, P., Roels, O., and Jaspers, E., Eds., Universa Press, Wetteren, Belgium, 1980, 427.
39. **Bowen, S. T., Sterling, G., and Barkan, J.**, The hemoglobins of *Artemia salina*. IV. A model for genetic control of Hb-1, Hb-2, and Hb-X, *Biochem. Gene.*, 15, 409, 1977.
40. **Ferry, J. and Bowen, S. T.**, Resolution of the subunits of *Artemia* hemoglobins on Triton-acid-urea polyacrylamide gels, *Comp. Biochem. Physiol.*, 71(B), 317, 1982.
41. **Ferry, J. A., Nichols, R. C., Condon, S. J., Stubbs, J. D., and Bowen, S. T.**, *Artemia* hemoglobins: increase in net synthesis of the beta-polypeptide (relative to the alpha-polypeptide) in hypoxia, *Biochem. Biophys. Acta*, 739, 249, 1983.
42. **Decleir, W., Vos, J., Bernaerts, F., and Van Den Branden, C.**, The respiratory physiology of *Artemia*, in *The Brine Shrimp Artemia*, Vol. II, Persoone, G., Sorgeloos, P., Roels, O., and Jaspers, E., Eds., Universa Press, Wetteren, Belgium, 1980, 137.
43. **Bowen, S. T.**, unpublished data.
44. **Trotman, C. N. A., Kirssansen, G. W., Kastner, B., and Tate, W. P.**, Occurrence and properties of the glycoprotein artemocyanin, in *Artemia Research and Its Applications*, Vol. 2, Sorgeloos, P., Bengtson, D. A., Decleir, W., and Jaspers, E., Eds., Universa Press, Wetteren, Belgium, 1987, 160.
45. **Sterling, G., Barkan, J., and Bowen, S. T.**, The hemoglobins of Artemia salina. V. Genetic variation in hemoglobin-1, hemoglobin-2, and hemoglobin-3, *Biochem. Genet.*, 15, 423, 1977.
46. **Gejardo, G. M. and Beardmore, J. A.**, Ability to switch reproductive mode in *Artemia* is related to maternal heterozygosity, *Mar. Ecol. Prog. Ser.*, 55, 191, 1989.
47. **Browne, R. A.**, Cross-breeding experiments between Mediterranean sexual populations of the brine shrimp, *Artemia*, *Biol. Bull.*, 165, 506, 1983.
48. **Browne, R. A.**, unpublished data.
49. **Dana, G. L.**, Comparative population ecology of the brine shrimp *Artemia*, M.Sc. thesis, San Francisco State University, CA, 1981.
50. **Thun, M. A. and Starrett, G. L.**, The effect of cold, hydrated dormancy and salinity on the hatching of *Artemia* cysts from Mono Lake, California, USA, in *Artemia Research and Its Applications*, Vol. 1, Sorgeloos, P., Bengtson, D. A., Decleir, W., and Jaspers, E., Eds., Universa Press, Wetteren, Belgium, 1987, 65.
51. **Lenz, P. H.**, Ecological studies on *Artemia*: a review, in *Artemia Research and Its Applications*, Vol. 3, Sorgeloos, P., Bengtson, D. A., Decleir, W., and Jaspers, E., Eds., Universa Press, Wetteren, Belgium, 1987, 89.
52. **Browne, R. A.**, Genetic and ecological divergence of sexual and asexual brine shrimp (*Artemia*) populations in the Mediterranean Basin, *Nat. Geogr. Res.*, 4, 310, 1988.
53. **Sherif, S. E. M.**, personal communication.
54. **Browne, R. A., Sallee, S. E., Grosch, D. S., Segreti, W. O., and Purser, S. M.**, Partitioning genetic and environmental components of reproduction and lifespan in *Artemia*, *Ecology*, 65, 223, 1984.
55. **Abreu-Grobois, F. A. and Beardmore, J. A.**, Genetic characterization of *Artemia* populations: an electrophoretic approach, in *The Brine Shrimp Artemia*, Vol. 1, Persoone, G., Sorgeloos, P., Roels, O. A., Jaspers, E., Eds., Universa Press, Wetteren, Belgium, 1980, 133.
56. **Abreu-Grobois, F. A. and Beardmore, J. A.**, Genetic differentiation and speciation in the brine shrimp, *Artemia*, in *Mechanisms of Speciation*, Barigozzi, C., Ed., Alan R. Liss, New York, 1982.

57. **Parker, E. D.,** Ecological implications of clonal diversity in parthenogenetic morphospecies, *Amer. Zool.*, 19, 753, 1979.
58. **Browne, R. A. and Hoopes, C. W.,** Genotype diversity and selection in asexual brine shrimp *(Artemia), Evolution*, in press.
59. **Parker, E. D. and Selander, R. K.,** The organization of genetic diversity in the parthenogenetic lizard *Cnemidophorus tesselatus, Genetics*, 84, 791, 1976.
60. **Harshamn, L. G. and Futuyma, D. J.,** The origin and distribution of clonal diversity in *Alsophila pometaria* (Lepidoptera: Geometridae), *Evolution*, 39, 315, 1985.
61. **Mitter, C., Futuyma, D. J., Schneider, J. C., and Hare, J. D.,** Genetic variation and host plant relations in a parthenogenetic moth, *Evolution*, 33, 777, 1987.
62. **Crease, T. J., Stanton, D. J., and Hebert, P. D. N.,** Polyphyletic origins of asexuality in *Daphnia pulex*. II. Mitochondrial DNA variation, *Evolution*, 43, 1016, 1989.
63. **Hoopes, C. W. and Browne, R. A.,** unpublished data.
64. **MacDonald, G. H. and Browne, R. A.,** Population dynamics of an asexual brine shrimp (*Artemia*) population, *J. Exp. Mar. Biol. Ecol.*, 133, 169, 1989.
65. **Browne, R. A. and Spencer, M. H.,** Intrapopulational differences in life-history traits of obligately parthenogenetic clones of brine shrimp, in *Artemia Research and Its Applications*, Vol. 1, Sorgeloos, P., Bengtson, D. A., Decleir, W., and Jaspers, E., Eds., Universa Press, Wetteren, Belgium, 1987, 115.
66. **Browne, R. A.,** unpublished data.
67. **Hedrick, P. W.,** *Genetics of Populations*, Jones and Bartlett, Boston, 1985, 629.
68. **Williams, G. C.,** *Sex and Evolution*, Princeton University Press, NJ, 1975, 201.
69. **Maynard Smith, J.,** *The Evolution of Sex*, Cambridge University Press, U.K., 1978.
70. **Muller, H. J.,** The relation of recombination to mutational advance, *Mutat. Res.*, 1, 2, 1964.
71. **Pamilo, P., Nei, M., and Li, W.,** Accumulation of mutations in sexual and asexual populations, *Genet. Res. Camb.*, 49, 135, 1987.
72. **MacDonald, G. H. and Browne, R. A.,** Differential inheritance of rare males in clones of the brine shrimp *Artemia parthenogenetica, Genetica*, 75, 47, 1987.
73. **Stefani, R.,** The origin of males in parthenogenetic populations of *Artemia salina, Riv. Biol.*, 57, 147, 1964.
74. **Browne, R. A. and Brownlee, D. H.,** unpublished data.
75. **Hebert, P. D.,** The population biology of *Daphnia* (Crustacea, Daphniidae), *Biol. Rev.*, 58, 387, 1978.
76. **Batuecas, B., Garesse, R., Calleja, M., Valverde, J., and Marco, R.,** Genome organization of *Artemia* mitochondrial DNA, *Nucleic Acid Res.*, 16, 6515, 1988.
77. **Avise, J. C., Arnold, J., Ball, R. M., Bermingham, E., Lamb, T., Neigel, J. E., Reeb, C. A., and Saunders, N. C.,** Intraspecific phylogeography: the bridge between population genetics and systematics, *Ann. Rev. Ecol. Syst.*, 18, 449, 1987.
78. **Marco, R., Garesse, R., Cruces, J., and Renart, J.,** personal communication.
79. **Cruces, J., Wonenburger, M. L. G., Diaz-Guerra, M., Sebastian, J., and Rehart, J.,** Identification of a satellite in the *Artemia* genome, in *Artemia Research and Its Applications*, Vol. 2, Decleir, W., Moens, L., Slegers, H., Jaspers, E., and Sorgeloos, P., Eds., Universa Press, Wetteren, Belgium, 1987, 403.
80. **Vaughn, J. C.,** DNA sequence organization in the genome of the brine shrimp, *Artemia salina*, in *Biochemistry of Artemia Development*, Bagshaw, J. C. and Warner, A. H., Eds., University Microfilm International, Ann Arbor, MI, 1979, 190.
81. **Vaughn, J. C., Whitman, D. J., Bagshaw, J. C., and Helder, J. C.,** Molecular cloning and characterization of ribosomal RNA genes from the brine shrimp, *Biochim. Biophys. Acta*, 697, 156, 1982.
82. **Medina, R., Gallego, M. E., Renart, J., and Cruces, J.,** unpublished data.

Chapter 10

ECOLOGY OF *ARTEMIA*

Petra H. Lenz and Robert A. Browne

TABLE OF CONTENTS

I. FIELD STUDIES

A. INTRODUCTION

The brine shrimp, *Artemia*, is a typical inhabitant of hypersaline lakes and ponds, which are characterized by communities with low species diversity and simple trophic structures.[1-3] These biological communities are often similar in structure, which has led to an underestimation of the diversity in physical and chemical characteristics of *Artemia* habitats (e.g., Por[3]). Furthermore, *Artemia* spp. show little morphological variation, suggesting little difference among populations. However, laboratory studies indicate that *Artemia* strains vary significantly with respect to their life-history traits, presumably as a result of selective pressures in their native habitats.[4]

B. *ARTEMIA FRANCISCANA*

The *Artemia* of the New World are bisexual and with the exception of one population, belong to the *A. franciscana* superspecies.[5-7] Although all *A. franciscana* populations are capable of interbreeding when reared in the laboratory in artificial media, several of them are geographically and chemically isolated, and gene flow is unlikely.[7] *Artemia* occur in lakes in Alberta and Saskatchewan, Canada (Little Manitou and Chaplin Lakes), in the U.S. Great Basin (Great Salt Lake, UT; Mono Lake, CA; Fallon Ponds, NV), in salterns along the west coast of the U.S. (San Francisco to San Diego, CA), on the Okanogan Plateau (Penley Lakes, WA), and in salt lake areas in Nebraska, Arizona, and New Mexico. A more complete list of findspots can be found in Vanhaecke et al.[8] and Bowen et al.[7] In Mexico they have been found in salterns along both the west and east coasts and in several inland salt wells.[9] *A. franciscana* also occurs naturally on many Caribbean Islands, such as in Puerto Rico, Curaçao, Bahamas, etc.[2,8] In South America, *A. franciscana* inhabit primarily coastal salterns such as Macau (Rio Grande do Norte, Brazil) and Cabo Frio (Rio de Janeiro, Brazil) but also coastal salt lakes (Boca Chica Salt Lake, Margarita Island, Venezuela) and inland salt lakes (Atacama Lake, Chile) and coastal hypersaline estuaries (Virilla, Peru). The Macau population was introduced in 1977 from a San Francisco Bay stock. Although the initial introduction was limited to a small area, *Artemia* populations have been established in many other salterns in Northeastern Brazil.[10] *A. franciscana* has also been introduced to Australia, New Zealand, Philippines, and Christmas Island. Natural populations occur on two Pacific Islands: Laysan Lagoon (Hawaii)[11,12] and San Salvador Island Lagoon (Galapagos Islands, Ecuador).[7]

1. North America

a. Water Chemistry

North American *Artemia* habitats range in salinity from 30 to 330 g/l and are characterized by diverse chemical compositions.[7] Cole and Brown[13] studied the anions in *Artemia* habitats and classified the lakes as chloride (the majority), sulfate (e.g., Penley Lakes, WA; Lake Chaplin, Saskatchewan, Canada), or carbonate (e.g., Sturgeon Lake, NB). Most lakes have one dominant anion (greater than 80%).[7] However, a few lakes have two dominant anions

(Green Pond, AZ: CO_3 and SO_4; Little Manitou, Saskatchewan, Canada: Cl and SO_4;[7] Abert Lake, OR: Cl and CO_3[14] or even three anions (Mono Lake, CA; Fallon Ponds, NV; and Jesse Lake, NB).[7] The individual *Artemia* populations vary considerably with respect to their tolerance for waters of different ionic compositions.[7] Ecological isolation among these populations occurs because of an intolerance for each other's natural habitat due to differences in lakewater chemistry.

A. franciscana is distinguished from other *Artemia* species by occurring in habitats with the largest diversity in water chemistry. The ecological consequences of this heterogeneity are not clear, and there may be only a limited effect on the *Artemia* population dynamics. The costs of osmoregulation can be expected to differ, and for the Mono Lake *Artemia* the upper salinity tolerance is 150 g/l, which is low compared to *Artemia* tolerances for chloride waters.[15] The different types of lake water may also favor different phytoplankton species. For example, *Dunaliella viridis* and *D. salina* are dominant in many chloride lakes, including Great Salt Lake,[16] whereas in Mono Lake *Nannochloris* dominates.[17] However, both *Dunaliella* and *Nannochloris* appear to be an adequate food source for *Artemia*.

b. Seasonality

The *Artemia* in most North American habitats are characterized by an interrupted growth season. In the permanent temperate lakes the interruption is caused by ~ 0°C temperatures in the winter, which preclude the survival of adults, juveniles, and nauplii.[14,16,18-21] In the temporary ponds, the interruption in the population cycle is caused by a combination of high salinities and ultimately by the drying of the pond (e.g., Fallon Ponds).[22]

The reseeding of the population occurs through the year via hatching of diapause cysts. After a major rain, a lower salinity layer may form on the surface, which may promote a period of increased hatching in populations with bouyant cysts. In Mono Lake, where cysts sink to the bottom of the lake, some hatching occurs year-round. However, peak hatching rates (up to 25,000 m^2/d) were observed in the spring.[23] The authors estimated that over 400,000 nauplii/m^2 hatched in a 48-day period between April and May. The standing crop of *Artemia* was nearly 20 times lower, suggesting high mortality rates for the newly hatched nauplii. Since food levels in the spring are abundant,[21,24] mortality must be due to another factor(s). During the remainder of the year, nauplii hatched from cysts do not appear to contribute significantly to the *Artemia* population dynamics. A similar situation may occur in other habitats, where nauplii, hatched after a summer rainstorm, may not be able to compete with an established adult population.

In Mono Lake, the first generation reaches adulthood in late May-early June after the lake has warmed to 15 to 18° C.[21] In other habitats (Great Salt Lake, Fallon Ponds), the first reproductive adults appear earlier, and this seems to be related to a faster warming of these lakes.[16,18,22] The appearance of adults is followed by a peak in nauplii, which arises from ovoviviparous reproduction.[20,21] These nauplii mature in July and August, when peak adult abundances are observed. Population recruitment is very low thereafter, and the population declines throughout the fall.[11,16,21] This two-generation population cycle has been observed consistently for the last 10 years in Mono Lake.[12,21,25,26] At least four generations were estimated for the warmer and shallower Fallon pond,[22] and Carpelan[27] estimated eight generations for San Francisco Bay. Several factors, including temperature and length of growing season, are probably important in determining the number of generations per year. More studies are needed before the number of generations can be predicted for any particular habitat.

c. Reproduction

The most detailed studies on female reproduction have been conducted on the Mono Lake population.[12,16,21] After an initial naupliar brood in June, females switch to oviparity.

After July, cyst production dominates. Mean brood sizes decline between June and July from over 100, down to 30 offspring. Brood sizes were significantly correlated with food levels, which are low during the summer. Estimates of total annual cyst production ranged between 2.5 and 7 million cysts/m^2 over a five-year period.[23] With spring adult densities ranging between 3000 and 50000/m^2, the actual proportion of cysts that become the first generation adults in the following year is between 0.1 and 2 percent. This scenario of a large number of cysts produced and a small fraction reseeding the population in the following season probably occurs in most *Artemia* populations.

d. Abundances and Limiting Factors

Peak adult *Artemia* densities occur in mid to late summer, and range between 1 and 10 individuals/1. Unusually high densities were reported for Little Manitou (1,300/1).[28] In Mono Lake, peak densities varied between 27,000 and 100,000/m^2 (2 and 7/l), and these differences were related to variations in spring adult abundances, and subsequent recruitment to the second generation.[21] Food availability and predation often limit population sizes in zooplankton. *Artemia* are indiscriminant filter feeders on particles ranging from 1 to 50 μm in diameter.[2] Their primary food source is phytoplankton; however, they also feed on benthic forms,[14] bacteria and detritus. Food limitation is suggested by the inverse relationship between *Artemia* densities and phytoplankton abundances in many habitats.[16-18] Female brood sizes in Mono Lake are correlated with food densities, and a similar relationship can be expected in other populations. It has been shown that *Artemia* can limit phytoplankton densities;[16,29] however, how phytoplankton densities affect *Artemia* population dynamics needs further study.

2. Caribbean and Latin America

a. Water Chemistry

The *Artemia* habitats in this region are primarily coastal, and thus contain chloride waters. *Artemia* are present at salinities ranging from 40 to > 300 g/l,[30,31] but appear to be limited to chloride waters. Bowen et al.[7] found that Puerto Rican *Artemia* can only tolerate chloride media, thus occupying a narrow chemical niche.

b. Seasonality

Few ecological studies have been conducted on *Artemia* populations in the Caribbean and Latin America. Seasonality in temporary lakes is driven by the salinity cycle, and in permanent lakes *Artemia* may occur year-round. In Boca Chica Salt Lake (Venezuela), where salinities range from 42 to 200 g/l and temperatures range from 25 to 35°C, *Artemia* are always abundant.[31] In the Antilles, *Artemia* are excluded from the saline ponds due to both drying and dilution of the lakes.[30] At the lower salinities, the ponds are occasionally invaded by predators (fishes and copepods). In their absence, *Artemia* will inhabit the ponds over the entire salinity range. However, when a predator population becomes established *Artemia* will be excluded from a pond even after the predator disappears at salinities of 80 g/l. This occurs because the cysts will not hatch at the higher salinities and *Artemia* will only become reestablished if the lake becomes diluted by rains. No studies have established the number of generations per year in these habitats, but due to the warm climate, multiple and overlapping generations would be expected.

c. Reproduction

The reproductive patterns in permanent vs. temporary lakes is expected to differ. The *Artemia* in temporary ponds need to produce cysts to ensure the reseeding of the pond (as occurs in temperate lakes). In the Antilles, females reproduced oviparously at the low salinities and brood sizes averaged 50 cysts.[30] Females switched to ovoviviparity at salinities

> 90 g/l, and naupliar broods of 15 offspring were produced. It is not clear whether this difference in brood sizes was due to food availability. Ovoviviparity predominates in permanent habitats[32] and in Boca Chica Salt Lake small brood sizes were observed (< 10 eggs).[31] In this population, adults tended to be small (<6 mm) and they comprised less than 5% of the population.

In Macau (Brazil) cyst production was high for the first five years after inoculation, then declined by over an order of magnitude.[10] Low nutrient concentrations and a resulting drop in primary productivity were suggested as possible causes for this decline. An alternative explanation would be that cyst production has been selected against in an environment with year-round conditions favorable for *Artemia*. Lack of oviparity has also been reported for Lake Grassmere and Laysan Lagoon (see below in Section 3c. Reproduction).

d. Abundances and Limiting Factors

Coastal salterns tend to be low in nutrients[1] and low in phytoplankton production.[32,33] Low adult *Artemia* densities were observed in Boca Chica Salt Lake (< 2/1), ESSA salterns, Mexico (< 1/1)[32] and Long Island, Bahamas ("rare").[33] Fertilization of the Long Island ponds with nitrogen and phosphorous, increased *Artemia* standing stocks and cyst production.[33]

3. Australia, New Zealand, and the Pacific

a. Water Chemistry

The *Artemia* habitats in this region are all coastal and therefore sodium chloride lakes, with a salinity range of at least 80 to 260 g/l.

b. Seasonality

In New Zealand, seasonal population studies have indicated that although temperatures decline to 5° C in the winter, *A. franciscana* are present year-round. During the winter, densities are low, and the population is dominated by adults and juveniles. Females are not reproductively active. As the temperature rises to ~ 15°C, reproduction begins, and the remaining season is characterized by many overlapping generations. Owing to the survival of adults through the winter, this population can be classified as having an uninterrupted population cycle.[34,35]

A similar situation appears to be present in the subtropical island population of Laysan Lagoon.[12] The population structure appeared relatively constant over the four-month period studied, suggesting a steady state with continuous low recruitment. The population was dominated by juveniles. This population structure is similar to the Venezulean population, but is in marked contrast to temperate *Artemia* populations, which are usually either dominated by nauplii or adults.[12,21,34-36]

c. Reproduction

Ovoviviparity predominates in Lake Grassmere[34] and in Laysan Lagoon.[12] Reproductive rates in the latter were low, with brood sizes smaller than four. No cyst production was observed in either the Lake Grassmere or the Laysan Lagoon populations, suggesting that cyst production has been selected against in these populations. This pattern is typical of *Artemia* populations with uninterrupted population cycles.

d. Abundances and Limiting Factors

From winter to summer, standing stock in Lake Grassmere increased from less than 1 g dry weight/m^3 (~ 9×10^{-3} adults/1) to 15 g dry weight/m^3 (equivalent to 0.125 adults/1).[35] These low densities may be due to low nutrient concentrations, which occur in Lake Grassmere[35] and other solar saltworks.[32,33] In winter, the population is limited by low temperatures. Mortality can result as well from artificial manipulation of conditions in solar saltworks.[34]

Australia has a number of high salinity lakes, which are characterized by an endemic salt lake fauna and flora differing from those of other continents.[37] The niche usually occupied by *Artemia* in salt lakes on other continents is taken up by *Parartemia*, another Anostracan.[38] The introduced *Artemia* are restricted to coastal saltworks, where *Artemia* and *Parartemia* compete with each other. *Parartemia* are found at the lower salinities, whereas *Artemia* dominate at the higher salinities. For a detailed discussion of the population dynamics of *Artemia* and *Parartemia*, see Chapter 14 by Williams and Geddes.

C. *ARTEMIA PARTHENOGENETICA*

Although the asexual *Artemia* are identified as a single species, *A. parthenogenetica,* this group is genetically heterogeneous. Chromosome numbers vary, and diploid (e.g., Salin de Giraud, France), triploid (Izmir, Turkey), tetraploid (Margherita di Savoia, Italy) and pentaploid (Burgas Pomerije, Bulgaria) strains have been reported.[39] In addition to differences in chromosome numbers, some populations of *A. parthenogenetica* are composed of a variety of distinct clones, with multiple clones occurring within a single saltwork (e.g., 65 in Salin de Giraud).[40]

A. parthenogenetica populations inhabit saline lakes and salterns in the coastal areas around the Mediterranean, Aegean, and Black Seas (see maps in References 41-43). They are also found in hypersaline lakes in the Canary Islands, along the Gulf of Aqaba and the potash works along the shore of the Dead Sea.[44,45] The *Artemia* populations in India are parthenogenetic[39] and Ahmadi[46] found *A. parthenogenetica* in Lake Shurabil (Iran). Asia is characterized by a number of arid regions with closed basins, which may have suitable *Artemia* habitats. Lists for the geographic location of *Artemia* populations have been compiled[2,8] albeit the listings for Asia are likely to be incomplete. It is not clear which of these populations are parthenogenetic, and both bisexual and parthenogenetic populations have been reported for China.[47,48] *A. parthenogenetica* was introduced to Australia, where it occurs in a few coastal salterns.[38] A parthenogenetic population has also been reported in Namibia, Africa.[8]

Exceptionally large *Artemia* were reported for a Greek parthenogenetic strain, where female lengths were reported up to 19 mm.[50] While differences in environmental conditions, mean adult age, and clone characteristics probably contribute to variation in female length, ploidy may be important as well. Amat[51] found that cyst and naupliar sizes were positively related to ploidy and this relationship may extend into the adult stage.

1. Water Chemistry

The coastal habitats are sodium chloride and *Artemia* are found in salinities ranging from 40 to 310 g/l.[36] *Artemia* were also reported in at least one Na-Mg-SO_4-(Cl) lake (Chiprana) in Northeastern Spain, but in general the *Artemia* habitats on the Iberic Peninsula are sodium chloride. Lake Didwana, an inland Indian salt lake, is also sodium chloride.[49] This is in contrast to the variety of water chemistries found on the North American continent. However, other inland Asian lakes may have varied water chemistries.

The Indian lakes and salterns have a salinity range that excludes *Artemia* on both extremes.[49,52] Salinities range from < 20 to 300 g/l, and *Artemia* occur between approximately 50 and 200 g/l. At the low salinities (less than 50 g/l) cysts hatch, but they are evidently excluded from the ponds by other zooplankters, probably through competition for food and predation.[49] Cysts from the Indian populations do not hatch below 13°C with an optimal hatching temperature of 30°C.[53]

2. Seasonality

The dynamics of one asexual diploid population has been studied at Salin de Giraud

(France).[36] The annual cycle begins in February when cysts begin to hatch. By late April this first generation has matured and a second peak of nauplii is observed. Five generations are estimated between February and early November, with a population peak in August. The fall decline correlated with an increase in salinity and a drop in temperatures.

3. Reproduction

At Salin de Giraud, females reproduced oviparously and ovoviviparously throughout the reproductive season.[36] The proportion of egg-bearing females fluctuated between 15 and 55%.[36] Periods of low percentages of gravid females seemed to correspond to periods of recruitment into the adult population. Brood sizes were high in May (100 offspring), declining to < 30 by June, presumably because of low food abundances.[36] Little is known about the reproductive characteristics of females in other populations.

4. Abundances and Limiting Factors

MacDonald and Browne[36] measured peak adult *Artemia* abundances of 30 individuals/l. At Salin de Giraud, population density is negatively correlated with food abundances, suggesting food limitation. A survey of the entire saltwork showed that *Artemia* densities increased with salinity between 40 and 240 g/l, and decreased with salinities > 240 g/l.[36] In India, peak *Artemia* densities varied among habitats: Bhargava et al.[49] reported up to 1300 individuals per liter (15% as adults), whereas Ramamoorthi and Thangaraj[52] report 2 and 15 individuals per liter for two coastal lagoons in South India (Tuticorin). More recently Ramanathan and Natarajan[54] reported peak densities of 25 to 30 individuals per liter for Tuticorin salterns. As in the Salin de Giraud, population densities are positively correlated with salinity and negatively correlated with phytoplankton densities.[49,52]

D. OTHER BISEXUALS: *ARTEMIA TUNISIANA*, *A. URMIANA*, AND *A. PERSIMILIS*

A. tunisiana is the most widely distributed among these three bisexual species. *A. tunisiana* is found in the Mediterranean region, more specifically in Spain, North Africa, and on several of the Mediterranean Islands, including Cyprus.[41] The distribution of this species overlaps with *A. parthenogenetica*, and the two species sometimes co-occur in a single salt lake.[55] *A. tunisiana,* however, is limited to lower latitudes (< 40°N).[41] The only location for *A. urmiana* is Lake Rezaieh, Iran (also known as Lake Urmia), and for *A. persimilis,* the Salinas Grandes de Hidalgo in Argentina.[56] The latter species is unique in that is has 44 chromosomes, instead of the usual 42 chromosomes found in all other diploid *Artemia*.

1. Water Chemistry

A. tunisiana has been collected primarily from sodium chloride lakes, as has *A. persimilis*. However, Wadi Natrun, Egypt where a population of *A. tunisiana* occurs (see Chapter 9 by Browne and Bowen) is a $NaCO_3$-Cl lake according to Hammer.[28]

2. Seasonality

A. tunisiana is cold adapted,[57] and when this species co-occurs with *A. parthenogenetica* the bisexual population predominates during the colder months.[55] Although little is known about the ecology of this species, it is possible that it is adapted to temporary ponds, which are filled during the winter months.[41,57]

3. Reproduction

Little is known about the reproductive characteristics of these bisexual species in their natural habitats. However, laboratory studies have shown that *A. tunisiana* prefer low temperatures (<23°C)[57] and relatively low salinities.[58] Cyst production dominates at a variety of temperatures.[57] Amat[55] reports that in Cadiz, *A. tunisiana* reproduces primarily by oviparity.

4. Abundances and Limiting Factors

With the exception of laboratory studies on reaction profiles to temperature[57] and salinity[58] little is known about abundances and limiting factors in the natural habitats.

II. LABORATORY STUDIES

A. REPRODUCTION AND LIFE-SPAN COMPARISONS: SINGLE ENVIRONMENTS

Browne et al.[4] have measured reproductive and life-span characteristics at 24°C and 90 g/l salinity for 12 populations encompassing four species grouping: *A. parthenogenetica, A. tunisiana, A. persimilis,* and *A. franciscana*. The four species are distinct with regard to reproductive and life-span characteristics. *A. tunisiana* populations consistently have the smallest broods and fewest total offspring and among the smallest number of broods per female, offspring per day per female, days between broods, female reproductive days, and length of female life-span. They also have the highest percent of offspring encysted and the longest female postreproductive period. In contrast, *A. parthenogenetica* populations have among the longest female life-spans, the longest recovery period between broods, and the highest ovoviviparity/cyst ratios. Interestingly, the two New World bisexual species, *A. persimilis* and *A. franciscana*, lie at opposite ends of the spectrum for many of the reproductive and life-span parameters. *A. franciscana* has the most offspring per brood, highest percent of cysts hatched, and longest male life-span. *A. franciscana* females also have the longest reproductive period and life-span among the bisexual populations. Alternately, *A. persimilis* has the fewest broods per female, shortest time between broods, fewest female reproductive days, and shortest female life-span.

A. parthenogenetica are at least twice as fecund as bisexual populations when calculated on a per-adult basis. However, when calculated per-adult female, *A. franciscana* females are as productive as *A. parthenogenetica* females in terms of daily rates and total offspring. The principal reproductive differences occur not between *A. parthenogenetica* and bisexual females, but between the New and Old World bisexual populations. *A. tunisiana* populations are characterized by steady, low reproductive output while *A. franciscana* populations show continually high output until senescence.

Variation among clones within a parthenogenetic population reared at static conditions is discussed subsequently in this chapter.

B. REPRODUCTION AND LIFE-SPAN CONDITIONS: MULTIPLE ENVIRONMENTS

Browne et al.[57] have measured fourteen life-history traits at 15, 24 and 30°C for 3 *Artemia parthenogenetica*, 2 *A. tunisiana*, and 1 *A. franciscana* populations. Data were also collected for the *A. franciscana* population at 27.5 and 32.5°C. *A. tunisiana* populations were found to have optimal reproductive performance at a lower temperature (15°C) compared to the 24°C optimum for *A. franciscana* and *A. parthenogenetica*. *A. tunisiana* generally appear to be poorly adapted to warm temperatures and they have the longest development times at all temperatures. *A. franciscana*, on the other hand, shows better adaptation to warm temperatures. The reproductive output of both bisexual species at 15°C exceeds that of the asexual populations. Thus, the asexual and bisexual species from the Old World appear to have different temperature optima, with *A. parthenogenetica* populations performing well at higher temperatures and *A. tunisiana* populations performing well at low temperatures.

Although Browne et al.[57] found that total female life-span increases linearly as temperature decreases, life-span components (especially the reproductive period) are not necessarily linearly related. The result is that the ratio between the reproductive period and total life-span shifts with temperature. For most populations, the ratio peaks at the mid-temperature of 24°. Both the absolute and relative fitness of each population (as measured by offspring production) change with temperature. For example, the *A. franciscana* population has relatively low offspring production at 15°C but has high production at 24°C, while *A. tunisiana* has relatively high production of 15°C but produces few offspring at 30°C. Thus, a population that performs well in one environment does not necessarily have the best output in other environments. For the *A. franciscana* and *A. tunisiana* populations, in particular, there appear to be physiological trade-offs between performance at low and high temperatures, with a relatively fixed optimal temperature range. *A. franciscana* stands out among the six populations in its reproductive performance at the extreme upper temperature range (27.5 to 32.5°C), producing more than five as many offspring as any other population at 30°C. However, performance is poor at 15°C. It would seem more than coincidental that the most apparent trade-off occurs in the populations that have the best performance at high (*A. franciscana*) and low (*A. tunisiana*) temperatures.

From the previously mentioned work,[57] it is clear that with regard to temperature *A. franciscana* is the most phenotypically plastic species, followed by *A. parthenogenetica*, and that *A. tunisiana* is the least plastic.

Mean survival times at 2 and 36°C have been estimated for adults from a number of *Artemia* populations.[58] The rank for increasing length of survival time at 2°C is *A. parthenogenetica* < *A. tunisiana* < *A. franciscana*, and at 36°C is *A. tunisiana* < *A. parthenogenetica* < *A. franciscana*. The Critical Thermal Maxima were also significantly higher for *A. franciscana* and significantly lower for *A. tunisiana* than for six other populations.[59]

Von Hentig[60] examined the Great Salt Lake population of *A. franciscana* at 10, 15, 20, and 30°C. Although different salinity levels were also incorporated into the experimental design, it is of limited use since the highest salinity level was 70 g/l. This is the lower limit of *Artemia's* natural habitat, and the optimal range for *A. franciscana* is above 70 g/l.[58] The reproductive output and life-span reported by Von Hentig[60] are much reduced from those reported for other *A. franciscana* populations,[4,57] probably due to less than optimal culture conditions. Even with these limitations, Von Hentig's[60] conclusions concerning the effects of temperature are generally in concordance with Browne et al.[57] Maturation rate and interclutch interval for the Great Salt Lake population were found to be negatively correlated with temperature, while life-span was greatest at 15°C. Total number of offspring was found to be positively correlated with temperature, although only a slightly larger number was found at 30° than 20°C. Von Hentig's[60] results may be limited by the fact that in the upper temperature range only 20 and 30°C were used, most likely missing the optimal mid-twenties temperature range found for the closely related San Francisco Bay population.[57]

Vanhaecke et al.[61] tested the response of 13 *Artemia* strains to various salinity-temperature combinations, with experimental temperatures ranging from 18 to 34°C and salinities from 5 to 120 g/1. The criterion used was nine-day survivorship. Hence, the tests are analogous to toxicity tests. They concluded that the interaction of salinity and temperature was negligible or very limited, with all populations showing high survival over a wide range of salinities (35 to 110 g/l). Their results showed that juveniles from most of the populations experienced high mortality at the upper end of the temperature range (30 to 34°C). They also reported that the two *A. tunisiana* populations tested were the least tolerant to high temperatures, with high mortalities occurring at 30°C.

Interestingly, some divergence in temperature response profiles appears to have taken place between the *A. franciscana* populations at San Francisco Bay and those at Macau (Brazil) and Barotac Nuevo (Philippines). The latter populations were derived from inoculated San Francisco Bay cysts but inhabit saltworks that experience higher temperatures.[62] Despite the relatively short time since inoculation, the Brazil and Philippines populations showed increased survivorship at temperatures above 30°C. This would argue for heritable adaptation to high-water temperature although plasticity cannot be ruled out completely.

Wear and Haslett[63] and Wear et al.[35] reported on the effects of temperature and salinity on the reproduction and life-span of a New Zealand population of *A. franciscana* (which was inoculated in the early 1950s). The results indicate that a temperature range of 20 to 28°C in salinities of 100 to 170 g/l are optimal. As with the Brazilian and Philippine populations derived from San Francisco Bay populations, some divergence appears to have taken place in the New Zealand population, where high salinity levels are maintained constantly and temperatures are high enough that a population can be maintained year-round. This is hypothesized by Wear et al.[35] to result in low cyst production and high levels of ovoviviparity.

Gilchrist[64] examined the temperature response of *Artemia*, although on a limited scale. She found that the diploid *A. parthenogenetica* population at La Palme in southern France had faster growth rates and reached sexual maturity earlier than the sexual *A. franciscana* from San Diego when reared at 25° and 140 g/l. Iwasaki[65] also looked at temperature effects, reporting a marked decrease in fecundity in both an *A. parthenogenetica* population from China and an *A. franciscana* population from California reared at 32°C.

The hypothesis that polyploid asexual *A. parthenogenetica* populations are better buffered against extreme environmental conditions than diploid asexuals[61,66-68] is not supported by the results of Browne et al.[57] Populations from Kutch, India (a triploid asexual) and Salin de Giraud, France (a diploid asexual) were found to have a very similar reproductive and life-span performances at 15, 24, and 30°C. Although the argument of superior buffering for polyploids may have merit, it may be balanced on a population basis by the ability of asexual diploids to spin off genetically and ecologically widely divergent clonal lines that can occur within the same population.[40,41,69] This option is not available to polyploid populations where clonal diversity is quite low or nonexistent.[39] Whether this is only applicable to the Salin de Giraud population, where clonal diversity is known to be very high (> 60 clones in the population)[40] or is also typical of other asexual diploid *Artemia* populations, remains to be explored. In order to fully test this hypothesis, additional populations and environmental conditions other than temperature need to be tested.

Research in progress[58] points to a lower optimal salinity range for *A. tunisiana* and a higher optimal salinity range for *A. parthenogenetica*. However, the plasticity for salinity tolerances is highest in *A. franciscana*, the same species with the greatest plasticity in temperature response.

In addition to temperature and salinity, the effects of food levels on life-history traits have been investigated.[70] *A. franciscana* populations were reared at high (HF) and low (LF = 10% of HF) levels. At LF levels, reproductive output per female was significantly lower than for HF females (232 vs. 1161, respectively). LF females had 48% fewer offspring per brood, 58% fewer broods, and a significant increase in interbrood interval. LF females also produced a higher proportion of offspring as cysts (35%) than HF females (17%), although these data should be interpreted with caution. Since LF females produce fewer broods, and initial broods typically have a higher proportion of offspring as cysts, the percent of offspring would be expected to be higher. LF levels also exacted a toll on life-span, with the effect more pronounced for males than females, with life-span decreasing by 30% and 23% respectively.

C. CORRELATIONS AMONG LIFE-HISTORY TRAITS

Artemia populations with a high portion of their offspring encysted were also found to have low reproductive output and short life-spans.[4] As with most correlations, cause and effect, if any, are unknown, However, habitat uncertainty and high encystment levels appear to be related, along with lower selective pressure for long life-spans. Many *A. tunisiana* populations are found in temporary habitats or habitats where abiotic factors are unpredictable. They also have a rate of encystment that approaches 100% and relatively short life-spans. In contrast, ovoviviparity dominates in *A. parthenogenetica* populations, which are often found in habitats (e.g., Old World commercial salinas) where abiotic factors such as temperature and salinity for each pond are human-regulated and remain relatively predictable from year to year.

Extremely low correlation values have been found between the length of the female prereproductive period and the other life-span or reproductive traits. For *Artemia*, no trade-offs occur between early onset of reproduction and increased death rates, either on an intrapopulation level[70] or interpopulation basis.[4] However, the length of time between broods does seem to play an important role. A longer recovery time between broods was significantly correlated with a greater number of broods, more offspring, and a longer female life-span.[4]

D. GENETIC AND ENVIRONMENTAL COMPONENTS OF LIFE-SPAN AND REPRODUCTION TRAITS

If an asexual population is monoclonal, variance levels for reproductive and life-span traits should be due almost entirely to environmental factors. Variance levels within sexual populations, however, are caused by both environmental and genetic variation. By comparing asexual to sexual population variance, partitioning of genetic from environmental effects is possible and trait heritability can be estimated.

Variance levels for *Artemia* life-history traits have been partitioned into genetic and environmental components.[4] Encystment rates and the percent of cysts that hatch appear to have a large environmental component of variance. Somewhat paradoxically, comparison of encystment rates among clones within an asexual population,[69] suggests a genetic factor as well. Both the interpopulational and interclonal evidence is in agreement concerning many other reproductive traits, including total production (total offspring per female) and reproductive rates (offspring per day per female and days between broods) all of which have large genetic components.

A genetic component, with regard to level of female heterozygosity controlling the level of encystment (at least in the Great Salt Lake population), has recently been suggested.[71]

E. ENCYSTMENT VS. OVOVIVIPARITY

Although cyst production has obvious advantages, it also entails a number of disadvantages. First, even in optimal laboratory conditions, cysts have a low hatch rate, ranging from 10 to 90%.[4] In the field even lower levels could occur. Second, there is at least circumstantial evidence that cyst production is more costly than ovoviviparous reproduction. A typical cyst brood is only 74% the size of ovoviviparous broods.[72] Von Hentig[60] determined that 22% of the dry mass of the cyst was utilized for encapsulation, which may explain smaller encysted brood size. The third disadvantage involves a delay in population growth. When conditions are favorable, ovoviviparously produced offspring would be expected to dominate future population growth. However, when conditions are drastically unfavorable, such as habitat dessication or when overwintering temperatures occur, only cysts will survive. Thus there is a trade-off between the advantages and disadvantages involved with the production of encysted and ovoviviparous offspring. One would predict relative environmental stability, with only occasional perturbations, in populations with a low percentage of encystment, or where cysts are produced late in life. Under these conditions, it would be of selective advantage to produce the majority of offspring ovoviviparously in order to maximize success in intraspecific competition. On the other hand, production of cysts early in a female's reproductive life and/or maintaining a high cyst/ovoviviparity ratio would be selectively advantageous if the environment tended towards instability.

For the *A. tunisiana* populations, where favorable conditions may occur sporodically, habitat encystment rates are not only very high but also most variable with regard to hatching.[4,57] For these populations, a direct analogy could be drawn between *Artemia* cysts and many desert seeds. As Cohen[73-75] has pointed out, seeds can be programmed to germinate in response to specific environmental factors, such that a fraction of the seeds germinates over subsequent years. In both cases, the "eggs" are placed into a number of different chronological "baskets."

F. REPRODUCTIVE COSTS

Costs of reproduction in *Artemia* were examined by postponing mating from 0 up to 40 d past sexual maturity.[70] Each reproductive regime was further subdivided into low and high feeding groups. Under low food, female life-span was negatively correlated with the length of the reproductive period. However, no such relationship was found among the four high food groups. Fertile females in the delayed reproduction groups gave birth to approximately the same total number of offspring as controls, not by lengthening life-span but by producing larger broods, with fewer days between broods. The results indicate that although some reproductive costs are unavoidable in *Artemia*, nutritional resources play a critical role in determining how extensively reproduction will affect life-span.

G. LIFE-HISTORY PARAMETERS OF "LABORATORY" VS. "WILD" POPULATIONS

Differences in reproductive and life-span characteristics between "wild" and "laboratory" San Francisco Bay cultures reared for 25 years under laboratory conditions, have been reported by Browne.[76] The number of generations that the laboratory population experienced was estimated to be from 50 to 75 over the 25 year period. Wild females produced more broods, more offspring per day, and had a longer reproductive period, resulting in a larger total number of offspring. However, the differences were not significant between the two populations, except for two parameters; "laboratory" males lived only 60% as long as "wild" males (55.5 d vs. 92.3 d) and the percent of offspring encysted was three times greater in laboratory populations (55% vs. 19%).

H. CLONAL VARIATION IN ASEXUAL POPULATIONS

Browne and Spencer[69] established nine clonal lines from single *A. parthenogenetica* females originating from Salin de Giraud, France. (For a discussion of the genetic aspects of multiclonality, see the Chapter 9 on "Taxonomy and Population Genetics" by Browne and Bowen). Offspring from each clonal line experienced identical standardized laboratory conditions. Significant differences were found between clonal groups for a number of life-history traits. The total number of offspring per female varied from 274 to 1162. The length of the reproductive period varied from 16.3 to 42.2 d. Other traits that were found to vary significantly were number of offspring per brood, number of broods per female, mean number of offspring per day, and the encystment-ovoviviparity ratio.

Differences between clones are significant and by no means trivial. For example, percent of offspring encysted varied from 0 to 95%.[40] Intrapopulational differences between clones for some traits approach those found for differences between populations belonging to the entire *Artemia* complex. In a study of 12 populations from three different species found on five continents,[4] the mean number of total offspring per female ranged from 112 to 1619. This is only slightly greater than the range of 274 to 1162 among clones. The mean number of broods per female ranged from 4.0 to 19.1 for the interpopulational study, and from 2.6 to 9.8 for the clonal groups. The interclonal range for the prereproductive period actually exceeded that found for the 12 populations. However, it should be pointed out that interclonal values are essentially measurements among replicated individuals whereas the values found among the twelve populations are differences among groups of individuals, with correspondingly dampened variability.

The existence of multiple clonal groups that differ widely in life history traits suggests that the evolutionary potential for *A. parthenogenetica* may approach that of sexual populations. Since genotypically different clones within the same population vary significantly in key fitness variables, the potential for change in asexual populations is greater than previously considered by most investigators. The existence of multiple clonal lines in *A. parthenogenetica* populations, along with the introduction of new genetic material via mutation (both single locus and polygenic traits) offers a considerable amount of genetic and phenotypic variability upon which natural selection can act.

I. INTERACTION BETWEEN SEXUAL AND ASEXUAL POPULATIONS

Most *Artemia* localities are limited to either asexual or sexual reproduction, with mixed species known only for two Spanish sites[55,57] and possibly Lake Urmia (Lake Rezaieh), Iran.[77] Even at these localities, the bisexual and parthenogenetic species appear to cycle temporarily, with the bisexual species dominating in the winter months when salinity and temperature are low and *A. parthenogenetica* populations prevailing in the summer months when salinity and temperature are relatively high.[78] Such partitioning would be expected based on temperature response profiles.[57] This temporal cycling and the biogeographical distribution are at least circumstantial evidence of the competitive interplay between sexual and asexual populations.

Competition experiments between bisexual and asexual *Artemia* have been reported.[79,80] Intra- and interspecific variability were incorporated into the experimental designs, as well as variation in food level. *A. franciscana* populations outcompeted *A. parthenogenetica* populations in 91% of the scorable trials. However, when *A. parthenogenetica* populations competed against the phylogenetically more closely related Old World sexual populations (*A. tunisiana*), bisexuals were eliminated in 98% of the scorable trials. This, along with the temperature and salinity response profiles of the populations, may explain why *A. parthenogenetica* populations dominate Mediterranean basin distribution sites.[41,42] In general, the competitive abilities of the *Artemia* populations (at least with regard to the environmental conditions in which they were tested) are *A. tunisiana* $<$ *A. parthenogenetica* $<$ *A. franciscana*. Interestingly, for the large majority of simple laboratory systems, competitive exclusion requires on the order of 10 to 100 generations,[81] but for *Artemia*, exclusion occurred in only 2 or 3 generations. This may be due to the almost complete niche overlap between *Artemia* species.

Bisexual-asexual competitive outcome may be influenced by a number of factors, including environmental conditions, geographic origins, and life-history traits of the competing populations.[57,70] One drawback to earlier experiments[79,80] is that they were conducted using only a limited range of environmental conditions. Experiments conducted in different conditions should be pursued. For example, temperature-response reaction profiles[57] reveal that the *A. tunisiana* populations have cold tolerance levels that exceed those of either *A. parthenogenetica* or *A. franciscana*.

An additional factor that may be important in the outcome of competitive experiments is the "weak link" that males represent. In competition experiments,[79,80] males died much sooner than females with low food conditions. Males were eliminated from the cultures in as little as ten days.[80] Males may be less efficient than females at food gathering and energy storage, or may swim at a faster rate than females and hence spend energy more rapidly. An alternate explanation is that the rapid disappearance of males from the *A. tunisiana*—*A. parthenogenetica* pairings[80] may represent elimination of the entire sexual population rather than just males.

Sperm robbing[79,82] may also play a role in the elimination of the sexual populations since mating males do not appear to discriminate between sexual and asexual females.[79] Asexual females may "capture" some males, depriving fertilization of some outbreeding females and ultimately decreasing the reproductive potential of sexual populations.

J. SUMMARY

The bisexual and parthenogenetic *Artemia* species are characterized by extensive genetic variation in life-history traits. Under laboratory conditions, longevity, fecundity, and age-at-first-reproduction vary over an order of magnitude among *Artemia* species and populations. Temperature/salinity tolerances and optima differ among species/populations, and genetic shifts in temperature optima can be selected for both in the laboratory and in the field. The experimental data suggest that *Artemia* spp. have been subjected to a large variety of selective pressures, as is reflected in their respective life-history characteristics.

Artemia inhabit hypersaline lakes and ponds, which differ in water chemistry, seasonality, biological community, and productivity. It can be expected that the life-histories of individual *Artemia* sp. and populations have evolved in response to the habitat characteristics. As an example, ovoviviparity appears to predominate in *Artemia* inhabiting lakes with an uninterrupted growth season (e.g., *A. franciscana* in Hawaii and New Zealand). In the Old World, temporal partitioning of the niche seems to have led to different temperature optima in *A. parthenogenetica* (summer) and *A. tunisiana* (winter). Less is known about the selective pressures acting on other life-history traits, and more field and laboratory work will be needed before we understand the evolution of life histories in *Artemia*.

ACKNOWLEDGMENTS

Support (to RAB) was provided by Whitehall Foundation Grant 86M14 and by an Archie Faculty Enrichment Grant. Daniel Hartline, Patrick Sorgeloos and Sarane Bowen made comments and suggestions on an earlier version of this review.

REFERENCES

1. **Lenz, P. H.,** Ecological studies on *Artemia*: a review, in *Artemia Research and Its Applications* Vol. 3, Sorgeloos, P., Bengtson, D. A., Decleir, W., and Jaspers, E., Eds., Universa Press, Wetteren, Belgium, 1987, 5.
2. **Persoone, G. and Sorgeloos, P.,** General aspects of the ecology and biogeography of *Artemia*, in *The Brine Shrimp Artemia*, Vol. 3, Persoone, G., Sorgeloos, P., Roels, O., and Jaspers, E., Eds., Universa Press, Wetteren, Belgium, 1980, 3.
3. **Por, F. D.,** A classification of hypersaline waters based on trophic criteria, *Mar. Ecol. Prog. Ser.*, 1, 121, 1980.
4. **Browne, R. A., Sallee, S. E., Grosch, D. S., Segreti, W. O., and Purser, S. M.,** Partitioning genetic and environmental components of reproduction and life-span in *Artemia*, *Ecology*, 65, 949, 1984.
5. **Bowen, S. T., Fogarino, E. A., Hitchner, K. N., Dana, G. L., Chow, H. S., Buoncristiani, M. R., and Carl, J. R.,** Ecological isolation in *Artemia*: population differences in tolerance of anion concentrations, *J. Crustacean Biol.*, 5, 106, 1985.
6. **Bowen, S. T., Hitchner, K. N., and Dana, G. L.,** *Artemia* speciation: ecological isolation, in *Vernal Pools and Intermittent Streams*, Jain, S. and Moyle, P., Eds., Institute of Ecology, University of California, Davis, Pub. No. 28, 102, 1984.
7. **Bowen, S. T., Buoncristiani, M. R., and Carl, J. R.,** *Artemia* habitats: ion concentrations tolerated by one superspecies, *Hydrobiologia*, 158, 1988, 201.
8. **Vanhaecke, P., Tackaert, W., and Sorgeloos, P.,** The biogeography of *Artemia*: an updated review, in *Artemia Research and Its Applications*, Vol. 1, Sorgeloos, P., Bengtson, D. A., Decleir, W., and Jaspers, E., Eds., Universa Press, Wetteren, Belgium, 1987, 129.

9. **Castro, T., Sanchez, L., and DeLara, R.,** Natural sources of brine shrimp (*Artemia*) in Mexico, in *Artemia Research and Its Applications,* Vol. 3, Sorgeloos, P., Bengston, D. A., Decleir, W., and Jaspers, E., Eds., Universa Press, Wetteren, Belgium, 1987, 153.
10. **Camara, M. R. and de Medeiros Rocha, R.,** *Artemia* culture in Brazil: an overview, in *Artemia Research and Its Applications,* Vol. 3, Sorgeloos, P., Bengtson, D. A., Decleir, W., and Jaspers, E., Eds., Universa Press, Wetteren, Belgium, 1987, 195.
11. **Caspers, H.,** On the ecology of hypersaline lagoons on Laysan Atoll and Kauai, Hawaii, with special reference to the Laysan duck, *Anas laysanensis* Rothschild, *Hydrobiologia,* 82, 261, 1981.
12. **Lenz, P. H. and Dana, G. L.,** Life cycle studies in *Artemia*: a comparison between a subtropical and a temperate population, in *Artemia Research and Its Applications,* Vol. 3, Sorgeloos, P., Bengtson, D. A., Decleir, W., and Jaspers, E., Eds., Universa Press, Wetteren, Belgium, 1987, 89.
13. **Cole, G. A. and Brown, R. J.,** The chemistry of *Artemia* habitats, *Ecology,* 48, 858, 1967.
14. **Comin, F. P. and Conte, P. A.,** Abundance and spatial distribution of *Artemia salina* in Lake Abert, Oregon, *Hydrobiologia,* 158, 237, 1988.
15. **Dana, G. L. and P. H. Lenz,** Effects of increasing salinity on an *Artemia* population from Mono Lake, California, *Oecologia,* 68, 428, 1986.
16. **Wirick, C. D.,** *Dunaliella-Artemia* plankton community of the Great Salt Lake, Utah, M. Sc. thesis, Utah University, Logan, 1972.
17. **Jellison, R. S. and Melack, J. M.,** Photosynthetic activity of phytoplankton and its relation to environmental factors in hypersaline Mono Lake, California, *Hydrobiologia,* 158, 237, 1988.
18. **Anderson, G. C.,** Some limnological features of a shallow saline meromictic lake, *Limnol. Oceanogr.,* 3, 1958, 259.
19. **Bradbury, J. P.,** Limnology of Zuni Lake, New Mexico, *Bull. Amer. Geol. Soc.,* 82, 379, 1971.
20. **Gillespie, D. M. and Stephens, D. W.,** Some aspects of plankton dynamics in the Great Salt Lake, Utah, in *Desertic Terminal Lakes,* Greer, D. C., Ed., Utah State University, Logan, 1977, 401.
21. **Lenz, P. H.,** Life-history analysis of an *Artemia* population in a changing environment, *J. Plankton Res.,* 6, 967, 1984.
22. **Dana, G. L.,** *Artemia* in temporary alkaline ponds near Fallon, Nevada with a comparison of its life-history strategies in temporary and permanent habitats, in *Vernal Pools and Intermittent Streams,* Jain, S. and Moyle, P., Eds., Institute of Ecology, University of California, Davis, Pub. No. 28, 1984, 115.
23. **Dana, G. L., Foley, C. J., Starrett, G. L., Perry, W. M., and Melack, J. M.,** 1988. *In situ* hatching of *Artemia monica* cysts in hypersaline Mono Lake, California, *Hydrobiologia,* 158, 183, 1988.
24. **Mason, D. T.,** Limnology of Mono Lake, California, *Univ. Calif. Pub. Zool.,* 83, 1, 1967.
25. **Lenz, P. H.,** Ecology of an alkali-adapted variety of *Artemia* from Mono Lake, California, USA, in *The Brine Shrimp Artemia,* Vol. 3, Persoone, G., Sorgeloos, P., Roels, O., and Jaspers, E., Eds., Universa Press, Wetteren, Belgium, 1980, 79.
26. **Dana, G. L.,** personal communication.
27. **Carpelan, L. H.,** Hydrobiology of the Alviso salt ponds, *Ecology,* 38, 375, 1957.
28. **Hammer, U. T.,** *Saline Lake Ecosystems of the World,* Dr. W. Junk Publishers, Dordrecht, Netherlands, 1986.
29. **Lenz, P. H.** Population studies on *Artemia* in Mono Lake, California, Ph.D. thesis, University of California, Santa Barbara, 230, 1982.
30. **Kristensen, I. and Hulscher-Emeis, T. M.,** Factors influencing *Artemia* populations in Antillean salines, *Stud. Fauna Curaçao, Caribbean Islands,* 39, 87, 1972.
31. **Scelzo, M. A. and Voglar, J. F.,** Ecological study of the *Artemia* population in Boca Chica Salt Lake, Margarita Island, Venezuela, in *The Brine Shrimp Artemia,* Vol. 3, Persoone, G., Sorgeloos, P., Roels, O., and Jaspers, E., Eds., Universa Press, Wetteren, Belgium, 1980, 115.
32. **Javor, B. J.,** Planktonic standing crop and nutrients in a saltern ecosystem, *Limnol. Oceanogr.,* 28, 153, 1983.
33. **Davis, J. S.,** Biological communities of a nutrient enriched salina, *Aquatic Bot.,* 4, 23, 1978.
34. **Wear, R. G. and Haslett, S. J.,** Studies on the biology and ecology of *Artemia* from Lake Grassmere, New Zealand, in *Artemia Research and Its Applications,* Vol. 3, Sorgeloos, P., Bengtson, D. A., Decleir, W., and Jaspers, E., Eds., Universa Press, Wetteren, Belgium, 1987.
35. **Wear, R. G., Haslett, S. J., and Alexander, N. L.,** Effects of temperature and salinity on the biology of *Artemia franciscana* Kellogg from Lake Grassmere, New Zealand, 2. Maturation fecundity and generation times. *J. Exp. Mar. Biol., Ecol.,* 98, 167, 1986.
36. **MacDonald, G. H. and Browne, R. A.,** Population dynamics of an asexual brine shrimp (*Artemia*) population, *J. Exp. Mar. Biol. Ecol.,* 133, 169, 1989.
37. **Geddes, M. C. and Williams, W. D.,** Comments on *Artemia* introductions and the need for conservation, in *Artemia Research and Its Applications,* Vol. 3, Sorgeloos, P., Bengston, D. A., Decleir, W., and Jaspers, E., Eds., Universa Press, Wetteren, Belgium, 1987, 19.

38. **Geddes, M. C.,** The Brine Shrimps *Artemia* and *Parartemia* in Australia, in *The Brine Shrimp Artemia,* Vol. 3, Persoone, G., Sorgeloos, P., Roels, O., and Jaspers, E., Eds., Universa Press, Wetteren, Belgium, 1980, 57.
39. **Abreu-Grobois, F. A. and Beardmore, J. A.,** Genetic differentiation and speciation in the brine shrimp *Artemia,* in *Mechanism of Speciation,* Barigozzi, C., Ed., Alan R. Liss, New York, 1982, 245.
40. **Browne, R. A. and Hoopes, C. W.,** Genotypic diversity and selection in asexual brine shrimp *(Artemia), Evolution,* in press.
41. **Browne, R. A.,** Ecological and genetic divergence of sexual and asexual brine shrimp (*Artemia*) populations from the Mediterranean basin, *Nat. Geogr. Res.,* 4, 548, 1988.
42. **Browne, R. A. and MacDonald, G. H.,** Biogeography of the brine shrmp, *Artemia:* distribution of parthenogenetic and sexual populations, *J. Biogeography,* 9, 331, 1982.
43. **Clark, L. S. and Bowen, S. T.,** The genetics of *Artemia salina.* VII. Reproductive isolation, *J. Hered.,* 67, 385, 1976.
44. **Amat, D. F.,** Differentiation in *Artemia* strains from Spain, in *The Brine Shrimp Artemia,* Vol. I, Persoone, G., Sorgeloos, P., Roels, O., and Jaspers, E., Eds., Universa Press, Wetteren, Belgium, 1980, 19.
45. **Goldschmidt, E.,** Fluctuation in chromosome number in *Artemia salina, J. Morph.,* 9, 111, 1952.
46. **Ahmadi, M. R.,** First report of *Artemia* occurrence in Shurabil Lake (Iran), in *Artemia Research and Its Applications,* Vol. 3, Sorgeloos, P., Bengtson, D. A., Decleir, W., and Jaspers, E., Eds., Universa Press, Wetteren, Belgium, 1987, 143.
47. **Yaneng, C.,** Observations on parthenogenetic and bisexual brine shrimp from the People's Republic of China, in *Artemia Research and Its Applications,* Vol. 1, Sorgeloos, P., Bengtson, D. A., Decleir, W., and Jaspers, E., Eds., Universa Press, Wetteren, Belgium, 1987, 227.
48. **Yaneng, C.,** New *Artemia* sibling species from PR China, *Artemia Newsl.,* 11, 41, 1989.
49. **Bhargava, S. C., Jakher, G. R., Saxena, M. M., and Sinha, R. K.,** Ecology of *Artemia* in Didwana Salt Lake (India), in *Artemia Research and Its Applications,* Vol. 3, Sorgeloos, P., Bengtson, D. A., Decleir, W., and Jaspers, E., Eds., Universa Press, Wetteren, Belgium, 1987, 127.
50. **Castritsi-Catherios, J., Christodoulopoulos, M., Aivatzidou, A., and Kiortsis, V.** Une source parthenogenetique d'*Artemia* provenant de l'ile de Milos de l'archiple, Grec, *Rapp. Comm. Int. Mer. Medit.,* 29, 63, 1985.
51. **Amat, F.,** Diferenciation y distribucion de las poblaciones de'*Artemia* de Espana. IV. Biometria de quistes y nauplices, *Investigaciones Pesqueras,* 46, 55, 1982.
52. **Ramamoorthi, K. and Thangaraj, G. S.,** Ecology of *Artemia* in salt pans of Tuticorin, South India, in *The Brine Shrimp Artemia,* Vol. 3, Persoone, G., Sorgeloos, P., Roels, O., and Jaspers, E., Eds., Universa Press, Wetteren, Belgium, 1980, 105.
53. **Lavens, P. and Sorgeloos, P.,** The cryptobiotic state of *Artemia* cysts, its diapause deactivation and hatching: a review, in *Artemia Research and Its Applications,* Vol. 3, Sorgeloos, P., Bengtson, D. A., Decleir, W., and Jaspers, E., Eds., Universa Press, Wetteren, Belgium, 1987, 26.
54. **Ramanathan, N. P. and Natarajan, P. P.,** Ecology of *Artemia* along the southeastern coast of India, in *Artemia Research and Its Applications,* Vol. 3, Sorgeloos, P., Bengtson, D. A., Decleir, W., and Jaspers, E., Eds., Universa Press, Wetteren, Belgium, 1987, 135.
55. **Amat, D. F.,** Zygogenetic and parthenogenetic *Artemia* in Cadiz sea-side salterns, *Mar. Ecol. Progress Series,* 13, 1983, 291.
56. **Halfer-Cervini, A. M., Piccinelli, M., Prosdocimi, T., and Baratelli-Zambruni, L.,** Sibling species in *Artemia* (Crustacea Branchiopoda), *Evolution,* 22, 373, 1968.
57. **Browne, R. A., Davis, L. E., and Sallee, S. E.,** Effects of temperature and relative fitness of sexual and asexual brine shrimp, *Artemia, J. Exp. Mar. Biol. Ecol.,* 124, 1, 1988.
58. **Browne, R. A. and Brownlee, D. H.,** unpublished data.
59. **Browne, R. A. and Dimock, G.,** unpublished data.
60. **Von Hentig, R.,** Einfluss von Salzgehalt und Temperatur auf Entwicklung, Wachstum, Fortpflanzung und Energiebilanz von *Artemia salina, Mar. Biol.,* 9, 145, 1971.
61. **Vanhaecke, P., Siddall, S. E., and Sorgeloos, P.,** International study on *Artemia.* XXXII. Combined effects of temperature and salinity on the survival of *Artemia* of various geographic origin, *J. Exp. Mar. Biol. Ecol.,* 80, 259, 1984.
62. **Vos, J., Leger, P., Vanhaecke, P., and Sorgeloos, P.,** Quality evaluation of brine shrimp, *Artemia* produced in Asian salt ponds, *Hydrobiology,* 108, 17, 1984.
63. **Wear, R. G. and Haslett, S. J.,** Effects of temperature and salinity on the biology of *Artemia franciscana* Kellogg from Lake Grassmere, New Zealand. 1. Growth and mortality, *J. Exp. Mar. Biol. Ecol.,* 98, 153, 1986.
64. **Gilchrist, B.,** Growth and form in the brine shrimp *Artemia salina, J. Morph.,* 91, 111, 1960.
65. **Iwasaki, T.,** Reproductive patterns in *Artemia* with special regard to food and temperature, *Zoology Magazine* (Tokyo), 85, 229, 1976.

66. **Artom, C.**, L'origine e l'evoluzione della'partenogenesi attraveso i differenti biotipi di una specie colletiva (*Artemia salina* L.) con speciale riferimento al biotipo diploide partenogenetico di Sete, *Academe d'Italia Rome Classe di Scienze Fisiche, Matematiche e Naturali, Memorie*, 2, 157, 1931.
67. **Chapman, J.**, The relative adaptive values of parthenogenesis and zygogenesis, M. Sc. thesis, San Francisco State College, 1968.
68. **Metalli, P. and Ballardin, E.**, Radiobiology of *Artemia*: radiation effects and ploidy, *Curr. Top. Radiat. Res. Quart.*, 7, 181, 1972.
69. **Browne, R. A. and Spencer, M. H.**, Intrapopulational differences in life-history traits of obligately parthenogenetic clones of the brine shrimp *Artemia*, in *Artemia Research and Its Applications*, Vol. 3, Decleir, W., Moens, L., Slegers, H., Sorgeloos, P., and Jaspers, E., Eds., Universa Press, Wetteren, Belgium, 115, 1987.
70. **Browne, R. A.**, The costs of reproduction in brine shrimp, *Ecology*, 63, 43, 1982.
71. **Gajardo, G. M. and Beardmore, J. A.**, Ability to switch reproductive mode in *Artemia* is related to maternal heterozygosity, *Mar. Ecol. Prog. Ser.*, 15, 191, 1989.
72. **Browne, R. A.**, Reproductive pattern and mode in the brine shrimp, *Ecology*, 61, 466, 1980.
73. **Cohen, D.**, Optimizing reproduction in a randomly varying environment, *J. Theor. Biol.*, 12, 119, 1966.
74. **Cohen, D.**, Optimizing reproduction in a randomly varying environment where a correlation may exist between the conditions at the time a choice has to be made and the subsequent outcome, *J. Theor. Biol.*, 16, 1, 1967.
75. **Cohen, D.**, A general model of optimal reproduction in a randomly varying environment, *J. Ecol.*, 56, 219, 1968.
76. **Browne, R. A.**, Divergence of demographic and reproductive variables over 25 years in laboratory and natural populations of the brine shrimp, *Artemia, Crustaceana* (Leiden), 45, 164, 1985.
77. **Sorgeloos, P.**, Two strains of *Artemia* in Urmia Lake (Iran), *Artemia Newsl.*, 13, 5, 1989.
78. **Perez, R.**, Cyst production of *Artemia* in salt ponds in southeastern Spain, in *Artemia Research and Its Applications*, Vol. 3, Sorgeloos, P., Bengtson, D. A., Decleir, W., and Jaspers, E., Eds., Universa Press, Wetteren, Belgium, 1987, 215.
79. **Browne, R. A.**, Competition experiments between parthenogenetic and sexual strains of brine shrimp, *Artemia salina, Ecology*, 61, 471, 1980.
80. **Browne, R. A. and Halanych, K. M.**, Competition between sexual and asexual *Artemia*: a reevaluation, *Crustaceana*, 57, 57, 1989.
81. **Miller, R. S.**, Patterns and process in competition, *Adv. Ecol. Res.*, 4, 1, 1967.
82. **Lynch, M.**, Destabilizing hybridization, general-purpose genotypes and geographic parthenogenesis, *Quart. Rev. Biol.*, 59, 257, 1984.

Chapter 11

USE OF *ARTEMIA* AS A FOOD SOURCE FOR AQUACULTURE

David A. Bengtson, Philippe Léger, and Patrick Sorgeloos

TABLE OF CONTENTS

I. INTRODUCTION

Although *Artemia* has been known to man for centuries, its use as a food for the culture of larval organisms apparently began only in the 1930s, when several investigators found that it made an excellent food for newly hatched fish larvae.[1-3] As aquaculture developed in the 1960s and '70s, the use of *Artemia* also became more widespread, due both to its convenience and to its nutritional value for larval organisms. The fact that *Artemia* dormant cysts can be stored for long periods in cans, then used as an off-the-shelf food requiring only 24 h of incubation makes them the most convenient, least labor-intensive, live food available for aquaculture. The nutritional value of *Artemia*, especially for marine organisms, is not constant, but varies both geographically and temporally.[4] During the last decade, however, both the causes of *Artemia* nutritional variability and methods to improve poor-quality *Artemia* have been identified.

This article covers both historical and present usage of *Artemia* in aquaculture, with particular emphasis on the nutritional aspects of this important food. The reader may also wish to consult other recent review articles[4-9] for more in-depth treatment of some of the subjects covered here.

II. *ARTEMIA* SUPPLY AND DEMAND

The history of *Artemia* cyst harvesting and commercialization reveals an interesting evolution. In the 1950s commercial supplies originated from two sources in the U.S., i.e., the salt pans in the Bay of San Francisco, California and the Great Salt Lake, Utah. *Artemia* cysts were marketed at a very low price (less than 10 U.S. dollars per kg) for the aquarium pet trade.[10] With the start of hatchery research in marine fish and crustacean farming in the early sixties, new marketing opportunities were created for *Artemia* cysts. San Francisco Bay, Great Salt Lake, and Chaplin Lake (Saskatchewan, Canada) *Artemia* were the first commercial strains used in aquaculture research and production.

Increased interests in aquaculture research and development around the globe, decreased harvests from the Great Salt Lake (because of weather conditions; see below) and Chaplin Lake, high import taxes in certain third world countries (e.g., 100% in Brazil in 1975),[11] and possible simulation of cyst shortages by certain commercial companies resulted in severe price rises of *Artemia* cysts. By the mid-seventies, wholesale prices ranged from 50 to 100 U.S. dollars per kg.

The dramatic impact of the aggravating cyst shortage on the expansion of hatchery aquaculture of marine fish and crustaceans was repeatedly underlined at international conferences.[12] Third world countries often could not afford to import the expensive cysts. The suggestion was made to abandon *Artemia* and look for alternative solutions to culture food for the early larval stages of fish and crustaceans.[13]

Fundamental and applied research with brine shrimp *Artemia* was initiated at the State University of Ghent in the early 70s. Sorgeloos claimed in 1976 that the cyst shortage was an artificial and temporary problem.[14] This was verified when commercial cyst production began from a number of new sources, including a number of third-world countries. This occurred with the help of international aid organizations, as well as many private companies. By 1980 a change in the *Artemia* cyst situation became obvious. New commercial products were available from Argentina, Australia, Brazil, Colombia, People's Republic of China, France, and Thailand. Prices dropped considerably and market competition was aggressive. A market survey estimated the annual cyst consumption by aquaculture hatcheries was about 60 metric tons.[15,16]

However, new problems arose. The hatching quality of commercial cyst products was inconsistent[17] and the nutritional value of *Artemia* varied among sources and even among batches.[16] Climatological conditions eliminated cyst production at a number of sites for several years. For example, flooding of the south arm of the Great Salt Lake, Utah, interrupted cyst production for over five years. On the other hand, dilution of the north arm resulted in optimal salinity levels that increased annual cyst harvests to several hundred tons during the mid-80s. By 1988 the increase in salinity in the south arm again resulted in new cyst production.[18] Finally, cyst production in large saltworks was high for a few years following inoculation (e.g., over 10 ton per year in the CIRNE saltworks at Macau, Rio Grande do Norte, Brazil, in the late 70s),[19] only to subsequently drop to insignificant levels due to lack of proper management, e.g., control of nutrient levels, manipulation of cyst induction, etc.

Currently, cyst production, availability, and price are relatively stable, due to efforts in *Artemia* research and development in the first half of the eighties,[4,20,21] application of improved methods for harvesting and processing of cysts, the introduction of *Artemia* cyst quality certificates, and the enhancement of cyst food value via bioencapsulation with particulate or emulsified enrichment diets. The price of cysts is now quality dependent, ranging from \$25 per kg for Great Salt Lake cysts (large nauplii, low levels of the essential fatty acid 20:5(n-3) to \$80 per kg for cysts characterized by high yield, synchronous hatching, and small nauplii which are pollution free and contain high levels of 20:5(n-3); i.e., >10 mg/g dry weight nauplii.

Since the mid-80s *Artemia* cyst consumption has increased to several hundred tons annually as a result of the worldwide expansion in commercial larviculture of marine fish, shrimp, and prawn. For several aquaculture candidate species it was only in recent years that a successful transition could be made from pilot into commercial larviculture. In view of the increased demands for fish-fry and shrimp postlarvae (considered to be the bottleneck in present-day aquaculture),[22] and the expected extension of the list of new commercially cultured species (e.g., mahi-mahi, grouper, halibut, etc.) *Artemia* cyst demand is expected to increase for years to come. For example, marine fish-fry production in the Mediterranean is expected to increase fivefold by 1992.[23] Although more use can be made of artificial diets as partial substitutes for *Artemia* nauplii,[24] complete substitution is not likely to occur because of nutritional reasons (e.g., palatability, availability, and digestibility), possible interference with water quality, and, last but not least, price difference.

Although present commercial cyst supplies largely meet demands, a larger diversity of commercial sources is urgently needed to ensure more, and especially predictable, quantities of a commercially available product. Today over 70% of all marketed cysts originate from the Great Salt Lake. An exceptionally wet winter in Utah might wipe out the *Artemia* population in the Great Salt Lake for an extended period. More commercial attention should be focused on the development of alternative/complementary sources, which, for example, are available for natural harvesting in the Soviet Union and the People's Republic of China.

Small-scale cyst production in man-made saltworks, although technically very successful in several countries in Southeast Asia and Latin America,[25] are not expected to contribute significantly to world cyst supplies. However, they provide interesting opportunities for local commercial developments, especially in specific third-world countries (e.g., Vietnam)[26] with restricted import opportunities, and where local availability of *Artemia* cysts is the first requirement in consideration of a viable hatchery industry.

Aside from the harvesting of cysts from large natural salt lake/pond systems it is anticipated that more integrated projects of man-managed *Artemia* production and use will be considered. Thanks to the discovery of the high food value of adult *Artemia* biomass as nursery and maturation diet for fish and crustaceans,[4,27,28] several fish/shrimp farming operations are planning to set up units for *Artemia* biomass production which could eventually be modified for cyst production in the event of an imminent cyst shortage.

TABLE 1
Procedure for Decapsulation of *Artemia* Cysts [6,7,30,32,55]

1. Hydrate cysts by placing them in fresh or salt water, bubbled in a separatory funnel for 2 h at 25°C.
2. Collect cysts on 125 μm mesh sieve, rinse, and transfer to hypochlorite solution.
3. The hypochlorite solution, made up of either liquid bleach NaOCl or bleaching powder $Ca(OCl)_2$ (in advance), should consist of 0.5 g active hypochlorite product (activity normally labeled on the package) per g of cysts, plus an alkaline product to keep pH > 10 (0.15 g technical grade NaOH for liquid bleach; either 0.67 g $NaCO_3$ or 0.4 g CaO for bleaching powder; per g of cysts), plus enough sea water to make up the final solution to 14 ml/g of cysts. (Dissolve bleaching powder first, before addition of alkaline product.)
4. Keeping the solution at 15—20° C by addition of ice to the surrounding water bath, add the hydrated cysts and stir for 5—15 min. Check the temperature regularly, since the reaction is exothermic; never exceed 40°C. When cyst color becomes grey (with powder bleach) or orange (liquid bleach), or when microscopic examination shows the chorions to be completely dissolved, remove the cysts, and rinse them with water on a 120 μm screen until the chlorine smell disappears.
5. Totally deactivate the hypochlorite by dipping the cysts (< 1 min) either in 0.1 N HCl or in 0.1% $Na_2S_2O_3$ solution, then rinse again with water.

III. FORMS OF *ARTEMIA* USED IN AQUACULTURE

A. DECAPSULATED CYSTS

The hard shell (chorion) that encysts the dormant *Artemia* embryo can be removed by chemical means in a procedure called decapsulation. This procedure requires hydration of the cysts, removal of the chorion with a hypochlorite solution, and washing and deactivation of the hypochlorite. The resulting decapsulated cysts can then be used immediately or dehydrated for storage. A detailed procedure for decapsulation can be found in Table 1. Decapsulation results in a naked, viable embryo that is approximately 210 to 270 μm in diameter,[29] depending on the strain of *Artemia* being used. The decapsulated cysts can then be hatched into Instar I nauplii or can be fed directly to aquaculture organisms.

The advantages of decapsulation are several. First, the embryos become disinfected during this procedure, so any bacteria associated with the cyst shells are not introduced into the culture tanks.[30] Second, the cyst shells are not introduced into the culture tanks and therefore cannot be ingested by the cultured organisms. Third, the decapsulated cysts contain a higher energy content when fed to a predator than do Instar I nauplii, because they do not have to expend energy in breaking out of the shell.[31] Fourth, the decapsulated cysts are smaller particles than Instar I nauplii (compare dimensions given above with the range of Instar I naupliar lengths of 420 to 520μm for various strains). Finally, if the decapsulated cysts are incubated for hatching as Instar I nauplii, the hatchability is improved, again because no shell breakout is required.[32]

The major disadvantage of decapsulated cysts is that they are nonmoving, nonbuoyant particles. Thus, some aquaculture organisms, like marine fish larvae, that could benefit most from their small size, bacteria- and shell-free nature, and high energy content might have difficulty ingesting them if a sufficient number of the particles is not kept in suspension by extra aeration or circulation of water in the tanks. Perhaps for this reason, the major usage of decapsulated cysts seems to be in penaeid shrimp hatcheries, where the postlarvae are more able to withstand extra aeration and can better capture particles at or near the bottom of the tanks.

B. NEWLY HATCHED NAUPLII

Nauplii in the Instar I-II stages are probably the form of *Artemia* most widely used in aquaculture. In order to optimize utilization of cysts for hatching into nauplii, it is useful to know as much as possible about the hatching characteristics of each batch. Percent hatch (the total percentage of the cysts that actually hatch), hatching efficiency (the number of

TABLE 2
Procedure for Hatching and Harvesting *Artemia*[55,61]

1. Place 5 g of cysts per liter of sea water (temperature 25-30°C; salinity >5 g/l) in a conical container. If low-salinity water (5-10 g/l) is used, or if large quantities of cysts are being hatched, buffer with up to 2 g technical grade $NaHCO_3$/l. pH must remain >8.
2. Add aeration to bottom of conical container to maintain oxygen levels above 2 mg/l until cysts have hatched.
3. Maintain a light level of 1000 lux, at least during the first 3—4 h of incubation.
4. When cysts have hatched, remove aeration and allow nauplii to settle for 5—10 min at bottom of container.
5. Remove nauplii onto a screen of ≤150 μm mesh and rinse them thoroughly. Similarly remove any more nauplii that settle at 5—10 min intervals.
6. Place in a separator box and use a light source to attract nauplii away from the unhatched cysts and debris; collect the nauplii with a pipet and transfer to clean sea water.

nauplii hatched per g of cysts), hatching rate, including T_0 (the time to first hatch) and T_{90} (the time to 90% hatch), and hatching output (dry weight of nauplii hatched per g of cysts) can be determined by the methods of Bruggeman et al.,[32] Sorgeloos et al.,[33] Vanhaecke and Sorgeloos,[34] and Vanhaecke and Sorgeloos,[17] respectively. For certain strains, especially those from the Canadian sulfate lakes, it might also be useful to know the difference in hatching percentage in 5 ppt sea water vs. 35 ppt and with decapsulated vs. untreated cysts.[17] Once these are known, routine hatching should be accomplished in containers with conical bottoms, with an external light source of at least 1000 lux to trigger the hatching mechanism[35] and aeration from the bottom of the cone. Specific procedures for hatching and harvesting *Artemia* nauplii are given in Table 2.

Knowledge of hatching characteristics is important due to the variability among batches found by Vanhaecke and Sorgeloos.[17] Percent hatch was found to vary from about 20 to 90% of the total cysts. This quality criterion for *Artemia* cysts obviously accounts for much of the price differences among *Artemia* batches. A batch with 80% hatch will clearly be worth more than a batch with 20% hatch. Number of nauplii hatching per gram of cysts can vary from <100,000 to >300,000. Time to first hatch was also found to vary from about 13 h to about 20 h, and time to 90% hatch from about 20 h to about 32 h. This is important because the practical aquaculturist normally wishes to harvest the nauplii and feed them to the cultured organisms as soon after hatching as possible. With each passing hour, the nauplii metabolize the yolk for their own energetic needs, therefore decreasing in energy content,[36] increasing in size and mobility, and making it more difficult for the predator to capture them.

Perhaps more important than hatching characteristics of the nauplii is the nutritional quality of those nauplii for the particular organism being fed. It should be stressed that, although *Artemia* nauplii are often a convenient, adequate food for short-term rearing of fish and invertebrate larvae, they may not be a nutritionally complete food for long-term rearing of those organisms. In particular, marine fish and invertebrate larvae appear to have different fatty-acid requirements than do freshwater organisms.[37] This subject will subsequently be covered in more detail; we simply remark that a batch of *Artemia* proven to be a good food for rearing freshwater organisms will not necessarily be good for marine organisms. Furthermore, there is no correlation between good hatching characteristics (e.g., >80% hatch) and good nutritional value. Nutritional value for marine organisms is another quality criterion that accounts for much of the difference in price among *Artemia* batches. Those batches that have an adequate fatty-acid composition for both freshwater and marine organisms can be worth two to three times as much as batches that are adequate only for freshwater organisms.

A prudent strategy for the aquaculturist is to buy a small quantity (e.g., one can) of a particular batch of cysts and test it for hatching and nutritional quality before procuring a large quantity. In some aquaculture operations, *Artemia* cysts are used at a rate of several kg per day. Thus, cysts may cost tens of thousands of U.S. dollars per year and the quality of the cysts should be verified before purchase or a quality certificate should be requested at purchase (as is commonly done now in Europe).

Newly hatched *Artemia* nauplii are normally fed immediately upon harvesting to the cultured organisms. They may be introduced into the culture tank all at once, or they may be metered in slowly, so that some food is always available in the water. The latter method has the inherent drawback that the nauplii lose nutritional value and increase in size while they are held. One solution is the cold storage of nauplii until they are introduced into the culture tanks.[38] If the nauplii are kept at about 4°C until they are added to the tank, their metabolism is slowed and they retain their nutritional value and newly hatched size for up to 48 h.

C. METANAUPLII

The term metanauplii is given to *Artemia* of Instar stages II through V, or roughly 2 to 5 d after hatching. Most nauplii will use up their yolk reserve and starve to death within 3 to 4 d after hatch, so successful utilization of metanauplii as food requires that they be offered food prior to being fed to predators. This has generally meant feeding them with algae so that the use of metanauplii in aquaculture has been limited. In recent years, however, the introduction of enrichment techniques (see below) has allowed for the much simpler culture of *Artemia* to the metanaupliar stages and these are now commonly used worldwide.

The main restriction on the use of metanauplii is their size (about 500 to 800 μm). Many fish and crustacean larvae are not able to ingest such large particles until several days to weeks after they have started to feed. For predators that are capable of ingesting metanauplii, however, the added nutritional value possibly obtained in algae-fed or enriched *Artemia* can make utilization of this form of *Artemia* especially worthwhile. Furthermore, the greater energy content obtained per particle ingested may reduce the energetic costs of feeding for the predator, as compared to the costs of feeding on newly hatched nauplii that may be smaller than the predator prefers.

D. JUVENILE AND ADULT *ARTEMIA*

These forms of *Artemia*, obtained through intensive *Artemia* culture in the laboratory or extensive culture in ponds, are primarily important as live food for penaeid shrimp aquaculture and are often referred to as *Artemia* "biomass". Methods for culture of *Artemia* have been published elsewhere.[7,13,39] Juvenile and adult *Artemia* can be harvested easily from an intensive culture system by siphoning or draining the tank. They can be harvested from a pond either by a large net pushed or towed by a boat or by a net placed in the flow of water at a pumping station between ponds.[40] The harvested *Artemia* may then be fed live to the predator,[41] frozen,[42] freeze-dried[40] for later usage, or made into a flake diet.[43,44]

Live *Artemia* biomass is apparently a good food both for the growth of penaeid shrimp juveniles in nursery ponds and for the maturation of adult brood stock in hatcheries.[20,41] The location of a salina or man-made *Artemia* production ponds in close proximity to an aquaculture facility can be beneficial to both *Artemia* producers and aquaculturists. For example, in Thailand small-scale salt producers also harvest live *Artemia* biomass as a by-product and sell it to fish and shrimp farmers, transporting it by truck.[41] In a demonstration project in the Philippines, salt, *Artemia*, aquaculture, and agriculture production were all integrated into a single business operation.[45]

Another popular use of live *Artemia* is as a food for tropical and ornamental pet fish.[14,42] The mode of transport for this application is by airplane, as well as by surface vehicles, since the quantities required are smaller.

E. FROZEN AND FREEZE-DRIED *ARTEMIA*

Artemia biomass can be frozen, freeze-dried, or made into a flake diet for storage and subsequent usage, with relatively little loss of nutritional composition. The frozen form is sold commercially by several companies, which often advertise in aquaculture trade magazines. Use of freeze-dried and flake diet forms has recently been suggested by Guimaraes and Lira do Rego[40] and Janata et al.,[43] although their effectiveness as food (compared to live *Artemia*) has not yet been completely demonstrated.

Frozen adult *Artemia* are normally used for culture of larval American lobsters (*Homarus americanus*), as well as other crustacean and fish species (see review in Léger et al.[4]) Frozen *Artemia* nauplii have also been used for culture of both crustacean[46,47] and fish[48,49] larvae. Some investigators have observed that frozen or freeze-dried nauplii could lead to reduced survival or growth of fish larvae,[50-53] however, perhaps due to the loss of essential nutrients during the thawing process.[54]

IV. FACTORS INFLUENCING THE EFFECTIVENESS OF *ARTEMIA* AS A FOOD FOR AQUACULTURE

A. CYST COLLECTION AND PROCESSING

The collection and processing of cysts is normally done by the commercial supplier and therefore is beyond the control of the aquaculturist. Variability in these procedures is likely to affect hatchability more than nutritional value. Optimal procedures for collection and processing have been given by Sorgeloos et al.[55] Hatchability of cysts is probably decreased the longer they remain on the shore of a pond, which is subject to alternations in drying and rehydration. The cysts should, if possible, be collected directly from the hypersaline pond water, e.g., through the use of collection barriers oriented perpendicular to the predominant wind direction.[55] The collected cysts should then be separated from debris, sand, etc. Optimal hatching will result when cysts are dried in a fluidized bed dryer and packaged in cans under either vacuum or nitrogen.[34]

B. STORAGE OF CYSTS

Cysts should be dehydrated before storage (moisture content preferably <5%), because a moisture content greater than 10% reduces hatchability.[6] Storage under vacuum or nitrogen is important because the presence of oxygen can cause formation of free radicals that reduce hatchability.[56] Properly packaged containers of *Artemia* cysts can be kept at room temperature or can be frozen;[6] however, if frozen, the cysts should be held for 1 week at room temperature before they are incubated in sea water.[57]

It should also be noted that storage of frozen brine shrimp (nauplii or adults) for extended periods in the presence of oxygen, similarly to any other aquaculture food, can result in the oxidation of the long-chain highly unsaturated fatty acids (possibly causing rancidity)[58] and the depletion of vitamin E due to its antioxidant properties.[59]

C. HATCHING OF CYSTS

Problems of hatching variability have been discussed above. A knowledge of the hatching characteristics of a particular batch of cysts and application of that knowledge to daily hatching procedures can greatly increase the effectiveness of usage of that batch in a hatchery. Some standard rules can be applied to maximize the hatching and naupliar survival. First, the container used for hatching should have a conical bottom into which aeration is introduced. Any container with a cylindrical or square bottom will have "dead spots" in which *Artemia* cysts or nauplii will accumulate and deplete the oxygen supply. Aeration in the bottom of a conical container will keep cysts and nauplii continually in suspension. Second, the aeration level should be sufficient to maintain a dissolved oxygen concentration of at least 2 mg/l (preferably 5 mg/l). Third, incubation should be carried out at a maximum concentration of 5 g cysts per 1 of sea water for small volumes (e.g., <20 l) decreasing to a concentration of 1 g cysts per l for large volumes (e.g., >100 l). Fourth, an external light source is needed, especially during the first few hours of incubation,in order to trigger the hatching mechanism. Fifth, technical grade sodium bicarbonate should be added at a maximum rate of 2 g/l in order to maintain the pH>8.0 throughout the incubation period. Finally, if optimal procedures are followed and cysts still do not hatch well, it may be possible to increase hatching percentage by treatment of the cysts with hydrogen peroxide (e.g., 5 min soaking of dry cysts in a 5% hydrogen peroxide solution).[6,60] For a more complete review of *Artemia* cyst dormancy and hatching, the reader is referred to the recent review article by Lavens and Sorgeloos.[6]

D. HARVESTING OF NAUPLII

Four aspects of harvesting nauplii primarily govern the effectiveness of those nauplii as food: (1) The harvest should take place as soon after T_{90} as possible to ensure maximum nutritional value of the nauplii.[31,36] (2) Nauplii should not be allowed to settle for too long in the bottom of the conical container (i.e., 5 to 10 min maximum), lest they die due to insufficient oxygen. (3) Nauplii should be collected on a submerged fine mesh (<150 μm) screen and rinsed well with minimal mechanical harm to the nauplii before being fed to cultured organisms, in order to prevent the introduction of hatching by-products like glycerol into the culture system. (4) Nauplii should be separated from hatching debris and any dead nauplii (e.g., with a separator box)[61] so that only freely swimming, live nauplii are introduced into the culture system. Feeding dead nauplii that become unavailable to the predator can reduce culture output[62] and foul the water.

E. BIOMETRICS

Difference among *Artemia* geographical strains with regard to size of cysts and nauplii has been well-studied by Vanhaecke and Sorgeloos.[29] Size appears to be genetically determined and cysts collected over time from a particular location are relatively similar in diameter. Cysts from San Francisco Bay and northeastern Brazil tend to be among the smallest available, and those from India, China, Italy, and France, among the largest. Finally, there is a high correlation between cyst diameter and naupliar length, as well as for several other biometric characteristics.

Size of nauplii is often not critical for crustacean larvae, which can capture and manipulate food particles with their feeding appendages. For fish larvae, which have no feeding appendages, the ability to engulf a prey particle in one bite is critical. Fish larvae that receive overly large *Artemia* nauplii may starve because they cannot ingest the particles. For at least one species, a high correlation exists between naupliar length of *Artemia* and larval fish mortality during the first five days after hatching.[63] Some fish species must be fed rotifers as a first food because nauplii from all *Artemia* strains are too large. In those cases, the size of nauplii will determine when the fish can be switched from a rotifer to an *Artemia* diet.

F. BIOCHEMICAL COMPOSITION

Many aspects of the biochemical composition of *Artemia* have been studied and the results will only be summarized here. The reader is referred to the review by Léger et al.[4] for a more extensive treatment of this subject.

The dry weight and individual energy content of Instar I *Artemia* nauplii are strongly related to the size of the cysts and nauplii,[31] as one might expect. Dry weights range from about 1.6 to 3.3 μg per nauplius and energy content from about 0.037 to 0.073 J per nauplius.[31]

Literature reports on proximate composition of nauplii vary greatly, ranging from 37 to 71% protein, 12 to 30% lipid, 11 to 23% carbohydrate, and 4 to 21% ash.[4] Values for adults are 50 to 69% protein, 2 to 19% lipid, 9 to 17% carbohydrate, and 9 to 29% ash.[4]

Fatty-acid composition of *Artemia* is probably the most studied biochemical component, due to the early observations of Watanabe et al.[64] demonstrating that the long-chain essential fatty acids in *Artemia* were especially variable. A recent review[4] of a large data base of published and unpublished fatty-acid profiles revealed that six fatty acids, 16:0, 16:1(n-7); 18:1(n-9); 18:2(n-6); 18:3(n-3); and 20:5(n-3), normally make up about 80% of the total fatty acids in an *Artemia* sample. The saturated and monoene fatty acids of that group, 16:0, 16:1(n-7), and 18:1(n-9), typically comprise 40 to 60% of the total fatty acids. However, it is clear from the work of Watanabe and his colleagues, as well as the work of the International Study on *Artemia*, and subsequent work on *Artemia* enrichment,[4] that the (n-3) series of fatty acids, 18:3(n-3) and 20:5(n-3), determine the nutritional effectiveness of the *Artemia* more than any other single biochemical component. Linolenic acid, 18:3(n-3), is an essential fatty acid for freshwater organisms, and eicosapentaenoic acid, 20:5(n-3), is an essential fatty acid for marine organisms.[37] If both fatty acids are lacking (a very rare phenomenon) the *Artemia* will probably be a poor food for both freshwater and marine aquaculture species. Normally *Artemia* samples contain either (a) 18:3(n-3) levels >20% of total fatty acids and 20:5(n-3) levels <5%, or (b) 18:3(n-3) levels <10% and 20:5(n-3) levels between 5 and 13%. Only the latter *Artemia* batches can be successfully fed to marine organisms, because marine organisms appear to require *Artemia* with 20:5(n-3) levels >5%, but both types can be fed to freshwater organisms.

The fatty acid composition of an *Artemia* batch is environmentally, not genetically, determined. Millamena et al.[65] and Lavens et al.[151] have demonstrated that the fatty-acid profiles of *Artemia* adults, as well as the cysts they produce, strongly reflect the fatty-acid profile of the diet that they were fed. *Artemia* cysts produced in large salt lakes with a single predominant algal species (e.g., Great Salt Lake, Utah) therefore have a more constant fatty-acid profile over time than do cysts from ponds in solar salt evaporation systems, in which algal species can change markedly both temporally and spatially. In recent years, cysts from Great Salt Lake (by the hundreds of metric tons) have contained high levels of 18:3(n-3) and low levels of 20:5(n-3). Thus, the majority of the world's cyst production has been adequate only for culture of freshwater organisms. This has strongly influenced world market prices and led to the commercialization of enrichment products (see below) so that food for marine organisms could be made available.

The amino acid composition of *Artemia* nauplii, by contrast, seems to be remarkably similar from strain to strain,[66] suggesting that it is not environmentally determined in the manner that the fatty acids are. Furthermore, the ten essential amino acids required for fish[67] are normally present in sufficient quantities in *Artemia* nauplii.[66]

The presence of several proteolytic enzymes in *Artemia* developing embryos and nauplii[4,68,69] and the measurement of proteolytic activity in *Artemia* nauplii[70] has led to speculation that the exogenous enzymes in the nauplii play a significant role in breakdown of those nauplii in the digestive tract of larval fish.[4,71] This has become an important question in view of the relatively low levels of digestive enzymes in many first-feeding fish larvae and the inferiority of prepared feeds (vs. live) for marine fish larvae.

Vitamin analyses have been carried out only on San Francisco Bay *Artemia* cysts[72] and adults.[73] Whether or not significant differences exist within or among strains is therefore unknown. The published data indicate that levels of niacin, pyridoxine, and riboflavin in *Artemia* are slightly less than recommended for cold-water fishes.[67]

The levels of several minerals in *Artemia* as reported in the literature have been summarized recently.[4] The mineral requirements of marine organisms are poorly known and may be met by the sea water that they ingest. The main concern about *Artemia* mineral composition is whether they meet the requirements of freshwater organisms in culture, about which our knowledge is also poor. A recent study of variability of 18 minerals and trace elements in *Artemia* cysts revealed that levels of selenium and manganese were the most variable and that selenium in some cases may not be present in sufficient quantity.[74]

V. THE NUTRITIONAL ENHANCEMENT OF *ARTEMIA*

A. SOLUTION TO THE PROBLEM OF NUTRITIONAL DEFICIENCIES

In the previous section it was concluded that the main factor affecting the nutritional value of *Artemia* as a food source for marine larval organisms was the content of the essential fatty acid 20:5(n-3). This highly unsaturated long-chain fatty acid, along with 22:6(n-3), have been proven to be essential for marine fish and crustaceans.[37,75-85] The content of 22:6(n-3) in freshly hatched *Artemia* is low, even nil, while the level of 20:5(n-3) is highly variable (see Table 3 and Léger et al.[4]). In view of the essential fatty acid (EFA) deficiency in *Artemia*, research has been conducted to increase the EFA content (review in Léger et al.[4]). Taking advantage of its continuous nonselective feeding behavior, *Artemia* may be fed any particulate diet when its particle size is below 50 μm.[86] Hence, a diet enriched with EFA 20:5(n-3) and 22:6(n-3) will produce nutritionally more adequate *Artemia*. Léger et al.[87] have demonstrated that feeding cod liver oil-coated micronized rice bran (CLO) fortified with 20:5(n-3) and 22:6(n-3), increased the level of both EFAs in San Pablo Bay *Artemia* (SPB); see Table 4. Prefeeding with rice oil-coated rice bran (RO, lacking 20:5(n-3) and 22:6(n-3)) did not markedly increase the level of either EFA in SPB *Artemia*. The nutritional value of EFA-enriched SPB *Artemia* for the marine crustaceans *Mysidopsis bahia* and *Penaeus stylirostris* was significantly better than the newly hatched SPB nauplii (Table 5). Prefeeding with the EFA-lacking diet (RO) did not enhance the nutritional quality of the nauplii. The enhancement of the dietary value of SPB *Artemia* nauplii was optimal with more refined (AA18) or more concentrated (SEC) enrichment diets (Table 4 and 5) which resulted in higher EFA levels in the enriched *Artemia*. A similar enhancement of the nutritional value of *Artemia* nauplii for marine fish larvae (red seabream) was demonstrated by Watanabe et al.[88] using algae and yeasts containing EFA (Table 6). The best culture results for both fish and shrimp were obtained when enriched *Artemia* nauplii contained not only 20:5(n-3), but also EFA 22:6(n-3); e.g., prefed with ω-yeast, CLO, AA18, and SEC (see Table 4 and 5). This probably explains why better culture results are usually obtained with natural zooplankton such as copepods (e.g., *Acartia* spp., *Eurytemora* spp., *Tigriopus* spp.) than with freshly hatched *Artemia* nauplii.[88-90] Marine zooplankton usually contain high levels of 22:6(n-3).[91] From these above experiments it is apparent that the nutritional value of *Artemia* is governed by its content of EFA (especially 20:5(n-3) and 22:6(n-3), generally referred to as (n-3)HUFA — Highly Unsaturated Fatty Acids), and that the nutritional value of *Artemia* can be enhanced by enriching the nauplii through prefeeding with EFA-rich or enriched diets.

B. ENRICHMENT TECHNIQUES

The finding that the nutritional value of *Artemia* can be improved by prefeeding with EFA, especially (n-3)HUFA-rich diets, has led to the development of a number of enrichment techniques and diets. Their application in marine larviculture is of particular interest because of the variable levels of 20:5(n-3) and the very low levels (if any) of 22:6(n-3) found in *Artemia* nauplii. Léger et al.[4] classified them into four groups: the British technique, with unicellular algae; the Japanese technique, with ω-yeast or emulsions plus baker's yeast; the French technique, with compound diets; and the Belgian technique, with microparticulate diets or self-emulsifying concentrates. Beside the enrichment diet used, the respective techniques vary with respect to hatching conditions, enrichment time (time between hatching and addition of enrichment diet), enrichment period, and temperature. The enrichment levels reported so far are summarized in Table VI of Léger et al.[4] and Tables 7 and 8 of this review.

TABLE 3
Variability in Content of 20:5(n-3) within Several Geographical Strains of *Artemia*

Artemia Geographical Strain	20:5(n-3) Content (range)
San Francisco Bay	0.3—13.3
Great Salt Lake - South Arm	2.7—3.6
Great Salt Lake - North Arm	0.3—0.4
Chaplin Lake (Canada)	5.2—9.5
Macau (Brazil)	3.5—10.6
Bohai Bay (China)	1.3—15.4

Note:: 20:5(n-3) Content given as area percent; i.e., content of 20:5(n-3) as a percentage of total fatty acids.

Data compiled from References 4 and 87.

TABLE 4
Content of (n-3)HUFAs in Enrichment Diets and *Artemia* Preparations

	20:5(n-3)		22:6(n-3)		(n-3)HUFA	
	Area %	mg/g	Area %	mg/g	Area %	mg/g
Enrichment diets						
CLO	8.0	6.3	10.0	5.2	20.9	14.2
RO	—	—	—	—	Tr	Tr
Artemia preparations						
SFB—newly hatched	9.3	11.8	0.2	0.3	11.5	14.6
SPB—newly hatched	0.5	0.5	—	—	0.7	0.8
SPB—starved 24 h	1.4	1.1	0.6	0.4	3.5	3.0
SPB—CLO-enriched 24 h	6.3	7.3	1.5	1.9	8.9	10.1
SPB—RO-enriched 24 h	0.9	0.8	—	—	1.9	1.9

Note: Data expressed both as area percent (i.e., (n-3)HUFA as a percentage of total fatty acids) and as mg (n-3)HUFA/g dry weight of tissue. Tr, trace; —, not detected Σ (n-3)HUFA > 20:3(n-3); CLO, cod liver oil; RO, rice oil; SFB, San Francisco Bay; SPB, San Pablo Bay.

Data from Reference 87.

TABLE 5
(n-3)HUFA Content of San Francisco Bay and San Pablo Bay *Artemia* Nauplii—Results of Culture Tests for *Mysidopsis bahia* and *Penaeus stylirostris* Fed on Those Diets

	SFB *Artemia*	SPB *Artemia*				
	Newly hatched	Newly hatched	RO-en-riched	CLO-en-riched	AA18-en-riched	SEC-en-riched
(n-3)HUFA content (area%)						
20:5(n-3)	9.3	0.2	0.9	6.3	8.2	9.9
22:6(n-3)	0.2	—	—	1.5	1.5	5.9
Σ (n-3)HUFA	11.4	0.7	1.9	8.9	10.6	17.8
Culture results (*M. bahia*)						
Survival (%)	93.3	62.0	60.0	75.0	92.5	95.8
Ind. length (μm)	5532	4587	4285	5029	5375	5254
Ind. dry weight (μg)	354	198	188	259	259	323
Culture results (*P. stylirostris*)						
Survival (%)	47.5	34.0	—	—	45.7	63.9
Ind. wet weight (mg)	1.8	1.7	—	—	2.0	2.7

Note: AA18 and SEC are commercial enrichment products; RO, rice oil; CLO, cod liver oil; Σ (n-3)HUFA >20:3(n-3).

Data from References 85, 87, and 106.

1. The British Technique

In this technique developed by Forster and Wickins[92] and Wickins,[93] algae (*Isochrysis galbana*) are used for enriching *Artemia* nauplii. A density of 10,000 nauplii per liter of sea water containing 300 cells/ml is applied during 24 h for improving the nutritional value of *Artemia* for prawn larvae (*Palaemon serratus*). No data are available on the enrichment levels obtained applying this technique. Using *I. galbana* (T-Iso), Sivertsen[94] obtained a maximum enrichment level of 5.5 mg/g (dry weight basis) after 12 h; this level dropped to 2.6 mg/g after 125 h. The disadvantages of using algae are that one has to maintain continuous cultures and that the (n-3)HUFA content in algae is variable.[95] Recently, Walford and Lam[96] proposed the use of microcapsules (AR 121, Frippak, United Kingdom) containing a high percentage of total lipids as a substitute for algae in enriching *Artemia* nauplii. Total (n-3)HUFA buildup in *Artemia* nauplii reached a maximum of 16.9% (percent of total fatty acids) after a 48-h enrichment. Rimmer and Reed[97] reported enrichment levels up to 12%, (20:5(n-3), 22:5(n-3), and 22:6(n-3) of total fatty acids) using Frippak microcapsules. They obtained better culture results with *Lates calcarifer* when using *Artemia* enriched in this way. Deru et al.,[98] using the same product during a 4-h enrichment, obtained a maximum (n-3)HUFA enrichment of 6.8% in *Artemia*. The dietary value for *Macrobrachium rosenbergi* larvae of these enriched *Artemia* was not significantly different from that of *Artemia* containing 6.0% (n-3)HUFA without enrichment.

2. The Japanese Technique

The "indirect method" as developed by Watanabe et al.[64,90,99] at first applied algae (*Chlorella minutissima*) for enrichment, similar to the British technique. Enrichment levels in *Artemia* obtained with this algae reached 15.5% (n-3)HUFA of total fatty acids. A similar procedure was adopted using the so-called ω-yeast as an alternative to algae. The (n-3)HUFA buildup after 24 h reached 13.8% of total fatty acids. This ω-yeast, an (n-3)HUFA-fortified baker's yeast, was developed by Imada et al.[100] and offers the advantage of having a less variable EFA content than does algae. The disadvantage of using this product is that it has to be used in a fresh, living condition. The "direct method" developed by Watanabe et al.[99,101] requires emulsified fish oils (e.g., cuttle-fish liver oil and pollock oil) and an (n-3)HUFA methylester mixture. The emulsion is fed to the *Artemia* nauplii which ingest and accumulate the droplets. Reported (n-3)HUFA levels attained with this method is 1.01% or 10.1 mg/g *Artemia* (wet or dry weight basis not stated). Watanabe and colleagues concluded that *Artemia* containing at least 0.3% (n-3)HUFA (dry or wet basis not stated) may be a satisfactory single feed for marine fish, *Pagrus major, Paralichthys* sp., and *Oplegnathus* sp.

3. The French Technique

Robin et al.[102] developed an enrichment technique applying a composed diet (*Spirulina* powder, yeast, amino acids, vitamins, cholesterol, and fish oil) during a 48-h prefeeding period. Later, Robin et al.[103] and Robin[104] improved this technique by splitting up the enrichment period in a 48-h prefeeding, followed by a short-term (30 min) enrichment bath, using a diet based on fish autolysate, fish oil, vitamins, and minerals. (n-3)HUFA buildup in *Artemia* reached 16 mg/g (dry weight basis). Higher levels up to 25 mg/g (dry weight basis) were reported by Robin et al.[105] through the use of higher dietary oil concentrations in the prefeeding diet. Better culture results were obtained when *Artemia*, enriched in this manner, were fed to seabass (*Dicentrarchus labrax*) and especially to turbot (*Scophthalmus maximus*). Recently, Sivertsen et al.[94] reported enrichment data obtained with a composed diet containing fish meals and oils. A maximum level of 10.6 mg/g (dry weight basis) was obtained in *Artemia* after a 12-h enrichment. When using micromilled pure squid meal, a maximum level of (n-3)HUFA was attained after 96 h (15.4 mg/g), while the level after a

TABLE 6
Content of (n-3) HUFAs in *Artemia* from Canada and San Francisco Bay (SFB) — Effect of Enriched SFB on Survival and Growth of Red Sea Bream Juveniles

	Canadian *Artemia*	San Francisco Bay *Artemia*		
	Newly hatched	Newly hatched	Fed *Chlorella* for 24 h	Fed ω-yeast for 24h
(n-3)HUFA content				
20:5(n-3)	5.2	1.6	3.2	3.4
22:6(n-3)	—	—	—	
Σ (n-3)HUFA	5.8	2.4	4.1	5.1
Fish culture test				
Survival (%)	68.4	43.4	66.8	86.4
Survival after activity test (%)	37.5	24.1	46.1	50.0
Final length (mm)	9.57	10.13	11.13	11.67

Note: Data from Reference 88.

TABLE 7
Summary of Enrichment Procedures for *Artemia* Nauplii, (n-3)HUFA Content in *Artemia*, and Results of Comparative Culture Tests with Enriched *Artemia*[a]

Artemia source	Hatching		Pre-enrichment			Enrichment			(n-3)HUFA content						Culture test		Ref.
									20:5(n-3)		22:6(n-3)		Σ (n-3)HUFA				
	t	T	Diet	t	T	Diet	t	T	%	mg/g	%	mg/g	%	mg/g	Species	Performance	
SPB (1628)			na			AA18, in hatching medium	46	30	—	2.2	—	3.1	—	9.6	*Dicentrarchus labrax*	+	147
GSL-NA			na			AA18, in hatching medium	46	30	—	3.4	—	2.3	—	6.7	*D. labrax*	+	147
GSL-NA	22	25	na			na			—	0.3	—	—	—	0.5	*D. labrax*	–	147
SPB (1628)	22	25	na			na			—	0.5	—	—	—	2.6	*D. labrax*	–	147
RAC	26	25	na			na			—	10.6	—	—	—	11.2	*D. labrax*	+	147
RAC	T_{90}	28	na			na			7.2	—	0.1	—	—	—	*Sparus aurata*	+	111
Bohai Bay (China)	T_{90}	28	na			na			—	—	—	—	—	—	*S. aurata*	+	111
GSL	T_{90}	28	na			na			0.3	—	—	—	—	—	*S. aurata*	–	111
SPB	T_{90}	28	na			na			0.5	—	—	—	—	—	*Sparus aurata*	–	111
GSL			na			AA18, in hatching medium	48	28	2.1	—	1.4	—	—	—	*S. aurata*	o	111
SPB						AA18, in hatching medium	48	28	1.8	—	2.6	—	—	—	*S. aurata*	o	111
RAC	T_{90}	28	na			na			7.2	—	0.1	—	—	—	*Penaeus serratus*	+	111
GSL	T_{90}	28	na			na			0.3	—	—	—	—	—	*P. serratus*	o	111
SPB	T_{90}	28	na			na			0.5	—	—	—	—	—	*P. serratus*	o	111
GSL			na			AA18, in hatching medium	48	28	2.1	—	1.4	—	—	—	*P. serratus*	±	111
SPB			na			AA18, in hatching medium	48	28	1.8	—	2.5	—	—	—	*P. serratus*	o	111
RAC	T_{90}	28	na			na			7.2	—	0.1	—	—	—	*P. adspersus*	+	111
GSL	T_{90}	28	na			na			0.3	—	—	—	—	—	*P. adspersus*	±	111
SPB	T_{90}	28	na			na			0.5	—	—	—	—	—	*P. adspersus*	±	111

TABLE 7 (continued)
Summary of Enrichment Procedures for *Artemia* Nauplii, (n-3)HUFA Content in *Artemia*, and Results of Comparative Culture Tests with Enriched *Artemia*[a]

Artemia source	Hatching		Pre-enrichment			Enrichment			(n-3)HUFA content						Culture test		Ref.
									20:5(n-3)		22:6(n-3)		Σ (n-3)HUFA				
	t	T	Diet	t	T	Diet	t	T	%	mg/g	%	mg/g	%	mg/g	Species	Performance	
GSL			na			AA18, in hatching medium	48	28	2.1	—	1.4	—	—	—	*P. adspersus*	+	111
SPB			na			AA18, in hatching medium	48	28	1.8	—	2.5	—	—	—	*P. adspersus*	±	111
Brazil (CIRNE)	48	24	na			na			—	—	—	—	6.8	9.2			105
Bohai Bay (China)	48	24	na			na			—	—	—	—	11.7	11.0			105
France	48	24	na			na			—	—	—	—	6.9	7.6			105
France	48	24	G	48	24	na			—	—	—	—	8.5	5.4			105
SFB (2557)	48	24	na			na			—	—	—	—	5.7	8.6			105
SFB (2557)	48	24	G	48	24	na			—	—	—	—	12.7	8.4			105
SFB (2557)	48	24	G	48	24	J			—	—	—	—	14.9	15.9			105
SFB (2557)	48	24	L	48	24	na			—	—	—	—	10.1	10.2			105
SFB (694)	48	24	na			na			—	—	—	—	4.0	5.8			105
SFB (694)	48	24	J	24	24	na			—	—	—	—	7.5	13.0			105
SFB (694)	48	24	G	48	24	na			—	—	—	—	12.2	9.0			105
SFB (694)	48	24	L	48	24	na			—	—	—	—	14.4	14.0			105
SFB (694)	48	24	M	48	24	na			—	—	—	—	16.6	22.3			105
GSL	48	24	na			na			—	—	—	—	4.0	5.2			105
GSL	48	24	G	48	24	na			—	—	—	—	8.9	9.0			105
GSL	48	24	G	48	24	J			—	—	—	—	10.9	12.3			105
GSL			na			na			—	0.5	—	0	—	—	*P. vannamei*	−	148
GSL			na			SELCO (after separation)	24	28	—	21.3	—	12.7	—	—	*P. vannamei*	+	148
AF	24	25	na			na			5.3	7.7	—	—	—	—	*D. labrax*	+	129
GSL	24	25	na			na			0.3	0.4	—	—	—	—	*D. labrax*	−	129

SFB	24	25	na			na			1.9	2.3	—	—	—	—	*D. labrax*	±	129
GSL	24	25	na			SELCO (after separation)	24	25	4.9	8.1	3.1	5.1	—	—	*D. labrax*	+	129
Bohai Bay (China)	48	25	na			na			13.0	13.0	—	—	—	—	*S. aurata*	±	130
Bohai Bay						SELCO (in hatching medium after decapsulation)	48	25	12.3	17.9	3.6	5.2	—	—	*S. aurata*	+	130
Bohai Bay	48	—	na			na			10.8	—	—	—	10.8	—			96
Bohai Bay	48	—	Baker's yeast	20	—	na			11.2	—	0.1	—	11.4	—			96
Bohai Bay	48	—	Baker's yeast	20	—	AR121 (Frippak, 5 million micro capsules/ml)	8	29	12.2	—	3.6	—	16.9	—			96
							24	29	12.6	—	1.2	—	14.4	—			96
							48	29	15.5	—	0.7	—	16.9	—			96
Aquarium products			starved	24	—	BOOSTER (Frippak) 0.3 g/l			7-9	—	1-2	—	—	—	*Lates calcarifer*	+	97
Aquarium products			starved	24	—	na									*L. calcarifer*	–	97
EG			na			na			—	3.2	—	—			*Hippoglossus hippoglossus*	–	94,149
EG						Squid meal 0.2 g/l/d	6	28	—	3.6	—	—			*H. hippoglossus*	±	94,149
							12	28	—	4.5	—	2.5					
							24	28	—	4.6	—	1.0					
							48	28	—	5.9	—	1.6					
							72	28	—	8.1	—	2.8					
							96	28	—	10.1	—	4.7					
							125	28	—	9.0	—	1.8					
EG						SUPERSELCO (0.3 g/l, 2x/d)	6	28	—	4.0	—	—			*H. hippoglossus*	o	94,149
							12	28	—	8.7	—	6.9					
							24	28	—	15.1	—	16.1					
							48	28	—	20.3	—	21.3					
							72	28	—	22.7	—	21.1					

TABLE 7 (continued)
Summary of Enrichment Procedures for *Artemia* Nauplii, (n-3)HUFA Content in *Artemia*, and Results of Comparative Culture Tests with Enriched *Artemia*[a]

Artemia source	Hatching		Pre-enrichment			Enrichment			(n-3)HUFA content						Culture test		Ref.
									20:5(n-3)		22:6(n-3)		Σ (n-3)HUFA				
	t	T	Diet	t	T	Diet	t	T	%	mg/g	%	mg/g	%	mg/g	Species	Performance	
EG						Squid meal plus SUPERSELCO (3:1)	6	28	—	3.2	—	—					94,149
							12	28	—	5.6	—	4.0					
							24	28	—	6.8	—	4.4					
							48	28	—	13.6	—	12.4					
							72	28	—	15.8	—	17.8					
							96	28	—	16.5	—	19.6					
							125	28	—	15.0	—	16.1					
EG						Special composed diet (SINTEF) 0.2 g/l/d	6	28	—	5.0	—	2.0					94,149
							12	28	—	6.5	—	3.4					
							24	28	—	5.9	—	1.7					
							48	28	—	5.9	—	2.3					
							72	28	—	6.8	—	2.2					
							96	28	—	6.6	—	1.6					
							125	28	—	6.3	—	1.1					
EG			na			*Isochrysis galbana*(T-Iso); 0.1 g/l/d	6	28	—	3.9	—	0.2					94,149
							12	28	—	3.8	—	0.9					
							24	28	—	3.6	—	0.5					
							48	28	—	2.6	—	0.5					
							72	28	—	2.3	—	0.6					
							96	28	—	2.2	—	0.5					
							125	28	—	2.0	—	0.4					
GSL	24	28	na			na									*Lates calcarifer*	–	132
GSL	24	28	na			SELCO after separation	24	28							*L. calcarifer*	+	132
GSL	24	28	na			na			3.0	3.4	0.3	0.4	3.4	3.9			107

GSL			na	SUPERSELCO after separation	6	28	7.4	11.7	5.1	8.1	13.4	21.7			
					12	28	10.3	18.4	6.7	12.2	18.0	32.7			
					18	28	13.4	25.0	9.5	18.0	24.5	46.1			
					24	28	16.3	31.7	11.4	22.1	30.2	56.0			
GSL	24	28	na	na			3.7	4.8	0.5	0.7	4.3	5.6	*Macrobrachium rosenbergii*	−	127
GSL				Emulsified coconut oil	24	28	2.3	4.0	0.4	0.8	2.8	4.9	*M. rosenbergii*	−	127
GSL				SUPERSELCO after separation	24	28	17.3	33.5	11.4	22.0	30.9	66.2	*M. rosenbergii*	+	127

Note: The symbols +, o, ±, −, refer to relative values; i.e., very good, good, average, and poor culture results, interpreted from the data given. Abbreviations: t, time period; T, temperature in °C; %, quantity of fatty acid as a percentage of total fatty acids (area%); mg/g, mg fatty acid methyl ester per g dry weight of tissue; Σ (n-3)HUFA, sum of (n-3) highly unsaturated fatty acids (note: this generally refers to the sum of (n-3) fatty acids with 20 or 22 carbons and 3 or more double bonds); na, no preenrichment or enrichment diet was applied; diets G, J, L, M, refer to Table 8; SPB, San Pablo Bay; SFB, San Francisco Bay; GSL-NA, Great Salt Lake, North Arm; RAC, Reference *Artemia* Cysts; AF and EG Artemia Systems, SA, commercial product. AA18, SELCO, and SUPERSELCO are commercial enrichment diets of Artemia Systems NV/SA, Belgium.

[a] These are additions to Table VI in Léger et al., *Oceanogr. Mar. Biol. Ann. Rev.*, 24, 521, 1986.

TABLE 8
Composition (%) of Preenrichment and Enrichment Diets for Which Abbreviations (G, J, L, M) are Given in Table 7

	Preenrichment diets (48 h)			Enrichment bath (0.5 h)
	G	L	M	J
Brewer's yeast	89.4	83.4	73.4	—
Cod liver oil	4	10	20	10
Fish autolysate	—	—	—	73
Choline chloride	2	1	1	4
DL-Methionine	1	2	2	2
Vitamin premix	—	3.6	3.6	—
Vitamin mineral premix	3.6	—	—	11

24-h enrichment was 5.9 mg/g. Replacing 25% of the squid meal ration with an emulsified enrichment diet (SUPERSELCO, Artemia Systems, Belgium) the values reached 38.0 mg/g and 12.0 mg/g, respectively. These authors also reported total amino acid contents in the enriched *Artemia*: 378.2 mg/g (dry weight) using SUPERSELCO during 24 h; 383.8 mg/g using squid meal; 419.4 mg/g using squid meal + SUPERSELCO; 413.9 mg/g using the compound diet; and 407.3 mg/g using *Isochrysis galbana*. The amino acid composition was not significantly different among enrichment diets.

4. The Belgian Technique

This technique consisted at first of prefeeding EFA-enriched microparticles, e.g., micronized rice bran coated with fish oils;[85] compound analog AA18.[85,106] *Artemia* enriched in this manner contained up to 15.1 mg (n-3)HUFA per gram (dry weight basis). Because of the complexity and cost for preparing these products, a new technology was applied for producing a more effective enrichment diet under the form of a self-emulsifying concentrate.[87] This diet is a self-dispersing complex mixture of mainly (n-3)HUFA sources, vitamins, carotenoids, and phospholipids. Upon dilution in sea water, finely dispersed stable microglobules are formed which are readily ingested by *Artemia* and which bring about EFA-enrichment levels which largely surpass the values reported in literature (e.g., 38.3 mg/g to 53.5 mg/g after 24 h and up to 87.6 mg/g after a 48-h enrichment; see Table VI of Léger et al.[4] and Tables 7 and 8 of this review). Different application regimes are proposed for applying the self-emulsifying concentrates to the hatching medium or after separation of the hatched nauplii.[85]

The high enrichment levels obtained with this technique are the result not only of an optimal diet composition and presentation but also of proper enrichment procedures; e.g., nauplii must be transferred or exposed to the enrichment medium as soon as possible before first feeding. In this way, the nauplii begin feeding immediately after the opening of the alimentary tract (Instar II stage). Furthermore, the size increase occurring after hatching during enrichment will be minimal. *Artemia* enriched according to previous procedures reach >900 μm, whereas here, higher enrichment levels are obtained in nauplii measuring 660 μm (12-h enrichment) to 790 μm (48-h enrichment).

Recently a dry form of the self-dispersing concentrate has been proposed with levels obtained in *Artemia* of between 30 mg/g and 60 mg/g, dry weight.

C. (n-3)HUFA REQUIREMENTS

Several studies have shown the beneficial effect of feeding (n-3)HUFA-enriched nauplii to several species of marine fish and crustaceans.[4,107] However, no study has clearly indicated the requirements for (n-3)HUFA for one species or another. Léger et al.[4] summarized I.S.A.

(International Study on *Artemia*) feeding tests with several strains of *Artemia* for several species. They concluded that high survival and growth of marine species are obtained with *Artemia* nauplii containing at least 4% 20:5(n-3) of total fatty acid methyl esters (or about 4 mg/g on a dry weight basis), while levels below 3% consistently yielded poor results. From their experiments Watanabe and colleagues concluded that *Artemia* containing at least 0.3% (n-3)HUFA (dry or wet basis not stated) may be a satisfactory single feed for marine fish. They added, however, that *Artemia* enrichment should always be applied since lipid contents in *Artemia* gradually decrease after hatching.[99]

Recently, it was found that (n-3)HUFA requirements may differ considerably among marine species. Halibut (*Hippoglossus hippoglossus*) and dolphin fish (*Coryphaena hippurus*) larvae appeared to require more than other species such as sea bream (*Sparus aurata*) and sea bass (*Dicentrarchus labrax*).[107] Significant increases in survival and growth were obtained only when fish were fed highly-enriched *Artemia* nauplii containing about 50 mg/g. Based on this knowledge, the Working Group on the Mass Rearing of Juvenile Fish (Mariculture Committee of the International Council for the Exploration of the Seas) in 1988, recommended the initiation of a series of studies in order to better identify the qualitative and quantitative requirements for (n-3)HUFA in marine larval fish. Three tasks are currently being conducted: (1) cataloging all existing information; (2) identifying (n-3)HUFA requirements through feeding tests using three enrichment diets containing different levels of (n-3)HUFA; (3) intercalibrating (n-3)HUFA determination methods by gas chromatography.

D. OTHER APPLICATIONS OF ENRICHMENT

Besides EFAs (especially (n-3)HUFA), other nutrients can be incorporated in *Artemia*. Watanabe et al.[8] reported on the accumulation of fat soluble vitamins (especially vitamin A and vitamin E) in *Artemia* during a short-term (9 h) enrichment period. Vitamin A levels increase from below 1 IU/g (wet weight basis) to over 16 IU/g within 9 h enrichment and vitamin E levels increase from below 20 mg/g (wet weight basis) to about 250 mg/g in freshly hatched and up to over 600 mg/g in starved metanauplii, 52 h old.

Besides nutrients, other components may be incorporated as well. Nin et al. (cited by Léger[107]) report on the use of *Artemia* nauplii for administering therapeutics to fish larvae. They demonstrated the use of *Artemia* nauplii enriched with chloramphenicol for *Cichlosoma nigrofasciatum*. Caudal tissue levels of chloramphenicol reached 3 to 4 ppm after 1 h feeding of 24-h enriched nauplii (Figure 1). Curative and preventive treatment of larvae through the food chain may be much more effective than treating the relatively large volume of culture water surrounding them.

The enrichment techniques and diets developed for *Artemia* may also be applied for adult *Artemia* and other live food sources such as rotifers and nematodes.[107-109]

VI. SPECIFIC APPLICATIONS OF *ARTEMIA* FOR FEEDING AQUACULTURE SPECIES

A. PENAEID SHRIMP CULTURE

Artemia is generally used for feeding the late larval and postlarval stages of penaeids. Freshly hatched nauplii are usually offered beginning with the first mysis stage, and sometimes earlier at the zoea-mysis molt. Some authors recommend introduction of *Artemia* even during the second zoea stage.[110] However, penaeids are usually fed algae prior to the *Artemia* and undergo a several-day weaning period when both foods are given. Thus, addition of *Artemia* too early in the life cycle can result in competition for the algae food between the uneaten *Artemia* and the penaeids. A convenient solution may be the early administration of killed nauplii or decapsulated *Artemia* cysts, as suggested by Mock et al.[46] and Wilkenfeld et al.[110]

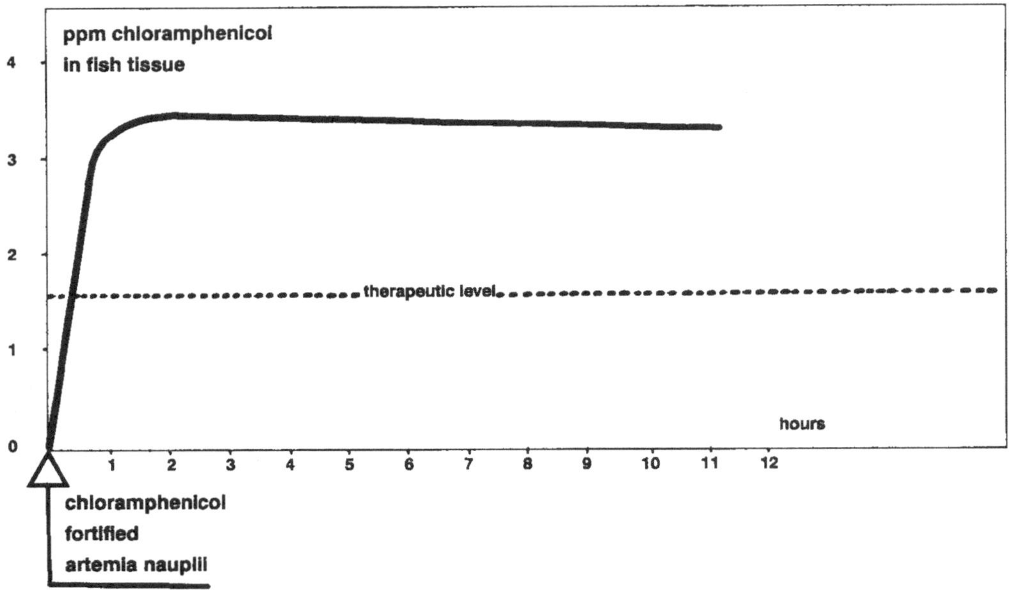

FIGURE 1. Chloramphenicol levels in fish larvae caudal tissue fed one single ration of 24 h-fortified *Artemia* nauplii.

Little information is available on the nutritional value of different strains of *Artemia* for penaeid shrimp. Léger et al.[85] demonstrated that San Francisco Bay (236 2016) *Artemia* nauplii performed significantly better than San Pablo Bay (1628) nauplii. The nutritional value of the latter could be improved by (n-3)HUFA enrichment. In addition, the (n-3)HUFA content of the diet (algae or substitutes) offered during the first developmental stages significantly affected the susceptibility of the larvae to the poor nutritional quality of the San Pablo Bay *Artemia* nauplii. For example, the negative effects of feeding these *Artemia* was aggravated when the animals had previously been fed an (n-3)HUFA-poor diet while no negative effects were noticed when an (n-3)HUFA-rich diet was administered during the earlier stages. Feeding enriched Great Salt Lake *Artemia* nauplii also resulted in better survival and growth in *Penaeus vannamei* postlarvae.[87] When these postlarvae were subsequently fed an artificial diet without transition from a live to a dry diet, better food acceptance, growth, and survival were recorded than in those postlarvae which were previously fed with (n-3)HUFA-poor *Artemia*.

Amat et al.[111] compared the nutritional value of different *Artemia* strains for *Penaeus kerathurus, P. serratus* and *P. adversus*. The best growth and survival of shrimp were obtained with *Artemia* strains containing the highest HUFA levels, although significant differences were not always detected. *P. adversus* larvae grew even better on *Artemia* containing only low levels of (n-3)HUFA (Great Salt Lake and San Pablo Bay *Artemia*). Ongrown *Artemia* have been used successfully for postlarval feeding of *Penaeus monodon*,[112-114] *P. kerathurus*,[115,116] *P. japonicus*,[40,117,118] and *P. aztecus*.[119] Yashiro[114] compared growth and survival in *P. monodon* postlarvae reared on *Artemia* of progressively larger size, fed with different diets (wheat flour, rice bran, and milled rice). Freshly hatched *Artemia* supplemented with cooked mussel meat were fed in the control treatment. Growth after 20 d was significantly better in the treatments fed ongrown *Artemia* (more than double the weight gain), while survival was best in the control treatment, though not significantly better than those shrimp fed *Artemia* grown on milled rice.

Guimaraes and Lira do Rego[40] successfully used ground and sieved freeze-dried and liquefied fresh *Artemia* biomass collected from the saltworks in Macau (Brazil) as a food source for larval and postlarval *P. japonicus* and *P. aztecus*. *Tetraselmis chuii* and baker's yeast were fed during the Nauplius V and Protozoea I stages only while *Artemia* biomass was offered from Protozoea I stage till harvest (PL VII—PL IX). Following this method, survival averaged 59% over a 3-year period.

B. FRESHWATER PRAWN, *MACROBRACHIUM* SP.

The culture of freshwater prawn larvae depends heavily on the use of *Artemia* nauplii, which is the most successful diet throughout the larval rearing period, and after one week usually in combination with prepared diets.[120-124] In contrast to penaeid shrimp, *Macrobrachium* can initially be fed with freshly hatched *Artemia* nauplii. Sick and Beaty[125] demonstrated that naupliar concentration should be higher than 0.1 nauplii/ml to assure proper ingestion. The same authors found that energy intake in *M. rosenbergii* was directly proportional not only to *Artemia* concentration but also to *Artemia* size. They demonstrated that Stage VIII prawn larvae attained a maximum energy ingestion of 0.0066 cal/mg animal dry weight/h when fed 0.7 mm *Artemia* metanauplii; 0.062 when fed 1.5 mm *Artemia* larvae; and 1.014 when fed 5.5 mm *Artemia* juveniles.

Only a few authors have studied the (n-3)HUFA requirements in *Macrobrachium* sp. D'Abramo and Sheen[126] demonstrated significantly better growth in juvenile prawn with a HUFA-supplemented experimental diet. Deru et al.[98] could not identify extrapolation of this HUFA requirement to the larval stages. They compared unfed *Artemia* nauplii containing 0.3% and 6.0% (n-3)HUFA, and 4-h enriched *Artemia* (Booster, Frippak, United Kingdom) containing up to 6.9% 20:5(n-3) + 22:6(n-3). Using highly enriched *Artemia* nauplii (30.9% (n-3)HUFA, 66.2 mg/g dry weight; SUPERSELCO, Artemia Systems Societé Anonyme, Belgium), Devresse and Rasowo[127] recently demonstrated significant differences in both growth and survival of *M. rosenberghii* larvae using (n-3)HUFA-fortified vs. control *Artemia* nauplii. The first postlarvae appeared on day 16 in the enriched *Artemia*-fed larvae and on day 19 in the control treatment. Metamorphosis was completed on day 22 and day 25, respectively.

C. MARINE FISH CULTURE

The larvae of several species of marine fish, e.g., bream, bass, and flatfish, can only be offered an *Artemia* diet after an initial week on a smaller prey, primarily the rotifer, *Brachionus plicatilis*. In contrast to crustacean larvae, marine fish larvae are cultured on *Artemia* for a much longer period of time, e.g., 20 to 40 d. *Artemia* cyst consumption is also among the highest in marine fish larviculture and ranges from 200 to 500 g per 1000 of the fry produced. *Artemia* biomass in live or frozen form is often applied as a transitional diet for the fry when weaned from a live to an inert food. With the development of better artificial diets, weaning can be successfully applied progressively earlier during the larval development of different fish species.

The variability of the nutritional value of *Artemia* nauplii as a food source for marine larval fish has been well documented.[12] Application of HUFA enrichment of the *Artemia* diet has had a significant effect in marine fish larviculture;[12] i.e., it resulted in increased survival and less variability in fish hatchery production. The latter is especially important since it was the missing link in the development of commercial production. Furthermore, the quality of the fry in terms of stress resistance, better pigmentation, reduced deformities, more successful swimbladder inflation, and increased vigor appears to be directly correlated with (n-3)HUFA enrichment of their larval diet.[128]

The survival of European sea bass (*Dicentrarchus labrax*) appears strongly correlated with the content of 20:5(n-3) in *Artemia* nauplii, while growth is highly correlated with 22:6(n-3) content. All larvae fed unenriched Great Salt Lake *Artemia* died within 35 d, while 25% of those fed (n-3)HUFA-enriched Great Salt Lake *Artemia* survived for 42 d.[129]

For good growth and survival in European sea bream (*Sparus aurata*), the food regime of rotifers and brine shrimp should contain high levels of both 20:5 and 22:6(n-3). Best resistance to stress conditions (i.e., activity test) was displayed by bream larvae fed 22:6(n-3)-enriched live feed.[130]

With turbot (*Scophthalmus maximus*) (n-3)HUFA enrichment has its most significant effect on larval pigmentation. However, survival remains too low and variability among experiments too high to evaluate the effect of dietary (n-3)HUFA levels on growth and survival.[131]

For the Pacific species, similar tendencies to those of the European species have been determined. Survival at metamorphosis and stress resistance (i.e., salinity shocks) in Asian sea bass (*Lates calcarifer*) is strongly correlated with the HUFA levels of its *Artemia* diet.[22] After 3 weeks culturing, milk fish (*Chanos chanos*) showed significant increases in growth (length and dry weight) when fed HUFA-fortified *Artemia*.[132] Similarly, rabbit fish larvae (*Siganus guttatus*) fed HUFA-rich *Artemia* displayed less mortality when disturbed than did controls fed HUFA-poor *Artemia*.[132]

Until early 1988, culture trials with mahi-mahi larvae (*Coryphaena hippurus*) had only been successful when they were fed natural copepods or other zooplankton. Culture tests with newly hatched *Artemia* did not prove successful.[133-135] In 1988 and 1989, significant progress in the larviculture of this fast-growing aquaculture candidate was achieved by various research groups in the U.S. and Australia. Feeding the larvae with (n-3)HUFA super-fortified *Artemia* resulted in more consistent larviculture outputs in terms of survival, larval growth, and health as compared to those cultured with the copepod *Calamoecia* or other zooplankton as food.[22,135]

Rearing of red drum larvae (*Sciaenops ocellatus*) is currently being studied in the U.S. Gulf Coast states. Rotifers are the first food, followed by *Artemia* nauplii,[136] and experiments on the effects of enrichment of both live foods are currently being conducted.[137]

D. FRESHWATER FISH CULTURE

Much freshwater fish culture is carried out in ponds with natural zooplankton as the larval food. The salmonids, perhaps the group cultured most widely on an intensive basis, have a relatively well-developed digestive tract at first feeding and are usually fed prepared diets immediately. Thus, *Artemia* is not used in the vast majority of freshwater fish culture. However, several species of freshwater fish are fed *Artemia*. Whitefish larvae (family Coregonidae) are often fed *Artemia* until they metamorphose and can be switched to a dry diet.[52] Hokanson and Lien[138] compared survival and growth of walleye larvae (*Stizostedion vitreum*) raised on diets of *Artemia*, natural zooplankton, and fish larvae and concluded that *Artemia* is the best choice for a first food. Barrows[139] suggested a 15-d feeding period on brine shrimp for walleye larvae prior to the feeding of traditional artificial diets. In the U.S. *Artemia* nauplii are increasingly used as a first food for striped bass larvae (*Morone saxatilis*).[140] Interestingly, although these fish are reared in fresh or very low-salinity water, recent evidence[141] suggests that they may have the fatty-acid requirements of a marine fish, which they become as adults. The larvae are typically fed *Artemia* from about 5 d post-hatch until about 20 d, then weaned onto an artificial diet until about 30 d, after which *Artemia* feeding ceases.[142]

A major drawback in feeding *Artemia* to freshwater organisms is that the *Artemia* die after 30 to 60 min in fresh water. Therefore, they are not continuously available to the predator, as they would be in marine systems, and must be fed intermittently every 2 to 3 h. A lot of the convenience of *Artemia* is thus lost, especially since dead *Artemia* may foul the water.

E. AQUARIUM FISH

As mentioned above, both live and frozen adult *Artemia* are used as food for species reared as pets by home aquarists. Cysts are also purchased by these users and hatched for feeding as nauplii. Survival, vigor, and pigmentation were reported to be significantly improved in several tropical species when (n-3)HUFA levels were increased.[150] Although the quantities used by a pet fish owner are small compared to those used by a commercial aquaculturist, there are many more of the former than the latter.

F. TOXICOLOGICAL RESEARCH

An increasing market for *Artemia* cysts is developing among researchers in aquatic toxicology. As governmental requirements for environmental testing increase, many more tests are being conducted with many more species. Standards now exist for the use of *Artemia* nauplii as food for those organisms[143] and the availability of reference *Artemia* diets[144-146] greatly assists in the rigorous conduct of that testing.

VII. CONCLUDING REMARKS

Artemia usage in aquaculture is currently undergoing great changes. As new aquaculture species are developed and as aquaculture production increases, the demand for *Artemia* continues to grow, especially in Asia, Latin America, and Europe. At the Second International Symposium on *Artemia* in 1985, projections were made that cyst sales would increase to 150—170 tons annually before the year 2000.[42] Only three years later, at the World Aquaculture Society meeting, participants estimated that 1988 consumption would already be 250 tons. As demand increases, prices are likely to increase also.

The development of products and techniques to enrich *Artemia* has meant that cysts from the world's major production source, Great Salt Lake, could be used for marine as well as freshwater species. This has helped ease the shortage of cysts that inherently have an adequate fatty-acid profile for marine organisms.

Two major questions exist at this point. First, will *Artemia* supply be able to keep up with demand? Research is currently being conducted to try to understand the processes in extensive production of *Artemia*, so that someday salinas and lakes might be biologically managed to optimize production. Second, can artificial formulated diets ever totally replace *Artemia* as a sole food for cultured larvae? Research is being conducted both on the development of diets and materials to bind or encapsulate them and on the physiological and nutritional requirements of larvae. It appears at this point that adequate diets will not be ready for several years, if ever. For the time being, *Artemia* will continue to be the major live food for larviculture and, therefore, a product that unites aquaculturists and researchers in both developed and developing countries.

ACKNOWLEDGMENTS

This work was supported by Cooperative Agreement CR-814721 between the U.S. Environmental Protection Agency and the University of Rhode Island representing Contribution Number X154 from the EPA Environmental Research Laboratory, Narragansett, RI. It was also supported by the Belgian National Science Foundation through FKFO Project *Artemia*, the Belgian Administration for Development Cooperation, the Belgian Ministry of Science Policy, and Artemia Systems NV/SA.

REFERENCES

1. **Seale, A.,** Brine shrimp (*Artemia*) as a satisfactory food for fishes, *Trans. Am. Fish. Soc.*, 63, 129, 1933.
2. **Gross, F.,** Notes on the culture of some marine plankton organisms, *J. Mar. Biol. Ass. U.K.*, 21, 753, 1937.
3. **Rollefsen, G.,** Artificial rearing of fry of seawater fish — preliminary communication, *Rapp. P.-v. Reun. Cons. Perm. Int. Explor. Mer*, 109, 133, 1939.
4. **Léger, P., Bengtson, D. A., Simpson, K. L., and Sorgeloos, P.,** The use and nutritional value of *Artemia* as a food source, *Oceanogr. Mar. Biol. Ann. Rev.*, 24, 521, 1986.
5. **Léger, P., Bengtson, D. A., Sorgeloos, P., Simpson, K. L., and Beck, A. D.,** The nutritional value of *Artemia*: a review, in *Artemia Research and Its Applications*, Vol. 3, Sorgeloos, P., Bengtson, D. A., Decleir, W., and Jaspers, E., Eds., Universa Press, Wetteren, Belgium, 1987, 357.
6. **Lavens, P. and Sorgeloos, P.,** The cryptobiotic state of *Artemia* cysts, its diapause deactivation and hatching: a review, in *Artemia Research and Its Applications*, Vol. 3, Sorgeloos, P., Bengtson, D. A., Decleir, W., and Jaspers, E., Eds., Universa Press, Wetteren, Belgium, 1987, 27.
7. **Sorgeloos, P., Bossuyt, E., Lavens, P., Léger, P., Vanhaecke, P., and Versichele, D.,** The use of brine shrimp *Artemia* in crustacean hatcheries and nurseries, in *Handbook of Mariculture*, Vol. 1, McVey, J. P., Ed., CRC Press, Boca Raton, FL, 1983, 71.
8. **Watanabe, T.,** The use of *Artemia* in fish and crustacean farming in Japan, in *Artemia Research and its Applications*, Sorgeloos, P., Bengtson, D. A., Decleir, W., and Jaspers, E., Eds., Universa Press, Wetteren, Belgium, 1987, 373.
9. **Watanabe, T., Kitajima, C., and Fujita, S.,** Nutritional values of live organisms used in Japan for mass propagation of fish: a review, *Aquaculture*, 34, 115, 1983.
10. **Schmidt, A. and Sanders, G.,** personal communication, 1979.
11. **Ferreira de Souza, T.,** personal communication, 1977.
12. **Sorgeloos, P.,** The use of the brine shrimp *Artemia* in aquaculture, in *The Brine Shrimp Artemia*, Vol. 3, Persoone, G., Sorgeloos, P., Roels, O., Jaspers, E., Eds., Universa Press, Wetteren, Belgium, 1980, 25.
13. **Conference recommendation,** FAO Technical Conference on Aquaculture, Kyoto, Japan, (May/June) *FAO Fish. Rep.*, 188, 1, 1976.
14. **Sorgeloos, P.,** The brine shrimp, *Artemia salina*: a bottleneck in mariculture, in *Advances in Aquaculture*, Pillay, T. V. R. and Dill, W. A., Eds., Fishing News Books, Farnham, England, 1979, 321.
15. **Anonymous,** Thai freshwater prawn and brine shrimp farming. Report on a study of economics, marketing and processing requirements, FAO Report THA:75:008:80/WP 17, 1980.
16. **Sorgeloos, P.,** Live animal food for larval rearing in aquaculture: the brine shrimp *Artemia*, in *Realism in Aquaculture: Achievements, Constraints, Perspectives*, Bilio, M., Rosenthal, H., and Sindermann, C. J., Eds., European Aquaculture Society, Bredene, Belgium, 1986, 199.
17. **Vanhaecke, P., and Sorgeloos, P.,** International Study on *Artemia*. XIX. Hatching data for ten commercial sources of shrimp cysts and reevaluation of the "hatching efficiency" concept, *Aquaculture*, 30, 43, 1983.
18. **San Francisco Bay Brand, Inc., and Sanders Brine Shrimp Co.,** personal communication.
19. **Camara, M. R. and De Medeiros Rocha, R.,** *Artemia* culture in Brazil: an overview, in *Artemia Research and Its Applications*, Vol. 3, Sorgeloos, P., Bengtson, D. A., Decleir, W., and Jaspers, E., Eds., Universa Press, Wetteren, Belgium, 1987, 195.
20. **Sorgeloos, P., Bengtson, D. A., Decleir, W., and Jaspers, E., Eds.,** *Artemia Research and Its Applications*, Vol. 1. Morphology, Genetics, Radiobiology, Strain Characterization, Toxicology, Universa Press, Wetteren, Belgium, 1987.
21. **Sorgeloos, P., Bengtson, D. A., Decleir, W., and Jaspers, E., Eds.,** *Artemia Research and Its Applications*, Vol. 3. Ecology, Culturing, Use in Aquaculture, Universa Press, Wetteren, Belgium, 1987.
22. **Sorgeloos, P.** Artemia Newsletter, No. 13, Artemia Reference Center, State University of Ghent, Belgium, 1989.
23. **Artemia Systems, SA,** Ghent, Belgium, unpublished personal communication.
24. **Sorgeloos, P.,** Artemia Newsletter, No. 12, Artemia Reference Center, State University of Ghent, Belgium, 1989.
25. **Sorgeloos, P.,** Brine shrimp *Artemia* in coastal saltworks: hydrobiological key to improved salt production; inexpensive food for vertically integrated aquaculture, in *Proc. International Symposium on Saltworks Conversion for Aquaculture*, D'Amelio, V. and Santulli, A., Eds., Libera Universita Trapani, 1987, 133.
26. **Vu Do Quynh,** Viet Nam. Coastal aquaculture in the Southern Provinces, *World Aquaculture*, 20(2), 1987.
27. **Sorgeloos, P.,** Artemia Newsletter, No. 11, Artemia Reference Center, State University of Ghent, Belgium, 1989.
28. **Sorgeloos, P.,** Artemia Newsletter, No. 10, Artemia Reference Center, State University of Ghent, Belgium, 1988.

29. **Vanhaecke, P. and Sorgeloos, P.,** International Study on *Artemia*. IV. The biometrics of *Artemia* strains from different geographical origin, in *The Brine Shrimp Artemia*, Vol. 3, Persoone, G., Sorgeloos, P., Roels, O., and Jaspers, E., Eds., Universa Press, Wetteren, Belgium, 1980, 393.
30. **Sorgeloos, P., Bossuyt, E., Laviña, E., Baeza-Mesa, M., and Persoone, G.,** Decapsulation of *Artemia* cysts: a simple technique for the improvement of the use of brine shrimp in aquaculture, *Aquaculture*, 12, 311, 1977.
31. **Vanhaecke, P., Lavens, P., and Sorgeloos, P.,** International Study on *Artemia*. XVII. Energy consumption in cysts and early larval stages of various geographical strains of *Artemia*, *Ann. Soc. R. Zool. Belg.*, 113, 155, 1983.
32. **Bruggeman, E., Sorgeloos, P., and Vanhaecke, P.,** Improvements in the decapsulation technique of *Artemia* cysts, in *The Brine Shrimp Artemia*, Vol. 3, Persoone, G., Sorgeloos, P., Roels, O., and Jaspers, E., Eds., Universa Press, Wetteren, Belgium, 1980, 216.
33. **Sorgeloos, P., Persoone, G., Baeza-Mesa, M., Bossuyt, E., and Bruggeman, E.,** The use of *Artemia* cysts in aquaculture: the concept of "hatching efficiency" and description of a new method for cyst processing, in *Proc. 9th Annual Meeting World Mariculture Society*, Avault, J. W., Jr., Ed., Louisiana State University, Baton Rouge, 1978, 715.
34. **Vanhaecke, P. and Sorgeloos, P.,** International Study on *Artemia*. XVIII. The hatching rate of *Artemia* cysts — a comparative study, *Aquacult. Eng.*, 1, 263, 1982.
35. **Vanhaecke, P., Cooreman, A., and Sorgeloos, P.,** International Study on *Artemia*. XV. Effect of light intensity on hatching rate of *Artemia* cysts from different geographical origin, *Mar. Ecol. Prog. Ser.*, 5, 111, 1981.
36. **Benijts, F., Van Voorden, E., and Sorgeloos, P.,** Changes in the biochemical composition of the early larval stages of the brine shrimp *Artemia salina* L., in *10th European Symposium on Marine Biology*, Vol. 1, Persoone, G., and Jaspers, E., Eds., Universa Press, Wetteren, Belgium, 1975, 1.
37. **Kanazawa, A., Teshima, S.-I., and Ono, K.,** Relationship between fatty acid requirements of aquatic animals and the capacity for bioconversion of linolenic acid to highly unsaturated fatty acids, *Comp. Biochem. Physiol.*, 63B, 295, 1979.
38. **Léger, P., Vanhaecke, P., and Sorgeloos, P.,** International Study on *Artemia*. XXIV. Cold storage of live *Artemia* nauplii from various geographical sources: potentials and limits in aquaculture, *Aquacult. Eng.*, 2, 69, 1983.
39. **Persoone, G., Sorgeloos, P., Roels, O., and Jaspers, E.,** *The Brine Shrimp Artemia*, Vol. 3, Ecology, Culturing, Use in Aquaculture, Universa Press, Wetteren, Belgium, 1980.
40. **Guimaraes, J. I. and Lira do Rego, F.,** The use of freeze-dried *Artemia* as food for penaeid shrimp larvae, in *Artemia Research and Its Applications*, Vol. 3, Sorgeloos, P., Bengtson, D. A., Decleir, W., and Jaspers, E., Eds., Universa Press, Wetteren, Belgium, 1987, 479.
41. **Tarnchalanukit, W. and Wongrat, L.,** *Artemia* culture in Thailand, in *Artemia Research and Its Applications*, Vol. 3, Sorgeloos, P., Bengtson, D. A., Decleir, W., and Jaspers, E., Eds., Universa Press, Wetteren, Belgium, 1987, 201.
42. **Lai, L. and Lavens, P.,** Workshop report: *Artemia* as a business perspective, in *Artemia Research and Its Applications*, Vol. 3, Sorgeloos, P., Bengtson, D. A., Decleir, W., and Jaspers, E., Eds., Universa Press, Wetteren, Belgium, 1987, 517.
43. **Janata, W. R., Bell, D. J., and Keller, P. D.,** The DSIR-*Artemia* harvesting programme, in *Artemia Research and Its Applications*, Vol. 3, Sorgeloos, P., Bengtson, D. A., Decleir, W., and Jaspers, E., Eds., Universa Press, Wetteren, Belgium, 1987, 221.
44. **Sumeru, S. U. and Nur, A.,** Preliminary test on the effect of *Artemia* flake diet on the growth and survival of *Penaeus monodon* postlarvae at various feeding rates, *Bull. Brackishwater Aquacult. Dev. Cent.*, 8, 94, 1986.
45. **Jumalon, N. A., and Ogburn, D. M.,** Nutrient flow and physicochemical profile studies of an integrated poultry-salt-*Artemia*-milkfish-sea bass-shrimp pond production system, in *Artemia Research and Its Applications*, Vol. 3, Sorgeloos, P., Bengtson, D. A., Decleir, W., and Jaspers, E., Eds., Universa Press, Wetteren, Belgium, 1987, 239.
46. **Mock, C. R., Fontaine, C. T., and Revera, D. B.,** Improvements in rearing larval penaeid shrimp by the Galveston Laboratory method, in *The Brine Shrimp Artemia*, Vol. 3, Persoone, G., Sorgeloos, P., Roels, O., and Jaspers, E., Eds., Universa Press, Wetteren, Belgium, 1980, 331.
47. **Mock, C. M., Revera, D. B., and Fontaine, C. T.,** The larval culture of *Penaeus stylirostris* using modifications of the Galveston Laboratory technique, in *Proc. 11th Annual Meeting World Mariculture Society*, Avault, J. W., Jr., Ed., Louisiana State University, Baton Rouge, 1980, 102.
48. **Gunkel, G.,** Laboratory experiments on the cultivation of young whitefish, *Coregonus fera* J.: food intake, gross growth-efficiency and growth of the fry, in *Cultivation of Fish Fry and Its Live Food*, Styczynska-Jurewicz, E., Backiel, T., Jaspers, E., and Persoone, G., Eds., European Mariculture Society, Bredene, Belgium, 1979, 211.

49. **Kentouri, M.,** Elevage des larves du loup (*Dicentrarchus labrax*) à l'aide d'organisme du zooplancton congelée: résultats préliminaires, *Aquaculture*, 21, 171, 1980.
50. **Hogendoorn, H.,** Controlled propagation of the African catfish, *Clarias lazera* (C and V.). III. Feeding and growth of fry, *Aquaculture*, 21, 233, 1980.
51. **Seidel, C. R., Schauer, P. S., Katayama, T., and Simpson, K. L.,** Culture of Atlantic silversides fed on artificial diets and brine shrimp nauplii, *Bull. Jap. Soc. Sci. Fish.*, 46, 237, 1980.
52. **Flüchter, J.,** Review of the present knowledge of rearing whitefish (Coregonidae) larvae, *Aquaculture*, 19, 191, 1980.
53. **Fuchs, J.,** Production de juveniles de sole (*Solea solea*) en conditions intensives. I. Le premier mois d'élevage, *Aquaculture*, 26, 321, 1981/1982.
54. **Grabner, M., Wieser, W., and Lackner, R.,** The suitability of frozen and freeze-dried zooplankton as food for fish larvae: a biochemical test program, *Aquaculture*, 26, 85, 1981.
55. **Sorgeloos, P., Lavens, P., Léger, P., Tackaert, W., and Versichele, D.,** *Manual for the Culture and Use of the Brine Shrimp Artemia in Aquaculture*, Artemia Reference Center, State University of Ghent, Ghent, Belgium.
56. **Crowe, J. H. and Clegg, J. S.,** *Anhydrobiosis*, Dowden, Hutchinson and Ross, Inc., Stroudsburg, PA, 1973.
57. **Kinne, O.,** Research cultivation, in *Marine Ecology*, Vol. 3, Part 2, Kinne, O., Ed., Wiley Interscience, New York, 1977, 579.
58. **Hardy, R. W.,** Diet preparation, in *Fish Nutrition*, 2nd ed., Halver, J. E., Ed., Academic Press, New York, 1989, 476.
59. **Halver, J. E.,** The vitamins, in *Fish Nutrition*, 2nd ed., Halver, J. E., Ed., Academic Press, New York, 1989, 32.
60. **Lavens, P., Tackaert, W., and Sorgeloos, P.,** International Study on *Artemia*. XLI. Influence of culture conditions and specific diapause deactivation methods on the hatchability of *Artemia* cysts produced in a standard culture system, *Mar. Ecol. Prog. Ser.*, 31, 179, 1986.
61. **Sorgeloos, P. and Persoone, G.,** Technological improvements for the cultivation of invertebrates as food for fishes and crustaceans. II. Hatching and culturing of the brine shrimp, *Artemia salina* L., *Aquaculture*, 6, 303, 1975.
62. **Ward, S. H.,** Feeding response of the mysid *Mysidopsis bahia* reared on *Artemia*, *Prog. Fish-Cult.*, 49, 29, 1987.
63. **Beck, A. D. and Bengtson, D. A.,** International Study on *Artemia*. XXII. Nutrition in aquatic toxicology - diet quality of geographical strains of the brine shrimp, *Artemia*, in *Aquatic Toxicology and Hazard Assessment: Fifth Conference*, Pearson, J. G., Foster, R. B., and Bishop, W. E., Eds., American Society for Testing and Materials, Philadelphia, PA, 1982, 161.
64. **Watanabe, T., Oowa, F., Kitajima, C., and Fujita, S.,** Nutritional quality of brine shrimp, *Artemia salina*, as a living feed from the viewpoint of essential fatty acids for fish, *Bull. Jap. Soc. Sci. Fish.*, 44, 1115, 1978.
65. **Millamena, O. M., Bombeo, R. F., Jumalon, N. A., and Simpson, K. L.,** The effect of various diets on the nutritional value of *Artemia* as feed for *Penaeus monodon* larvae, in *Book of Abstracts*, 16th Annual Meeting World Mariculture Society, Orlando, FL, 1985, 21.
66. **Seidel, C. R., Kryznowek, J., and Simpson, K. L.,** International Study on *Artemia*. XI. Amino acid composition and electrophoretic protein patterns of *Artemia* from five geographical locations, in *The Brine Shrimp Artemia*, Vol. 3, Persoone, G., Sorgeloos, P., Roels, O., and Jaspers, E., Eds., Universa Press, Wetteren, Belgium, 1980, 375.
67. **National Research Council,** *Nutrient Requirements of Cold-Water Fishes*, National Academy Press, Washington, D.C., 1981.
68. **Ezquieta, B. and Vallejo, C. G.,** *Artemia* trypsin-like proteinase - a developmentally regulated proteinase, in *Artemia Research and Its Applications*, Vol. 2, Decleir, W., Moens, L., Slegers, H., Sorgeloos, P., and Jaspers, E., Eds., Universa Press, Wetteren, Belgium, 1987, 53.
69. **Perona, R., Ezquieta, B., and Vallejo, C. B.,** The degradation of yolk in *Artemia*, in *Artemia Research and Its Applications*, Vol. 2, Decleir, W., Moens, L., Slegers, H., Sorgeloos, P., and Jaspers, E., Eds., Universa Press, Wetteren, Belgium, 1987, 105.
70. **Dabrowski, K. and Glogowski, J.,** Studies on the proteolytic enzymes of invertebrates constituting live fish food, *Hydrobiologia*, 52, 171, 1977.
71. **Dabrowski, K. R.,** The role of proteolytic enzymes in fish digestion, in *Cultivation of Fish Fry and Its Live Food*, Styczynska-Jurewicz, E., Backiel, T., Jaspers, E., and Persoone, G., Eds., European Mariculture Society, Bredene, Belgium, 1979, 107.
72. **Stults, V. J.,** The nutritional value of brine shrimp cysts: the encysted eggs of *Artemia salina*, Ph.D. thesis, Michigan State University, East Lansing, MI, 1974.
73. **Gallagher, M. and Brown W. D.,** Composition of San Francisco Bay brine shrimp *(Artemia salina)*, *Agric. Food Chem.*, 23, 630, 1975.

74. **Cowgill, U. M., Emmel, H. W., Boggs, G. U., Murphy, P. G., Gersich, F. M., Takahashi, I. T., and Bengtson, D. A.**, Variations in chemical composition of *Artemia* cysts from three geographical locations, in *Artemia Research and Its Applications*, Vol. 1, Sorgeloos, P., Bengtson, D. A., Decleir, W., and Jaspers, E., Eds., Universa Press, Wetteren, Belgium, 1987, 173.
75. **Owen, J. M., Adron, J. W., Sargent, J. R., and Cowey, C. B.**, Studies on the nutrition of marine flatfish. The effect of dietary fatty acids on the tissue fatty acids of the plaice *Pleuronectes platessa*, *Mar. Biol.*, 13, 160, 1972.
76. **Owen, J. M., Adron, J. W., Middleton, C., and Cowey, C. B.**, Elongation and desaturation of dietary fatty acids in turbot, *Scophthalmus maximus*, and rainbow trout, *Salmo gairdneri*, *Lipids*, 10, 528, 1975.
77. **Yone, Y. and Fujii, M.**, Studies on nutrition of red sea bream. XI. Effect of (n-3) fatty acid supplement in a corn oil diet on growth rate and feed efficiency, *Bull. Jap. Soc. Sci. Fish.*, 41, 73, 1975.
78. **Cowey, C. B., Owen, J. M., Adron, J. W., and Middleton, C.**, Studies on the nutrition of marine flatfish. The effect of different dietary fatty acids on the growth and fatty acid composition of turbot *(Scophthalmus maximus)*, *Brit. J. Nutr.*, 36, 479, 1976.
79. **Guary, J. C., Kayama, M., Murakami, Y., and Ceccaldi, H. J.**, The effects of a fat-free diet and compounded diets supplemented with various oils on moult, growth and fatty acid composition of prawn, *Penaeus japonicus* Bate, *Aquaculture*, 7, 245, 1976.
80. **Gatesoupe, J., Girin, M., and Luquet, P.**, Recherche d'une alimentation artificielle adaptée a l'élevage des stades larvaires des poissons. II. Application a l'élevage larvaire du bar et de la sole, *Actes Colloq. C.N.E.X.O.*, 4, 59, 1977.
81. **Kanazawa, A., Teshima, S.-I., and Tokiwa, S.**, Nutritional requirements of prawn — VII. Effects of dietary lipids on growth, *Bull. Jap. Soc. Sci. Fish.*, 43, 849, 1977.
82. **Yone, Y.**, Essential fatty acids and lipid requirements in marine fish, in *Dietary Lipids in Aquaculture*, Japanese Society for Scientific Fisheries, Eds., Koseisha-Koseikaku, Tokyo, 1978, 43.
83. **Léger, C., Gatesoupe, F. J., Metailler, R., Luquet, P., and Frémont, L.**, Effect of dietary fatty acids differing by chain lengths and ω series on the growth and lipid composition of turbot *Scophthalmus maximus* L., *Comp. Biochem. Physiol.*, 64B, 345, 1979.
84. **Bell, M. V., Henderson, R. J., Pirie, B. J. S., and Sargent, J. R.**, Effects of dietary polyunsaturated fatty acid deficiencies on mortality, growth, and gill structure in the turbot, *Scophthalmus maximus*, *J. Fish Biol.*, 26, 181, 1985.
85. **Léger, P., Bieber, G. F., and Sorgeloos, P.**, International Study on *Artemia*. XXXIII. Promising results in larval rearing of *Penaeus stylirostris* using a prepared diet as algal substitute and for *Artemia* enrichment, *J. World Maricult. Soc.*, 16, 354, 1985.
86. **Dobbeleir, J., Adam, N., Bossuyt, E., Bruggeman, E., and Sorgeloos, P.**, New aspects of the use of inert diets for high density culturing of brine shrimp, in *The Brine Shrimp Artemia*, Vol. 3, Persoone, G., Sorgeloos, P., Roels, O., and Jaspers, E., Eds., Universa Press, Wetteren, Belgium, 1980, 165.
87. **Léger, P., Naessens-Foucquaert, E., and Sorgeloos, P.**, International Study on *Artemia*. XXXV. Techniques to manipulate the fatty acid profiled in *Artemia* nauplii, and the effect on its nutritional effectiveness of the marine crustacean *Mysidopsis bahia* (M.), in *Artemia Research and Its Applications*, Vol. 3, Sorgeloos, P., Bengtson, D. A., Decleir, W., and Jaspers, E., Eds., Universa Press, Wetteren, Belgium, 1987, 411.
88. **Watanabe, T., Oowa, F., Kitajima, C., and Fujita, S.**, Relationship between dietary value of brine shrimp *Artemia* salina and their content of (n-3) highly unsaturated fatty acids, *Bull. Jap. Soc. Sci. Fish.*, 46, 35, 1980.
89. **Fukusho, K.**, The food value of *Artemia* to the percoid fish larvae, *Oplegnathus fasciatus*, *Aquaculture*, 21, 71, 1974.
90. **Kuhlman, D., Quantz, G., and Witt, U.**, Rearing of turbot larvae (*Scophthalmus maximus* L.) on cultured food organisms and post-metamorphosis growth on natural and artificial food, *Aquaculture*, 23, 183, 1981.
91. **Watanabe, T., Arakawa, T., Kitajima, C., Fukusho, K., and Fujita, S.**, Nutritional quality of living feed from the viewpoint of essential fatty acids for fish, *Bull. Jap. Soc. Sci. Fish.*, 44, 1223, 1978.
92. **Forster, J. R. M. and Wickins, J. F.**, Experiments on the culture of the prawn *Palaemon serratus*, ICES Mariculture Committee, E:13, 1967.
93. **Wickins, J. F.**, The food value of brine shrimp, *Artemia salina* L. to larvae of the prawn, *Palaemon serratus* Pennant, *J. Exp. Mar. Biol. Ecol.*, 10, 151, 1972.
94. **Sivertsen, A., Rodriguez, J. R., Reitau, K. I., and Olsen, Y.**, Chemical composition of Artemia fed different diets, S.T.F. 21, F 90-27. SINTEF, Trondheim, Norway, 31, 1989.
95. **Scott, A. P. and Middleton, C.**, Unicellular algae as a food for turbot (*Scophthalmus maximus* L.) larvae — the importance of dietary long-chain polyunsaturated fatty acid, *Aquaculture*, 18, 227, 1979.
96. **Walford, J. and Lam, T. J.**, Effect of feeding with microcapsules on the content of essential fatty acids in live foods for the larvae of marine fishes, *Aquaculture*, 61, 219, 1987.
97. **Rimmer, M. A. and Reed, A.**, Effects of nutritional enhancement of live prey organisms on survival of barramundi/seabass (*Lates calcarifer*) larvae, in *Book of Abstracts*, Conference on Advances in Tropical Aquaculture, Feb/March, Tahiti, 1989.

98. **Deru, J., McColgan, S., Ahmed, K., and Jones, D. A.**, Requirement for dietary HUFA in the larval culture of *Macrobrachium rosenbergii* and other caridean larvae, in *Book of Abstracts*, Aquaculture, '89, February 1989, Los Angeles, CA.
99. **Watanabe, T., Ohta, M., Kitajima, C., and Fujita, S.**, Improvement of dietary value of brine shrimp *Artemia salina* for fish larvae by feeding them on (n-3) highly unsaturated fatty acids, *Bull. Jap. Soc. Sci. Fish.*, 48, 1775, 1982.
100. **Imada, O., Kageyama, Y., Watanabe, T., Kitajima, C., Fujita, S., and Yone, Y.**, Development of a new yeast as a culture medium for living feed used in the production of fish seed, *Bull. Jap. Soc. Sci. Fish.*, 45, 955, 1979.
101. **Watanabe, T., Tamiya, T., Oka, A., Hirata, M., Kitajima, C., and Fujita, S.**, Improvement of dietary value of live foods for fish larvae by feeding them on (n-3) highly unsaturated fatty acids and fat-soluble vitamins, *Bull. Jap. Soc. Sci. Fish.* 49, 471, 1983.
102. **Robin, J. H., Gatesoupe, F. J., and Ricardez, R.**, Production of brine shrimp (*Artemia salina*) using mixed diet. Consequences on rearing of sea bass larvae *(Dicentrarchus labrax)*, *J. World Maricult. Soc.*, 12, 119, 1983.
103. **Robin, J. H., Gatesoupe, F. J., Stephan, G., Le Delliou, H., and Salaun, G.**, Méthodes de production de filtreurs-proies et amélioration de leur qualité alimentaire, *Oceanis*, 10, 497, 1984.
104. **Robin, J. H.**, unpublished data, 1988.
105. **Robin, J. H., LeMilinaire, C., and Stephan, G.**, Production of *Artemia* using mixed diets: control of fatty acid content for marine fish larvae culture, in *Artemia Research and Its Applications*, Vol. 3, Sorgeloos, P., Bengtson, D. A., Deceir, W., and Jaspers, E. Eds., Universa Press, Wetteren, Belgium, 1987, 437.
106. **Léger, P.**, unpublished data, 1985.
107. **Léger, P.**, De nutritionele waarde van *Artemia* in de aquacultuur, Ph.D. thesis, State University of Ghent, Ghent, Belgium, 1989.
108. **Gatesoupe, J.**, Nutritional and antibacterial treatments of live food organisms: the influence on survival, growth rate and weaning success of turbot *(Scophthalmus maximus)*, *Ann. Zootech.*, 31, 353, 1982.
109. **Triviño-Guerrero, A.**, unpublished data, 1987.
110. **Wilkenfeld, J. S., Lawrence, A. L., and Kuban, F. D.**, Survival, metamorphosis and growth of penaeid shrimp larvae reared on a variety of algal and animal foods, *J. World Maricult. Soc.*, 15, 31, 1984.
111. **Amat, F., Hontoria, F., and Navarro, J. C.**, International study on *Artemia*. XLIV. Preliminary nutritional evaluation of different *Artemia* nauplii as food for marine fish and prawn larvae, in *Artemia Research and Its Applications*, Vol. 3, Sorgeloos, P., Bengtson, D. A., Decleir, W., and Jaspers, E., Eds., Universa Press, Wetteren, Belgium, 1987, 425.
112. **Millamena, O. M., Bombeo, R. F., Jumalon, N. A., and Simpson, K. L.**, Effects of various diets on the nutritional value of *Artemia* sp. as food for the prawn *Penaeus monodon*, *Mar. Biol.*, 98, 217, 1988.
113. **Bombeo, R. F.**, Growth and survival of *Penaeus monodon* and *Chanos chanos* fry fed with *Artemia* singly or in combination with an artificial diet, in *Artemia Research and Its Applications*, Vol. 3, Sorgeloos, P., Bengtson, D. A., Decleir, W., and Jaspers, E., Eds., Universa Press, Wetteren, Belgium, 1987, 465.
114. **Yashiro, R.**, The effect of *Artemia* fed with different diets on the growth and survival of *Penaeus monodon* Fabricius postlarvae, in *Artemia Research and Its Applications*, Sorgeloos, P., Bengtson, D. A., Decleir, W., and Jaspers, E., Eds., Universa Press, Wetteren, Belgium, 1987, 447.
115. **Rodriguez, A. M.**, Expériences de ponte et d'élevage de larves et de post-larves de crevettes *Penaeus kerathurus* (Forskal, 1779), *Etud. Rev. CGPM*, 55, 49, 1976.
116. **San Feliu, J. M., Munor, F., Amat, F., Ramos, J., Pena, J., and Sanz, A.**, Cultivo experimental de larvas de crustaceos y peces en tanques, *Inf. Tecn. Inst. Invest. Pesq.*, 36, 3, 1976.
117. **Palmegiano, G. B. and Trotta, P.**, *Artemia salina* as pabulum for growing penaeids under laboratory conditions, *Aquacult. Eng.*, 2, 173, 1983.
118. **Trotta, P., Villani, P., Palmegiano, G. B., Forneris, G., and Sarra, C.**, Laboratory-grown *Artemia* as reference food for weaning fish fry and shrimp postlarvae, in *Artemia Research and Its Applications*, Vol. 3, Sorgeloos, P., Bengtson, D. A., Decleir, W., and Jaspers, E., Eds., Universa Press, Wetteren, Belgium, 1987, 459.
119. **Flores Tom, A.**, Preliminary production results of *Artemia* to be used in local shrimp farming in La Paz (Mexico), in *Artemia Research and Its Applications*, Vol. 3, Sorgeloos, P., Bengtson, D. A., Decleir, W., and Jaspers, E., Eds., Universa Press, Wetteren, Belgium, 1987, 283.
120. **White, D. B. and Stickney, R. R.**, A manual of flatfish rearing, Tech. Rept. 730-7, Georgia Marine Science Center, Skidaway Island, 1973.
121. **Dugan, C. C., Hagood, R. W., and Frakes, T. A.**, Development of spawning and mass larval rearing techniques for brackish-freshwater shrimps of the genus *Macrobrachium* (Decapoda, Palaemonidae), Publ. No. 12, Florida Marine Research Lab., Florida Dept. Nat. Res., St. Petersburg, 1975.
122. **Hanson, J. A. and Goodwin, H. L.**, *Shrimp Farming in the Western Hemisphere*, Dowden, Hutchinson and Ross, Stroudsburg, PA, 1977.

123. **Murai, T. and Andrews, J. W.**, Comparison of feeds for larval stages of the giant prawn (*Macrobrachium rosenbergii*), in *Proc. 9th Annual Meeting World Maricult. Soc.*, Avault, J. W., Jr., Ed., Louisiana State University, Baton Rouge, 1978, 189.
124. **Corbin, J. S., Fujimoto, M. M., and Iwai, T., Jr.**, Feeding practices and nutritional considerations for *Macrobrachium rosenbergii* culture in Hawaii, in *Handbook of Mariculture*, Vol. 1, McVey, J. P., Ed., CRC Press, Boca Raton, FL, 1983, 391.
125. **Sick, L. V. and Beaty, H.**, Culture techniques and nutrition studies for larval stages of the giant prawn, *Macrobrachium rosenbergii*, Tech. Rep. No. 74-5, Georgia Marine Science Center, Skidaway Island, 1974.
126. **D'Abramo, L. R. and Sheen, S. S.**, Essential fatty acid requirements of juvenile freshwater prawn *Macrobrachium rosenbergii*, in *Book of Abstracts*, Aquaculture '89, February 1989, Los Angeles, CA.
127. **Devresse, B. and Rasowo, J.**, unpublished data, 1989.
128. **Sorgeloos, P., Léger, P., and Lavens, P.**, Improved larval rearing of European and Asian sea bass, sea bream, mahi-mahi, siganid and milkfish using enrichment diets for *Brachionus* and *Artemia*, *World Aquaculture*, 19, 78, 1988.
129. **Franicevic, V., Lisac, D., Buble, J., Léger, P., and Sorgeloos, P.**, Etude internationale sur l'*Artemia*. XLII. L'effet de la valeur nutritionelle de l'*Artemia* sur la croissance et la survie des larves du loup (*Dicentrarchus labrax* L.) dans une écloserie commerciale, in *Production in Marine Hatcheries*, Loix, B., Ed., FAO-MEDRAP, Rovinj-Zadar, Yugoslavia, 1987, 139.
130. **Lisac, D., Franicevic, V., Vejmelka, Z., Buble, J., Léger, P., and Sorgeloos, P.**, International Study on *Artemia*. XLIII. The effect of live food fatty acid content on growth and survival of sea bream *(Sparus aurata)* larvae, Paper presented at the conference Ichthyopathology in Aquaculture, October 1986, Inter-University Center, Dubrovnik, Yugoslavia.
131. **Stottrup, J.**, personal communication, 1989.
132. **Dhert, P.**, Aquaculture Department of the Southeast Asian Fisheries Development Center, Manila, Philippines personal communications.
133. **Sorgelóos, P.**, *Artemia* Newsletter, No. 8, Artemia Reference Center, State University of Ghent, Belgium, 1988.
134. **Schmid, R.**, personal communication, 1989.
135. **Nel, S.**, personal communication, 1989.
136. **Holt, G. J.**, Studies of larval red drum (*Sciaenops ocellatus* feeding: diet development, in *Book of Abstracts*, Aquaculture '89, Los Angeles, CA.
137. **Holt, G. J.**, personal communication, 1989.
138. **Hokanson, K. E. F. and Lien, G. J.**, Effects of diet on growth and survival of larval walleyes, *Prog. Fish-Cult.*, 48, 250, 1986.
139. **Barrows, R.**, Diets and culture techniques for larval walleye and muskellunge, in *Book of Abstracts*, Aquaculture '89, February, Los Angeles, CA.
140. **Kraueter, J. N. and Woods, C., III**, Culture of striped bass and its hybrids: first feeding to six months, in *Hybrid Striped Bass Culture: Status and Perspective*, Hodson, R., Smith, T., McVey, J., Harrell, R., and Davis, N., Eds., University of North Carolina Sea Grant, Chapel Hill, 1987, 23.
141. **Webster, C.**, personal communication, 1988.
142. **Friedman, B.**, personal communication, 1988.
143. **American Society for Testing and Materials**, Standard E1203: Standard practice for using brine shrimp nauplii as food for test animals in aquatic toxicology, in *Annual Book of Standards*, Vol. 11.04, ASTM, Philadelphia, PA, 1989, 796.
144. **Sorgeloos, P.**, Availability of Reference *Artemia* cysts, *Mar. Ecol. Prog. Ser.*, 3, 363, 1980.
145. **Bengtson, D. A., Beck, A. D., and Simpson, K. L.**, Standardization of the nutrition of fish in aquatic toxicological testing, in *Nutrition and Feeding in Fish*, Cowey, C. B., Mackie, A. M., and Bell, J. G., Eds., Academic Press, London, 1985, 431.
146. **Collins, G. and Bengtson, D. A.**, Characterization of Reference *Artemia* III for marine toxicological studies, in preparation.
147. **Van Ballaer, E., Amat, F., Hontoria, F., Léger, P., and Sorgeloos, P.**, Preliminary results on the nutritional evaluation of (n-3)-HUFA-enriched *Artemia* nauplii for larvae of the sea bass, *Dicentrarchus labrax*, *Aquaculture*, 49, 223, 1985.
148. **Léger, P., Chamorro, R., and Sorgeloos, P.**, Improved hatchery production of postlarval *Penaeus vannamei* through application of innovative feeding strategies with an algal substitute and enriched *Artemia*, in *Book of Abstracts*, 18th Annual Meeting World Aquaculture Society, Guayaquil, Ecuador, 1987.
149. **Bolla, S.**, unpublished data, 1989.
150. **Cooper, E.**, Evaluation of Selco for improving the nutritional value of the shrimp *Artemia salina*, *Freshwater Mar. Aquarium*, 4, 121, 1988.
151. **Lavens, P., Léger, P., Sorgelóos, P.**, Manipulation of the fatty acid profile in *Artemia* offspring using a controlled production unit, in *Aquaculture—a Biotechnology in Progress*, De Pauw, N., Jaspers, E., Ackefors, H., and Wilkins, N., Eds., European Aquaculture Society, Bredene, Belgium, 1989, 731.

Chapter 12

SEMI-INTENSIVE CULTURING IN FERTILIZED PONDS

Wim Tackaert and Patrick Sorgeloos

TABLE OF CONTENTS

I. INTRODUCTION

Artemia biomass and cysts can be produced in intensive as well as semi-intensive and extensive conditions. Intensive production is performed in indoor tank systems under completely controlled conditions (see Chapter 12). Semi-intensive and extensive production refer to the culture of *Artemia* in outdoor conditions. The former is performed in small and managed salt pond systems (mostly seasonal solar saltponds); e.g. with some degree of control over salinity, water retention time, and feed availability. The latter (extensive production) consists of the harvesting of mostly natural *Artemia* populations from large biotopes with year-round high salinity conditions such as large solar salt operations or salt lakes.

Since *Artemia* is highly susceptible to predation, a major prerequisite to semi-intensive and extensive production is the availability of brine of sufficiently high salinity that is free from fish and invertebrates. In most saltpond locations/natural biotopes, this situation is reached at salinities from >80 to 100 g/l, although situations have been reported where fish and insects were still present at salinities of >100 to 130 ppt.[1]

Natural populations of *Artemia* are widely distributed over the five continents in a variety of isolated biotopes such as inland salt lakes, coastal lagoons and especially coastal salterns associated with commercial solar salt production.[2,3] A recent list of natural *Artemia* sites compiled by Vanhaecke et al.[3] extends to over 350 localities.

In most natural *Artemia* populations, densities are low, mainly as a result of food limitation due to a low nutrient content of the water. The few exceptions which have higher productivity are those biotopes (mostly large solar saltworks) which are located in highly eutrophic areas (e.g., near population centers, estuaries, or mangrove areas) such as the Leslie Saltworks in the San Francisco Bay, and the salinas of the Bohai Bay in China.[4] Production estimates for a few natural *Artemia* biotopes are given in Table 1. As a result of their generally low productivity, most of the natural biotopes offer opportunities for extensive harvesting of *Artemia* biomass only. Production of cysts in these biotopes, especially when the local ecological conditions are fairly stable, occurs only occasionally or is erratic. Moreover, since the quality of *Artemia* cysts differs from strain to strain and even from harvest to harvest[5,6] it is imperative to determine its nutritional value for specific application in aquaculture, prior to commercial development.

A schematic outline of a typical solar saltwork is given in Figure 1. Sea water flows over a series of successive ponds in which salinity gradually increases as sea water evaporates. During this process, salts with low solubility precipitate as carbonates and later as gypsum (see Figure 2). Finally, when the sea water has evaporated to approximately one tenth of its original volume, mother brine is transferred to the crystallizers where pure sodium chloride is deposited. *Artemia* is only found in the evaporation ponds of intermediate salinity, i.e., from approximately 90 g/l (≃ upper tolerance level of predators) to approximately 200 to 250 ppt. At elevated salinities *Artemia* die as a result of either starvation because of increased energy associated with hyperosmoregulatory physiology[7] and/or increased toxicity of the brine due to drastic changes in ionic composition caused by gradual precipitation and enrichment of different salts.

Very extended solar salt industries (i.e., highly mechanized operations of hundreds to thousands of hectares each with individual evaporation ponds of up to hundreds of hectares) are localized in climatic zones with high evaporation rates and restricted rainfall (e.g., various regions in Australia, South America, Mexico, United States, China, southern France, and southeast Italy) and are usually in production (though not necessarily harvesting) on a year-round basis. In contrast, numerous small cottage-scale units for artisanal solar salt production are in production in the tropical-subtropical belt only during a restricted period of the year; i.e., cycles of 3 to 6 months during the dry season when conditions are favorable for making salt. The large operations can only be managed with regard to *Artemia* presence,

TABLE 1
Production Estimates for Natural *Artemia* Populations

Site	Country	Maximum production	Period	Ref.
Lake Rezaiyeh	Iran	1.2 adults/l		68
Sivash Salt Lakes	U.S.S.R.	400/l		69
Slagbaai	Bonaire, Netherlands Antilles	200—360/l	Oct.—June	70
Mono Lake	California	4 adults/l		
		12 nauplii/l	June—Sept.	71
		400/l	Aug.—Sept.	72
Great Salt Lake	Utah	10/l		73
		100—200g dw/m^2	per year	74
Salin de Giraud	Camargue, France	10-100/l	March—Oct.	75
		0.02-0.2 g/l ww		
		16,000/m^2		76
Long Island Salina	Bahamas	25-100/l	May—Sept.	51
Alviso Salt Ponds	California	13 g/m^3 dw	summer	77
San Francisco Bay saltponds	California (harvest)	5 kg/ha ww	per week	78
Crimea Salt Lakes	U.S.S.R.	250 kg/ha	October	79
		3000 kg/ha	June	
Burgas-Pomorije saltworks	Bulgaria	2.75 g/l adults ww		
		0.93 g/l juveniles	June—Sept.	80
		0.5 g/l nauplii		
Lake Grassmere	New Zealand	4.2 g dw/m^3	Nov.—May	81,82
		4.9 g dw/m^3	Nov.—April	81,82

[a] Some data compiled from Persoone and Sorgeloos.[83]

especially to control the opportunistic dispersion of *Artemia*[3] when better hydrobiological conditions are required for solar salt production.[4,8-10] As a result, the big-pond systems can be tapped for extensive harvesting only. The small units are much more versatile and provide realistic possibilities for human-managed production of *Artemia* biomass and cysts through inoculation with selected strains, through salinity control, pond retention times, food availability, predator presence, etc. This chapter will cover the basic principles and strategies for semi-intensive production of *Artemia* in seasonal saltponds.

II. SEASONAL PRODUCTION OF *ARTEMIA* IN SMALL SOLAR SALTFARMS

A. TYPICAL CHARACTERISTICS OF SEASONAL SALTFARMS

Seasonal saltfarms generally have an area of less than a few hectares, while individual ponds can be as small as a few hundred m^2. The ponds are usually shallow with a water depth of <10 cm. Sea water of mangrove or estuarine origin is usually supplied by tidal inflow, although some farms use pumps, windmills, and/or manually operated waterscoopers to permit better manipulation of water levels.

During winter (e.g., in China, southern Spain, and Sicily) or monsoon season (e.g., in Central America and Southeast Asia) salt production is abandoned. In monsoon climates, salt evaporation ponds are eventually converted into paddy fields or shrimp/fish ponds (e.g., Southeast Asia). In some farms deep brine reservoirs (pickle ponds) are available for storage of the brine which remains available at the end of the dry season (e.g., Philippines and Indonesia); at the onset of the dry season, brine is pumped back into the evaporation ponds and allows a faster start-up of salt production.

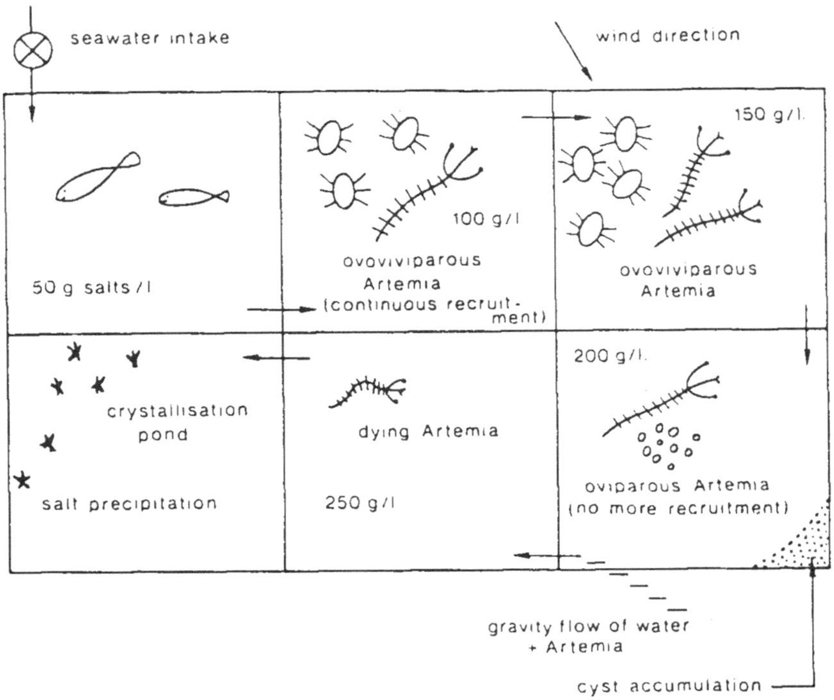

FIGURE 1. Schematic diagram of a solar salt operation with natural occurrence of *Artemia*. (From Sorgeloos, P., Léger, P., Lavens, P., and Tackaert, W., *Aquacult. Dév. Cahiers Ethologie Appliquée*, 7, 43, 1987. With permission.)

B. POND MODIFICATIONS

Successful production of *Artemia* in seasonal solar-salt fields involves minor pond modifications to increase the water depths in the future *Artemia* ponds (in the salinity range of 100 to 180 g/l) to a minimum level of about 40 to 50 cm (preferably 70 to 100 cm). High water depths are essential not only to prevent lethal high temperature conditions for *Artemia*, but also to promote the development of phytoplankton which through its shading effect inhibits the development of phytobenthos. In contrast to phytoplankton, the latter is undesirable because it is too large to be ingested by *Artemia*, and as it starts to float, it may further reduce evaporation and eventually contaminate the harvests of *Artemia* biomass and cysts as well as the salt. Since *Artemia* is a planktonic organism, deep ponds can also sustain a larger production per surface area than shallow ponds. High water depths and associated phytoplankton result in a dark coloration of the brine which is also beneficial for salt production since it enhances the evaporation efficiency through increased absorption of solar heat.[9] Higher water depths can easily be achieved by digging an inner perimeter ditch and using the soil from the ditches for heightening the dikes (Figure 3). However, the water levels in the *Artemia* ponds will than be higher than in the nonmodified upstream evaporation ponds, which implies that the brine has to be relifted (by pump or windmill) into the first *Artemia* pond from where it further gravitates into the following downstream ponds.

C. PREPARATION OF PONDS

Prior to *Artemia* production, it is recommended that ponds be completely emptied to expose bottom soil for a period of one to two weeks, followed by raking the upper layer of the soil to enhance mineralization of accumulated organic matter. Fish left in remaining mud holes may be killed by the use of rotenone or tea-seed cake or by application of lime in

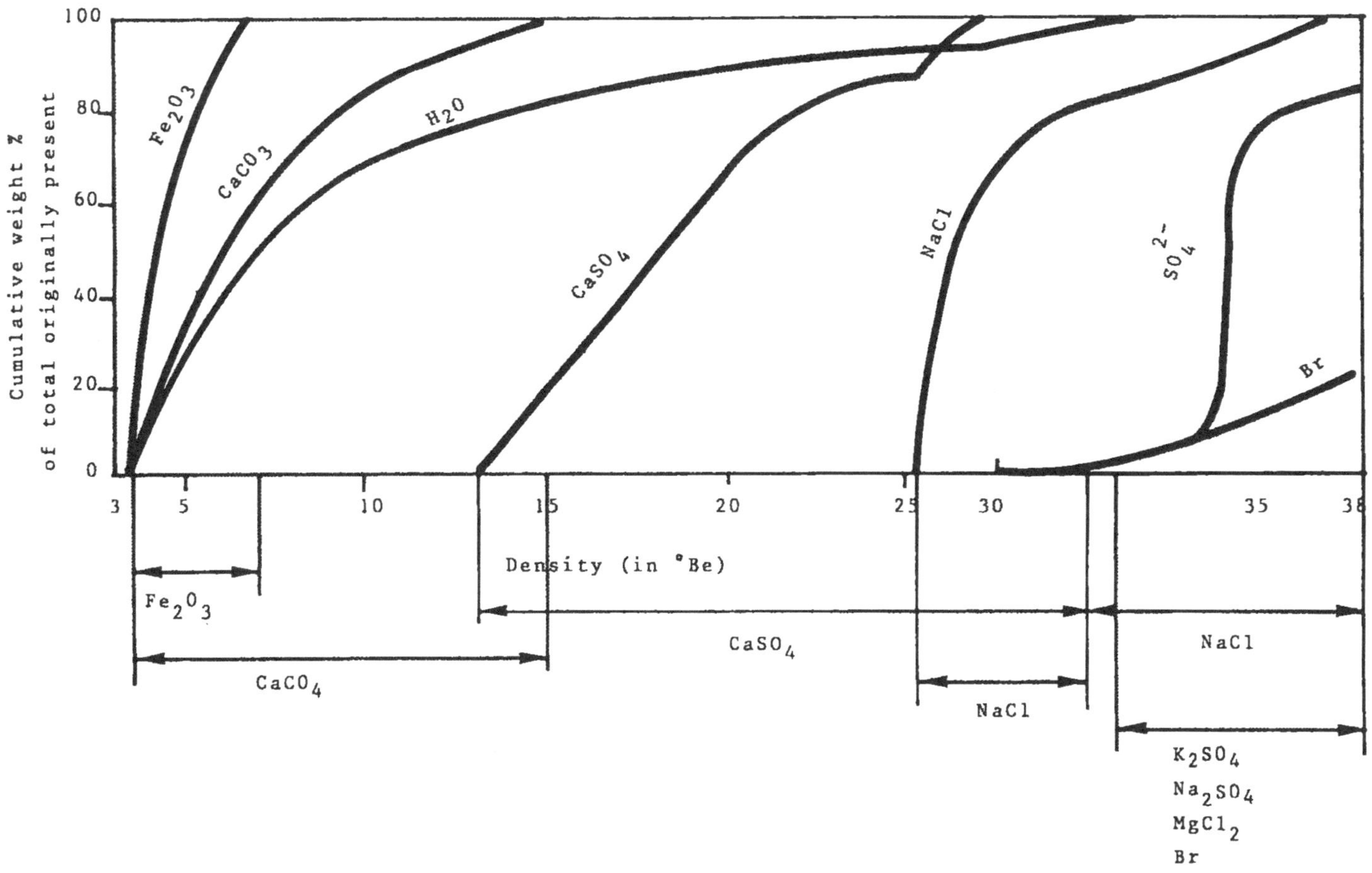

FIGURE 2. Deposition of salts during concentration of sea water (from Bradley, personal communication).[87]

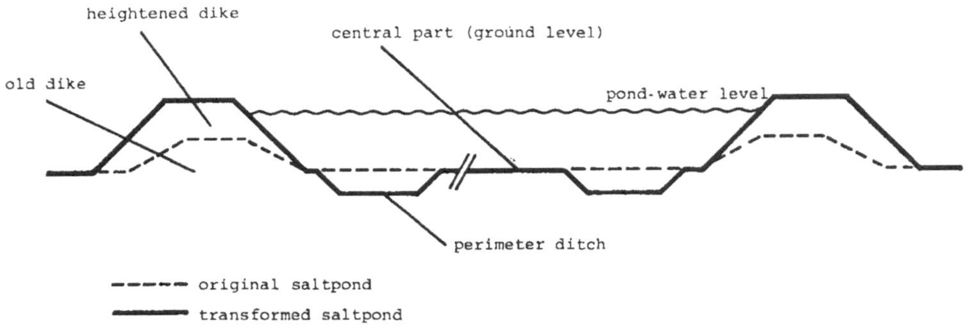

FIGURE 3. Longitudinal section through a modified *Artemia* pond. (Modified from Tackaert, W., Léger, P., Lavens, P., and Sorgeloos, P., *El cultivo del Camaron, Langostino y Congrejo en el Mundo: Bases y Tecnologias* (The Aquaculture of Shrimp, Prawn, and Crawfish in the World: Basics and Technologies), Chavez Justo, C. and Sosa Nishizaki, O., Eds., McGraw-Hill, Mexico, 1989. With permission.)

combination with ammonium sulphate.[11,12] Coastal saltfields may be located in mangrove areas associated with acid sulphate soils containing pyrite. Upon exposure to air, pyrite is oxidized to iron oxides and sulphuric acid, especially in newly excavated ponds. The release of the acid entails very low pH values, resulting in aluminium and iron leaching from the soil. The latter conditions are unfavorable for most aquatic organisms including *Artemia*; phytoplankton production is also inhibited through stripping of phosphorus from the water.[13] In a number of cases, soil acidity may be visually observed; i.e., air-exposed soils turn yellow to brownish-red. Addition of lime neutralizes the acid soil conditions, allowing for a better bioavailability of nutrients, which in turn enhances phytoplankton growth and production of *Artemia*. Lime, furthermore, aids in decomposition of pond muds, long term buffering of pH (which is essential in ponds where heavy organic fertilization is applied), and in the killing of undesirable fish eggs and pests through the toxic action of caustic components. There are several forms of lime: (1) calcium oxide, CaO, or quicklime has a neutralizing efficiency of 173% $CaCO_3$, and is used for fast action in ponds with very low pH, in the range of 3.5 to 5.0; (2) calcium hydroxide, $Ca(OH_2)$, or hydrated lime has a neutralizing efficiency of 135% $CaCO_3$ and also acts quickly to increase the soil pH; (3) calcium carbonate, $CaCO_3$, or agricultural lime (ground limestone) acts relatively slow and therefore may be used for long-term acidity control. CaO and $Ca(OH)_2$ may be used in newly excavated ponds while $CaCO_3$ is used in older, more stabilized ponds. To raise the pH by 0.1 unit, about 500 kg of $CaCO_3$ is applied per hectare.[13] Lime is applied to dry pond bottoms, although $CaCO_3$ at low rates not exceeding 400 kg/ha, may also be used in ponds filled with water. Best results are obtained by spreading lime over the entire bottom or surface. Dikes should also be limed to prevent acidic runoff in case of rainfall.

D. INTAKE OF SEA WATER AND INCREASE OF REQUIRED SALINITIES AND WATER DEPTHS

Most artisanal saltfarms are designed to permit intake of sea water by tide. Some are provided with sluices in order to obtain the maximum water level in the reservoir as close to the high-tide level as possible. Even then, and especially in the case where modified *Artemia* ponds are operated at higher water depths, it is rare that the brine can travel through the entire system by gravity alone. Pumps or windmills (see Figure 4) are required to supplement or take the place of the tidal gates. While water depths in the reservoir should always be as high as possible (to maximize supply of the brine to the downstream ponds), evaporators and *Artemia* ponds should initially only be filled to a level of 10 to 15 cm, in

FIGURE 4. Windmill used in seasonal solar salt farming in Thailand. (From Sorgeloos, P., Lavens, P., Léger, P., Tackaert, W., and Versichele, D., *Manual for the Culture and Use of Brine Shrimp Artemia in Aquaculture*, Artemia Reference Center, State University of Ghent, Belgium, 1986. With permission.)

order to ensure maximum evaporation and to create high water temperatures harmful to predators. Predators should furthermore be avoided by screening the intake water upon filling the ponds. If the pH of the water in the modified *Artemia* ponds is still lower than 7.0, it is advisable to wash out the remaining acidity by repeated flushing of the pond with sea water. If the pH is higher than 7.0, low water levels in the *Artemia* ponds are allowed to evaporate until a salinity of about 100 g/l is reached; i.e., high salinity in combination with high temperatures obtained in thin waterlayers will eliminate copepods and other predators. At this point, gradual intake of sea water is restarted at a rate to maintain the salinity in the *Artemia* ponds around 100 g/l and be continued until the desired levels in the *Artemia* ponds (50 cm or more) are reached. While some predators (e.g., *Cyprinodon variegatus* and *Aphanius fasciatus*) are able to adapt to gradually increasing salinities, they will not resist the severe salinity shock created by the above practice of water intake. As soon as the *Artemia* ponds are filled to their maximum level, the rate of sea water intake is adjusted to maintain these depths. Consequently a density gradient typical for a normal salt operation will be established in the successive *Artemia* ponds and evaporators. Between the *Artemia* ponds, the brine is preferentially bottom-drawn not only to prevent temperature and salinity stratification but also to enhance distribution and release of organic matter which accumulates on the bottom.

E. FERTILIZATION

1. General Requirements

Before introducing *Artemia* in the ponds, enough particulate food should be present in the water to guarantee high population productivity. Water with a green-brown color and a transparency of less than 20 cm mostly contains high concentrations of organic detritus particles and/or algae that can be used as food by the *Artemia*. This is generally the case in saltfarms associated with mangrove areas or eutrophic estuaries. The availability of the nutrients from mangrove or estuaries can be maximized by pumping at low tide.[14] In this situation of good food availability, fertilization is not required, at least not before the introduction of the *Artemia* nauplii. Water with only a slight coloration and a high transparency (>30 cm) is not productive enough and requires fertilization to increase the availability of natural food, 3 to 7 days prior to inoculation of *Artemia*.

Since the goal is to stimulate phytoplankton and not phytobenthos, it is essential to apply the fertilizer only to ponds already filled to maximum water levels. In flow-through systems, it is best to fertilize the low salinity ponds as we have often experienced difficulties in initiating a phytoplankton bloom when fertilizing high salinity ponds; i.e., chemical interactions limit the nutrient availability for a restricted number of algal species. In the latter systems, phytoplankton-rich water is ultimately drained into the high salinity ponds. Two kinds of fertilizers or a combination of both can be used: (1) organic fertilizers, such as dried chicken manure, and (2) inorganic fertilizers (commercial products used in local agriculture) with a high nitrogen and phosphorus content. Generally inorganic fertilizers stimulate phytoplankton growth more rapidly, while organic fertilizers act more slowly but provide a long-range effect since they first have to be degraded by bacterial action to release plant nutrients. In addition, some organic fertilizers such as dried and ground chicken manure will easily disperse into the water column. Since they still contain up to 20% protein, they may also act as a direct food source for *Artemia*. Organic products are cheaper than inorganic fertilizers but much more bulky, and therefore involve more labor in their use. Moreover, organic fertilizers, especially when not properly distributed, may accumulate and decay at the pond bottom and create anaerobic zones resulting in oxygen deficiency, acidity, and toxicity through production of hydrogen sulphides.

Optimal rates of application are difficult to predict since they will vary from location to location due to climatic differences and quantitative/qualitative fertility of the local soil and water. Morales[86] reported that minimum concentrations of nitrogen and phosphorus in the water in order to obtain blooming of phytoplankton should be 1 to 2 mg/l and 0.1 mg/l, respectively. Different fertilization programs also favor different types of food in the ponds; e.g., plankton is favored by a high ratio of nitrogen to phosphorus whereas organic manures, which are usually high in phosphorus, enhance the growth of undesirable filamentous algae. The fertilization program in *Artemia* ponds needs to be adjusted to optimize the availability of phytoplankton. The following doses can be recommended as a guideline. These concentrations have proven to be effective but other application rates and combinations of both organic and inorganic fertilizers are not excluded.

2. Organic Fertilizers

Best results to date have been obtained with chicken manure. Although cow and goat dung have been successfully used in some cases, chicken manure, in contrast to some other manures, such as from cattle (low nitrogen to phosphorus ratio: 1.5; high contents of undigested insoluble material) is more effective in inducing a phytoplankton bloom since it has a relatively high N/P ratio (3.5) and good dispersibility, providing a larger surface for bacterial breakdown. In addition, it does not accumulate on the pond bottom since it is highly soluble. Van der Zanden[15] reported a significant increase in the development of phytobenthos or "lab-lab" whenever cow dung was used.

Chicken manure needs to be dried and sieved for removal of debris, bran, feathers, etc., and it is preferable to grind it to increase its availability as a direct food source for *Artemia*. Nonetheless, good results have been reported by Jumalon et al.[11] when using chicken manure suspension (1:1 ratio of manure to sea water).

Dry chicken manure is applied at rates of 0.5 to 1.25 ton/ha at the start with dressings of 100 to 200 kg every 2 to 3 days.

3. Inorganic Fertilizers

- Combination of 100 kg/ha mono-ammonium-phosphate (N:P:K ratio expressed in weight percentage of 16:20:0) and 50 kg/ha ammonium-nitrate (33:0:0); weekly dressings of 50 and 25 kg/ha, respectively.
- Combination of 50 kg/ha di-ammonium-phosphate (18:46:0) and 50 kg/ha urea (44:0:0); weekly dressings of 25 and 20 kg/ha, respectively.
- 100 kg/ha urea (44:0:0); weekly dressings of 40 kg/ha (e.g., in ponds where natural levels of phosphorus are high).

Recently a by-product from the industrial production of monosodium glutamate derived from cassava or sugar cane molasses has been succesfully used as a cheap fertilizer for *Artemia* ponds. Application rates of up to 2500 l/ha have been found very effective in inducing dense phytoplankton blooms. In view of its acidity the effluents of the monosodium glutamate fermentation should be applied in small quantities but on a frequent basis.

Both organic and inorganic fertilizers should be very evenly spread over the pond surface. Slow dissolving pelleted fertilizers (e.g., 16:20:0) are first made into concentrated solutions (prepared overnight) or are placed on a platform in areas of active water flow (e.g., near the brine intake), 15 to 20 cm below the water surface in order to ensure a more even release and mixing and prevent any trapping of nutrients into the soil.

Although the fertilizer is generally applied directly to the *Artemia* ponds, several farmers are now working with separate "food production ponds" from which they feed the *Artemia* ponds; these may be low salinity ponds, 50 to 80 g/l, which are not necessarily integrated in the brine circuit. Availability/production of phytoplankton in these ponds is maximized through intake of "green water" from fish/shrimp ponds or through supplemental fertilization with fecal droppings from a vertically integrated poultry farm.[11,12]

F. *ARTEMIA* STRAIN SELECTION

In view of the high degree of genetic variation[16] associated with the diversity of biochemical, biometrical, and physiological characteristics[5,17] found among strains of *Artemia* the selection of the strain best adapted to the particular ecological conditions (especially temperature regimes) of the saltfarm and/or most suitable to its later application in aquaculture farms, is very important. Strain selection can be based on the available data of growth and production performance,[18,19] reproductive characteristics,[20] anion concentration tolerance,[21] and especially temperature/salinity tolerance.[22] In addition, whenever possible, a comparative bioassay culture test should be performed in closely simulated conditions using the untreated brines of the habitat as culture medium. Strain selection might also be restricted by the intended application of the produced *Artemia* in local aquaculture; e.g., if small nauplii are needed for the production of the early larval stages of fish and shrimps, a strain producing small cysts and nauplii is to be preferred.[23] On the other hand, if local aquaculture is primarily interested in *Artemia* biomass as a nursery/weaning or shrimp maturation diet, then a strain showing good growth and survival, as well as dominant ovoviviparous reproduction characteristics will be the most interesting.

G. INOCULATION PROCEDURES

Inoculation of *Artemia* should be performed as early as possible in the brine circuit where no predators are found (i.e., usually at salinity levels of around 100 g/l). In flow-through systems with short pond-retention times, downstream ponds at higher salinity need not be inoculated since they will be gradually stocked with *Artemia* drained from the inoculated pond. Although Jumalon and Robles[24] reported an optimum *Artemia* production at an inoculation density of 50 nauplii per l, Vu Do Quynh and Nguyen Ngoc Lam[25] reported faster growth and maturation as well as higher fecundity at densities <20 nauplii per l. From our observations it is clear that small inocula (10 to 20 nauplii per l) are generally as effective (at least under normal temperature conditions) and more economical than large inocula since they exhibit a faster population increase when compared to the latter.

The quantity of cysts needed to obtain the number of nauplii needed for inoculation (and taking into account a 30% mortality at stocking) is calculated from the pond volume and the hatching efficiency of the selected cyst batch.[26] Cysts are preferably hatched close to the ponds. Optimal hatching conditions[26] are often difficult to achieve under field situations; nonetheless, the following aspects should be taken into consideration:

- Shaded or roof-covered area should be used to prevent excessive heating of the hatching containers by direct sunlight.
- Transparent, preferably funnel-shaped hatching containers, should be used. If flat-bottomed containers have to be used, they should be equipped with several aeration lines on their bottom so as to provide good mixing and aeration of the water.
- Hatching containers should be illuminated during the night to provide the essential light stimulus,[27] especially when cysts are incubated in late afternoon or evening.
- Air should be supplied from a blower, compressor, or several aquarium pumps powered by the net current, a generator, or batteries.
- Clean 20 to 35 ppt sea water should be used, filtered through a fine filter cloth (100 μm).
- Sodium bicarbonate should be added at a rate of 1 g/l hatching medium so as to buffer the sea water.
- Cyst densities should not exceed >1 g/l, especially under suboptimal hatching conditions.

It is essential to harvest the nauplii in the first instar stage. This is determined from subsamples taken at regular intervals or from hatching rate and synchrony data for the given strain or batch.[28] Older instar stages will not survive the salinity shock when transferred from natural sea water into 100 g/l of salt water. After hatching, the nauplii should be screened over a 125 μm filter, thoroughly washed and transferred to clean sea water or pond water at half their hatching density. They are now ready for inoculation into the pond. If the pond is not within walking distance from the hatching site, aeration should be provided during transport (battery-operated pump or oxygen tank) to prevent mortality. If transport takes several hours, it is best to cool the nauplii container to 0—5°C using cooled sea water or adding ice bags. At these low temperatures the naupliar metabolism and motoric activity are strongly reduced without affecting their viability.[29] Moderate aeration, e.g., with a battery-operated aquarium pump, has to be provided to keep the motionless animals in suspension. In this way about 100 million nauplii can be successfully transported for a period of several hours in a 20 liter plastic bag packed in a cooled styrofoam box. The best time of the day to inoculate a pond is during late evening when the water temperature is low and will continue to drop until early morning. When the nauplii have been transported at low temperature it is essential to allow the temperature in the containers to rise so that the animals can resume their motoric activity before they are introduced in the pond. Under heavy wind conditions it is important to siphon the nauplii on the leeward side of the pond to avoid having them driven on shore by heavy wave action.

H. FACTORS AFFECTING POPULATION GROWTH

During the first days after inoculation it is very difficult to see the nauplii since they have lost their distinct orange color and tend to concentrate in the deeper parts of the pond. It is only when the nauplii have grown into adults that one may evaluate if the inoculation has been successful. In fertilized ponds, operated under optimal conditions (i.e., temperatures within the tolerance range for the selected strain; pH between 7.5 and 8.5; intermediate salinity levels of 100 to 150 g/l; and presence of sufficient quantities of particulate feed, especially phytoplankton) sexual maturity may already be attained 7 to 10 days after inoculation.[25,30] Under these conditions the parental and the first generations will generally reproduce by ovoviviparity, resulting in a fast increase of the population. The size to which the population will grow is determined by the carrying capacity of the pond. The principal factors which affect this carrying capacity are pond depth, food availability (determined by the concentration of nutrients in the water), and the frequency of water intake which will improve the water quality and result in an extra nutrient influx and better mixing of the nutrients which accumulate on the pond bottom. Aside from quantity, quality of the planktonic algal population may also affect the population growth of *Artemia*. Green algae (e.g., *Tetraselmis* and *Dunaliella*) and diatoms (e.g., *Chaetoceros, Navicula,* and *Pleurosigma*) are a much better food for *Artemia* than planktonic filamentous blue-green algae (e.g., *Lyngbya* and *Oscillatoria*). The latter are too big for *Artemia* to be ingested and clog their thoracopods resulting in starvation of the *Artemia*. Filamentous blue-green algae might predominate in stagnant waters (since green algae and diatoms settle) and in conditions of high concentrations of organic matter, high pH, and low CO_2 levels.[31-33] In this regard, Jumalon and Ogburn[33] found a high correlation between the arrest of water intake rich in CO_2, the collapse of green algal populations, and the development of planktonic filamentous cyanophytes. This implies that regular water intake is important in pond management; i.e., aside from affecting the amount of food, continuous flow also stimulates blooming of particular phytoplankton species more suitable for *Artemia* development.

I. FACTORS CONTROLLING CYST PRODUCTION

Although the factors controlling the mode of reproduction are not fully understood, oviparity in *Artemia* is generally considered to be induced by environmental stress.[34,35] In pond systems, cyst production is often observed when the population is exposed to high salinities (e.g., when the *Artemia* reach the high salinity ponds). Salinity shocks have also been found effective in switching the population toward cyst production[36,37] i.e., abrupt lowering or raising of the salinity through rapid intake of brine of a very different density or at heavy rainfall. In addition, low oxygen concentrations or considerable fluctuations in dissolved oxygen levels reportedly induce oviparous reproduction in *Artemia*.[34,38] This condition stimulates the synthesis of hemoglobin and the consecutive excretion by the brown shell gland of its metabolic end product, hematin, which is the main constituent of the cyst shell of *Artemia*.[39-41] Oxygen stress in pond conditions may be accomplished by raising the salinity and/or by increasing the rate of fertilization to induce blooms of phytoplankton, creating extensive diurnal fluctuations in dissolved oxygen.

Several authors postulated that the frequency of oviparity in *Artemia* is not only correlated with environmental stress, but is also influenced by the geographical strain of *Artemia* which has been used for inoculation.[20,42,43] In this regard, Gajardo and Beardmore[44] suggested that encystment in *A. franciscana* is under genetic control and is associated at least in part with the levels of heterozygosity found in the females. In view of the considerable interstrain differences in the distribution of heterozygosity,[16] care must be taken in selecting a strain showing high heterozygosity levels; e.g., a strain inhabiting a variable and stressful environment, if cyst production is preferred. Even then, pond management should be directed toward creation of stress conditions to retain the genetic variability in the population and consequently prevent a decline in the cyst production.

It has been observed on several occasions in tropical habitats that newly introduced populations initially exhibit a high rate of oviparous reproduction, followed, however, by a drastic decrease in cyst production as soon as the population has become fully established (adapted to the new environment?) and/or the biotope has become completely stabilized, e.g., in Brazil,[45] Thailand,[12] and Vietnam.[46] This phenomenon was recently observed for the second time in Macau, Brazil: reappearance of cyst production at the end of 1987 and beginning of 1988 was associated with transformation of the local shrimp farm into evaporation ponds for salt production presenting a "new biotope" for *Artemia*. Berthélémy-Okazaki and Hedgecock[47] presumed that this decline in cyst production may be due to harvesting of the cysts leading to a removal of the genotypes predisposed towards oviparity from the population. They suggested that cyst production could possibly be revived by reinoculation with a highly oviparous strain.

J. MONITORING OF ENVIRONMENTAL CONDITIONS AND FOOD PRODUCTION IN PONDS

A basic prerequisite for correct pond management implies regular evaluation of the environmental conditions of the ponds. The physico-chemical parameters to be monitored include (1) dissolved oxygen, readily measured with an oxygen electrode or by Winkler titration;[48] (2) brine densities at the water surface and the pond bottom measured with a refractometer or with Baumé scale hydrometers (see conversion tables for brine density and degrees Beaumé, as well as corrections for temperature in Tables 2 and 3); (3) pH-values using a pH meter; (4) air and water temperatures at water surface and pond bottom with a minimum-maximum thermometer; and (5) water depth, read from a depth gauge.

Since maintenance of a healthy phytoplankton population is considered to be one of the most important keys for a successful *Artemia* production, regular monitoring of nutrient levels and associated standing crops of phytoplankton are important for proper pond management, i.e., rate of water intake, fertilization dressings, and biomass harvesting. The concentration of the major nutrients should be regularly controlled to ensure that no deficiency causing inhibition of algal growth is developing and/or that the ratio of N/P is not becoming too low. This involves analysis of reactive inorganic phosphate (the major form of phosphorus required for algal cells), reactive nitrate, nitrite, and ammonia, preferably by standard colorimetric procedures.[48,49] The phytoplankton population should be analyzed at least once a week. Phytoplankton densities may be determined from a representative sample by direct microscopic counting using a counting chamber. Whenever possible, species composition and cell size of the algal population should be determined, as the first may directly affect the nutritional value of the *Artemia* produced,[5] and the latter determines whether the algal cells (especially when forming chains or colonies) are small enough (≤ 50 μm) for ingestion by *Artemia*.[50] Records of phytoplankton species commonly found in nutrient enriched salinas were published by Davis[51] and Wongrat.[52] Other parameters to follow in respect to phytoplankton densities are water turbidity, the concentration of chlorophyl, phytoplankton dry weight, and primary productivity. Turbidity may be measured using a colorimeter or a Secchi disk. Procedures for measuring dry weight, chlorophyl, and primary productivity are described in Vonshack,[53] Strickland and Parsons,[48] and Boyd.[54] In situations where the water

TABLE 2
Conversion Table for Various Units of Salinity

Density (g/ml)	Degree Beaumé (°Be)	Salinity (g/l)	Density	Degree Beaumé	Salinity	Density	Degree Beaumé	Salinity	Density	Degree Beaumé	Salinity
1.020	2.8	28.6	1.061	8.4		1.102	13.4		1.141	17.8	
1.021	3.0		1.062	8.5		1.103	13.5		1.142	17.9	
1.022	3.1		1.063	8.7		1.104	13.6		1.143	18.0	
1.023	3.3		1.064	8.8		1.105	13.7		1.144	18.1	
1.024	3.4		1.065	8.9		1.106	13.8		1.145	18.2	
1.025	3.6		1.066	9.0		1.107	14.0		1.146	18.3	
1.026	3.7		1.067	9.2		1.108	14.2		1.147	18.5	
1.027	3.8		1.068	9.3		1.109	14.3		1.148	18.6	
1.028	4.0		1.069	9.4		1.110	14.4	159.5	1.149	18.7	
1.029	4.1		1.070	9.5	99.4	1.111	14.5		1.150	18.8	222.1
1.030	4.2	42.4	1.071	9.6		1.112	14.6		1.151	19.0	
1.031	4.4		1.072	9.7		1.113	14.7		1.152	19.1	
1.032	4.5		1.073	9.9		1.114	14.9		1.153	19.2	
1.033	4.7		1.074	10.0		1.115	15.0		1.154	19.3	
1.034	4.8		1.075	10.1		1.116	15.1		1.155	19.4	
1.035	4.9		1.076	10.2		1.117	15.2		1.156	19.5	
1.036	5.0		1.077	10.3		1.118	15.3		1.157	19.6	
1.037	5.1		1.078	10.5		1.119	15.4		1.158	19.7	
1.038	5.3		1.079	10.6		1.120	15.5	175.1	1.159	19.8	
1.039	5.4		1.080	10.7	114.1	1.121	15.6		1.160	19.9	237.8
1.040	5.5	56.4	1.081	10.8		1.122	15.7		1.161	20.0	
1.041	5.7		1.082	11.0		1.123	15.8		1.162	20.2	
1.042	5.8		1.083	11.1		1.124	15.9		1.163	20.3	
1.043	6.0		1.084	11.2		1.125	16.0		1.164	20.4	
1.044	6.1		1.085	11.3		1.126	16.2		1.165	20.5	
1.045	6.2		1.086	11.5		1.127	16.3		1.166	20.6	
1.046	6.4		1.087	11.6		1.128	16.4		1.167	20.7	
1.047	6.5		1.088	11.7		1.129	16.5		1.168	20.8	
1.048	6.6		1.089	11.8		1.130	16.6	190.6	1.169	20.9	
1.049	6.7		1.090	11.9	128.6	1.131	16.7		1.170	21.0	253.7
1.050	6.8	70.6	1.091	12.0		1.132	16.8		1.171	21.1	
1.051	7.0		1.092	12.1		1.133	16.9		1.172	21.2	
1.052	7.2		1.093	12.3		1.134	17.0		1.173	21.3	
1.053	7.3		1.094	12.4		1.135	17.1		1.174	21.4	
1.054	7.5		1.095	12.5		1.136	17.3		1.175	21.5	
1.055	7.6		1.096	12.6		1.137	17.4		1.176	21.6	
1.056	7.7		1.097	12.7		1.138	17.5		1.177	21.7	
1.057	7.9		1.098	12.8		1.139	17.6		1.178	21.8	
1.058	8.0		1.099	13.0		1.140	17.7	206.3			
1.059	8.1		1.100	13.1	144.0						
1.060	8.2	84.9	1.101	13.2							

TABLE 3
Temperature Corrections (to 20°C) For Density Readings of Concentrated Sea Water

		Density range (g/ml at 20°C)					
Temp. (°C)		from 1.00 to 1.05	1.05 1.10	1.10 1.15	1.15 1.20	1.20 1.25	1.25 1.30
10	Subtract	.002	.002	.003	.003	.003	.003
11	correction	.002	.002	.003	.003	.003	.003
12	from	.001	.002	.002	.003	.003	.002
13	measured	.001	.002	.002	.003	.002	.002
14	density	.001	.001	.002	.002	.002	.002
15		.001	.001	.002	.002	.002	.002
16		.001	.001	.001	.002	.002	.001
17		.001	.001	.001	.001	.001	.001
18		—	.001	.001	.001	.001	.001
19		—	—	.001	.001	—	—
20		—	—	—	—	—	—
21		—	—	.001	.001	.001	—
22	Add	.001	.001	.001	.001	.001	.001
23	correction	.001	.001	.001	.002	.002	.001
24	to measured	.001	.002	.002	.002	.002	.002
25	density	.002	.002	.003	.003	.003	.002
26		.002	.002	.003	.003	.003	.003
27		.003	.003	.004	.004	.004	.004
28		.003	.003	.004	.005	.005	.004
29		.004	.004	.005	.005	.005	.005
30		.004	.004	.005	.006	.006	.006
31		.004	.005	.006	.006	.006	.006
32		.005	.006	.006	.007	.007	.007
33		.005	.007	.007	.007	.007	.007
34		.006	.007	.007	.008	.008	.008
35		.006	.007	.008	.008	.008	.008

contains little or no silt, a good correlation is found between the latter parameters and the water turbidity, the measure of which still is the easiest and most rapid way to determine changes in phytoplankton densities.

K. MONITORING OF *ARTEMIA* PRODUCTION PERFORMANCE

Precise estimates of *Artemia* densities are difficult to make because of the heterogenous distribution in ponds which is influenced by wind, water temperature, light, pond depth, etc.[55,56,57] Nevertheless, rough estimates of *Artemia* densities, among other field data, may provide a valuable tool to assess the rate of biomass harvesting. Water samples should be taken at weekly intervals from fixed stations scattered throughout the pond, in the ditch as well as in the central part or along fixed transects. Sampling is preferentially done as early as possible in the morning when *Artemia* are more uniformly distributed.[24,25] Samples may be taken with a specific sampler[24] or with a variety of containers such as beakers, buckets, etc. When using the latter sampling procedure, it is advisable to thoroughly mix the water column so as to stir up bottom-dwelling *Artemia*.[58] The number of samples to be taken from a pond depends on the distribution of the *Artemia* as well as on the volume of the sample and the abundance of *Artemia*. This may be estimated by calculating the coefficient of variance (CV, variance/mean density), i.e., the lower the CV value, the more precise the sampling.

Of essential importance in pond management (e.g., rate of biomass harvesting, fertilizer

TABLE 4
Examples of Population Composition in *Artemia* Pond at Various Time Intervals (A through H)

Sampling time	Nauplii	Juveniles	Preadults	Adults	Cysts
A	+ +	−	−	−	−
B	−	−	+	+ +	−
C	+ +	−	−	+	−
D	+	+	+	+	−
E	+	−	−	+ +	−
F	−	−	−	+ +	−
G	−	−	−	+ +	+
H	+	+	+	+	+

addition, pond retention time, etc.) is the population composition, which provides valuable information on the population dynamics of *Artemia* in the ponds. The population composition should be analyzed from representative samples taken at weekly intervals from several places in the pond (e.g., combined with the density sampling). Samples containing large numbers of animals may be subsampled until they contain 200 to 300 animals. The *Artemia* are categorized into five classes: cysts and/or nauplii (Instar I—IV), juveniles (Instar V—VII, larvae with developing thoracopods), preadults (adult size but not yet reproductively active), and adults. These classes may be distinguished under a dissection microscope or by pouring the *Artemia* over three successive filter screens with mesh sizes of about 500, 375, and 125 μm which respectively retain adults and preadults; juveniles; and nauplii and cysts. The adults and preadults can easily be separated by eye. The (relative) presence of each *Artemia* class is expressed as percentage of the total number of *Artemia* counted in the plankton sample or is evaluated as follows: − absent; + present; + + dominant presence. Of further interest is the evaluation of the reproductive activity of the females; i.e., empty or full broodsacs and nauplii or cysts bearing. Table 4 shows a typical example of the population composition in an *Artemia* pond at various time intervals. The population changes over one-week intervals, from A through D, reveal a very healthy population; e.g., inoculation (A), growth up to preadult and adult stage (B), first generation of nauplii released (C), and continuous reproduction and good growth conditions (D). However, a population composition remaining for consecutive weeks (E) eventually evolves into a situation which reveals food-limiting conditions (F); e.g., initial algal concentrations are still sufficient for the adults to ensure reproductive activity but too limited for the nauplii which have a lower feeding efficiency than adults; subsequently (F) food becomes too scarce even for the adults. When oviparity is the dominant mode of reproduction no population recruitment is observed (G); in heavily fertilized ponds a mixed reproductive activity is often observed (H).

L. STRATEGIES FOR CULTURE MAINTENANCE

The information collected from the monitoring programs is used to make appropriate decision about pond management. Optimal conditions for biomass production are at the lower salinity levels (100 to 150 g/l) and under conditions of very regular food availability. When transparency levels are high (>30 cm), and pond nutrient levels become undetectable or fall below levels found in the intake water, fertilizer dressing or intake of nutrient-rich water (e.g., from a feed production pond) should be considered. During temperature/salinity stratification (which causes lethal high temperatures for *Artemia*) or when planktonic blue-green algae become dominant, the bottom flow of water from pond to pond should be maximized. Selection of desirable phytoplankton species and/or the prevention of the development of undesirable cyanophytes which have the ability to fix atmospheric nitrogen may also be aided by supplemental fertilization with specific nutrients; e.g., through application of nitrogen fertilizer during conditions where nitrogen is limiting but phosphorus is abundant, or through application of silicate in low salinity ponds to enhance the growth of diatoms, which are rich in the desired highly unsaturated fatty acids.

Sustained population growth also entails regular harvesting of the biomass so as not to exceed the carrying capacity of the pond. Insufficient harvesting may lead to a complete removal of the food (even in ponds with high nutrient levels) due to the high grazing pressure of increasing numbers of *Artemia*, eventually leading to a collapse of the population. Similarly, over-harvesting may reduce the grazing pressure on planktonic algal blooms which may deleteriously affect the salt production.[4,9] Ideally, the rate of biomass harvesting should approach the maximum sustainable yields. Since a precise estimation of the latter is impossible, the rate of biomass harvesting should be assessed from estimation of density, population composition, fertility parameters, and phytoplanktonic standing crop (e.g., estimated by water turbidity levels). In situations where densities increase over time with a population composition showing all *Artemia* classes represented, combined with a dominant ovoviviparous reproduction, biomass is being renewed at a high rate. In this case and when transparency levels increase, indicating that the primary production rate cannot sustain the grazing pressure of the *Artemia* population, frequent harvesting of biomass is recommended. If on the other hand, densities are decreasing and/or the population consists mostly of preadults and adults showing low fecundity and/or dominant oviparous reproduction, harvesting should cease.

In Section II.I. of this chapter we described the factors which induce cyst production. Production of cysts may naturally occur in the high salinity ponds. It may also be induced, even at low salinity levels, by increasing the rates of fertilization or by applying through salinity shocks of 10 to 20 g/l rapid intake of low salinity brines. Oviparity, however, should only be induced when the population density is sufficiently high since during conditions of dominant cyst production, recruitment will be inhibited, eventually resulting in a gradual decrease of the population due to constant mortality.

M. HARVESTING AND QUALITY CONTROL

Produced cysts float at the surface and accumulate along the windward side of the pond. They can be easily collected from the water surface with a double-screen dip net (Figure 5). Cysts should be harvested as soon as possible after production (accumulation) in order to ensure maximum recovery and hatching quality because:

- Cysts washed ashore are difficult to harvest as well as to clean; in addition they will dry and may eventually become airborne.
- Cysts may be exposed to temperatures >40°C which are lethal for hydrated cysts.[26]
- Cysts may be exposed to repeated hydration/dehydration cycles (rainfall, high humidity) and lose part of their energy reserves[59] resulting in a decreased hatchability.[26]

In order to prevent the cysts from being washed ashore and to facilitate harvesting, the windward pond corner or side should be steepened or lined with a cyst barrier, e.g., corrugated plastic. When winds develop heavy waves, foam is built up in which cysts are trapped and then blown away. In this case wave breakers should be installed in two or more rows parallel to the cyst barrier (Figure 6).

Harvested cysts may undergo an on-site cleaning by washing the harvested product with saturated brine or water from the pond over screens with different mesh widths (e.g., 1000, 500, 125 μm) in order to remove debris larger and smaller than the cysts. This wet-dry product stored in brine or mixed with crude salt (for proper dehydration) may be an acceptable product for local use, provided it is consumed within a few weeks. For the production of high quality cysts with optimal hatchability, maximum purity, and storability, the following

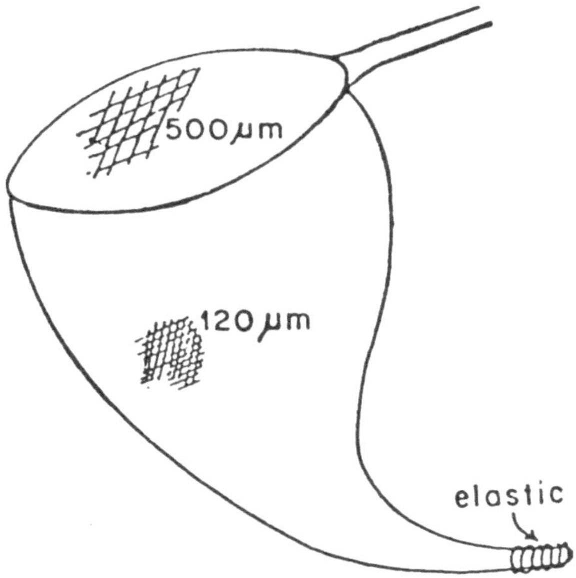

FIGURE 5. Double screen dip net for cyst harvesting. (From Sorgeloos, P., Lavens, P., Léger, P., Tackaert, W., and Versichele, D., *Manual for the Culture and Use of Brine Shrimp Artemia in Aquaculture,* Artemia Reference Center, State University of Ghent, Belgium, 1986. With permission.)

FIGURE 6. Floating bamboo poles used as a wave breaker for the harvesting of *Artemia* cysts. (From Sorgeloos, P., Lavens, P., Léger, P., Tackaert, W., and Versichele, D., *Manual for the Culture and Use of Brine Shrimp Artemia in Aquaculture,* Artemia Reference Center, State University of Ghent, Belgium, 1986. With permission.)

additional processing steps are however required: (1) density separation in brine to remove heavy debris in the same size range of the cysts; (2) rapid washing in fresh water to remove salt; (3) density separation in fresh water to separate full cysts from empty cysts and other small light debris — this step should not take longer than 15 minutes in order to prevent elevated hydration levels which may initiate cysts metabolism; (4) removal of excess water by squeezing or centrifuging the cysts; and (5) drying in order to reduce the water level in the cysts below the critical level of 10% (preferably between 2 and 5%) in order to arrest the metabolic activity in the cysts. Optimal cyst quality in terms of hatching efficiency, hatching rate, and energy content is obtained by fast and homogeneous drying of all cysts at temperatures just below 40°C.[84] Among the different drying techniques, optimal results are obtained when the cysts are kept in continuous movement in the drying air; i.e., each cyst is dried individually at the same time. This may be accomplished in a fluidized bed dryer (Figure 7) or a rotary dryer (Figure 8). If the latter equipment is not available, cysts may also be dried on drying racks on which the cysts are spread in thin layers of uniform thickness (few mm only). The drying racks are placed in the open air, protected from the sun to avoid temperatures higher than 40°C, or in a temperature-controlled room or oven at 35 to 38°C provided with good air exchange. Homogeneous drying is enhanced by granulating the cysts upon distribution on the drying racks through a 3 mm screen and by regular brushing (initially every h) of the cysts. Table 5 shows the effect of drying conditions on the hatching efficiency of cysts in Lavalduc, France. For more details with regard to cyst processing and drying, we refer to Sorgeloos et al.[26]

For long-term storage, cysts should be packed in air tight containers (cans) under vacuum or nitrogen.

Adult biomass may be harvested manually with a dip net (in small ponds) or with a conical net (see Figure 9) which can be towed over the entire pond. In highly eutrophic ponds optimal catches are made during the early morning after a calm night, when the dissolved oxygen concentration in the ponds is so low that the *Artemia* concentrate in very dense "blow-ups" in the upper water layers where they perform so-called surface respiration.[26] Biomass may also be harvested by static nets installed at the pond gates where active brine-flows drain large numbers of *Artemia*. The end part of the harvesting nets should have a small mesh size (<100 μm) to prevent extrusion of the animals. The nets must be harvested at intervals of less than 1 h since the *Artemia* accumulated at the end of the filter sac are exposed to anaerobic conditions which they cannot survive for more than 2 h. The harvesting efficiency of static nets can be improved by harvesting during the night using light attraction to take advantage of the positive phototactic behavior of the brine shrimp. Harvested biomass may be temporarily stocked (up to one week) in nylon screen cages (e.g., 1.5 × 2.0 × 0.5m) with a mesh width of 800 μm, which are suspended in the culture ponds (see Figure 10). For long distance live transport, the *Artemia* biomass may be packed at densities of 100 g/l in plastic bags filled with one-third cooled pond water (5 to 10°C) and two-third oxygen at atmospheric pressure. The bags are placed in a styrofoam box together with a few bags of ice (see Figure 11).

If not used directly, *Artemia* biomass should be frozen or dried after thorough washing with fresh water. Since *Artemia* is extremely prone to decomposition (due to proteolytic enzyme activity) it is essential to freeze the animals when still alive. In order to ensure optimal quality, the biomass should be frozen as quickly as possible in a blast or plate freezer (−25°C or lower) in thin layers (maximum 1 cm thick) or in small ice cube trays. A properly frozen product when thawed in water, yields only intact animals and does not pollute the water by leaching of body fluids (see Figure 12 for quality control).

N. SPECIFIC MODIFICATIONS FOR FURTHER OPTIMIZATION OF *ARTEMIA* PRODUCTION

Since pond production of *Artemia* requires the availability of high salinities to exclude

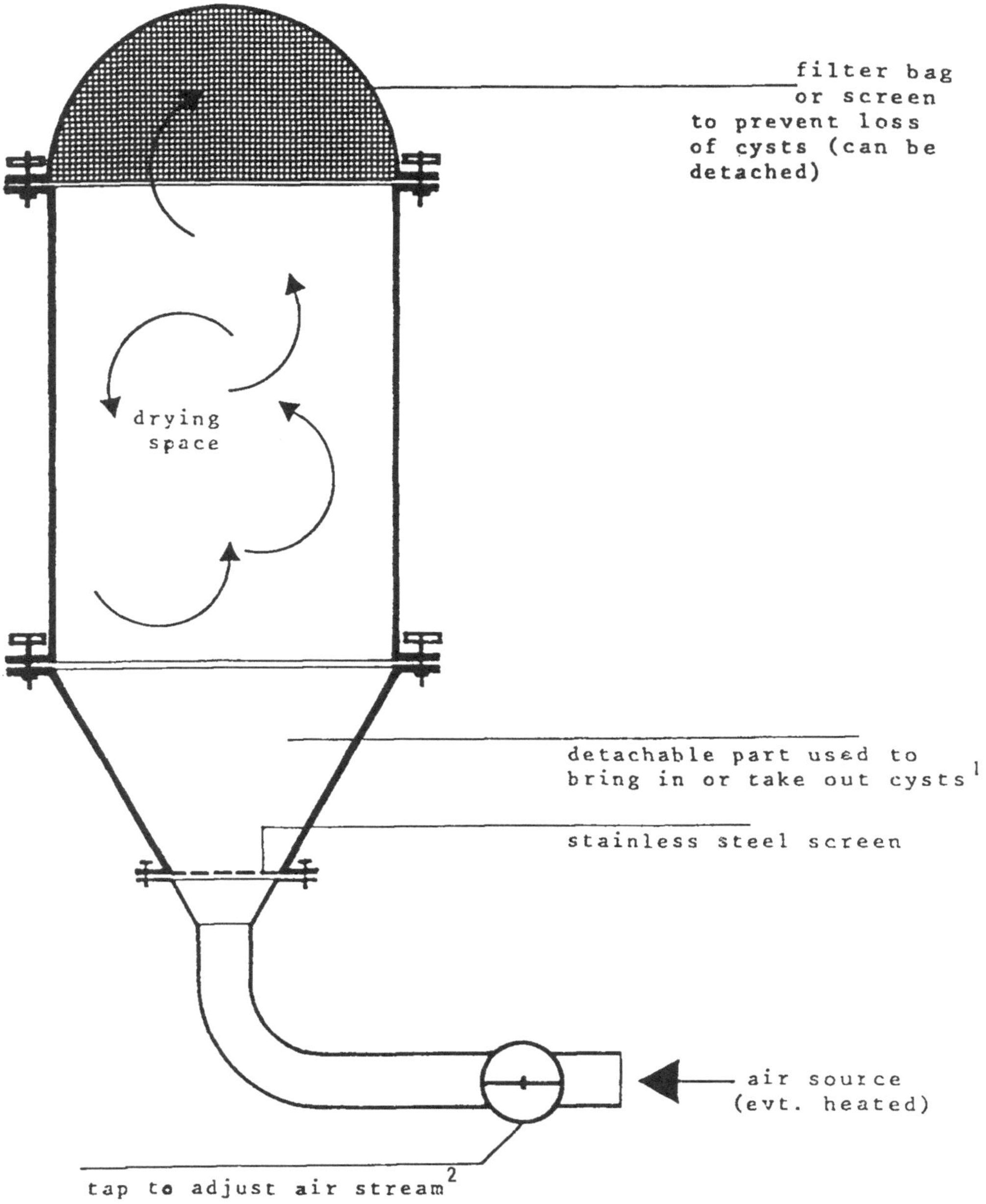

[1]conical shape results in differences in air pressure and assures better mixing of the cysts

[2]more pressure is needed at the start (heavy cysts containing much water before dehydration) than at the end (light cysts with low water content) to keep the cysts suspended in the drying chamber.

FIGURE 7. Schematic drawing of fluidized bed dryer for *Artemia* cysts. (From Sorgeloos, P., Lavens, P., Léger, P., Tackaert, W., and Versichele, D., *Manual for the Culture and Use of Brine Shrimp Artemia in Aquaculture*, Artemia Reference Center, State University of Ghent, Belgium, 1986. With permission.)

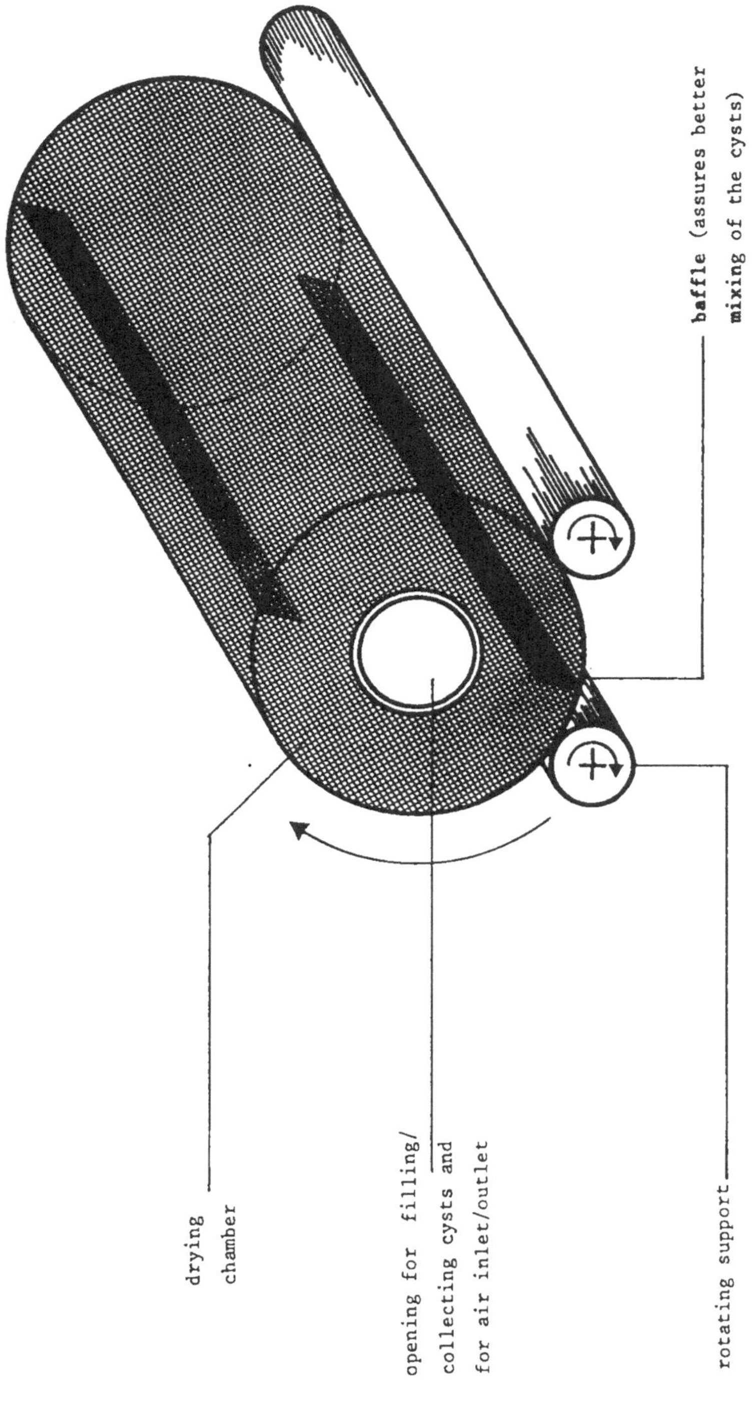

FIGURE 8. Schematic drawing of rotary dryer for Artemia cysts. (From Sorgeloos, P., Lavens, P., Léger, P., Tackaert, W. and Versichele, D., *Manual for the Culture and Use of Brine Shrimp Artemia in Aquaculture*, Artemia Reference Center, State University of Ghent, Belgium, 1986. With permission.)

TABLE 5
The Effects of Drying Conditions on the Hatching Efficiency of Cysts from Lavalduc, France[a]

Drying conditions			Hatching efficiency (nauplii/g cysts)	
Method	Temp. (°C)	Thickness of cysts layer (cm)	X[b]	SD[b]
Oven dryer				
	30	1.5	69,120	9,760
	30	0.5	149,600	10,240
			(154,120)	(7,600)
	38	1.5	150,880	7,200
	38	0.5	181,360	9,600
			(179,200)	(10,100)
Fluidized-bed dryer	35		182,400	6,400
			(181,960)	(6,920)
Control (unprocessed cysts)			178,640	(8,840)

[a] Data compiled from Sorgeloos et al.[26]
[b] In parentheses, data for same cysts but after 1-month storage under vacuum.

predators, the production season in monsoon climates is basically limited to the dry season only. Nevertheless, a significant extension of the production season may be realized by the installation of overflow devices (e.g., PVC turndown pipes and level controlled gates) in order to allow decanting of stratifying layers of rain water, combined by rigorous control of predators (e.g., through screening of intake water by means of a bag screen or semi-circular screen mounted to or surrounding the gate/pump). Year round production of biomass has been successfully applied in both the Philippines[11] and Thailand[12] at salinities of 60 to 80 and 70 to 90 g/l, respectively.

In farms having brine reservoirs, salinity control during the rainy season may further be facilitated by the recirculation of surplus brine from these reservoirs into the *Artemia* ponds. In addition, this practice allows for maximal water exchange (essential for good phytoplankton production) and salinity manipulation (e.g., salinity shocks for the induction of cyst production). Another specific modification beneficial for *Artemia* production under high temperature conditions involves the installation of shading platforms (e.g., made of coconut fronds). De los Santos et al.[60] reported that *Artemia* tend to concentrate under this shade to escape lethally high temperatures occurring on sunny days.

III. PRODUCTION FIGURES OF *ARTEMIA* IN FERTILIZED PONDS

Table 6 shows production figures of *Artemia* cysts and biomass in different man-managed saltfarms. Although the most successful farms yield 10 to 20 kg dry weight (dw) cysts and/or 100 to 375 kg wet weight (ww) biomass/ha/month, there is considerable variation from farm to farm, mainly as a result of differences in farm management. A survey of salt *cum* *Artemia* farms in Thailand in 1983,[61] revealed that poor production in the Samut Sakorn and Phetburi area (see Table 6) were correlated with low pond water depths, inappropriate local conditions such as acidity of the soil, and insufficient fertilization. Poor farm management including lack of pumping and application of cow dung instead of the previously

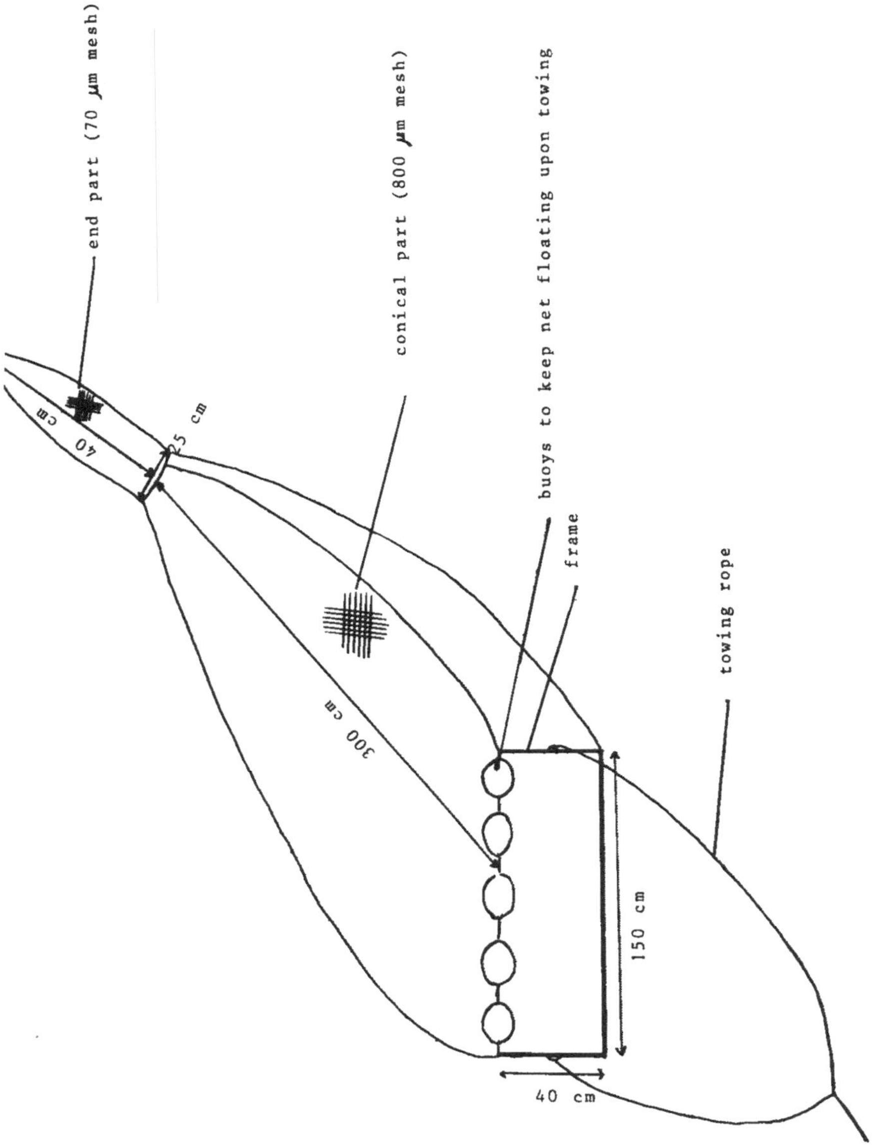

FIGURE 9. Conical harvesting net for *Artemia* biomass.

FIGURE 10. Storage net for *Artemia* biomass produced in seasonal salt ponds in S. E. Asia. (from Sorgeloos, P., Lavens, P., Léger, P., Tackaert, W., and Versichele, D., *Manual for the Culture and Use of Brine Shrimp Artemia in Aquaculture,* Artemia Reference Center, State University of Ghent, Belgium, 1986. With permission.)

used chicken manure resulted in the development of lab-lab and overall food-limiting conditions. These were also responsible for the sharp decrease in cyst productions in Vinh Chau, Vietnam (area not specified) in 1988 (29.1 kg ww) as compared to 1987 (120 kg ww).[15] On the other hand, Vu Do Quynh and Nguyen Ngoc Lam[25] found that the introduction of a flow-through type management in Cam Ranh Bay, Vietnam improved the cyst yield from 1.4 — 6.8 to 8.6 kg of dw cysts/ha/month.

Recently, biomass has been the product of preference (especially in Thailand) largely because it is easier to master than cyst production, which in most farms has remained inconsistent. Biomass production furthermore offers new local marketing opportunities; e.g., during the dry season in Thailand more than 3000 kg of locally produced biomass is being harvested and consumed on a daily basis as a starter feed in shrimp nursing.[85]

IV. SOCIO-ECONOMIC ASPECTS AND BENEFITS OF SALT *CUM ARTEMIA* PRODUCTION

Seasonal solar-salt production as practiced in Southeast Asia, Central America, East Africa, etc. is a labor-intensive activity generating employment for thousands of families. Its profitability, however, is usually limited, largely because of the low yields and the poor quality of the salt produced, owing to the small scale and artisanal manufacturing practices of this type of salt operation. In Viet Nam, for example, the mean annual income/worker in 1987/1988 was equivalent to 30 kg of rice.[15] In Thailand, the revenue of solar salt production has drastically decreased due to competition from rock-salt mining.[12] In fact, in many countries (e.g., Thailand, Panama, and Costa Rica) hundreds of those family-operated saltfarms are being abandoned for socio-economic reasons.

The profitability of these seasonal saltfarms can be considerably improved by integrating

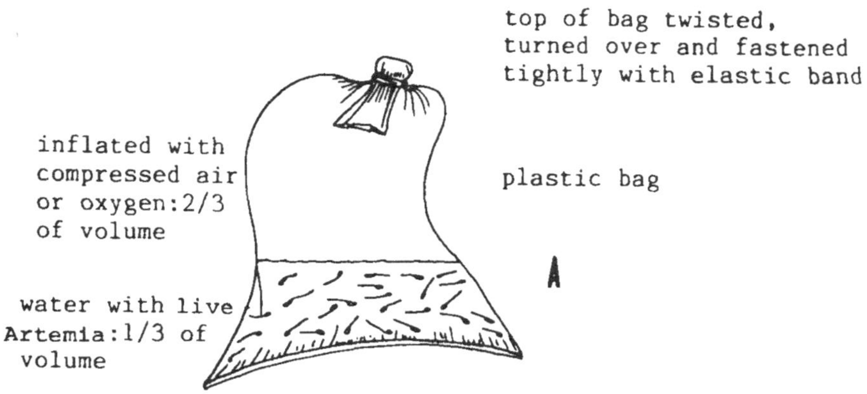

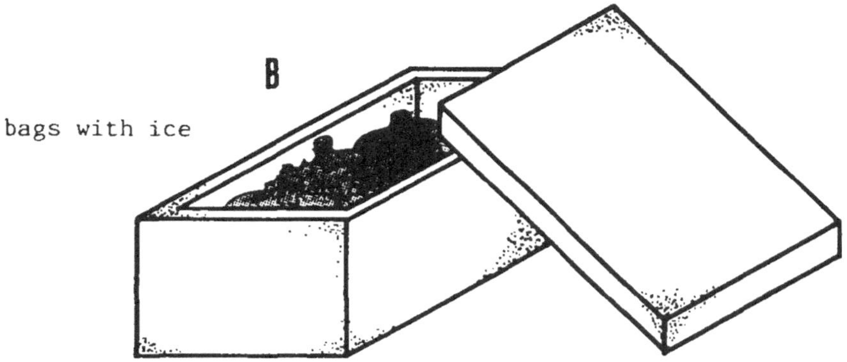

FIGURE 11. Live transport of *Artemia*—transport bag (A) and styrofoam box (B). (From Sorgeloos, P., Lavens, P., Léger, Ph., Tackaert, W. and Versichele, D., *Manual for the Culture and Use of Brine Shrimp Artemia in Aquaculture*, Artemia Reference Center, State University of Ghent, Belgium, 1986. With permission.)

Artemia production with solar-salt production. Based on a survey of five salt *cum Artemia* farms in Thailand, Vanhaecke[61] estimated that the total cost required for pond modification and operation of one ha *Artemia* pond was about 2040 U.S. dollars. Assuming a production of 180 kg ww cysts and 500 kg ww biomass/ha (extrapolated from average production figures of farms that adopted proper pond modification and good biological management) at average market prices of $16 and $4.4/kg respectively, the average benefits from *Artemia* production amounted to $3040 and $3870 in the first year and $3870 in the following years of operation. This represents an additional income almost triple of that derived from salt.[12]

During recent years salt *cum Artemia* production has become a profitable business. The latest data for Thailand[62] reveal annual revenues from *Artemia* biomass production of over $14,000 (for average production yields of 260 to 375 kg/ha/month on a year round basis and wholesale prices of about $4/kg). Integrated *Artemia* production is not only attractive from a socio-economic point of view, it also stimulates the development of local aquaculture (especially in those countries which do not have hard currency for importing *Artemia* cysts, e.g., Vietnam, Bangladesh, etc.) through the local availability of cheap *Artemia* products.

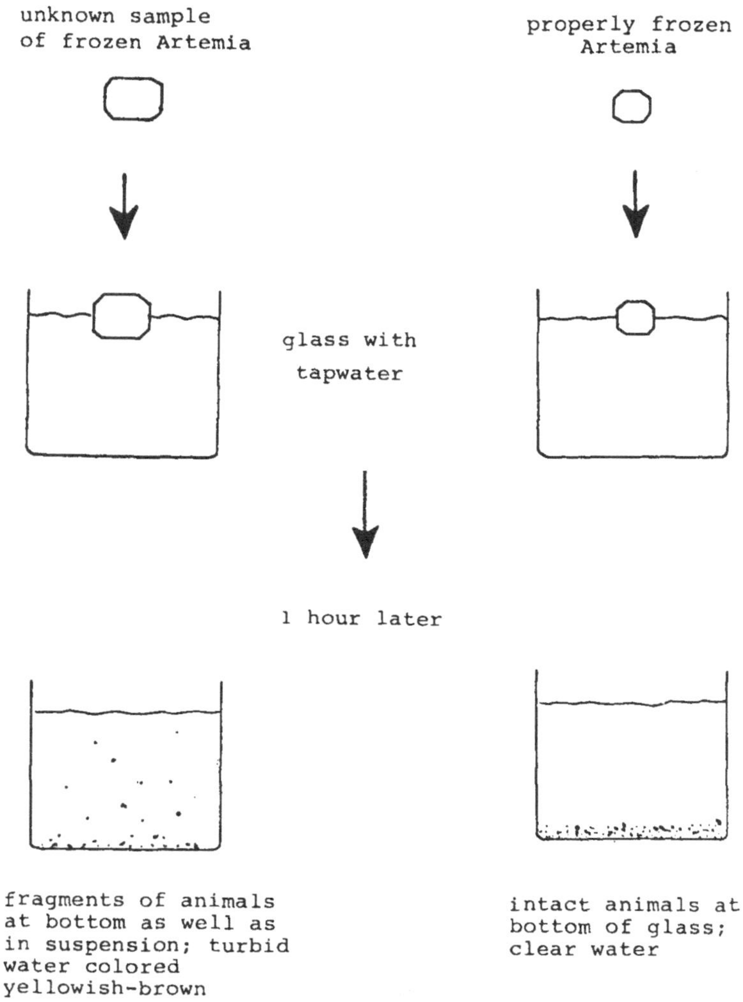

FIGURE 12. Quality control of *Artemia* biomass. (From Sorgeloos, P., Lavens, P., Léger, P., Tackaert, W., and Versichele, D., *Manual for the Culture and Use of Brine Shrimp Artemia in Aquaculture,* Artemia Reference Center, State University of Ghent, Belgium, 1986. With permission.)

TABLE 6
Production of *Artemia* Cysts and Biomass in Man-Managed Saltfarms

Country	Location	*Artemia* production (kg)		Ref.
		Cysts	Biomass	
Thailand	Chonburi	23.1 (dw/ha/month)	52.5 (ww/ha/month)	61
Thailand	Samut Sakorn	5.2 (dw/ha/month)	61.7 (ww/ha/month)	61
Thailand	Phetburi	3.0 (dw/ha/month)	27.2 (ww/ha/month)	61
Thailand	Cha-Choengsao	17.5 (dw/ha/month)	14.4 (ww/ha/month)	61
Thailand	Samut Songkram	15.3 (dw/ha/month)	51.5 (ww/ha/month)	61
Thailand		25.0 (ww/ha/month)	—	12
Thailand	Tambon Klong Tamru, Chonburi	—	260—375 (ww/ha/month)	62
Thailand	Cha-Choengsao	5.0 (ww/ha/month)	—	58
Philippines	Barotac Nuevo	5.0—18.6 (dw/ha/month)	29.4 (ww/ha/month)	60
Philippines	Negros Oriental	20.0 (dw/ha/month)	2000—7000 (standing crop/ha)	11
Viet Nam	Cam Ranh Bay	1.4-6.8-8.6 (dw/ha/month)	—	25
Viet Nam	Vinh Chau	3.2-3.4 (dw/ha/month)	—	15
Viet Nam	Vung Tau	5.0 (dw/ha/month)	—	63
China	Xuwen County	74.6 (dw/ha/year)	—	64
Peru	Virrilla	35.0 (ww/ha/month)	0.06 (standing crop/m^3)	65
Indonesia	Madura Island	38.0 (dw/ha/month)	—	66
Jamaica	Portland Cottage	8.2 (dw/ha/month)	—	67

REFERENCES

1. **Sorgeloos, P.,** *Artemia Newsletter,* No. 11, Artemia Reference Center, State University of Ghent, Belgium, 1989.
2. **Browne, R. A. and MacDonald, G.,** Biogeography of the brine shrimp, *Artemia*: distribution of parthenogenetic and sexual populations, *J. Biogeogr.,* 9, 331, 1982.
3. **Vanhaecke, P., Tackaert, W., and Sorgeloos, P.,** The biogeography of *Artemia*: an updated review, in *Artemia Research and Its Applications,* Vol. 3, Sorgeloos, P., Bengtson, D. A., Decleir, W., and Jaspers, E., Eds., Universa Press, Wetteren, Belgium, 1987, 129.
4. **Sorgeloos, P.,** Brine shrimp *Artemia* in coastal saltworks — hydrobiological key to improved solar salt production — inexpensive source of food for vertically integrated aquaculture, in *Atti del Convegnio Internationale "Conversione delle saline in acquacoltura",* D'Amelio, V. and Santulli, S., Eds., Libera Universa Trapani, Italy, 1988, 133.
5. **Léger, P., Bengtson, D. A., Simpson, K. L., and Sorgeloos, P.,** The use and nutritional value of *Artemia* as a food source, *Oceanogr. Mar. Biol. Ann. Rev.,* 24, 521, 1986.
6. **Léger, P., Bengtson, D. A., Sorgeloos, P., Simpson, K. L., and Beck, A. P.,** The nutritional value of *Artemia*: a review, in *Artemia Research and Its Applications,* Vol. 3, Sorgeloos, P., Bengtson, D. A., Decleir, W., and Jaspers, E., Eds., Universa Press, Wetteren, Belgium, 1987, 357.
7. **Croghan, P. C.,** The mechanism of osmotic regulation in *Artemia salina* (L.): The physiology of the branchiae, *J. Exp. Biol.,* 35, 234, 1958.
8. **Davis, J. S.,** Importance of microorganisms in solar salt production, in *4th Symp. Salt,* Coogan, A. H., Ed., The Northern Ohio Geological Society, Inc., Cleveland, Ohio, 1977.
9. **Davis, J. S.,** Experience with *Artemia* at solar saltworks, in *The Brine Shrimp Artemia,* Vol. 3, Persoone, G., Sorgeloos, P., Roels, O., and Jaspers, E., Eds., Universa Press, Wetteren, Belgium, 1980, 51.
10. **Jones, A. G., Ewing, C. M., and Melvin, M. V.,** Biotechnology of solar saltfields, *Hydrobiologia,* 82, 391, 1981.
11. **Jumalon, N. A., Estenor, D. G., and Ogburn, D. M.,** Commercial production of *Artemia* in the Philippines, in *Artemia Research and Its Applications,* Vol. 3, Sorgeloos, P., Bengtson, D. A., Decleir, W., and Jaspers, E., Eds., Universa Press, Wetteren, Belgium, 1987, 232.
12. **Tarnchalanukit, W. and Wongrat, L.,** *Artemia* culture in Thailand, in *Artemia Research and Its Applications,* Vol. 3, Sorgeloos, P., Bengtson, D. A., Decleir, W., and Jaspers, E., Eds., Universa Press, Wetteren, Belgium, 1987, 202.

13. **Boyd, C. E.,** Lime requirements and application in fish ponds, in *Advanced in Aquaculture,* FAO Technical Conference on Aquaculture, Kyoto, Japan, 1976, 120.
14. **Haxby, R. E. and Tackaert, W.,** Workshop report: role of *Artemia* in solar salt operation, in *Artemia Research and Its Applications,* Vol. 3, Sorgeloos, P., Bengtson, D. A., Decleir, W., and Jaspers, E., Eds., Universa Press, Wetteren, Belgium, 1987, 291.
15. **Van der Zanden, J. J. G.,** Third report on the activities on the culture of *Artemia salina* and *Macrobrachium rosenbergii* in Can Tho and Vinh Chau in Southern Vietnam, Dutch Committee for Science and Technology for Vietnam, 1988.
16. **Abreu-Grobois, A.,** A review of the genetics of *Artemia,* in *Artemia Research and Its Applications,* Vol. 1, Sorgeloos, P., Bengtson, D. A., Decleir, W., and Jaspers, E. Eds., Universa Press, Wetteren, Belgium, 1987, 61.
17. **Vanhaecke, P.,** Vergelijkende studie van diverse geografische rassen van het pekelkreeftje, *Artemia* ter verbetering van zijn gebruik in de aquakultuur, Ph.D. thesis, State University of Ghent, Belgium, 1983.
18. **Vanhaecke, P. and Sorgeloos, P.,** International Study on *Artemia.* XIV. Growth and survival of *Artemia* larvae of different geographical origin in a standard culture test, *Mar. Ecol. Progr. Ser.,* 3, 303, 1980.
19. **Vanhaecke, P. and Sorgeloos, P.,** International Study on *Artemia.* XLVII. The effect of temperature on cyst hatching, larval survival and biomass production for different geographical strains of brine shrimp *Artemia,* spp., *Ann. Soc. Zool. Belg.,* 119, 7, 1989.
20. **Browne, R. A., Sallee, S. E., Grosch, D. S., Segreti, W. O., and Purser, S. M.,** Partitioning genetic and environmental components of reproduction and lifespan in *Artemia, Ecology,* 65, 949, 1984.
21. **Bowen, S. T., Fogarino, E. A., Hitchner, K. N., Dana, G. L., Chow, H. S., Buoncristiani, M. R., and Carl, J. R.,** Ecological isolation in *Artemia:* population differences in tolerance of anion concentrations, *J. Crust. Biol.,* 5, 106, 1989.
22. **Vanhaecke, P., Sidall, S. E., and Sorgeloos, P.,** International Study on *Artemia.* XXXII. Combined effects of temperature and salinity on the survival of *Artemia* of various geographical origin, *J. Exp. Mar. Biol. Ecol.,* 80, 259, 1984.
23. **Vanhaecke, P. and Sorgeloos, P.,** International Study on *Artemia.* IV. The biometrics of *Artemia* strains from different geographical origin, in *The Brine Shrimp Artemia,* Vol. 3, Persoone, G., Sorgeloos, P., Roels, O., and Jaspers, E., Eds., Universa Press, Wetteren, Belgium, 1980, 393.
24. **Jumalon, N. A. and Robles, R. E.,** Sampling and stocking density studies for *Artemia* production in ponds, in *Proc. 1st Int. Warm Water Aquacult. Conf. 1983* (Crustaceans), Brigham Young University, Hawaii campus, 1983, 188.
25. **Vu Do Quynh and Nguyen Ngoc Lam,** Inoculation of *Artemia* in experimental ponds in Central Vietnam: an ecological approach and a comparison of three geographical strains, in *Artemia Research and Its Applications,* Vol. 3, Sorgeloos, P., Bengtson, D. A., Decleir, W., and Jaspers, E., Eds., Universa Press, Wetteren, Belgium, 1987, 253.
26. **Sorgeloos, P., Lavens, P., Léger, P., Tackaert, W., and Versichele, D.,** *Manual for the Culture and Use of the Brine Shrimp Artemia in Aquaculture,* Artemia Reference Center, State University of Ghent, Belgium, 1986.
27. **Sorgeloos, P.,** First report on the triggering effect of light on the hatching mechanism of *Artemia salina* dry cysts, *Mar. Biol.,* 22, 75, 1973.
28. **Vanhaecke, P. and Sorgeloos, P.,** International study on *Artemia.* XVIII. The hatching rate of *Artemia* cysts - a comparative study, *Aquacult. Eng.,* 1, 263, 1982.
29. **Léger, P., Vanhaecke, P., and Sorgeloos, P.,** International Study on *Artemia.* XXIV. Cold storage of live *Artemia* nauplii from various geographical sources: potentials and limits in aquaculture, *Aquacult. Eng.,* 2, 69, 1983.
30. **Tunsutapanich, A.,** Cyst production of *Artemia salina* in salt ponds in Thailand, in *Giant Prawn Farming,* New, M. G., Ed., Elsevier Scientific, Amsterdam, Netherlands, 1982, 233.
31. **Fogg, G. E.,** The comparative physiology and biochemistry of the blue-green algae, *Bacterial. Rev.,* 20, 148, 1956.
32. **Oloffson, J. A.,** The role of microlayers in controlling phytoplankton productivities, in *Developments in Hydrobiology.* Vol 2., Hypertrophic Ecosystems, Barica, J. and Mur, L. R., Eds., Dr. W. Junk, The Hague, Netherlands, 1980, 83.
33. **Jumalon, N. A. and Ogburn, D. M.,** Nutrient flow and physico chemical profile studies of an integrated poultry - salt - *Artemia* - milkfish - seabass - shrimp pond production system, in *Artemia Research and Its Applications,* Vol. 3, Sorgeloos, P., Bengtson, D. A., Decleir, W., and Jaspers, E., Eds., Universa Press, Wetteren, Belgium, 1987, 239.
34. **Lavens, P. and Sorgeloos, P.,** Controlled production of cysts under standard conditions in a recirculating culture system, *Aquacult. Eng.,* 3, 221, 1984.
35. **Lavens, P.,** Intensieve productie en kwaliteitsonderzoek van *Artemia* adulten en hun nakomelingen, Ph.D. thesis, State University of Ghent, Belgium, 1989.
36. **Sorgeloos, P.,** *Artemia Newsletter,* No. 11, Artemia Reference Center, State University of Ghent, Belgium, 1989.

37. **Tarnchalanukit, W.,** personal communication.
38. **Versichele, D. and Sorgeloos, P.,** Controlled production of *Artemia* cysts in batch cultures, in *The Brine Shrimp Artemia* Vol. 3, Persoone, G., Sorgeloos, P., Roels, O., and Jaspers, E., Eds., Universa Press, Wetteren, Belgium, 1980, 231.
39. **Dutrieu, J.,** Quelques observations biochemiques et physiologiques sur le développement d'*Artemia salina* Leach, *Arch. Zool. Exp. Gén.*, 99, 1, 1960.
40. **Gilchrist, B. M. and Green, J.,** The pigments of *Artemia*, *Proc. R. Soc.*, (Ser. B), 152, 118, 1960.
41. **Fautrez, J. and Fautrez-Firlefyn, N.,** Contribution à l' étude des glandes coquilières et des coques de l' oeuf d'*Artemia salina*, *Arch. Biol.*, 82, 42, 1971.
42. **Metalli, P. and Ballardin, E.,** Radiobiology of *Artemia*: radiation effects and ploidy, *Curr. Top. Radiat. Res. Q.*, 7, 181, 1972.
43. **Dhert, Ph.,** Vergelijking van de kwalitatieve en kwantitatieve kenmerken van de voortplanting bij diverse *Artemia* rassen, M.Sc. thesis, State University of Ghent, Belgium, 1986.
44. **Gajardo, G. M. and Beardmore, J. A.,** Ability to switch reproductive mode in *Artemia* is related to maternal heterozygosity, *Mar. Ecol. Progr. Ser.*, 55, 191, 1989.
45. **Camara, M. R. and De Medeiros Rocha, R.,** *Artemia* culture in Brasil: an overview, in *Artemia Research and Its Applications*, Vol. 3, Sorgeloos, P., Bengtson, D. A., Decleir, W., and Jaspers, E., Eds., Universa Press, Wetteren, Belgium, 1987, 195.
46. **Van der Zanden,** personal communication.
47. **Berthélémy-Okazaki, N. J. and Hedgecock, D.,** Effect of environmental factors on cyst formation in the brine shrimp Artemia, in *Artemia Research and Its Applications*, Vol. 3, Sorgeloos, P., Bengtson, D. A., Decleir, W., and Jaspers, E., Eds., Universa Press, Wetteren, Belgium, 1987, 167.
48. **Strickland, J. D. and Parsons, T. R.,** *A Practical Handbook of Seawater Analysis*, Bull. Fish. Res. Board Can., 1972, 167.
49. **Grasshoff, K.,** *Methods of Seawater Analysis*, Chemie Verlag, New York, 1976.
50. **Dobbeleir, J., Adam, N., Bossuyt, E., Bruggeman, E., and Sorgeloos, P.,** New aspects of the use of inert diets for high density culturing of brine shrimp, in *The Brine Shrimp Artemia*, Vol. 3, Persoone, G., Sorgeloos, P., Roels, O., and Jaspers, E., Eds., Universa Press, Wetteren, Belgium, 1980, 165.
51. **Davis, J. S.,** Biological communities of a nutrient enriched salina, *Aquat. Bot.*, 4, 23, 1978.
52. **Wongrat, L.,** Biological analysis of *Artemia* from salt and salt *cum Artemia* farm, in *Artemia Newsletter*, No. 4, Artemia Reference Center, State University of Ghent, Belgium, 1987.
53. **Vonshack, A.,** Laboratory techniques for the cultivation of microalgae, in *CRC Handbook of Microalgal Culture*, Richmond, E., Ed., CRC Press, Boca Raton, FL, 1986, 117.
54. **Boyd, C. E.,** *Water Quality in Warmwater Fish Ponds*, Alabama Agricultural Experiment Station, Auburn University, Auburn, Alabama, 1988.
55. **Marchant, R. and Williams, W. D.,** Population dynamics and production of brine shrimp, *Austral. J. Mar. Freshwater Res.*, 28, 417, 1977.
56. **Lenz, P. H.,** Ecological studies on *Artemia*: a review, in *Artemia Research and Its Applications*, Vol. 3, Sorgeloos, P., Bengtson, D. A., Decleir, W., and Jaspers, E., Eds., Universa Press, Wetteren, Belgium, 1987, 5.
57. **Wear, R. G. and Haslett, S. J.,** A minimal strategy for assessing *Artemia* biomass harvestable from production salinas, in *Artemia Research and Its Applications*, Vol. 3, Sorgeloos, P., Bengtson, D. A., Decleir, W., and Jaspers, E., Eds., Universa Press, Wetteren, Belgium, 1987, 217.
58. **Vos, J. and Tunsutapanich, A.,** Detailed report on *Artemia* cyst inoculation in Bangpakong, Chachoengsao Province, FAO/UNDP Field Document, THA/75/008, 1979.
59. **Rakowicz, M.,** Notes on *Artemia salina*, in *Prawn Farming Systems Techniques*, Food Institute, East-West Center, Honolulu, HA, 1975.
60. **De los Santos, C., Sorgeloos, P., Lavina, E., and Bernardino, A.,** Successful inoculation of *Artemia* and production of cysts in man-made salterns in the Philippines, in *The Brine Shrimp Artemia*, Vol. 3, Persoone, G., Sorgeloos, P., Roels, O., and Jaspers, E., Eds., Universa Press, Wetteren, Belgium, 1980, 159.
61. **Vanhaecke, P.,** Report mission to Thailand, Feb. 22 — March 13, Artemia Reference Center, State University of Ghent, Belgium, 1983, unpublished.
62. **Tarnchalanukit, W. and Wongrat, L.** personal communication, 1989.
63. **Tackaert, W.,** unpublished data.
64. **Tang Senming,** personal communication, 1989.
65. **Salgado Leu, A.,** Preliminary trials of extensive culture of *Artemia* in Peru, in *Artemia Research and Its Applications*, Vol. 3, Sorgeloos, P., Bengtson, D. A., Decleir, W., and Jaspers, E., Eds., Universa Press, Wetteren, Belgium, 1987, 289.
66. **Yap, W.,** personal communication, 1989.
67. **Henry, M.,** personal communication, 1989.
68. **Parker, G. H.,** Lake Urmi, *Amer. Natural.*, 315, 1900.

69. **Gun'ko, A. F.,** Methods of increasing the output of sturgeon hatcheries by controlling the temperature regime during incubation and by using *Artemia* as food for young sturgeon, *Tr. Azovsk. Nauchn. Issled Inst. Rybn. Khoz*, 5, 73, 1962 (in Russian).
70. **Rooth, J.,** The flamingos in Bonaire (Netherland Antilles), Habitat, diet and reproduction of *Phoenicopterus ruber*. Uitgave Natuurwetenschappelijke Studiekring in Suriname en de Nederlandse Antillen, Utrecht, The Netherlands, 41, 1965.
71. **Mason, D. T.,** Limnology of Mono Lake, California, *U. Calif. Public. in Zool.*, 83, 1, 1967.
72. **Lenz, P. H.,** Ecology of an alkali-adapted variety of *Artemia* from Mono Lake, California (U.S.A.), in *The Brine Shrimp Artemia*, Vol. 3, Persoone, G., Sorgeloos, P., Roels, O., and Jaspers, E., Eds., Universa Press, Wetteren, Belgium, 1980, 79.
73. **Wirick, C. D.,** *Dunaliella-Artemia* plankton community of the Great Salt Lake, Utah, M.Sc. thesis, University of Utah, 1972.
74. **Stephens, D. W. and Gillespie, D. M.,** Phytoplankton production in the Great Salt Lake Utah, and a laboratory study of algal responses to enrichment, *Limnol. Oceanogr.*, 21, 74, 1976.
75. **Isenmann, P.,** Observations sur la mouette pygmée *(Larus mimutus)* en Camargue de 1971 à 1974, *Terre et Vie*, 29(1), 77, 1975.
76. **Britten, R. H. and Johnson, R.,** An ecological account of a Mediterranean salina: the Salin de Giraud, Camargue (S. France), Biol. Conserv., 42, 185, 1987.
77. **Carpelan, L. H.,** Hydrobiology of Alviso salt ponds, *Ecology*, 38, 375, 1957.
78. **Baker, M. J.,** Autoecology of *Artemia:* factors influencing hemoglobin synthesis and cyst production, M.Sc. thesis, San Francisco State College, CA, 1966.
79. **Voronov, P. M.,** Seasonal abundance and biomass of *Artemia salina* and its eggs in saline lakes of the Crimea, in *Problems of Rational Marine Fisheries and Reproduction of Marine Fish and Shellfish*, Proceedings of All-Union Research Institute of Marine Fisheries and Oceanography (VNIRO), 94, 170, 1973 (in Russian, English abstract).
80. **Lüdskanova, J.,** Die Entwicklung von *Artemia salina* L. in den Teichen der Salzgarten von Burgas und Pomorije, *Arch. Hydrobiol.*, 74, 473, 1974.
81. **Haslett, S. J. and Wear, R. G.,** Biomass estimation of *Artemia* at Lake Grassmere, Marlborough, New Zealand, *Aust. J. Mar. Freshwater Res.*, 36, 537, 1985.
82. **Wear, R. G. and Haslett, S. J.,** Studies on the biology and ecology of *Artemia* from Lake Grassmere, New Zealand, in *Artemia Research and Its Applications*, Vol. 3, Sorgeloos, P., Bengtson, D. A., Decleir, W., and Jaspers, E., Eds., Universa Press, Wetteren, Belgium, 1987, 101.
83. **Persoone, G. and Sorgeloos, P.,** General aspects of the ecology and biogegography of Artemia, in *The Brine Shrimp Artemia*, Vol. 3. Persoone, G., Sorgeloos, P., Roels, O., and Jaspers, E., Eds., Universa Press, Wetteren, Belgium, 1980, 51.
84. **Vanhaecke, P. and Sorgeloos, P.,** International Study on *Artemia*. XVIII. The hatching rate of *Artemia* cysts - a comparative study, *Aquacult. Eng.*, 1, 263, 1982.
85. **Sorgeloos, P.,** Booming of shrimp hatcheries in Thailand, in *Artemia Newsletter*, No. 11, Artemia Reference Center, State University of Ghent, Belgium, 1989.
86. **Morales, J.,** personal communication, 1989.
87. **Bradley, H. L.,** personal communication, 1987.

Chapter 13

PRODUCTION OF *ARTEMIA* IN CULTURE TANKS

Patrick Lavens and Patrick Sorgeloos

TABLE OF CONTENTS

I. INTRODUCTION

Compared to semi-intensive and intensive pond production techniques, super-intensive tank culture systems for *Artemia* offer two main advantages. First, there is no restriction with regard to production site or time; i.e., the culture procedure does not require highly saline waters or specific climatological conditions. Although *Artemia* in its natural environment only occurs at high salinities where no predators can survive (usually >100 g/l), they can thrive in natural sea waters, artificial sea waters, or saline well waters. Their physiological performance in terms of growth efficiency is even best in the salinity range 30—80 ppt.[1] Furthermore, optimal growth temperature can be maintained by the use of thermal effluents, or simply by heaters. The second advantage is that the controlled production can be performed with very high densities of brine shrimp, e.g., several thousand animals per liter vs. a maximum of a few hundred animals per liter in outdoor culture ponds. As a consequence very high production yields per volume of culture medium can be expected with tank rearing systems. However, it is obvious that in view of its operational sophistication and its energy and feed demand, this type of super-intensive *Artemia* culture yields a much more expensive *Artemia* product than with pond production. Nonetheless, they are a valuable alternative since more applications with *Artemia* culture in tanks occur in aquaculture as well as in the aquarium pet industry.[2,3]

Depending on the objectives and opportunities, different culture procedures for super-intensive *Artemia* production may be applied. In this regard we have divided this chapter into five main topics. The first part gives detailed information on abiotic and biotic culture conditions which may affect production yields. The second topic covers the design of upscaled culture systems and the general culture procedures. Disease control is explained in the third part, followed by harvesting techniques and production yields in the fourth topic. Last, we deal with the utilization of *Artemia* in aquaculture.

II. ABIOTIC AND BIOTIC CULTURE CONDITIONS

A. PHYSICO-CHEMICAL CULTURE CONDITIONS

1. Temperature and Salinity

Growth and survival of *Artemia* are greatly influenced by temperature and salinity.[4] Both parameters should be considered together since a reciprocal interaction has been detected for aquatic invertebrates in general[5] and is confirmed for *Artemia*.[6] Although it is generally accepted that brine shrimp are eurythermal as well as euryhaline, differences in survival occur among different strains. Nonetheless Vanhaecke et al.[7] state that for most strains (except e.g., in Chaplin Lake, Canada) a common range of preference of 19—25°C and 35—110 ppt can be detected in which mortality is less than 10%. Above 30°C temperatures soon become lethal for most strains. This is also the case for salinities under 20 g/l. The optimal range of temperature varies per strain if biomass production or food conversion efficiency is considered as the criterion: European bisexual strains and the population from San Francisco Bay, California (SFB) prefer the lower range (20—25°C) whereas parthenogenetic species and the Great Salt Lake, Utah (GSL) strain prefer the higher range (25—30°C).[8] See also Table 1.

2. Ionic Composition of the Culture Media

Besides natural sea water or brine, several artificial media with different ionic composition are frequently used in indoor installations for brine shrimp production. Examples of the composition of such media are given in Table 2. Sometimes *Artemia* grow better in these culture media than in natural sea water.[3,9] It is not even essential to use complex formulas; e.g., Dietrich and Kalle — media prepared with only six technical salts — proved

TABLE 1
The Effect of Temperature on Different Production Parameters for Various Geographical Strains of *Artemia*[a]

Geographical strain	Temperature (°C)					
	20	22.5	25	27.5	30	32.5
San Francisco Bay, CA						
Survival (%)	97	97	94	91	66	J
Biomass production (%)[b]	75	101	100	94	88	-
Specific growth rate[e]	0.431[h]	0.464[e]	0.463[ef]	0.456[f]	0.448[g]	-
Food conversion[d]	3.89[f]	3.35[c]	3.64[f]	3.87[f]	4.15[g]	-
Great Salt Lake, UT						
Survival (%)	77	85	89	89	87	88
Biomass production (%)	69	104	122	128	135	78
Specific growth rate	0.392[i]	0.437[g]	0.454[f]	0.460[ef]	0.465[e]	0.406[h]
Food conversion	3.79[g]	2.90[f]	2.65[ef]	2.52[e]	2.40[e]	4.14[h]
Chaplin Lake, Canada						
Survival (%)	72	75	77	65	50	-
Biomass production (%)	78	102	108	106	90	-
Specific growth rate	0.422[g]	0.452[ef]	0.459[e]	0.456[e]	0.437[fg]	-
Food conversion	3.42[e]	3.00[e]	3.03[e]	3.11[e]	3.72[e]	-
Manaure, Colombia						
Survival (%)	94	94	91	93	86	77
Biomass production (%)	52	102	112	110	104	105
Specific growth rate	0.379[g]	0.451[f]	0.462[e]	0.460[ef]	0.454[ef]	0.455[cf]
Food conversion	5.82[f]	3.43[e]	3.36[e]	3.40[e]	3.62[g]	3.56[e]
Buenos Aires, Argentina						
Survival (%)	96	90	94	91	80	-
Biomass production (%)	93	113	115	98	96	-
Specific growth rate	0.462[f]	0.430[e]	0.485[e]	0.468[f]	0.465[f]	-
Food conversion	2.78[e]	2.66[e]	2.81[e]	3.29[f]	3.38[f]	-
Larnaca, Cyprus						
Survival (%)	98	94	89	62	3	-
Biomass production (%)	61	94	95	83	0	-
Specific growth rate	0.384[f]	0.430[e]	0.432[e]	0.417[g]	0[h]	-
Food conversion	4.29[f]	3.23[e]	3.43[e]	3.93[f]	-	-
Shark Bay, Australia						
Survival (%)	97	96	97	92	68	-
Biomass production (%)	48	67	84	75	66	-
Specific growth rate	0.332[g]	0.369[f]	0.393[e]	0.381[ef]	0.367[f]	-
Food conversion	6.49[g]	5.35[ef]	4.57[e]	5.08[e]	5.81[fg]	-
Tientsin, PR-China						
Survival (%)	95	94	91	93	84	54
Biomass production (%)	41	61	80	92	85	16
Specific growth rate	0.299[g]	0.343[f]	0.371[e]	0.387[e]	0.378[e]	0.208[h]
Food conversion	7.22[g]	5.42[f]	4.46[ef]	3.84[e]	4.22[ef]	22.04[h]

[a] data compiled from Vanhaecke *et al.*[8]
[b] expessed as % recorded for the *Artemia* reference strain (San Francisco Bay, batch 288-2596) at 25°C after 9 days culturing on a diet of *Dunaliella* cells.

[c] specific growth rate

$$k = \frac{\ln \frac{W_t}{W_o}}{T}$$

with T = duration of experiment in days (= 9)

[d] food conversion

$$= \frac{F}{W_t - W_o}$$

F = µg dry weight *Dunaliella* offered as food
W_t = µg dry weight *Artemia* biomass after 9 days culturing
W_o = µg dry weight *Artemia* biomass at start of experiment

[e–i] means with the same superscript are not significantly different at the $P < 0.05$ level.
J not analyzed.

TABLE 2
Artificial Sea Waters Used for Tank Production of *Artemia*[a]

Dietrich and Kalle[b]		Instant Ocean		ARC	
Solution A					
NaCl	: 23.9	Cl^-	: 18.4	NaCl	: 31.08
$MgCl_2.6H_2O$	: 10.83	Na^+	: 10.22	$MgCl_2$	: 6.09
$CaCl_2$ anhydric	: 1.15	SO_4^{2-}	: 2.516	$CaCl_2$	: 1.53
$SrCl_2.6H_2O$	: 0.004	Mg^{2+}	: 1.238	KCl	: 0.97
KC1	: 0.682	K^+	: 0.39	$MgSO_4$	: 7.74
KBr	: 0.099	Ca^{2+}	: 0.37	$NaHCO_3$	: 1.80
		HCO_3^-	: 0.142		
H_2O distilled	: 856	H_2O	: 1000	H_2O	: 1000
Solution B					
$NaSO_4.10H_2O$	: 9.06				
$NaHCO_3$	: 0.02				
NaF	: 0.0003				
H_3BO_3	: 0.0027				
H_2O distilled	: 1000				

[a] Data compiled from Lavens [3]; all data in g.

[b] Solutions A and B are prepared separately, then mixed and strongly aerated for 24 h.

to be as good as complete artificial seawater formulas. Moreover, culture tests with GSL *Artemia* in modified ARC (artificial sea waters) showed that KCl can be eliminated, and $MgCl_2$ and $MgSO_4$ reduced without affecting production characteristics. Ca^{++} concentrations >20 ppm are essential for chloride-habitat *Artemia* populations whereas carbonate-habitat strains prefer Ca^{++} concentrations <10 ppm in combination with low levels of Mg^{++} [10]

3. pH, Oxygen Concentration, and Illumination

In the literature, it is generally accepted that the pH tolerance for *Artemia* ranges from 6.5 to 8.[11,12]

With regard to oxygen, only concentrations <2 ppm can limit the production of biomass:[13] i.e., *Artemia* has the ability to adapt quickly to varying levels of O_2 by producing different hemoglobins with specific affinities for O_2.[13] For optimal production, however, O_2 concentrations >2 ppm are suggested.[14] High levels of O_2 (>5 ppm), on the other hand, will result in the production of pale animals, possibly with a lower individual dry weight, and which may therefore be less perceptible and attractive for the predators.

Little is known about the effect of light upon growth. Some studies postulate, without much proof, that *Artemia* is less active in darkness, thus consuming less energy for its metabolism and in this way allowing a faster growth.[15-17]

4. Water Quality

The quality of the culture medium can be affected by particles as well as soluble substances. Although no exact data on maximal concentrations are available, it is generally accepted that high levels of suspended solids will affect production characteristics, either by their interference at food uptake and propulsion by *Artemia*, or by bacterial growth and, consequently, oxygen demand. Furthermore, soluble waste products such as food proteins give rise to toxic nitrogen compounds (e.g., NH_3-N, NO_2^--N, and NO_3^--N) through oxidative deamination, mineralization, and nitrification processes. Ammonia is also excreted by the ammonotelic brine shrimp;[18] excretion rates, however, are a function of density, stage, and strain of *Artemia*, as well as temperature and food quality/quantity. The tolerance levels in *Artemia* for NH_3-NH_4^+-N, respectively NO_2^--N, and NO_3^--N, seem to be much higher than in other water organisms: e.g., acute and chronical toxicity tests with GSL brine shrimp larvae showed no significant effect on survival (LC_{50}) nor growth for concentrations up to 1000 ppm NH_4^+, respectively 320 ppm NO_2^-.[3,19] It is therefore very unlikely that N-components will interfere directly with the *Artemia* cultures. Nevertheless, the presence of soluble substances should be restricted as much as possible because they are an ideal substrate for bacteria.

B. FEEDING CONDITIONS

Artemia is a continuous, nonselective, particle-filtering organism. The morphology and operation of the filter-feeding apparatus and alimentary canal of nauplii and (pre-)adults are described in detail in the chapter in this volume on morphology by Criel.

The coupling of propulsion, respiration, and filtration by the thoracopods results in a practically continuous filter feeding. Ingestion, however, can be interrupted at high particle concentrations: i.e., mandibles and maxillae stop their action, and the food bolus accumulated behind the labrum is rejected into the medium by the first pair of thoracopods.[20,21]

Furthermore, there is no selection in size or sorting of particles ingested by the brine shrimp.[1,22] Although maximal particle sizes that can be ingested by the naupliar and post-naupliar stages have not been exactly determined, diameters should not exceed 50 to 70 μm.

Artemia is an obligate particle feeder, at least with regard to fulfilling its protein and carbohydrate needs.[23] Micronutrients can be ingested under a soluble form only if concentrations in the culture medium are sufficiently high.

In spite of the above-mentioned characteristic of being a continuous filter feeder, various factors may influence the feeding behavior of *Artemia* by affecting the filtration rate, ingestion rate, and/or assimilation. As illustrated in Table 3, factors which may interfere are the quality and quantity of the food offered, the developmental stage of the larvae, and the culture conditions. More detailed information about these processes is given in Lavens[3] and in Coutteau and Sorgeloos.[30] In general we can make the following conclusions:

- A relatively wide range of particle densities is optimal for maximal ingestion; the minimal and maximal concentrations of this range, however, are very specific for each culture condition. Although the influence of these factors on the feeding activity of *Artemia* is only important when they measure extreme values, a combination of sub-optimal conditions that may occur in intensive culture systems might result in a limited rate of food uptake and consumption. This will eventually lead to low biomass production yields.
- The filter-feeding activity of brine shrimp is not very efficient in the early larval stages when only the second antennae contribute to filtration activity. Maximal filtration capacity is attained when all the thoracopods are functional.
- Artemia seems to be able to meet its protein requirements by regulating the food intake and assimilation as a function of the chemical composition of the diet offered.
- The food digestion process is dependent on the retention time in the gut which is determined by the ingestion rate, the enzyme activity, and the digestibility of the food.

C. SUITABLE DIETS

The nutritional requirements of brine shrimp can be summarized as follows: (1) the optimal protein: carbohydrate ratio is situated between 1:3 and 1:5;[31] (2) the essential amino acids are probably the same as for Crustacea in general; (3) exogenous nucleotides and sterols are essential;[32,33] (4) the essential vitamins are thiamine, nicotinamide, Ca-pantothenate, pyridoxine, riboflavin, folic acid, and putrescine;[23] (5) highly unsaturated fatty acids are not essential for growth but stimulate reproduction.[3,34]

TABLE 3
Filtration Rates and Ingestion Rates of *Artemia*[a]

Type of Food (diameter, dimensions or volume)	*Artemia* characteristics								
	Strain	Stage	Size (mm)	Temp. (°C)	C_i	C_f	I_{max}	F_{max}	Ref.
Yeast	—	Juvenile		20	1200		10.08		24
Yeast	—	Adult			1200		33.60		
Phaeodactylum (113 μm³)	GSL		0.5	20	220	160			20
Phaeodactylum	GSL		1		180	90			
Phaeodactylum	GSL		2		210	100			
Phaeodactylum	GSL		5		120	55			
Phaeodactylum	GSL		7.5		145	70			
Phaeodactylum	GSL		10		160	15	9.6		
Dunaliella (345 μm³)	GSL		10				3.6		
Chlorella (31 μm³)	GSL		10						
Sand (3-40 μm³)	GSL		9	20		1000	9.6		1
Phaeodactylum	GSL		9			50	13		
Chlamydomonas (10-15 μm × 6-9 μm)	SFB		5	28		120	7		26,27
Chaetoceros (46 μm³)	—		11	20	400	400	14.4	408	21
Dunaliella (251 μm³)	—		11		30	30	0.96	408	21
Chlamydomonas (1017 μm³)	—		5	25	170		0.31		33
Chlorella (14 μm³)	—		5		850		20.4		
Phaeodactylum	—	Nauplius		20			0.8		29
Phaeodactylum	—	Juvenile						17	
Phaeodactylum	—	Preadult						35-37	
Phaeodactylum	—	Adult						60-70	

Note: C_i, critical cell concentration for maximal ingestion (10^3 cells/ml); C_F, critical cell concentration for maximal filtration (10^3 cells/ml); I_{max}, maximal ingestion rate (10^6 cells/24 h/animal); F_{max}, maximal filtration rate (ml/24 h/animal); —, not specified.

[a] Data compiled from Lavens, P., Ph.D. thesis, 1989. With permission.

Subtle differences in nutritional requirements can be detected among the different geographical strains: e.g., the bisexual GSL *Artemia* grew optimal on the artificial test medium of Provasoli and Pintner[34] but the asexual population from Commacchio (Italy) was unable to perform well.

An enormous diversity of live and non-living diets is available as a potential food source for the successful culture of brine shrimp, e.g., live and dried algae, bacteria and yeasts (single cell proteins), and by-products from agricultural crops. Two specific characteristics of *Artemia* are correlated with this broad range in suitable diets: on the one hand the nonselective filtration behavior which facilitates the use of any type of food provided that the particle size is sufficiently small, and on the other hand the possibility to take up and digest exogenous microflora as part of the diet. Bacteria and protozoans which develop easily in the *Artemia* cultures are indeed able to biosynthesize essential nutrients as they use the supplied brine shrimp food as a substrate; in this way they compensate possible deficiencies in the diet composition. As early as 1956, Gibor[35] observed that *Artemia* grew better on live algae cultured under xenic as compared to axenic conditions. This effect is confirmed for diets based on dried algae or yeasts and fed to Anostracans or other invertebrates.[36-39] This phenomenon is even more critical with regard to the utilization of waste products from agricultural crops which on their own are probably nutritionally inadequate.

The interference by bacteria makes it difficult to identify nutritionally adequate diets as such, since growth tests are difficult to run under axenic conditions. As a consequence nutritional composition of the diet does not play the most critical role in the selection of foods suitable for high density culture of brine shrimp. More important criteria are

- Availability and cost
- Particle-size composition
- Digestibility
- Consistency in composition among different batches and storage capacity
- Solubility
- Food conversion efficiency (FCE)
- Buoyancy
- Possible interactions with the applied culture technology (e.g., clogging of the particles, fouling, etc.)

1. Live Micro-Algae

Many publications report on the culture success of *Artemia* fed with different algal species.[23] It is, however, useless to list all suitable algal species here as their nutritional composition (e.g., protein content) may fluctuate as a function of production conditions.[40,41] Moreover, in most cases the results obtained are from small culture trials and never from upscaled systems.

Some algal species are unsuitable as a diet for *Artemia* as a consequence of characteristics other than the nutritional composition: e.g., an indigestible, thick cell wall (*Chlorella*[42] or *Stichococcus*[35]), production of gelatinous substances which interfere with the ingestion mechanism *(Coccochloris*[44]*)*, or secretion of toxic components (Dinoflagellates).[23]

Good examples of intensive *Artemia* productions using an algal diet are the St. Croix Artificial Upwelling Project at the U.S. Virgin Islands,[45] and a wastewater treatment system of the Dow Chemical Co. in Texas.[44] The main restriction, however, in using this food source is its limited availability. Mass culturing of suitable algae for *Artemia* is economically not realistic so that their use only can be considered in those places where the algal production is an additional feature of the main activity. In this regard, the effluents from intensive culture ponds of fish or shrimp, or from tertiary treatment plants may offer interesting possibilities for the super-intensive production of brine shrimp.[43]

2. Dried Algae

Various commercially available algal meals have been evaluated as a potential diet for *Artemia* culturing. In most cases they give satisfactory growth performance, especially when water quality conditions are kept optimal. To date, good culture success has been reported with dried *Scenedesmus*,[47] *Spirulina*,[48-50] *Gracilaria*,[51] and *Dunaliella salina*.[104] Drawbacks in the use of these feeds are their high cost (>$12/kg), as well as their high fraction of water soluble components which cannot be ingested by the brine shrimp but interferes with the water quality of the culture medium.

3. Bacteria and Yeasts

Single cell proteins (SCP) have several characteristics which make them an interesting alternative for micro-algae, not only for *Artemia* but also for other filter-feeding organisms, e.g., phyllopods, rotifers, bivalves, and penaeid shrimp:[30]

- The cell diameter is mostly smaller than 20 μm.
- The nutritional composition is fairly complete.[52]
- The rigid cell walls prevent the leakage of water soluble nutrients in the culture medium.
- Products are commercially available at acceptable cost (e.g., commonly used in cattle feeds[53,54]).

Nevertheless, these advantages do not guarantee culture success. Although good performances are reported for the small lab cultures of brine shrimp that are fed fresh or instant baker's yeast, or alcohol yeast, upscaling tests revealed highly variable production yields. When using different types of marine yeast, survival rates upon reaching the adult stage varied between 10 and 80%.[55,56] James et al. reported yields up to 7 kg of live weight *Artemia* when feeding marine *Candida* spp.[57] With *Rhodotorula*, *Saccharomyces* and *Kluyveromyces* species, low productions were obtained in raceway systems[58] or under flow-through conditions[3], due to low survival and growth performance after day ten. The problems that arise when feeding a yeast mono-diet are often assigned to nutritional deficiencies of the yeast diet[38,59,60] and therefore should be met by supplementation with other diets. Robin et al. detected that the addition of DL-methionine, choline, or a vitamin premix to methanol yeast resulted in a significant improvement of the *Artemia* growth.[61] Experiments in a flow-through installation where mono- and mixed diets containing *Kluyveromyces* yeast ("Protibel", France) were administered to high density cultures of GSL *Artemia* confirmed this hypothesis.[62] As can be seen from Figure 1 the high mortality that starts on day 10 could be overcome and better growth rates could eventually be obtained when 50% of the "Protibel" diet is supplemented with soybean pellets or corn (maize) bran. Biomass productions increased from 5 to >10 kg/m^3 and food conversion efficiencies remained very high as well, especially during the first 10 days (FCE = 0.6; in comparison FCE for micronized diets averages 1.0).[3] Analogous data were obtained by other researchers using mixed diets of yeast and waste products from agricultural crops.[61,63-65] These results indicate that SCP indeed could be interesting ingredients of diets used in the intensive production of *Artemia*.

In some trials SCP digestibility by the *Artemia* can be a problem as well: i.e. observations on feces of *Artemia* fed on baker's yeast revealed that they were packed with undigested cells. Complete removal of the complex and thick yeast cell wall by enzymatic treatment, or supplementation of the diet with live algae significantly improved assimilation rate and growth rate of the brine shrimp.[66] These experiments clearly demonstrated that the suitability of SCP (including micro-algae) is primarily related to the digestibility of the product. A digestibility treatment for fresh baker's yeast has recently been developed: the resulting diet can be successfully applied for the complete or partial substitution of microalgae in the diet of *Artemia* or mollusc larvae.[66-68]

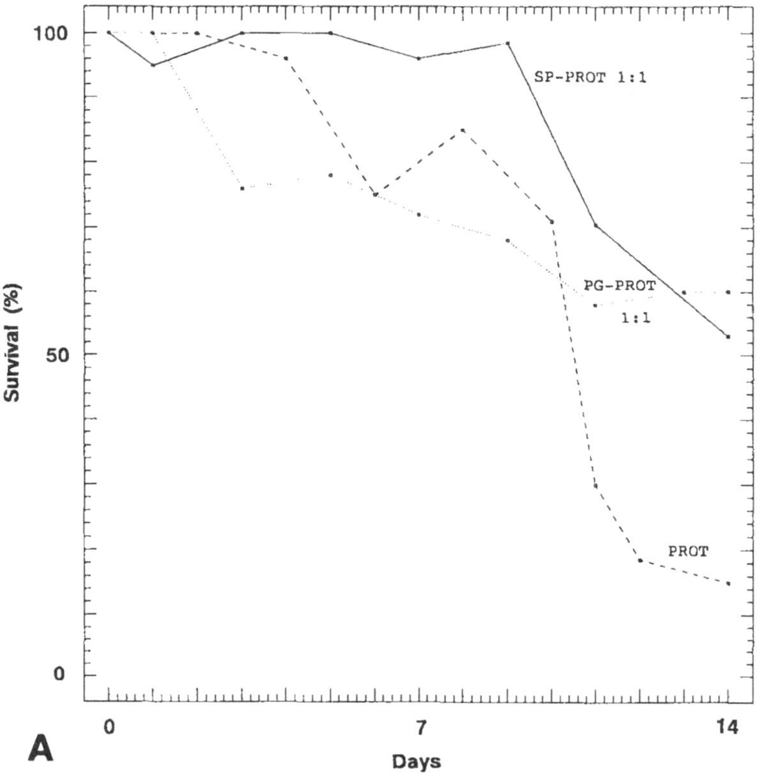

FIGURE 1. Survival (A) and growth (B) of Great Salt Lake *Artemia* cultured on Protibel-based diets. PROT, Protibel; SP, soybean pellets; PG, corn (maize) bran. (Modified from Lavens, P., Ph.D. thesis, 1989. With permission.)

According to D'Agostino[23] and Seki[69] live bacteria as a sole diet for *Artemia* seem to be unsuitable. Yasuda and Taga,[70] however, obtained good growth and survival when feeding them *Acinobacter* spp. Dried microbial detritus harvested from chitinovorous bacteria cultured on waste products, also supported good growth, but there was a low survival rate of the brine shrimp.[71]

4. Waste Products from the Food Industry

Nonsoluble waste products from agricultural crops or from the food-processing industry, e.g., rice bran, corn (maize) bran, soybean pellets, lactoserum, and sugarcane molasses appear to be a very suitable food source for high-density culturing of *Artemia*.[22] Their main advantages are the low cost and the worldwide availability. In most cases, however, these commercially available feeds do not meet the particle-size requirements for *Artemia* and an extra treatment is needed. When man power is cheap, a manual preparation which consists of blending the product in sea water followed by squeezing through a 70 μm filter bag can be used to obtain particle sizes less than 70 μm. Since this feed suspension cannot be stored, this manual method is restricted to daily preparation of the feed. This technique is successfully applied for brine shrimp culture in developing countries.[64,72] In order to reduce the manual labor in food preparations and to obtain a storable, dry product with the appropriate physical properties, mechanical techniques for dry grinding and processing need to be used. Of the different techniques examined to date, satisfactory results were only achieved with the relatively expensive micronization process: ± 80% (by weight) of the ground product has a particle size smaller than 70 μm.[3] Using micronized rice bran for *Artemia* culturing, food

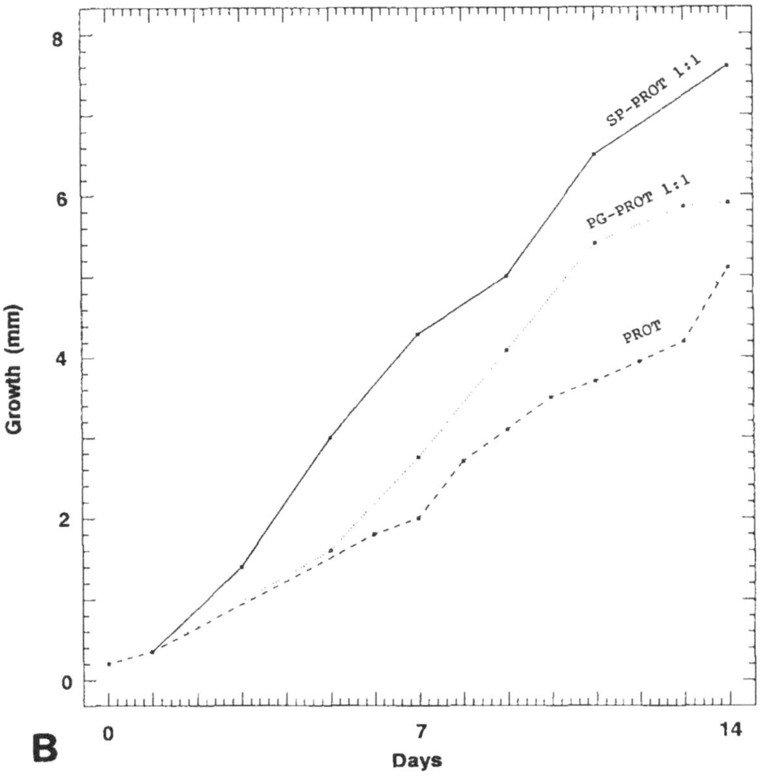

FIGURE 1B (continued)

conversion efficiencies of about 1 (ww *Artemia*/dw food) could be obtained with final productions up to 5 kg/m^3 or 25 kg/m^3 for batch[73] or flow-through conditions,[74] respectively.

Although many different waste products have been screened at the Artemia Reference Center (ARC) for application in intensive *Artemia* production, optimal results have only been obtained with diets consisting of a mixture of two ingredients: e.g., 1:1 mixtures of corn (maize) and rice bran or corn bran and soybean pellets (Figure 2). Comparison of the production data even show that there are no major differences in survival (± 60%) or body length (± 5.7 mm) among the feeding trials. This demonstrates that the biochemical composition of the diets is not of primary importance; e.g., the protein content is fluctuating between 18% and 40%. This is most probably due to the influence of microorganisms as previously explained. Comparison of the results of super-intensive production tests run at the ARC in the past decade and using waste products as diet, indicate that a high-fiber content in the food may be beneficial.

In conclusion, it may be stated that waste products from the food industry may be a good alternative to algae and SCP in the production of high densities of *Artemia*. Depending on the culture technique applied final yields can be in the order of 12 to 20 kg wet weight *Artemia*/m^3,[74,75] which is relatively close to the figures obtained with live algae.[45]

5. Microcapsules

Two publications mention the application of artificial diets under the form of microcapsules as a diet for *Artemia* culturing.[76,103] Such diets have the advantage that they should be more nutritionally complete and that no breakdown products pollute the culture medium. However, only low survival rates (20—35%) and moderate growth rates were attained after 18 d of culture.

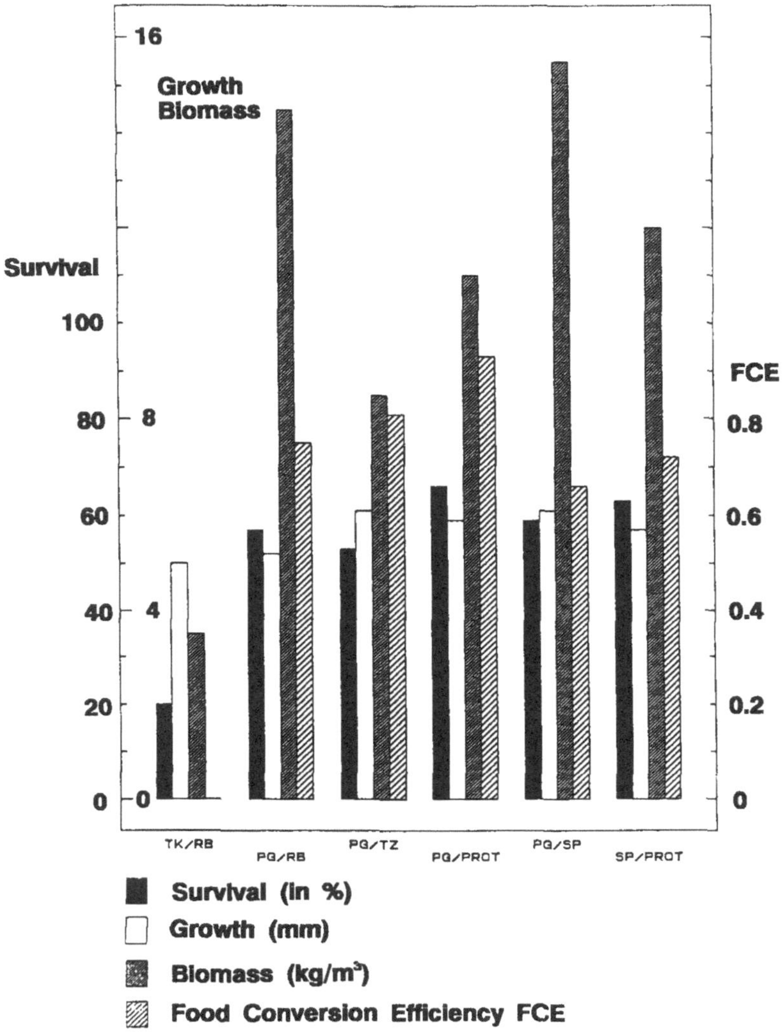

FIGURE 2. Production characteristics of super-intensive *Artemia* cultures fed different mixed diets. RB, rice bran; TK, wheat flour; PG, corn (maize) bran; PROT, Protibel yeast; SP, soybean pellets. (Modified from Lavens, P., Ph.D. thesis, 1989. With permission).

D. STRAIN SELECTION

Strain selection is not only important as a function of culture conditions (e.g., temperature; see Table 1), production yields can also vary from strain to strain. A comparative study with different *Artemia* strains fed two types of food (live *Dunaliella* algae and rice bran) demonstrated that differences in growth and survival may be important (see Tables 1 and 4). Under standard laboratory conditions GSL and Galerazamba, Colombia (GZ) brine shrimp produced about 50% more biomass than the reference strain from San Francisco Bay (SFB) origin. These results from small-scale experiments are confirmed in intensive production systems: highest biomass productions were always attained with GSL and GZ-populations cultured either in raceway systems[77] or in flow-through installations.[3,78]

E. DENSITY OF *ARTEMIA*

The growth rate of *Artemia* is not much affected by the density of animals in the culture

TABLE 4
Survival and Growth of Different *Artemia* Strains Cultured under Standard Laboratory Conditions with *Dunaliella* (DUN) or Rice Bran (RB) as Food for a Period of 7 Days[a]

Geographical strain	Survival at Day 7 (%)		Growth expressed as % recorded for reference strain San Francisco Bay (288-2596)	
	DUN diet	RB diet	DUN diet	RB diet
Larnaca Salt Lake (Cyprus)	70[c]		88[c]	
Santa Pola (Spain)	76[c]		88[c]	
Salin du Giraud (France)	66[c]		90[c]	
China (unknown locality)	84	83	91[c]	
Shark bay (Australia)	90(84)[b]	51[c]	95(96)	89[c]
San Francisco Bay (CA batch No 236-2016)	98		96	
Macau (Brazil, harvested in March 1978)	84(96)	90	96(98)	102
Barotac Nuevo (Philippines)	86	86	97	103
Tientsin (China)	96	86	98	
San Francisco Bay (CA batch No 933-235)	94		99	
Aigues Mortes (France)	90		99	
Izmir (Turkey)	96	88	101	108
Margherita di Savola (Italy)	96	88	102	99
San Felix (Spain)	74[c]		102	
Macau (Brazil, harvested in May 1978)	94		103	
Bonaire (Netherlands Antilles)	66[c]	90	104	104
San Pablo Bay (CA batch No 1628)	90	92	105	106
Port Araya (Venezuela)	90	81	107	112
Eilat (Israel)	92[c]		107	
Lavalduc (France)	70[c]	93	109	96
Adelaide (Australia)	88(88)		113[c](113)	
Manaure (Colombia)	90	85	115[d]	129[d]
Bahia Salinas (Puerto Rico)	88	74[c]	122[d]	122[d]
Great Salt Lake (UT harvested in 1977)	94	86	119[d](125)	115[d]
Buenos Aires (Argentina)	72[c]	80	126[d]	124[d]
Galerazamba (Colombia)	98	87	126[d]	133[d]
Great Salt Lake (UT harvested in 1966)	66[c]		127[d]	
Chaplin Lake (Canada)	88	77	130[d]	130[d]

[a] Data compiled from Vanhaecke and Sorgeloos [9] and Vanhaecke *et al.*[8]
[b] In parenthesis result of replicate test.
[c] significantly different from the strain at the 0.01 level.
[d] significantly different from the strain at the 0.05 level.

medium. Depending on the culture technique applied, inoculation densities up to 5,000 larvae/l for batch culture,[73] 10,000 for closed flow-through culture,[75] and 18,000 for open flow-through culture[45] can be maintained without affecting survival and growth. These so-called maximum densities are in fact not critical as there is no real effect on behavior as is the case for most other anostracans, copepods, etc. However, each culture has its maximum carrying capacity: above these optimal densities, culture conditions become suboptimal (e.g., water quality deterioration and lower individual food availability). This results in reduced growth and increased mortalities during the second culture week (Table 5). Below the maximum carrying capacity there is no effect on biomass production but length and individual weight are significantly higher at lower densities.[3,73]

TABLE 5
Production Characteristics of Great Salt Lake *Artemia* Cultures as a Function of Different Animal Densities at Start of Culture[a]

Density (indiv. per ml)	Survival (%)			Growth (mm)			Biomass (in g wet weight/l)	Food Conversion Efficiency (ratio g dry food: g wet weight biomass)
	day 7	day 10	day 14	day 7	day 10	day 14	day 14	day 14
5	82	77	58	3.3	4.7	6.1	15.5	0.66
8	66	50	56	3.2	4.8	5.2	14.5	0.75
10	95	73	72	1.8	2.9	3.5	11.6	0.67
13	62	32	28	-	-	-	8.0	1.13
15	55	41	34	2.0	2.9	3.8	10.0	1.17
18	58	37	29	-	-	-	5.8	1.34
40	74	12	-	1.5	2.3	-	-	-

[a] Data compiled from Lavens.[3]

III. TECHNOLOGICAL ASPECTS OF MASS CULTURING

Although high-density culturing of brine shrimp is relatively simple compared to other aquatic organisms, several technological conditions are needed to ensure a functional culture system. First, an efficient agitation of the culture medium is essential for an even distribution of *Artemia* in the water column (and consequently a maximal utilization of the total culture volume), for a homogenous suspension of the food particles, and for a sufficient supply of oxygen in the medium. This agitation should not interfere with survival or growth. This is determined by the shape and surface/volume ratio of the tank. Different circulation systems have been described: paddle-wheels in shallow basins,[80] air-lines or aeration collars in high cylinders, rectangular tanks or cylindroconical recipients,[45,47,81] and air-water-lifts operating in a raceway concept.[77]

Optimal results have been obtained with rectangular tanks equipped with either an air-water-lift operating system (volumes up to 10 m^3) (Figure 3), or a central standpipe and aeration collar for tank sizes of maximum 1 m^3 (Figure 4).

The construction and operation of a raceway tank is simple. The most important parameter for the configuration of the tank is the height/width ratio, which should be close to 1/2 (Figure 5). If axial blowers are used for aeration, the water depth should not exceed 1.2 m to assure optimal water circulation. The corners of the tank may be curved to prevent dead zones where sedimentation can take place. A partition, to which air-water-lifts (AWLs) are fixed, is installed in the middle of the tank to assure a combined horizontal and vertical movement of the water which results in a screw-like flow pattern. AWLs are constructed with PVC pipes and elbows; the lower part of the tube which stands on the bottom of the raceway is cut at an angle of 45°, and the elbow's outflow is half-submerged (Figure 6).

An aeration line from a central air-distribution cylinder extends almost to the bottom of each AWL in order to assure optimal water lift effects. Although the agitation by this type of circulation systems is very intense, neither survival nor production of the brine shrimp are negatively affected.[14]

Second, the culture system should be designed in such a way that the most critical parameter, i.e., water quality, can be maintained as optimal as possible. This means that the concentration of particles and soluble metabolites should remain minimal so as to prevent toxicity problems, proliferation of micro-organisms, and interferences with the filter-feeding

FIGURE 3. Air-water-lift operated raceway for super-intensive *Artemia* production following the stagnant culture principle (culture volume of 5 m^3).

FIGURE 4. 300-l *Artemia* culture tank equipped with standpipe and aeration collar for super-intensive flow-through culturing.

system of the brine shrimp. Flow-through culture techniques are suitable for maintaining optimal water quality since water is (semi-)continuously renewed, whereas stagnant systems (batch) need a particle separation unit in order to reach sub-optimal levels of water quality.

Of the various techniques used for growing *Artemia* we will only describe those configurations that have already proven to be suitable for commercial exploitation or for inte-

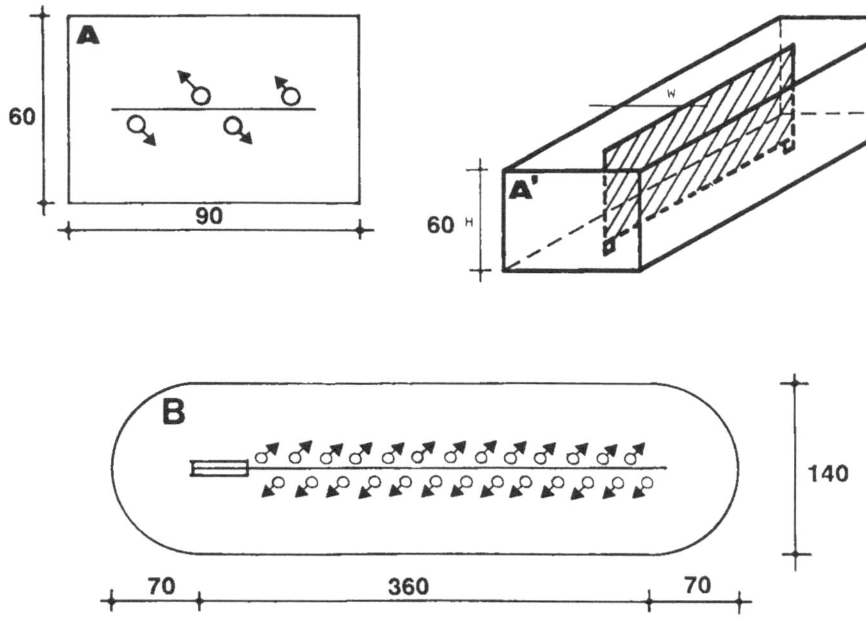

FIGURE 5. Schematic views and dimensions of raceway systems for *Artemia* culturing. (A) 300-l tank: diameter AWL 4 cm, ± 7 l air/min/AWL;(B) 5 m^3 tank: 80 cm water depth, diameter AWL 5 cm, ± 16 l air/min/AWL. (Modified from Bossuyt, E. and Sorgeloos, P., *The Brine Shrimp Artemia*, Vol. 3., Persoone, G., Sorgeloos, P., Roels, O., and Jaspers, E., Eds., Universa Press, Wetteren, Belgium, 1980, 133. With permission.)

gration into existing aquaculture activities. The type of installation will be subject to local conditions, production needs, and investment possibilities.

A. FLOW-THROUGH CULTURE TECHNIQUES

It is obvious that a discontinuous or continuous renewal of culture water, with consequent dilution of particulate and dissolved metabolites, will result in optimal culture conditions and thus high production capacities. Application of a flow-through culture technique, however, is limited to those situations where large volumes of sufficiently warm sea water (or brine) are available at relatively low cost — areas with (sub-)tropical climate or thermal effluents from power plants, desalination plants, and geothermal wells — and/or where large quantities of food are available, e.g., effluents from artificial upwelling projects, tertiary treatment systems, and intensive grow-out ponds of shrimp. These restrictions can also be overcome by the integration of a recirculation system which mechanically and biologically purifies the culture effluent so that finally it can be recycled to the *Artemia* tanks.

The most important and critical equipment in flow-through culturing is the special filter system used for efficient removal of excess culture water and metabolites without loosing the brine shrimp from the culture tank. Furthermore, it is essential that these filter-units operate without clogging for periods of more than 24 h.

Initially, standpipe filters were used which consist of a cylindrical PVC-frame around which an interchangeable nylon screen cylinder is fixed.[45,78] An aeration collar, positioned at the bottom of the filter, ensures a continuous friction of a curtain of air bubbles against the sides of the filter screen which results in an efficient reduction of filter-mesh clogging (Figure 7A). This filter has proven to be suitable for culture systems where live algae are fed to the *Artemia* population.[45]

However, when working with micronized waste products, we experienced much faster

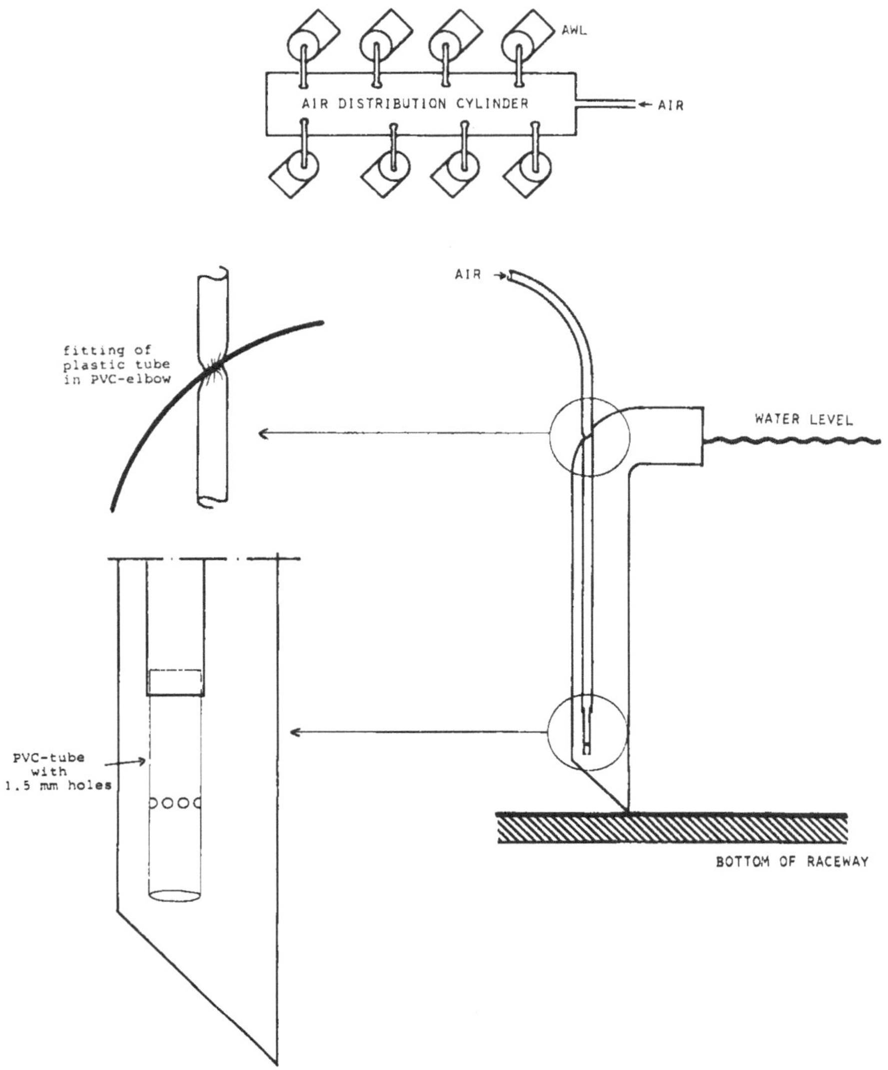

FIGURE 6. Details on the installation of AWL in raceways. (Modified from Bossuyt, E. and Sorgeloos, P., *The Brine Shrimp Artemia*, Vol. 3., Persoone, G., Sorgeloos, P., Roels, O., and Jaspers, E., Eds., Universa Press, Wetteren, Belgium, 1980, 133. With permission.)

clogging of the filter screen, especially for the smaller mesh sizes, and the design of the standpipe had to be adapted.[74] A PVC-frame was constructed with all stand-up sides inclined under an angle of 10° and over which the filter-screen bag of a specific mesh size (depending on the size of the *Artemia*) can be fitted (Figure 7B). In this arrangement, the rising air bubbles have a very intensive contact with the screen, ensuring a more efficient cleaning.[74] The upper part of the filter bag which is positioned just above and underneath the water level is made of smooth nylon cloth to prevent any trapping of the cultured larvae that are foamed-off by the effect of the strong aeration.

A new type of cylindrical filter system (Figure 8) has recently been introduced for still more efficient removal of the effluents from brine shrimp culture tanks.[75] It consists of a welded-wedge screen cylinder, made of stainless steel and vertically placed in the center of the culture tank. Its bottom and top part extend into PVC-rings, and an aeration collar is

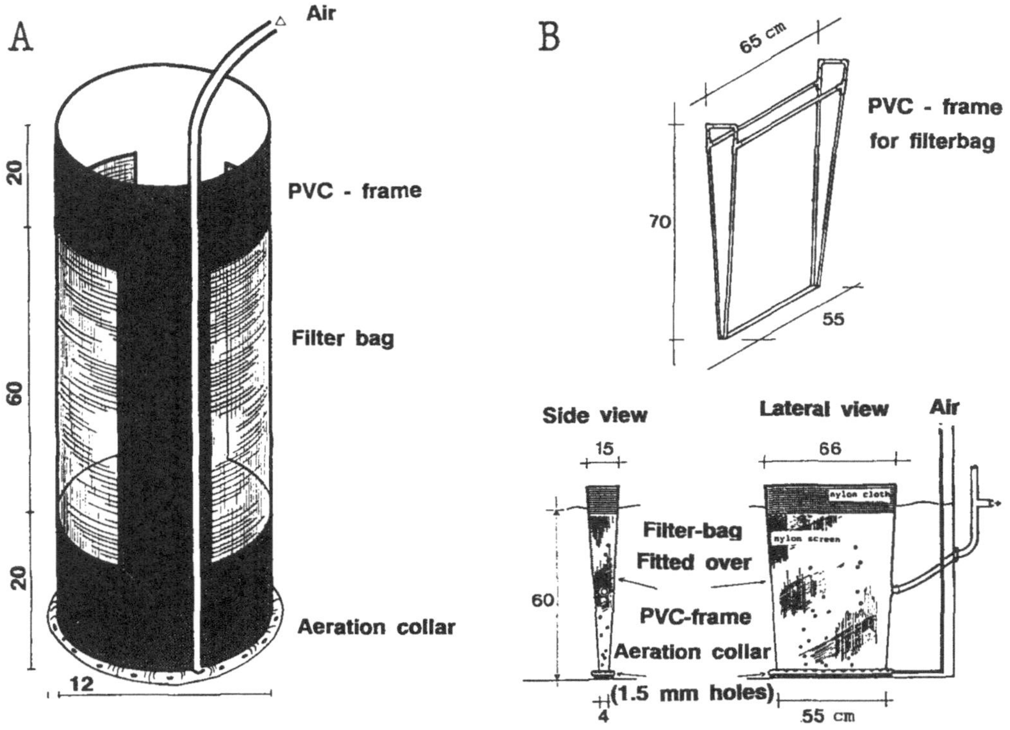

FIGURE 7. Schematic views and dimensions of filter systems used in flow-through culturing of *Artemia*. (Modified from Brisset, P., Versichele, D., Bossuyt, E., De Ruyck, L., and Sorgeloos, P., *Aquacult. Eng.*, 1, 115, 1982; Sorgeloos, P., Lavens, P., Léger, P., Tackaert, W., and Versichele, D., *Manual for the Culture and Use of Brine Shrimp Artemia in Aquaculture*, Artemia Reference Center, State University of Ghent, Belgium, 1986. With permission.)

fixed to the lower end of the filter. This welded-wedge system has several advantages in comparison to the filter-screen types:

- Oversized particles with an elongated shape can still be evacuated through the slit openings.
- The specific V-shape of the slit-openings creates specific hydrodynamic effects as a result of which filter particles that are only slightly smaller than the slit-opening itself are sucked through.
- Possible contact points between particles and filter are reduced to two instead of four, which consequently reduces the clogging probability.

Comparative tests in the ARC flow-through recirculation system[3,75] revealed that this new filter can be operated autonomously for much longer periods than with traditional mesh filters. Proportionally smaller welded-wedge filters can be used, leaving more volume for the animals in the culture tank. Furthermore, they are very cost effective since they do not wear out.

As brine shrimp grow, the filters are progressively switched for one with a larger mesh—or slit-opening—in order to achieve a maximal evacuation of molts, feces, and other waste particles from the culture tanks. A set of filters covering a 14-d culture period should consist of approximately six different slit/mesh-openings starting from 120 μm to 150 μm up to 350 μm and 550 μm, respectively (Table 6).

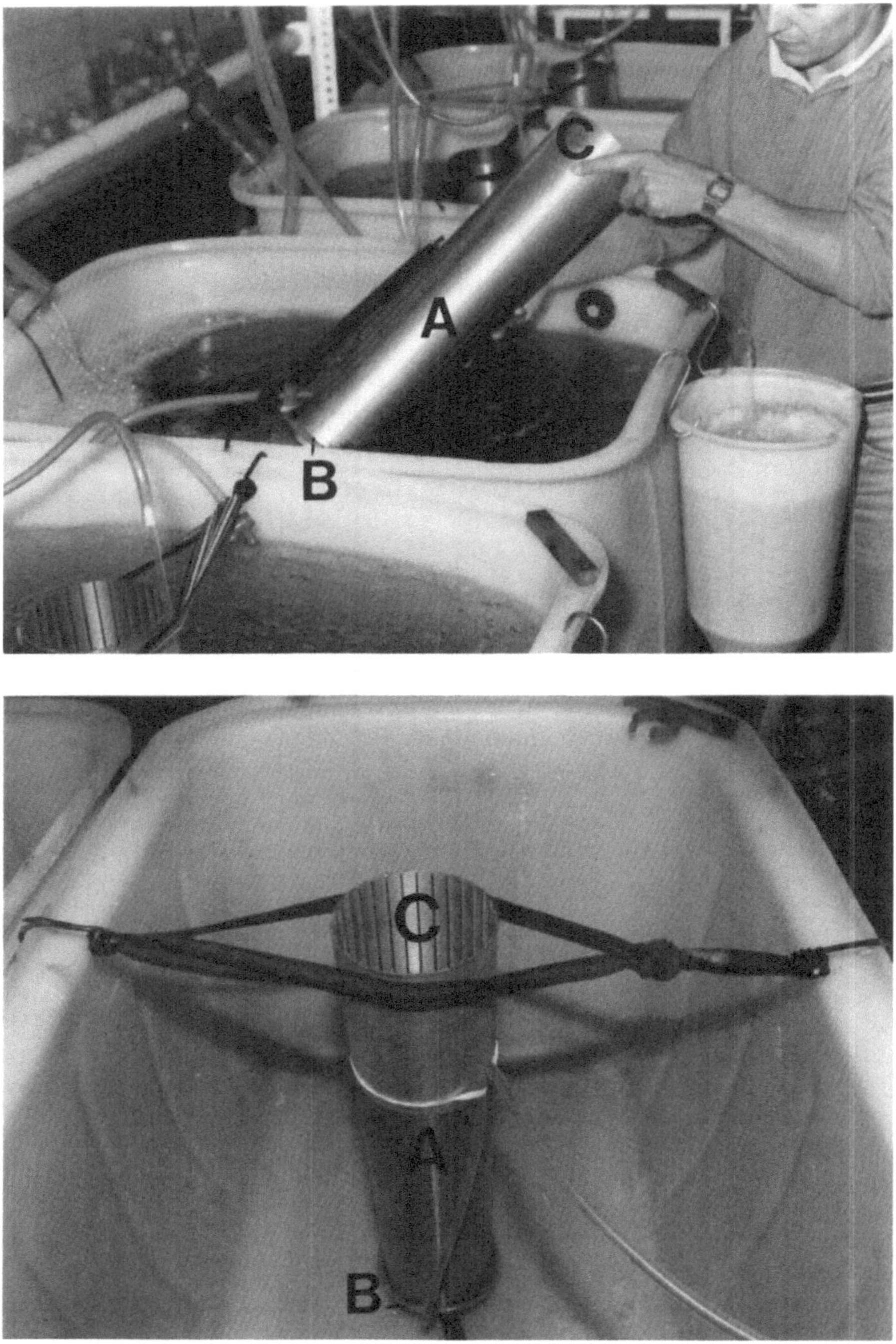

FIGURE 8. Overview of new filter system for use in flow-through *Artemia* culturing; (A) stainless-steel welded-wedge screen cylinder, (B) aeration collar, (C) upper PVC-ring.

The water-flow intake, and thus the hydraulic retention time in the culture tank, need to be adjusted in relation to the filter size used and the overall culture conditions in order to reach an optimal compromise between efficient evacuation of wastes and minimal food losses. An optimal regime applied for a 300-l rectangular tank (see Figure 4) equipped with welded-wedge filters of 65 cm high and 14 cm in diameter is outlined in Table 6. The

TABLE 6
Typical Example of Rate of Exchange of Filters, Culture Water Renewal and Feeding Regime Applied in a 300 L Super-Intensive *Artemia* Culture[a]

Culture day	Slit opening of filter (μm)	Flow rate (l/hr)	Retention time in culture tank (hr)	Feeding regime (period between distributions of 2.5 g each) (min)	Food added per day (g)
1	120	80	3.75	36	100
2	150	100	3	30	120
3 - 4	200	100	3	24	150
5 - 7	250	150	2	20	180
8 - 9	300	150	2	20	180
10 - 12	350	200	1.5	15	240
13 - 14	350	300	1	12	300

[a] Data compiled from Lavens.[3]

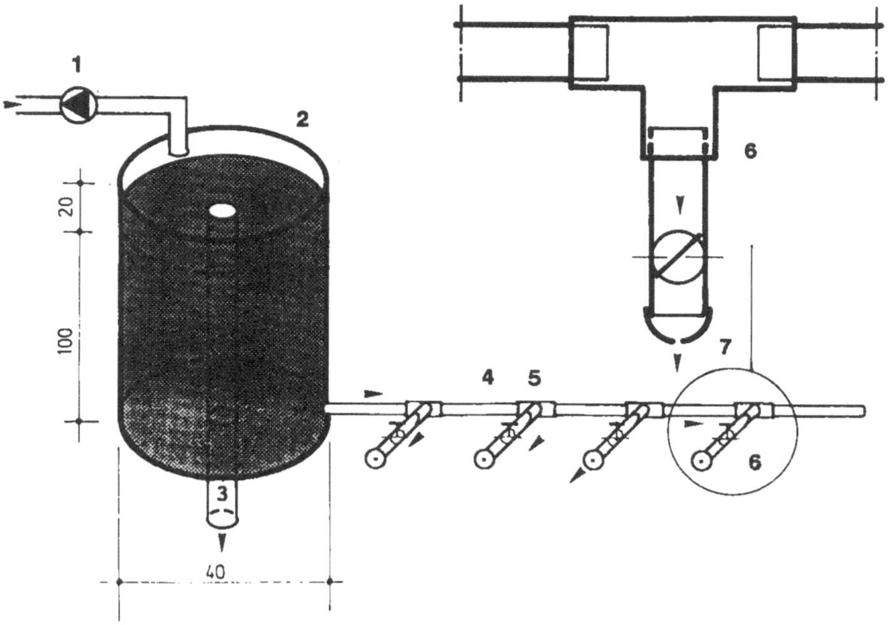

FIGURE 9. Schematic view of a water-inlet system assuring precise control of the flow rates to the culture tanks. (1) water pump; (2) constant head tank with (3) overflow; (4) supply pipe with taps (5); (6) PVC-caps with calibrated outlet opening (7). (Modified from Sorgeloos, P., Lavens, P., Léger, P. Tackaert, W., and Versichele, D., *Manual for the Culture and Use of Brine Shrimp Artemia in Aquaculture*, Artemia Reference Center, State University of Ghent, Belgium, 1986. With permission.)

installment of the flow rate can easily be done by using interchangeable PVC-caps which fit onto the PVC-water-supply pipe connected to a constant head tower, and which have a calibrated outlet (Figure 9). Depending on the size of the installed outlets a preset flow rate can be administered.

Since the continuous water addition will dilute the amount of food in the culture tank (provided no food is taken in with the influent water), fresh food needs to be added frequently in order to maintain optimal food particle concentrations for growth. This can be achieved with water pumps connected to the feed tank and activated by a preset electronic timer. A properly functioning system for food distribution is outlined in Figure 10; its main component

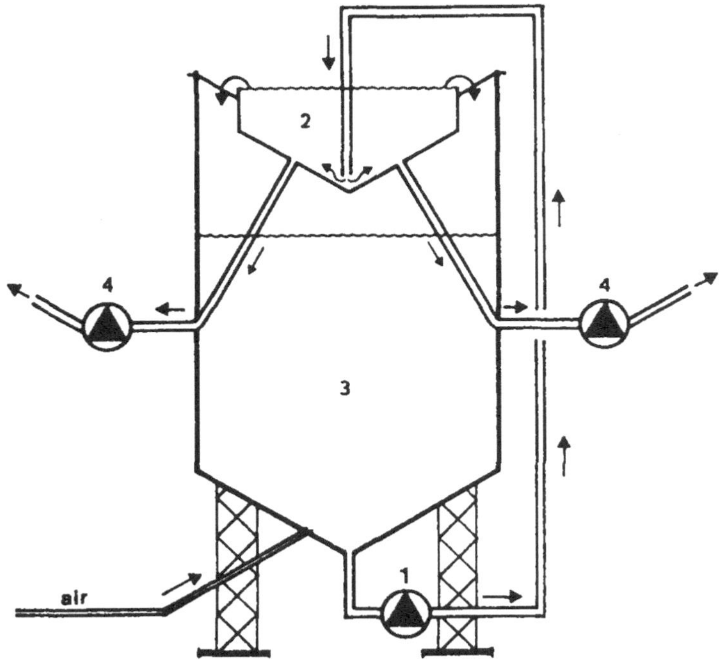

FIGURE 10. Schematic view of food distribution system for use in super-intensive *Artemia* culturing. Food is continuously pumped (1) into the constant head tank (2), the overflow of which drains into the main reservoir (3); individually activated by electronic timers, pumps (4) distribute feed suspension to the respective culture tanks. (From Lavens, P., Baert, P., De Meulemeester, A., Van Ballaer, E., and Sorgeloos, P., *J. World Aquacult. Soc.*, 16, 498, 1985. With permission.)

is a small constant head which assures equal food additions at any moment of the day. Pumping rates are adjusted daily to maintain constant levels of food in the culture tank (Table 6).

Using agricultural by-products as the diet source, the transparency of the culture medium can be used as a very useful parameter for the determination of the optimal food dose.[22,77] For the corn (maize) bran/soybean pellet (4:1) diet, transparency levels of 20 to 25 cm appear to be optimal for juvenile *Artemia*.[75] The transparency measurements can easily be performed with a modified Secchi-disk.[77] A simple electronic transparency meter (Figure 11) can be connected to the effluent tube of the culture tank to monitor possible overfeeding conditions. When the food level drops below a preset critical value, the apparatus will shut off the food distribution pump. Accurate functioning of the transparency meter is ensured by the continuous water flow through the apparatus, together with a regular air injection close to the light emitter, which limits sedimentation and clogging.

However, for monitoring adult cultures of brine shrimp a different feeding strategy should be applied. Molting rates in older *Artemia* gradually decrease as a result of which the setae of the thoracopods become easily clogged with food, especially at high particle densities. This affects respiration and ingestion, eventually resulting in decreased survival rates. Once animals reach the preadult stage, best production yields are obtained by gradually switching from a transparency controlled food distribution (see above) to a feeding scheme of about 10% (dry weight feed to live weight *Artemia*) per day distributed on a semi-continuous basis (every 5 min).[82]

Overfeeding can be prevented by the installation of a simple alarm system which controls

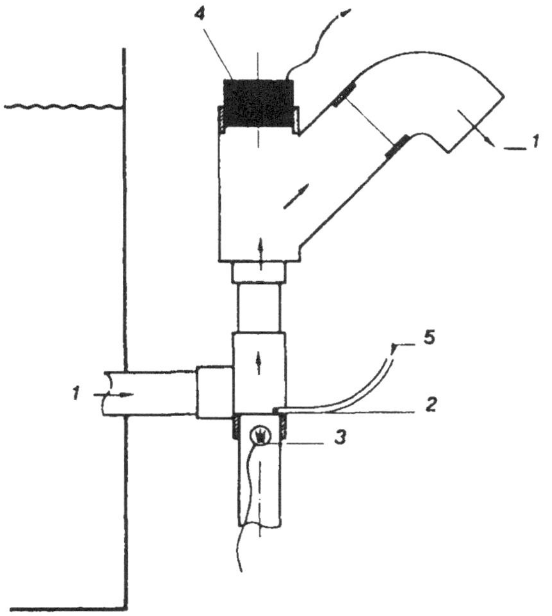

FIGURE 11. Schematic view of electronic transparency meter connected to the outlet of a super-intensive flow-through *Artemia* culture. (1) drain of culture water effluent; (2) transparent PVC-plate; (3) light bulb; (4) light receptor connected to electronic detector; (5) air injection. (From Lavens, P., Baert, P., De Meulemeester, A., Van Ballaer, E., and Sorgeloos, P., *J. World Aquacult. Soc.*, 16, 498, 1985. With permission.)

the water flow to the culture tanks eventually deactivating the automatic food distribution in case of reduced flow rates.

Recently, a very simple semi-flow-through system has been successfully developed by Bombeo et al.[83] The low flow rates applied (retention times of minimum 12 h) are easily adjusted by means of a siphon with a selected diameter which drains the stock tank by gravity. This latter tank holds culture water and food, and needs to be filled only once a day. This technique involves minimal sophistication and appears to be very predictable in production yields which are in between those obtained in batch and flow-through systems.

B. CLOSED SYSTEMS FOR FLOW-THROUGH CULTURING

When only limited quantities of warm sea water are available, high-density flow-through culturing of *Artemia* can be maintained via a recirculation unit. Although there has been considerable research for developing recirculation systems, the construction of a predictable culture system for marine organisms in closed circuits today is still more an art than a science.[84-85] The following description is an example of an operational recirculating system for *Artemia* culture effluent, as developed at the ARC[75] (Figure 12).

First, the effluent is drained into a small decantation tank (120-1) in which a spiral flow pattern is created by a rolled PVC-plate. After this initial purification step, water flows into a rotating biological contactor (RBC) for biological treatment, i.e., bacterial degradation of soluble organic wastes via oxidative deamination, and nitrification into nitrate. Biodisc purification has been selected for our pilot installation because of its stable operation under conditions of fluctuating hydraulic and organic loadings. It consists of three units of 40 sand-blasted PVC-discs (100 cm in diameter) which rotate at 6 rpm in the interconnected

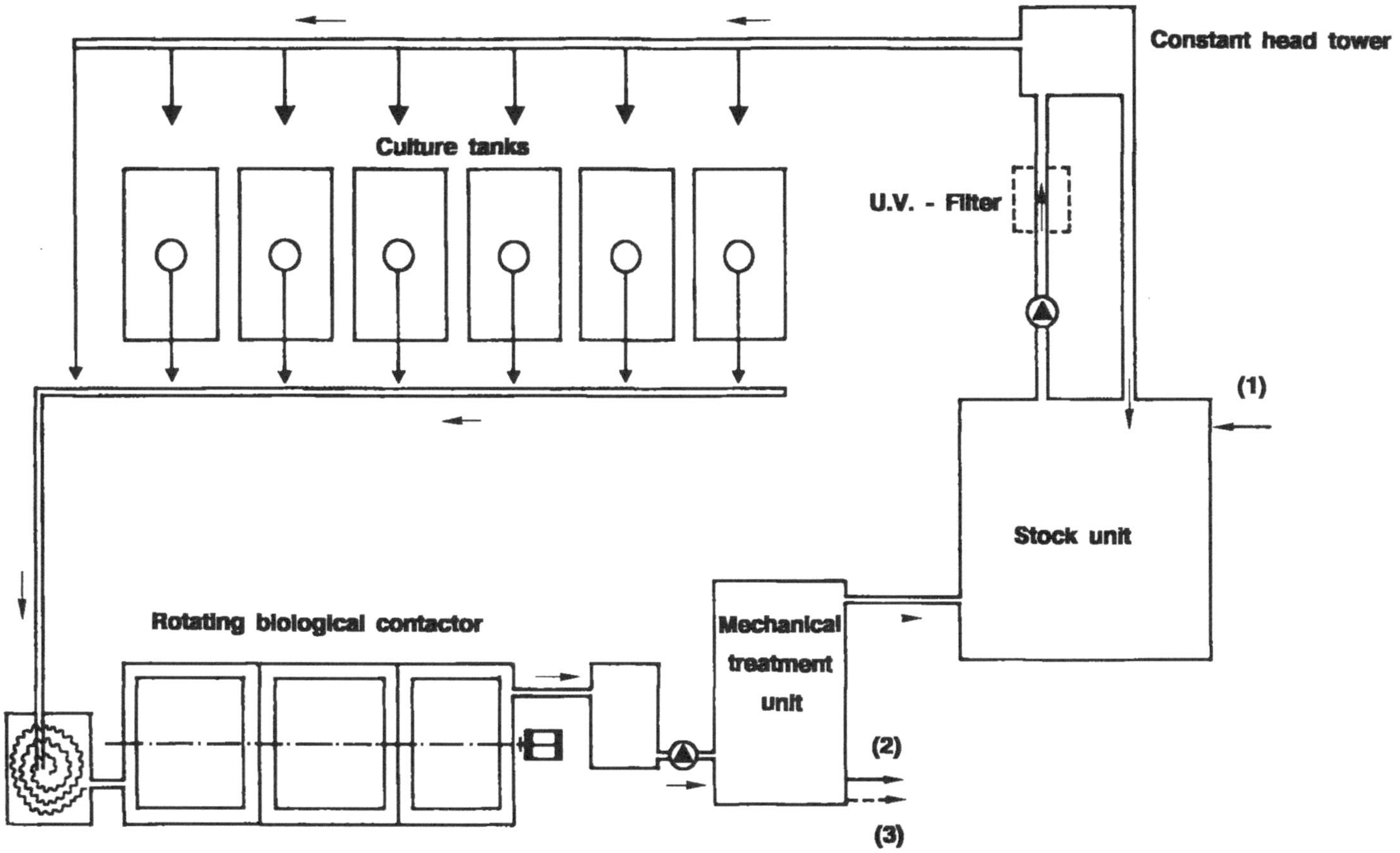

FIGURE 12. Schematic diagram of set-up for closed flow-through culturing of *Artemia*. (1) fresh seawater addition; (2) drain of water; and (3) suspended solids. (From Lavens, P., Baert, P., De Meulemeester, A., Van Ballaer, E., and Sorgeloos, P., *J. World Aquacult. Soc.*, 16, 498, 1985. With permission.)

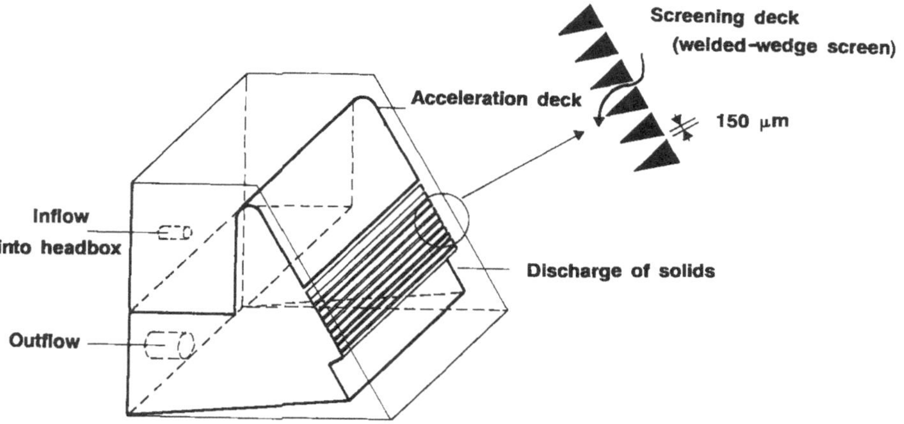

FIGURE 13. Cross-flow sieve system for concentrating suspended particles from *Artemia* culture effluent. (Modified from Bossuyt, E., and Sorgeloos, P., *The Brine Shrimp Artemia*, Vol. 3., Persoone, G., Sorgeloos, P., Roels, O., and Jaspers, E., Eds., Universa Press, Wetteren, Belgium, 1980, 133. With permission.)

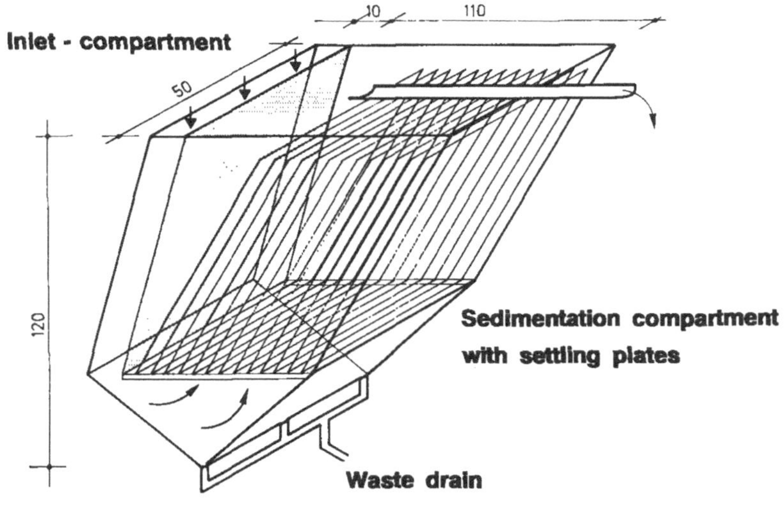

FIGURE 14. Schematic view and dimensions (in cm) of plate separator (650 l) used for primary treatment of 2 m^3 *Artemia* raceway. (Modified from Sorgeloos, P., Lavens, P., Léger, Ph., Tackaert, W., and Versichele, D., *Manual for the Culture and Use of Brine Shrimp Artemia in Aquaculture*, Artemia Reference Center, State University of Ghent, Belgium, 1986. With permission.)

compartments. Total effective surface for bacterial growth is 190 m^2 which is more than sufficient for the purification of 1800-l of culture water with an organic load of about 30 mg/l BOD_5.

The biodisc effluent is pumped over a cross-flow sieve[77] (200 cm^2) with a slit opening of 150 μm (Figure 13). The concentrated sludge is subsequently drained in a plate separator, while the purified water is directly returned to the reservoir. In this way the volume of the plate separator could be reduced 650-l. Its construction is a modification of the plate separator developed by Mock et al.[86] and Bossuyt and Sorgeloos.[77] A schematic drawing and the characteristics of the decantor system are given in Figure 14. It consists of an inclined tank subdivided into a small inflow and a large settling compartment interconnected by an open space at the conical-shaped bottom part. In the settling compartment 20 sand-blasted PVC-plates with a total surface area of 5.54 m^2 are mounted in an inclined position (60°) oriented in the direction of the water flow through the separator tank. A drain fixed on top of the plates finally evacuates the clean water to the stock tank.

An optimal retention time of ~15 min assures maximal sedimentation of the waste particles. Once every three days the accumulated sludge should be drained by siphoning from the bottom of the decantor.

A last step in the purification process may be disinfection. We installed a UV unit, 4 × 50 W. However, there is no absolute proof of its effectiveness in the culture of *Artemia* (or other organisms), nor does it result in a completely sterile medium.[87] Finally, part of the water should be renewed at regular time intervals. In our system we apply an arbitrarily choosen 25% culture water exchange on a weekly basis.

Using this recirculation system for our *Artemia* culture unit of six 300-l tanks, water quality could be maintained at acceptable levels; i.e., the biological and mechanical treatments result in an effluent containing less than 5 ml/l BOD_5[75] and 22.5 mg/l TOC[3], which accounts for a removal percentage >80%. The levels for NH_4^+-N and NO_2^--N remained <10.9 ppm and 0.6 ppm, respectively, which is far below the tolerance limits for *Artemia*.[3] Levels for suspended solids did not exceed 380 mg/l.

C. STAGNANT CULTURE SYSTEMS

Of the various techniques that have been tested for growing brine shrimp larvae in stagnant culture systems, the air-water-lift operated raceway has proven to be the most suitable.[77] With the exception of the water exchange, the same culture conditions can be maintained as described for flow-through systems. However, automation is more difficult, and monitoring, especially with regard to feeding conditions, more critical.

The first days of culture do not pose any problem and optimal feeding levels can easily be maintained. However, on the fourth day the water quality deteriorates quickly and waste particles — e.g., fecal pellets, food aggregates, and exuviae — physically hamper food uptake by the *Artemia* and interfere with the transparency measurements. The only solution is to install a primary treatment by means of a plate separator (see above) connected to the raceway. This also implies the use of filters to retain the *Artemia* in the culture tanks. Clogging of these filters is less problematic here than in flow-through culturing since no extra water is added and not overflow risks exist. Flow rates and plate separator dimensions are function of the volume of the raceway; optimal particle removal is assured when the culture medium passes over the plate separator at least eight times a day.

Besides the elimination of particulate wastes, it may also be desirable to partially remove the soluble fraction from the culture medium, especially when feeding products that are rich in proteins or which contain a high fraction of soluble material. An easy technique to incorporate in the batch culture design is foam fractioning.[88] A schematic drawing of the foam tower that we used for the final treatment of the effluent of a 2 m^3 raceway is provided in Figure 15.

IV. CONTROL OF INFECTIONS

When experimenting with micronized agricultural by-products as feed for *Artemia* cultures we occasionally found heavy losses of preadults due to infections with filamentous *Leucothrix* bacteria.[3,75] According to Solangi et al.[89] these infections especially occur in nutrient rich media. The *Leucothrix* colonies fix on the exoskeleton, by preference on the thoracopods, and only become visible from the Instar V/VI stage onwards. The brine shrimp suffer physically, as the movements of their thoracopods become affected and filtration rates consequently are reduced. At the upper limit, growth and molting are arrested, overfeeding of the tanks occur, resulting in a collapse of the *Artemia* culture.

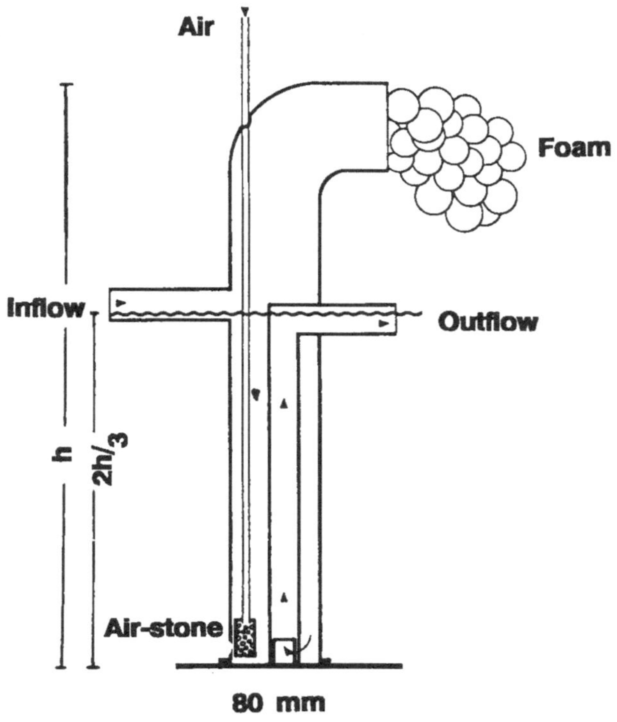

FIGURE 15. Schematic diagram of a foam separator. (From Sorgeloos, P., Lavens, P., Léger, P., Tackaert, W., and Versichele, D., *Manual for the Culture and Use of Brine Shrimp Artemia in Aquaculture*, Artemia Reference Center, State University of Ghent, Belgium, 1986. With permission).

Solangi et al. recommend the application of terramycin;[89] however, antibiotics cannot be used in recirculation systems as they will affect the biological treatment unit. We have explored other techniques that are frequently used in urban waste water treatment to eliminate filamentous microorganisms. Frequent additions of low doses of peroxide (up to 50 ppm H_2O_2) proved to be effective in reducing or eliminating the *Leucothrix* populations without affecting the brine shrimp. The most practical solution appeared to be an increase in salinity from 35 to 50—60 g/l together with a higher water renewal rate of 25% instead of 10% on a weekly basis.[3,75]

A second observed disease in *Artemia* cultures is the so-called "black disease"; i.e., part of the animals show black spots, especially at their extremities (thoracopods and antennae). (See Figure 16). According to Hernandorena[90,91] this disease consists of the detachment of the epidermis from the cuticula, and is caused by a dietary deficiency which interferes with the lipid metabolism. In high density culturing of *Artemia* using agricultural by-products as food source, we observed the black disease when water quality deteriorates (probably interfering with the composition of the bacterial population and consequently the diet composition) and/or when feeding rates are not optimal. Improving these conditions doesn't save the affected animals but appears to avoid further losses.

V. PRODUCTION YIELDS

A. HARVESTING AND PROCESSING TECHNIQUES

Harvesting of high-density cultures of *Artemia* can be facilitated by taking advantage of the surface respiration behavior of the animals. When the aeration in the culture tank,

FIGURE 16. Microscopic view of *Artemia* preadult with "black disease" spots on its thoracopods.

together with the flow-through and the automatic feeding are interrupted, oxygen levels in the water drop quickly and all waste particles sink to the bottom; after about 30 min the *Artemia* respond to the oxygen depletion by concentrating at the water surface where they perform surface respiration (i.e., moving the thoracopods at the interface air-water). The concentrated population, free from suspended solids, can easily be scooped with a net of an appropriate mesh size. At low animal densities brine shrimp can be harvested by draining the complete culture over a sieve which should be partially submerged under water.

After a thorough wash in fresh water or sea water, the harvested *Artemia* can be offered as a highly suitable live food for freshwater, as well as marine predators. The salinity of the culture water is of no concern as *Artemia* are hypo-osmoregulator, i.e., their body fluids have a constant and low-salt content of about 9 g/l, as a result of which they are an acceptable food for freshwater animals as well. Furthermore, when transferred into freshwater *Artemia* will continue to swim for another 5 h after which time they eventually die as a result of osmoregulatory stress. In sea water, they remain alive without feeding for several days.

Transportation of live brine shrimp can be done in plastic bags containing cooled sea water and inflated with oxygen. Details about this technique as well as the freezing of *Artemia* biomass are described in Chapter 12 on pond production.

B. PRODUCTION YIELDS

Figure 17 provides a summary of average production data expressed in *Artemia* survival and length, obtained in the different culture systems described in this chapter. After 2 weeks of culturing, preadult or adult *Artemia* with an average length of 5 mm or more can be harvested. Under flow-through culture conditions significant mortality occurs in the first week, probably because the early naupliar stages are more sensitive than the juvenile or preadult stages. This mortality can be reduced by offering a better food, e.g., when feeding live *Chaetoceros* in the St. Croix upwelling system, almost no mortality occurred during the initial culture period.[45,78] No significant differences in survival are noticed between open or closed (with recirculation of the culture medium) flow-through cultures. In stagnant culture installations, on the other hand, survival decreases continuously into the second week which can be explained by the deterioration of the water quality.

Average production yields in terms of ww biomass harvested after 2 weeks culture, amount to 5 kg, 15 kg, and 25 kg *Artemia*/m^3 of tank volume for batch production[73] and flow-through systems using micronized feeds,[74,75] or live algae, respectively.[45] These differences in production figures are primarily the result of differences in maximum stocking density at the start, and final survival at the end of the culture trial.

VI. NUTRITIONAL QUALITY AND USE OF TANK-PRODUCED *ARTEMIA* JUVENILES AND ADULTS IN AQUACULTURE

The opportunities for feeding ongrown *Artemia* to numerous fish and crustacean larval species are well documented in literature.[25] However, until recently, these applications were never utilized at the industrial level because of the limited availability of live or frozen biomass, its high cost, and its variable quality.[2] The possibilities for constructing intensive and super-intensive production facilities for brine shrimp in or nearby the aquaculture farm have resulted during the last decade in increased interests with regard to *Artemia* biomass. Although tank production of *Artemia* is far more expensive than pond production, its advantages for application are manifold:

- Year-round availability of ongrown *Artemia*, independent of climate or season;
- Specific stages (juveniles, preadults, adults) can be harvested with regard to possible size preferences of the predator;
- Quality of the *Artemia* can be better controlled (e.g., nutritionally and free from diseases).

The nutritional quality of *Artemia* juveniles and adults produced in super-intensive systems is analogous to natural product, except for its lipid content. Compared to Instar I nauplii, the protein content of adult *Artemia*, independent of its rearing conditions or food, is appreciably higher, i.e., 60% instead of 47%, and especially richer in essential amino acids.[16,93] (See Tables 7 and 8.) On the other hand, the lipid profile is a reflection of the diet offered to the *Artemia* cultures; i.e., when the animals are cultured on a HUFA-deficient diet (e.g., all agricultural by-products tested so far) their biomass at harvest will contain only small amounts of the fatty acids 20:5(n-3) and 22:6(n-3) (Table 8). However, this does not restrict their application since high levels of essential fatty acids can easily be attained in the *Artemia* biomass by applying the simple bio-encapsulation technique of Léger et al.[97] i.e., in less than 1 h the digestive tract of the brine shrimp can be filled with a HUFA-

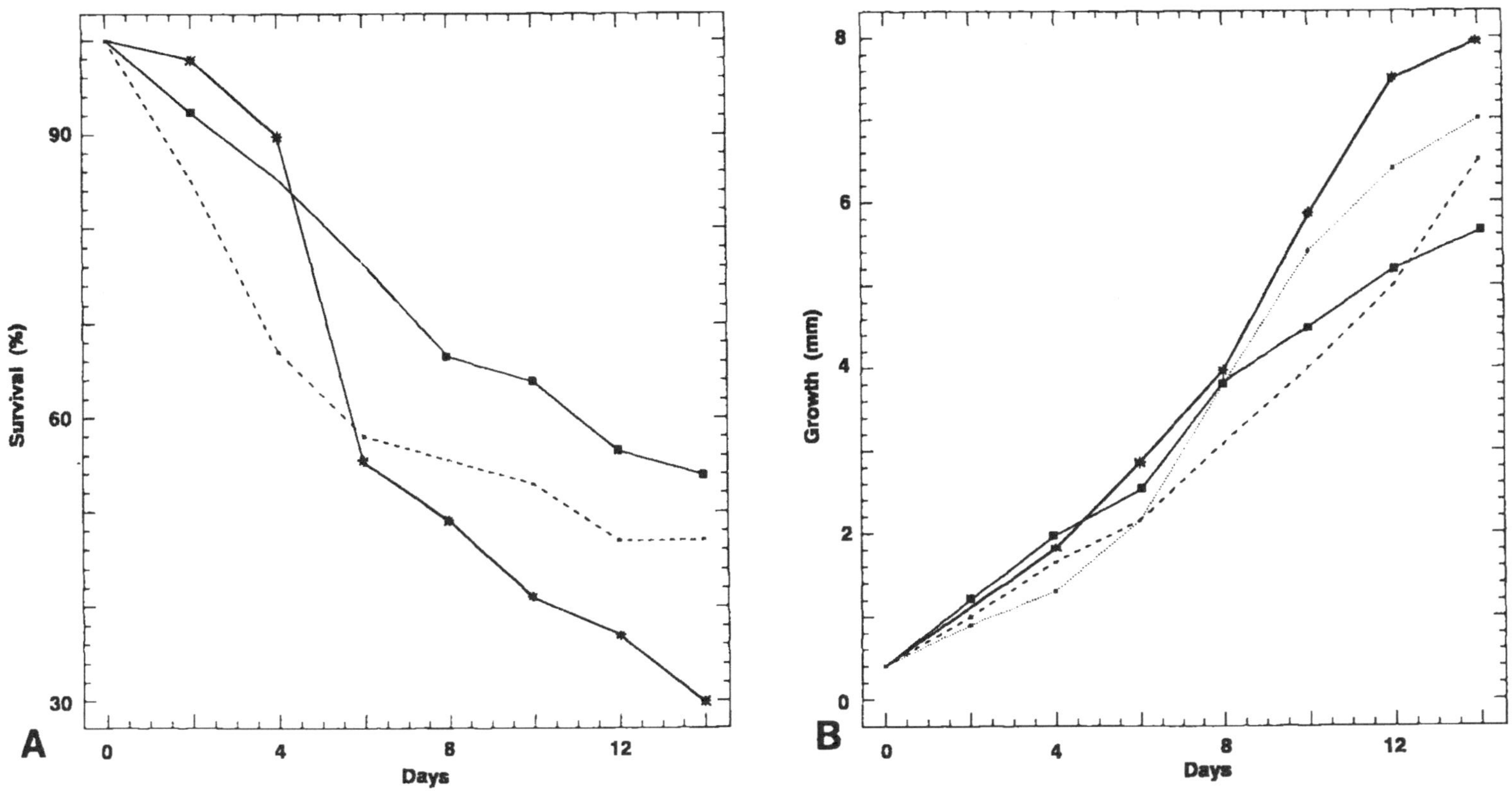

FIGURE 17. Survival (A) and growth (B) of Great Salt Lake *Artemia* cultured in different super-intensive culture systems. ---- 3501 open flow-through culture on rice-bran diet; ····· 2001 open flow-through culture on *Chaetoceras* diet; ▪——▪ 3001 closed flow-through culture on mixed diet of soybean waste and corn (maize) bran; *——* 2 m^3 stagnant raceway culture on rice-bran diet. (Modified from Lavens, P., Ph. D. thesis, 1989. With permission.)

TABLE 7
Comparison of the Biochemical Composition of Nauplii and of Preadults Harvested from Super-Intensive Cultures[a]

Artemia characteristics	Proteins	Lipids	N-free extract	Ash	Reference
Instar I - nauplii	41.6 - 47.2	20.8 - 23.1	10.5 - 22.7	9.5	Von Hentig[6], Claus *et al.*[94] Schauer *et al.*[95]
Preadults cultured on corn bran/soybean pellets in a flow-through system	55.2	17.6	10.4	16.8	Abelin[100]
Preadults cultured on rice bran in a stagnant raceway system	56.5	19.5	-	9.0	Dobbeni[96]

[a] All data for the Great Salt Lake strain and expressed in % of dry weight.

enrichment product, e.g. a marine fish-oil emulsion, boosting the HUFA content in the *Artemia* from a low level of 3 mg/g dw up to levels of >16 mg/g. In addition, this enrichment technique offers opportunities to quickly incorporate high amounts of essential nutrients, pigments, or therapeutics in ongrown *Artemia* for specific applications with different predators.

The use of *Artemia* juveniles and adults is not only advantageous from a nutritional point of view but also for energetic reasons; i.e., when feeding on large *Artemia* instead of freshly hatched nauplii, predator larvae need to chase less prey organisms per unit of time to meet their food requirements. This improved energy balance may result in faster growth, a higher developmental rate, and/or an improved physiological condition as has been demonstrated in *Macrobrachium,*[98] *Scophthalmus,*[99] *Sparus,*[100] and *Lates*[101] larviculture. For the latter two species, the introduction of ongrown *Artemia* as a hatchery/nursery food resulted in significant savings of *Artemia* cysts of about 60% and, consequently, a significant reduction in the total larval feed cost.

The aquarium pet shop industry offers expanding marketing opportunities for live *Artemia* biomass produced in regional super-intensive culture systems. Today, over 95% of the more than 3000 metric tons of *Artemia* biomass that are marketed in this sector[2] are sold frozen since they are harvested from a restricted number of natural sources and live transportation to other continents is cost prohibitive. Singapore, for example, already experiences a bottleneck where the local tropical aquarium industry is threatened by a shortage of live foods.[102] In the last decade several super-intensive *Artemia* farms have been established, e.g., in the U.S., France, U.K., and Australia, to supply local demand. Depending on the selected culture technology and implantation facilities, production costs are estimated at 2.5 to 12 US$/kg live weight *Artemia* with wholesale prices varying from $25 to $100.

ACKNOWLEDGMENTS

Research at the Artemia Reference Center on this subject has been supported by the Belgian National Science Foundation, the Belgian Ministry of Science Policy, the Belgian Administration for Development Cooperation, and Artemia Systems NV/SA.

TABLE 8
Fatty Acid and Amino Acid Profiles of Great Salt Lake Preadults, Cultured under Flow-Through Conditions on a Mixed Diet of Corn (Maize) Bran and Soybean-Pellets[a]

A. FATTY ACIDS[(b)]

saturated	16.20
14:0	0.70
15:0	0.50
16:0	9.10
17:0	0.70
18:0	5.20
19:0	
20:0	
24:0	
unsaturated	46.70
14:1	1.20
14:2	
15:1	0.30
16:1w7 + 16:1w9	4.30
16:2	
16:3	0.30
17:1	
18:1w7 + 18:1w9	18.30
18:2	15.90
18:4	0.10
19:4	
20:1	4.00
21:5	0.30
22:1	2.00
24:1	
unsaturated w3	9.00
18:3	
20:3	0.30
20:4	
20:5	2.80
22:3	0.90
22:4	
22:5	0.40
22:6	4.60
unsaturated w6	0.40
18:3	
20:3	
20:4	
22:4	0.40
22.5	

B. AMINO ACIDS[(c)]

essential	26.94
tryptophan	
lysine	4.23
histidine	1.30
arginine	2.69
threonine	2.42
valine	3.20
methionine	0.71
isoleucine	2.96
leucine	4.52
tyrosine	2.16
phenylalaline	2.75
non essential	25.71
asparagine	5.82
serine	2.63
glutamine	7.54
proline	3.29
glycine	2.68
alanine	3.61
cysteine	0.14

[a] Data compiled from Abelin.[100]

[b] Values expressed in mg/g dry matter.

[c] Values expressed in g/100 g dry matter.

REFERENCES

1. **Reeve, M. R.**, The filter-feeding of *Artemia*. II. In suspensions of various particles, *J. Exp. Biol.*, 40, 207, 1963.
2. **Lai, L. and Lavens, P.**, Report of workshop: *Artemia* as a business perspective, in *Artemia Research and Its Applications*, Vol. 3, Sorgeloos, P., Bengtson, D. A., Decleir, W. and Jaspers, E., Eds., Universa Press, Wetteren, Belgium, 1987, 517.
3. **Lavens, P.**, Intensieve produktie en kwaliteitsonderzoek van Artemia adulten en hun nakomelingen, Ph.D. thesis, State University of Ghent, Belgium, 1989.
4. **Wear, R. G. and S. J. Haslett**, Effect of temperature and salinity on the biology of *Artemia franciscana* Kellogg from Lake Grassmere, New Zealand. 1. Growth and mortality, *J. Exp. Mar. Biol. Ecol.*, 98, 153, 1986.
5. **Kinne, O.**, Temperature. 3.3 Animals. 3.31 Invertebrates, in *Marine Ecology, A Comprehensive, Integrated Treatise on Life in Oceans and Coastal Waters*. Vol I. Part 1, Wiley Interscience, New York, 1970, 407.
6. **Von Hentig, R.**, Einfluss von Salzgehalt und Temperatur auf Entwicklung, Wachstum, Fortpflanzung und Energielbilanz von *Artemia salina*, Thesis. University of Hamburg, Federal Republic of Germany, 1970, 136.
7. **Vanhaecke, P., Siddall, S. E., and Sorgeloos, P.**, International Study on *Artemia*. XXXII. Combined effects of temperature and salinity on the survival of *Artemia* of various geographical origin, *J. Exp. Mar. Biol. Ecol.*, 80, 259, 1984.
8. **Vanhaecke, P. and Sorgeloos, P.**, International Study on Artemia. XLVII. The effect of temperature on cyst hatching, larval survival and biomass production for different geographical strains of the brine shrimp *Artemia* spp., *Ann. Soc. Zool. Belg.*, 119, 7, 1989.
9. **Vanhaecke, P. and Sorgeloos, P.**, International study on Artemia. XIV. Growth and survival of *Artemia* larvae of different geographical origin in a standard culture test, *Mar. Ecol. Progr. Ser.*, 3, 303, 1980.
10. **Bowen, S. T., Buoncristiani, M. R., and Carl, J. R.**, *Artemia* habitats: ion concentrations tolerated by one superspecies, *Hydrobiology*, 158, 201, 1988.
11. **Provasoli, L. and D'Agostino, A. S.**, Development of artificial media for *Artemia salina*, *Biol. Bull.*, 136, 434, 1969.
12. **McShan, M., Trieff, N. M., and Grajcer, D.**, Biological treatment of wastewater using algae and *Artemia*, *J. Water Poll. Contr. Fed.*, 46, 1742, 1974.
13. **Decleir, W., Vos, J., Bernaerts, F., and Van den Branden, C.**, The respiratory physiology of *Artemia*, in *The Brine Shrimp Artemia*, Vol. 2, Persoone, G., Sorgeloos, P., Roels, O., and Jaspers, E., Eds., Universa Press, Wetteren, Belgium, 137, 1980.
14. **Platon, R. R.**, Scale-up studies on the culture of brine shrimp *Artemia* fed with rice bran, Ph.D. thesis, University of British Columbia, Vancouver, Canada, 1985.
15. **Sorgeloos, P.**, The influence of light on the growth rate of larvae of the brine shrimp, *Artemia salina* L., *Biol. Jaarboek Dodonaea*, 40, 317, 1972.
16. **Sorgeloos, P.**, Het gebruik van het pekelkreeftje *Artemia* spec. in de aquakultuur, thesis, State University of Ghent, Belgium, 1979.
17. **Royan, J. P.**, Effect of light on the hatching and growth of *Artemia salina*, *Mahasagar Bull. Nat. Inst. Oceanogr.*, 9, 83, 1976.
18. **Moffett, D. F. and Fisher, W. S.**, Ammonia production rates of *Artemia salina* under various culture conditions, *J. Can. Fish. Res. Bd*, 35, 1643, 1978.
19. **Chen, J. C., Chen, K. J. and Liao, J. M.**, Joint action of ammonia and nitrite in *Artemia salina* nauplii, in *Book of Abstract*, 19th Annual Meeting World Aquacultural Society, Honolulu, Hawaii, 26, 1988.
20. **Reeve, M. R.**, The filter feeding of *Artemia*. I. In pure cultures of plant cells, *J. Exp. Biol.*, 40, 195, 1963.
21. **Yanase, R. and Shiraishi, K.**, The filtering and feeding rates of the brine shrimp, *Artemia salina*, *Annual Report Iwate Medical University*, School of Liberal Arts and Sciences, 7, 27, 1972.
22. **Dobbeleir, J., Adam, N., Bossuyt, E., Bruggeman, E., and Sorgeloos, P.**, New aspects of the use of inert diets for high density culturing of brine shrimp, in *The Brine Shrimp Artemia*, Vol. 3, Persoone, G., Sorgeloos, P., Roels, O., and Jaspers, E., Eds., Universa Press, Wetteren, Belgium, 165, 1980.
23. **D'Agostino, A. S.**, The vital requirements of *Artemia*: physiology and nutrition, in *The Brine Shrimp Artemia*, Vol. 2, Persoone, G., Sorgeloos, P., Roels, O., and Jaspers, E., Eds., Universa Press, Wetteren, Belgium, 55, 1980.
24. **Sushchenya, L. M.**, Quantitative data on nutrition and energy balance in *Artemia salina* (L.)., *Akad. Nauk. SSR Doklady, Biol. Sci. Sect*, 143, 1205, 1962.
25. **Léger, P., Bengtson, D. A., Simpson, K. L., and Sorgeloos, P.**, The use and nutritional value of *Artemia* as a food source, *Oceanogr. Mar. Biol. Ann. Rev.*, 24, 521, 1986.
26. **Nimura, Y.**, Biology of the brine shrimp, *Bull. Jpn. Soc. Sci. Fish.*, 33, 690, 1967.
27. **Nimura, Y.**, Retarded growth of *Artemia salina* by overfeeding, *Bull. Jpn. Soc. Sci. Fish.*, 46, 681, 1980.

28. **Sushchenya, L. M.,** Food rations, metabolism and growth of crustaceans, in *Marine Food Chains,* Steele, J. H., Ed., Oliver and Boyd, Edinburgh, U.K., 127, 1970.
29. **Braun, J. G.,** The feeding of *Artemia* on *Phaeodactylum tricornutum,* in *The Brine Shrimp Artemia,* Vol. 2, Persoone, G., Sorgeloos, P., Roels, O. and Jaspers, E., Eds., Universa Press, Wetteren, Belgium, 197, 1980.
30. **Coutteau, P. and Sorgeloos, P.,** Feeding of the brine shrimp *Artemia* on yeast: Effect of mechanical disturbance, animal density, water quality, and light intensity, in *Aquaculture Europe '89, Short Communications and Abstracts,* Billard, R. and De Pauw, N., Eds., European Aquaculture Society, Special Publication No. 10, Bredene, Belgium, 75, 1989.
31. **Conklin, D. E. and Provasoli, L.,** Biphasic particulate media for the culture of filter-feeders, *Biol. Bull.,* 154, 47, 1978.
32. **Hernandorena, A.,** *Artemia* nutrition, in *Proceedings of an International Conference on Aquacultural Nutrition: Biochemical and Physiological Approaches to Shellfish Nutrition,* Pruder, G. D., Langdon, C. J., and Conklin, D. E., Eds., Louisiana State University, Baton Rouge, 166, 1983.
33. **Dadd, R. H.,** Arthropod nutrition, in *Chemical Zoology,* Vol. 5, Florkin, M. and Sheer, B. T., Eds., Academic Press, New York, 35, 1970.
34. **Provasoli, L. and Pintner, I. J.,** Biphasic particulate media for the parthenogenetic *Artemia* of Sète, in *The Brine Shrimp Artemia,* Vol. 2, Persoone, G., Sorgeloos, P., Roels, O., and Jaspers, E., Eds., Universa Press, Wetteren, Belgium, 321, 1980.
35. **Gibor, A.,** Some ecological relationships between phyto- and zooplankton, *Biol. Bull.,* 11, 230, 1956.
36. **Hirayama, K. and Watanabe, K.,** Fundamental studies on physiology of rotifer for its mass culture. 4. Nutritional effect of yeast on population growth of rotifers, *Bull. Jpn. Soc. Sci. Fish.,* 39, 1129, 1973.
37. **Hirayama, K., and Funamoto, H.,** Supplementary effect of several nutrients on nutritive deficiency of baker's yeast for population growth of the rotifer *Brachionus plicatilis, Bull. Jpn. Soc. Sci. Fish.,* 49, 505, 1983.
38. **Douillet, P.,** Effect of bacteria on the nutrition of the brine shrimp *Artemia,* fed on dried diets, in *Artemia Research and Its Applications,* Vol. 3, Sorgeloos, P., Bengtson, D. A., Decleir, W., and Jaspers, E. Eds., Universa Press. Wetteren, Belgium, 295, 1987.
39. **Langis, R., Proulx, D., de la Noue, J., and Couture, P.,** Effects of a bacterial biofilm on intensive *Daphnia* culture, *Aquacult. Eng.,* 7, 21, 1988.
40. **Landau, M., Bolis, C., and Miyamoto, G.,** A method for the production of the brine shrimp, *Artemia salina* Leach, in a manure-based system, *Agricult. Wastes,* 15, 79, 1986.
41. **Becker, E. W.,** Nutritional properties of microalgae: potentials and constraints, in *CRC Handbook of Microalgal Mass Culture,* CRC Press, Boca Raton, FL, 339, 1986.
42. **Sick, L. V.,** Nutritional effect of five species of marine algae on the growth, development, and survival of the brine shrimp, *Artemia salina, Mar. Biol.,* 35, 69, 1976.
43. **Zmora, O. B.,** in *Artemia Newsletter,* Sorgeloos, P., Ed., Artemia Reference Center, State University of Ghent, Belgium, 6, 1987.
44. **Takano, H.,** Rearing experiments of the brine shrimp on a diatom diet, *Bull. Tokai Reg. Fish. Res. Lab.,* 52, 1, 1967.
45. **Tobias, W. J., Sorgeloos, P., Bossuyt, E., and Roels, O. A.,** The technical feasibility of mass-culturing *Artemia salina* in the St. Croix "artificial upwelling" mariculture system, *Proc. World Maricult. Soc.,* 10, 203, 1979.
46. **Milligan, D. J., Quick, J. A., Hill, S. E., Morris, J. A., and Hover, R. J.,** Sequential use of bacteria, algae and brine shrimp to treat industrial wastewater at pilot plant scale, in *The Brine Shrimp Artemia,* Vol. 3, Persoone, G., Sorgeloos, P., Roels, O., and Jaspers, E., Eds., Universa Press, Wetteren, Belgium, 193, 1980.
47. **Sorgeloos, P.,** High density culturing of the brine shrimp, *Artemia salina, Aquaculture,* 1, 385, 1973.
48. **Cognie, D.,** Production d'Artemia salina Leach en grand volumes recycles, in *10th European Symposium on Marine Biology,* Vol. 1, Persoone, G. and Jaspers, E., Eds., Universa Press, Wetteren, Belgium, 35, 1976.
49. **Person-Le Ruyet, J.,** Elevage larvaire d'*Artemia salina* (Branchipode) sur nourriture inerte: Spirulina maxima (Cyanophycee), *Aquaculture,* 8, 157, 1976.
50. **Paulsen, C.,** *Artemia* culture, in *Annual Report 1977 of the Oceanic Institute,* Waimanalo, Hawaii, 11, 1978.
51. **Klein, V. L. M. and Ibarra, Su, J.,** Different rations types in the feeding of *Artemia salina* in laboratory conditions, in *Book of Abstracts,* First Interamerican Congress on Aquaculture, September 14-21, Salvador, Brazil, 68, 1986.
52. **Vandamme, E. J. and Demey, L.,** De Productie van "Single Cell Protein", *Het Ingenieursblad,* 49, 89, 1980.
53. **Peppler, H. J. and Stone, C. W.,** Feed yeast products. An update of the properties of feed yeasts and their applications, *Feed Management,* 1976.

54. **Litchfield, J. H.,** Microbial protein production, *Bioscience*, 30, 387, 1980.
55. **Shimaya, M., Kanazawa, A., and Kashiwada, K.,** Studies in the utilization of marine yeast I. Culture of *Artemia* and *Daphnia* by marine yeast, *Mem. Fac. Fish., Kagoshima Univ.*, 16, 34, 1967.
56. **Kawano, T., Kojima, H., Ohosawa, H., and Maringa, K.,** Feed and method of aquianimals cultivation, Patent Number 3,939,379. February 17, 1976.
57. **James, C. M., Abu-Rezeq, T. S., and Dias, P.,** Production of *Artemia* biomass for feeding marine fish larvae, in *Artemia Research and Its Applications*. Vol. 3, Sorgeloos, P., Bengtson, D. A., Decleir, W., and Jaspers, E., Eds., Unviersa Press, Wetteren, Belgium, 349, 1987.
58. **Johnson, D. A.,** Evaluation of various diets for optimal growth and survival of selected life stages of *Artemia*, in *The Brine Shrimp Artemia*, Vol. 3, Persoone, G., Sorgeloos, P., Roels, O., and Jaspers, E., Eds., Universa Press, Wetteren, Belgium, 185, 1980.
59. **Urban, E. Jr. and Langdon, C.,** Reduction in costs of diets for the american oyster, *Crassostrea virginica* (Gmelin), by the use of non-algal supplements, *Aquaculture*, 38, 277, 1984.
60. **Hirayama, K.,** A consideration of why mass culture of the rotifer *Brachionus plicatilis* with baker's yeast is unstable, *Hydrobiol.*, 147, 269, 1987.
61. **Robin, J. H., Gatesoupe, F. J., and Ricardez, R.,** Production of *Artemia salina* using mixed diet: Consequences on rearing of sea-bass larvae *(Dicentrarchus labrax), J. World Maricult. Soc.*, 12, 119, 1981.
62. **Lavens, P., De Meulemeester, A., and Sorgeloos, P.,** Evaluation of mono- and mixed diets as food for intensive culturing of *Artemia*, in *The Brine Shrimp Artemia*, Vol. 3, Persoone, G., Sorgeloos, P., Roels, O., and Jaspers, E., Eds., Universa Press, Wetteren, Belgium, 309, 1987.
63. **Nimmannit, S., and Assawamunkong, S.,** Study on yeast as feed for marine organisms (brine shrimps), in *Proceedings of Living Aquatic Resources*, Chulalongkorn University, Thailand, 21, 1985.
64. **Talloen, M.,** The use of locally available food for the mass-culture of the brine shrimp *Artemia salina*, in *Bulletin of the Jepara Brackishwater Aquaculture Development Center*, 1978.
65. **James, C. M., and Makkeya, B. A.,** Production of rotifers, *Brachionus plicatilis*, brine shrimp, *Artemia salina* and copepods for aquaculture, in *Annual Research Report*, Kuwait Institute for Scientific Research, 103, 1981.
66. **Coutteau, P., Lavens, P., and Sorgeloos, P.,** The use of yeast as single-cell protein in aquacultural diets, in *Proceedings of the Congress on Biotechnology*, Ghent, 1989.
67. **Lavens, P., Coutteau, P., Sorgeloos, P., and Vandamme, E.,** Aliment pour aquaculture, Patent: 231 592 D. 12213 SBM. 15, 1988.
68. **Albentosa, M., Naessens, E., Léger, P., Coutteau, P., Lavens, P., and Sorgeloos, P.,** Promising results in the seed culturing of the Manila clam *Tapes semidecussata* with a manipulated yeast product as a partial substitute for algae, in *Aquaculture Europe '89, Short Communications and Abstracts*, Billard, R. and De Pauw, N., Eds., European Aquaculture Society, Special Publication No. 10, Bredene, Belgium, 7, 1989.
69. **Seki, H.,** Studies on microbial participation to food cycle in the sea I. Participation in the microcosm at static conditions, *J. Oceanogr. Soc.* Jpn. 20, 122, 1964.
70. **Yasuda, K. and Taga, N.,** A mass-culture method for *Artemia salina* using bacteria as food, *Mer, Bull. Soc. Fr. Jpn. Oceanogr.*, 18, 55, 1980.
71. **Suhair, A. H., James, C. M., and Makkeya, B. A.,** Preliminary analysis on the use of chitin-derivated microbial detritus for the mass culture of *Artemia*, in *Technical Report of the Kuwait Institute for Scientific Research*, 11, 1982.
72. **Sorgeloos, P.,** De invloed van abiotische en biotische faktoren op de levenscyclus van het pekelkreeftje, *Artemia salina* L., Ph.D. thesis, State University of Ghent, Belgium, 1975.
73. **Bossuyt, E. and Sorgeloos, P.,** Batch production of adult *Artemia* in 2 m^3 and 5 m^2 air-water-lift operated raceways, in *Book of Abstracts of the World Conference on Aquaculture*, Venice, Italy, 17, 1981.
74. **Brisset, P. J., Versichele, D., Bossuyt, E., De Ruyck, L., and Sorgeloos, P.,** High density flow-through culturing of brine shrimp *Artemia* on inert feeds - preliminary results with a modified culture system, *Aquacult. Eng.*, 1, 115, 1982.
75. **Lavens, P., Baert, P., De Meulemeester, A., Van Ballaer, E., and Sorgeloos, P.,** New developments in the high density flow-through culturing of brine shrimp *Artemia*, *J. World Maricult. Soc.*, 16, 498, 1986.
76. **Jones, D. A., Munford, J. G., and Gabbott, P. A.,** Microcapsules as artificial food particles for aquatic filter feeders, *Nature*, 247, 233, 1974.
77. **Bossuyt, E. and Sorgeloos, P.,** Technological aspects of the batch culturing of *Artemia* in high densities, in *The Brine Shrimp Artemia*, Vol. 3, Persoone, G., Sorgeloos, P., Roels, O., and Jaspers, E., Eds., Universa Press, Wetteren, Belgium, 133, 1980.
78. **Tobias, W. J., Sorgeloos, P., Roels, O. A., and Sharfstein, B. A.,** International Study on *Artemia*. XIII. A comparison of production data of 17 geographical strains of *Artemia* in the St. Croix artificial upwelling-mariculture system, in *The Brine Shrimp Artemia*, Vol. 3, Persoone, G., Sorgeloos, P., Roels, O., and Jaspers, E., Eds., Universa Press, Wetteren, Belgium, 383, 1980.

79. **Abreu-Grobois, F. A., Briseno-Duenas, R., Herrera, M. A., and Malagon, M. L.,** A model for growth of *Artemia* cultures based on food ration dependent gross growth efficiencies, *Hydrobiol.*, in press.
80. **Dohse, H.,** Das Artemium. Die kleine "Lebendfutterfabrik", *Die Dtsche Aquar. Terrar. Zeitschr.*, 24, 413, 1971.
81. **Person-Le Ruyet, J.,** Techniques d'élevage en masse d'un rotifère (Brachionus plicatilis Muller) et d'un crustacé branchiopode (*Artemia salina* L.), in *Tenth European Symposium on Marine Biology*, Vol. 1, Persoone, G. and Jaspers, E., Eds., Universa Press, Wetteren, Belgium, 331, 1976.
82. **Lavens, P. and Sorgeloos, P.,** Design, operation, and potential of a culture system for the continuous production of *Artemia* nauplii, in *Artemia Research and Its Applications*. Vol. 3, Sorgeloos, P., Bengtson, D. A., Decleir, W., and Jaspers, E., Eds., Universa Press, Wetteren, Belgium, 339, 1987.
83. **Bombeo, R. F., Dhert, P., and Lavens, P.,** Design of a simple culture technique for the production of *Artemia* juveniles or adults, in preparation.
84. **Chiba, K.,** Present status of flow-through and recirculation systems and their limitations in Japan, in *Aquaculture in Heated Effluents and Recirculation Systems*, Vol. II, Tiews, K., Ed., Heeneman Verlaggesellschaft, Berlin, West Germany, 342, 1981.
85. **Tiews, K.,** *Aquaculture in Heated Effluents and Recirculation Systems*, Vol. II, Tiews, K., Ed., Heeneman Verlaggesellschaft, Berlin, West Germany, 1981.
86. **Mock, C. R., Ross, L. R., and Salser, R. B.,** Design and preliminary evaluation of a closed system for the shrimp culture, in *Proc. 8th Ann. Meeting World Maricult. Soc.*, Avault, J. W., Jr., Ed., Louisiana State University, Baton Rouge, 335, 1977.
87. **Rosenthal, H.,** Ozonization and sterilization, in *Aquaculture in Heated Effluents and Recirculation Systems*, Vol. I, Tiews, K., Ed., Heeneman Verlaggesellschaft, Berlin, West Germany, 219, 1981.
88. **Wheaton, F. W.,** *Aquacultural Engineering*, Wiley Interscience, New York.
89. **Solangi, M. A., Overstreet, R. M., and Gannam, A. L.,** A filamentous bacterium on the brine shrimp and its control, *Gulf Res. Rep.*, 6, 275, 1979.
90. **Hernandorena, A.,** Nutrition, croissance et metabolisme d' *Artemia salina* (L.), Ph.D. thesis, University of Paris, France, 154, 1974.
91. **Hernandorena, A.,** An increased dietary tryptophan requirement induced by interference with purine interconversion in *Artemia*, in *Artemia Research and Its Applications*. Vol. 2, Decleir, W., Moens, L., Slegers, H., Sorgeloos, P., and Jaspers, E., Eds., Universa Press, Wetteren, Belgium, 233, 1987.
92. **Sorgeloos, P., Lavens, P., Léger, P., Tackaert, W., and Versichele, D.,** *Manual for the Culture and Use of Brine Shrimp Artemia in Aquaculture*, Artemia Reference Center, State University of Ghent, Belgium, 1986.
93. **Ronsivalli, P. C. and Simpson, K. L.,** The brine shrimp *Artemia* as a protein source for humans, in *Artemia Research and Its Applications*. Vol. 3, Sorgeloos, P., Bengtson, D. A., Decleir, W., and Jaspers, E., Eds., Universa Press, Wetteren, Belgium, 503, 1987.
94. **Claus, C., Benijts, F., Vandeputte, G., and Gardner, W.,** The biochemical composition of the larvae of two strains of *Artemia salina* (L.) reared on two different algal foods, *J. Exp. Mar. Biol. Ecol.*, 36, 171, 1979.
95. **Schauer, P. S., Johns, D. M., Olney, C. E., and Simpson, K. L.,** International Study on *Artemia*. IX. Lipid level, energy content and fatty acid composition of the cysts and newly hatched nauplii from five geographical strains of *Artemia*, in *The Brine Shrimp Artemia*, Vol. 3, Persoone, G., Sorgeloos, P., Roels, O., and Jaspers, E., Eds., Universa Press, Wetteren, Belgium, 365, 1980.
96. **Dobbeni, A.,** personal communication, 1983.
97. **Léger, P., Naessens-Foucquaert, E., and Sorgeloos, P.,** International Study on *Artemia*. XXXV. Techniques to manipulate the fatty-acid profile in *Artemia* nauplii, and the effect on its nutritional effectiveness for the marine crustacean *Mysidopsis bahia* (M.), in *Artemia Research and Its Applications*. Vol. 3, Sorgeloos, P., Bengtson, D. A., Decleir, W., and Jaspers, E., Eds., Universa Press, Wetteren, Belgium, 411, 1987.
98. **Sick, L. V. and Beaty, H.,** Development of formulated foods designed for *Macrobrachium rosenbergii* larval and juvenile shrimp, in *Proc. 6th Ann. Meeting World Mariculture Society*, Avault, J. W., Jr. and Miller, R., Eds., Louisiana State University, Baton Rouge, 557, 1975.
99. **Purdom, C. E. and Preston, A.,** A fishy business, *Nature*, 266, 396, 1977.
100. **Abelin, P.,** personal communication, 1988.
101. **Dhert, P., Bombeo, R. F., and Lavens, P.,** Optimization of a semi flow-through system for the super-intensive culture of the brine shrimp *Artemia*, in preparation.
102. **Shim, K. F.,** Mass production of *Moina* in Singapore using pig waste, *World Aquacult.*, 19, 59, 1988.
103. **Kanazawa, A., Teshima, S. I., Sasada, H., and Rahman, S. A.,** Culture of the prawn larvae with micro-particulate diets, *Bull. Jpn. Soc. Scient. Fish.*, 48, 195, 1982.
104. **Pinxt, R.,** Australia, personal communication.

Chapter 14

ANOSTRACANS OF AUSTRALIAN SALT LAKES, WITH PARTICULAR REFERENCE TO A COMPARISON OF *PARARTEMIA* and *ARTEMIA*

W. D. Williams and M. C. Geddes

TABLE OF CONTENTS

I. INTRODUCTION

Artemia is the most widely encountered and the best-known anostracan genus of saline inland waters, but it is not the only genus of Anostraca to occur in such environments. As Hammer[1] noted, several other genera occur in inland saline waters of salinity below ~50 g/l, namely *Parartemia* (Branchipodidae), *Branchinecta* (Branchinectidae) and *Branchinella* and *Branchinectella* (Thamnocephalidae). Above 50 g/l, however, only *Parartemia* and *Artemia* have been recorded.

Parartemia is endemic to Australia where it is widely encountered and common. Indeed, although *Artemia* is recorded from Australia, it is almost entirely restricted to coastal, man-made solar ponds into which it may have been introduced.[2] Australia, therefore, as in so many other respects involving its aquatic fauna, stands apart from other continents in the nature of the anostracan fauna found in saline lakes; the native Anostraca of inland salt waters in Australia belong either to the family Branchipodidae and the endemic genus *Parartemia*, or the Thamnocephalidae and the genus *Branchinella*.

Other chapters in this book emphasize the scientific and commercial importance of *Artemia*. The commercial importance of *Artemia* has also attracted attention in Australia where it is used for solar pond management and is of interest to aquaculturalists.[3] *Parartemia*, in marked contrast, has attracted only passing commercial interest.[4] On the other hand, it has attracted considerable scientific interest, with a few features of its biology at least as well, if not better known, than those of *Artemia*. On the whole, however, studies of *Parartemia* have generally not attracted a great deal of attention beyond Australia.

Given the overall aims of the present book, it is appropriate that it contains a review of what is known about the Anostraca of Australian inland saline waters, especially *Parartemia*, emphasizing comparisons with *Artemia*. Since most studies of *Parartemia* (and halophilic *Branchinella* species) have addressed matters of mainly scientific interest, our own stress, consequently, is more upon basic biological features than upon features of direct commercial interest. Whether species of *Parartemia* have greater or lesser commercial values *vis-à-vis* *Artemia* species remains to be determined; nevertheless, it seems strange that *Artemia* has been the focus of commercial interest even in Australia, whereas *Parartemia*, as noted, has received little attention in this respect. Our attempt to provide a comprehensive bibliography concerning *Parartemia* may help correct this imbalance.

II. TAXONOMY

Three of the eight described families of Anostraca are known definitely to occur in Australia. The Artemiidae Grochowski 1896 is represented by one or more 'species' of *Artemia* Leach, 1819, the Branchipodidae Simon, 1886, by eight endemic species of *Parartemia* Sayce, 1903, and the Thamnocephalidae Simon, 1886, by twenty endemic species of *Branchinella* Sayce, 1903. *Parartemia* is itself endemic, but *Branchinella*, although most diverse in Australia, is also found in southern Africa, Europe, Asia, and North America. Australian records of *Streptocephalus archeri* Sars, 1896 (Streptocephalidae Daday, 1910), probably refer to a species of *Branchinella*.[5,6] There are no Australian records of the only other anostracan family known to include species found in inland saline waters, the Branchinectidae Daday, 1910. Species of *Artemia* and *Parartemia* are found only in saline waters; *Branchinella* occurs in fresh to moderately saline waters according to species.

Major taxonomic works concerning Australian Anostraca are those of Sars,[7] Sayce,[8] Linder,[5] and Geddes.[6,9,10] They provide a comprehensive if not complete basis for taxonomic knowledge of Anostraca in Australian saline lakes. A variety of morphological features has been used to distinguish taxa, but those most used include (in males) the second antennae, frontal appendage (when present), and penes, and in both sexes, the phyllopod structure and form of the genital segments.

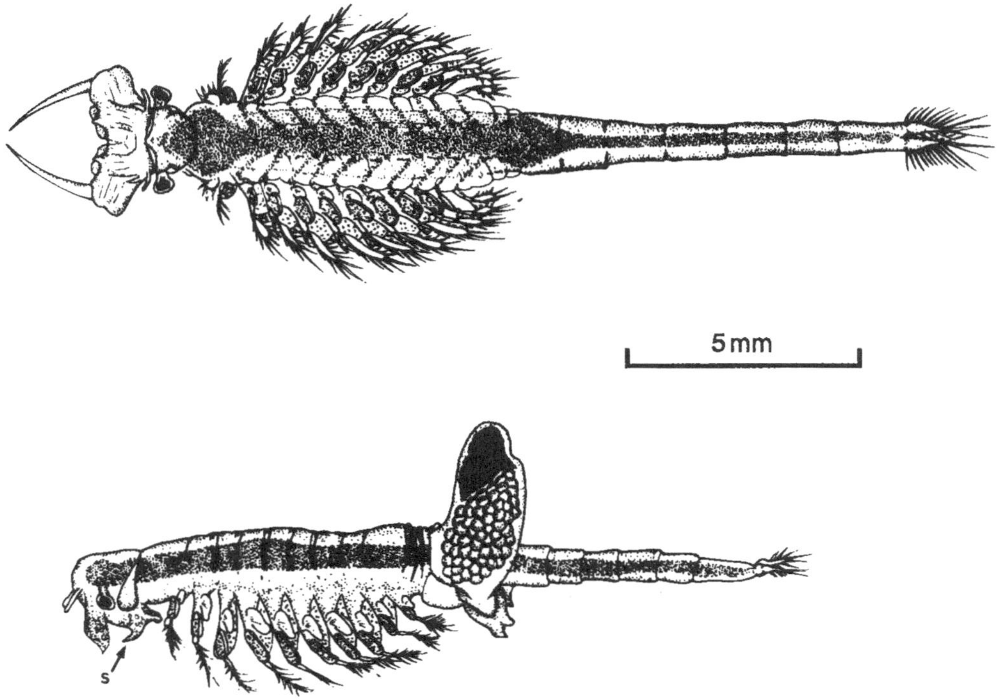

FIGURE 1. Whole animal views of *Parartemia zietziana*. Upper, male; lower, female; s, spine on labrum. (After Geddes, M. C., *Hydrobiologia*, 81, 169, 1981.)

Early taxonomic keys[5] were based entirely upon male characters, but as male specimens are sometimes rare, and penis structure in any case is difficult to study, later keys also made use of female characters. Since no generally accessible key to species of *Parartemia* exists, and to facilitate further study of Anostraca in Australian inland saline waters, we provide a key to *Artemia*, to those species of *Branchinella* tolerant of moderately saline waters in Australia, and to all species of *Parartemia*. All Australian species of *Branchinella* may be readily identified using the key of Geddes.[6] A few species of *Branchinella* not included in the key may occasionally be found in slightly saline waters (e.g., *B. affinis* has been collected at a salinity of 4.8 g/l).[11]

In the key, characters for both sexes are used in generic determinations, but mostly male characters are used for identification of halophilic species of *Branchinella* and only female characters for identification of *Parartemia* species. Females of *P. longicaudata* Linder, 1941 are unknown and so this species is omitted from the key. It may be noted, however, that male specimens of *Parartemia* can usually be identified, but their diagnostic characters are not as readily recognizable as those of females. The key can only be used with mature (adult) specimens, and is designed to facilitate ecological and physiological studies rather than provide definitive taxonomic identifications. Given the strong likelihood that further species of both *Branchinella* and *Parartemia* remain for discovery (and that three new species of *Parartemia* await description), definitive identifications can only be made from the taxonomic literature.

The best-known species of *Parartemia*, *P. zietziana*, a taxon widespread and common in southeastern Australia, is illustrated in Figure 1.

KEY TO SPECIES OF ANOSTRACA FOUND IN AUSTRALIAN INLAND SALINE WATERS

1. Basal parts of penes rigid; male without frontal appendage; 1 anterior seta on endite 3 of phyllopod in form of long curved spine; telson reduced and often not distinct from final abdominal segment; found only in saline waters .... (*Artemia* and *Parartemia*) 5
   - Basal part of penes soft and flexible; male with or without frontal appendage; anterior setae on endite 3 much shorter than posterior setae; telson well developed and demarcated; found in fresh or moderately saline waters .... (*Branchinella*) 2
2. Frontal appendage of male present, either biramous in form or found as a low ridge .... freshwater species of *Branchinella* (see Geddes[6])
   - Frontal appendage of male absent .... halophilic species of *Branchinella** 3
3. Large outgrowths from basal joints of male second antenna about equal in length to apical joint of second antenna .... *B. nichollsi* Linder[4]
   - No outgrowths from basal joints of male second antenna 4
4. Basal joints of second male antenna short and thick, with medial swelling covered with fine teeth, and united for more than half their length; phyllopods with elongated endite reaching further than exite on middle legs; female with lateral spines on genital segments .... *B. compacta* Linder, 1941
   - Basal joints of second male antenna slender, with no ornamentation and united for scarcely one quarter their length; phyllopods with exite longer than endite and with distinct notch in pre-epipodite; female without lateral spines on genital segment ....*B. simplex* Linder, 1941
5. Apical parts of penes without spines; no spine on labrum of female ...... *Artemia* sp.
   - Apical parts of penes with spines; spine on labrum of female............... .... (*Parartemia*) 6
6. Integument of abdomen smooth.. 7
   - Integument of abdomen roughened or granulated 8
7. Egg sac consisting of a single ventral pouch.................................. .... *P. cylindrifera* Linder, 1941
   - Egg sac with large, upwardly directed lateral lobes (Figure 1)............. ....*P. zietziana* Sayce, 1903
8. No reduction of hindmost thoracic segments................*P. extracta* Linder, 1941
   - Hindmost thoracic segments reduced and scarcely ornamented.................. 9
9. Egg sac consisting of a simple ventral pouch .... *P. minuta* Geddes, 1973
   - Egg sac with variously shaped lateral lobes 10
10. Eleventh thoracic segment completely coalesced with genital segments .... *P. contracta* Linder, 1941
    - Eleventh thoracic segment not coalesced with genital segments................ 11
11. With large dorsal swelling on thoracic segments 8 and 9 .... *P. serventyi* Linder, 1941
    - With lateral coherent swellings extending from segments 6 to 9 and 9 to 11 ... .*P. informis* Linder, 1941

* viz., waters with salinities >3 g/l.

Little of rigor can be written concerning the taxonomic relationships of the Australian species of *Branchinella* and *Parartemia*. However, *Branchinella*, judging from its geographical distribution, is clearly an old genus, and may have originated in Australia given its species diversity there. According to Linder,[5] *Parartemia* is most closely related to *Branchipodopsis* Sars, 1898, a genus found in arid regions of Africa and Asia where, however, all its species live in fresh waters.

Whatever the taxonomic relationships, the most notable taxonomic features of the Australian anostracan fauna is its low generic diversity and high species endemism.[12] For the Anostraca of highly saline waters in particular, a further notable feature is the high degree of morphological differentiation between species of *Parartemia vis-à-vis* the low degree of such differentiation in *Artemia* species. The extent of genetic homogeneity within morphotypes of *Parartemia*, moreover, has yet to be ascertained. Perhaps this will reveal even further diversity.

As for the taxonomic relationships of *Artemia* 'species' in Australia, one 'species' is similar to forms found in India, France, U.S.S.R., and Japan: it is parthenogenetic and occurs on Rottnest Island and near Port Hedland in Western Australia,[13] and may be related to populations found elsewhere in the same state as well as near Adelaide.[14] A sexual form of perhaps more recent introduction occurs in Queensland and may have come from San Francisco Bay populations.[15,16] Additional strains may also have been introduced, so that the overall situation, as noted recently,[16] is confused.

III. GEOGRAPHICAL DISTRIBUTION AND EVOLUTION

The distribution of halophilic species of *Branchinella* in Australia is indicated in Figure 2, and of all species of *Parartemia* and *Artemia* in Figure 3. Leaving aside the distribution of *Artemia*, Figures 2 and 3 reflect a common biogeographical feature of Australian anostracans in both fresh and saline waters: most species have rather limited geographical ranges, but a few have extensive ones. Thus, *B. simplex* and *B. compacta* occur in relatively small regions in southwestern and southeastern Australia, respectively; *B. nichollsi*, on the other hand, is recorded from Western Australia, Victoria, and Queensland. The related freshwater species, *B. australiensis*, is even more widely distributed. Likewise, species of *Parartemia* are mostly found in geographically restricted regions, but one species, *P. cylindrifera*, is found in both southwestern Western Australia and southeastern South Australia, and another, *P. minuta*, in Lake Eyre[17] and Lake Buchanan, Queensland.[18]

This pattern of distribution has been explained, at least in part for *Parartemia*, as a reflection of the way species have evolved.[19] (A similar explanation may well apply to the analogous geographical pattern displayed by species of *Branchinella*, but this is not explored further here.) Two basic hypotheses are involved: (1) Evolution only occurs in predictably filled localities; episodic water bodies have little or no value as evolutionary loci; (2) The most important environmental determinant within a salt lake is the seasonal (predictable) temporal pattern of interaction between salinity, temperature, and the presence of water. Over a considerable range, salinity per se is not important.

If these hypotheses are considered within the context of palaeoclimatic changes in Australia, two events will be obvious. First, regions of evolutionary importance, i.e., regions of predictably filled lakes, underwent geographical migration: in times of greater aridity than the present, they moved away from the center of Australia, in particular southwards, and became smaller in extent; in times of less aridity than the present, they moved towards the center of Australia and became larger in extent (Figure 4). Over geological times, such areal regression and extension occurred on numerous occasions, and in Western Australia, since at least mid-Tertiary times. Second, areas of predictably-filled salt lakes moved between climatic belts, and thus expanded into or regressed from regions where rainfall was mainly

FIGURE 2. Geographical distribution of halophilic species of *Branchinella* in Australia. C, *B. compacta*; N, *B. nichollsi*; S, *B. simplex*. (Partly after Geddes, M. C., *Aust. Mus. Mem.*, 18, 155, 1983.)

a summer or winter occurrence. Thus, certain areas of evolutionary importance experienced fundamental differences in the temporal patterns of interaction between salinity, temperature, and occurrence of water.

It is argued that these events, combined with the geographical separation of populations resulting from the closed nature of endorheic drainage basins, the poor dispersal powers of most species of *Parartemia* (whose cysts, unlike those of most species of *Artemia*, sink), and the restriction of *Parartemia* species to astatic water bodies, have largely driven evolution in *Parartemia*. The occurrence in Western Australia of environmentally similar localities in close proximity to each other, each with a different species of *Parartemia*, can be seen as a reflection of this evolutionary scenario. The maintenance of unispecific populations can be explained on the basis that once a particular genotype (species of *Parartemia*) dominated a particular locality, it had a genetic advantage over invasive genotypes (other species of *Parartemia*) and excluded them.

Exceptions to this evolutionary scenario appear to be *P. minuta, P. cylindrifera,* and *P. contracta*. The first, as noted, is recorded from localities as far apart as Lake Eyre and Lake Buchanan, the second, from localities in both Western Australia and South Australia, and the third, from acid salt lakes in Western Australia.

For *P. minuta*, perhaps selection was not towards adaptation to local environmental conditions (reinforced by poor powers of dispersal) as in the case of most species of *Parartemia*, but towards a more general adaptation enabling colonization of episodically filled salt lakes of central Australia. In this way, *P. minuta* could take advantage of opportunities offered by the otherwise presumably vacant ecological niches in such waters. To do this, of course, good dispersal mechanisms would be needed.

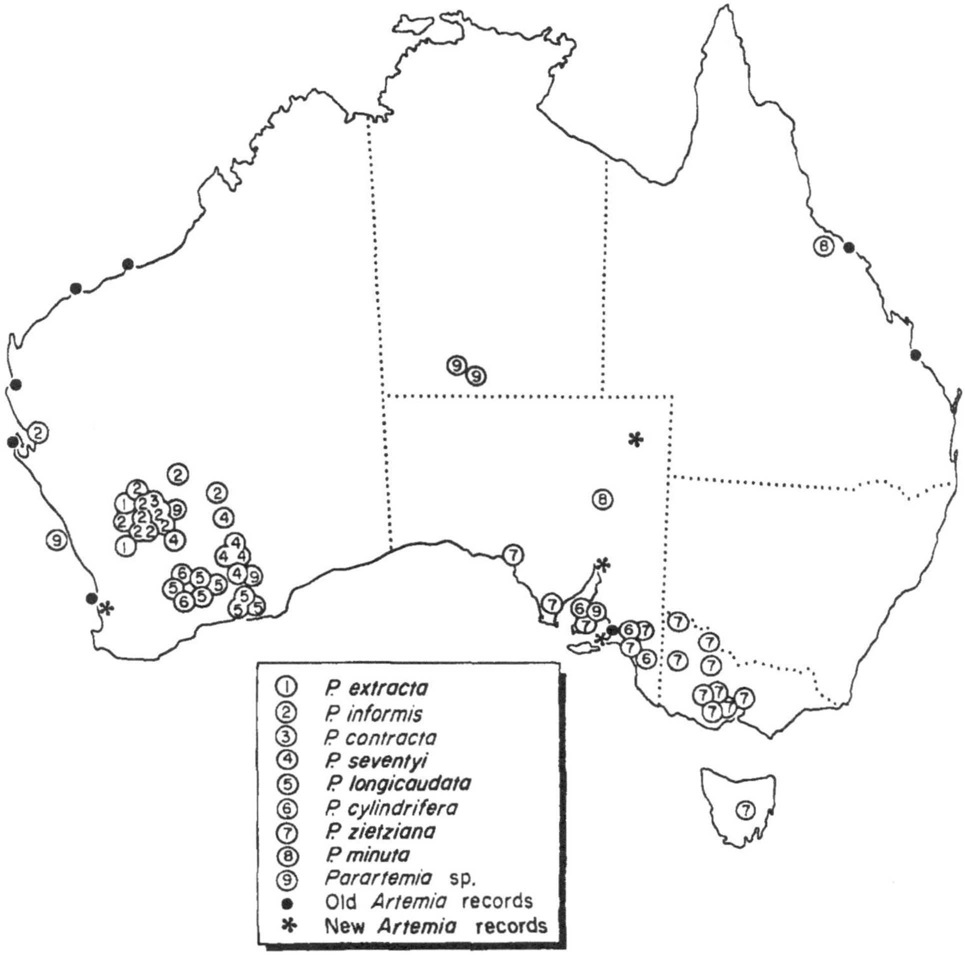

FIGURE 3. Geographical distribution of *Parartemia* spp. and *Artemia* in Australia. (After Geddes, M. C. and Williams, W. D., in *Artemia Research and Its Applications*, Vol. 3, Universa Press, Wetteren, Belgium, 1987, 19.)

For *P. cylindrifera*, perhaps tolerance of both high and very low (~3 g/l) salinities (see De Deckker and Geddes[20]) gives it an advantage over congeners, and has enabled an eastward migration in waters of only moderate salinity. Other species of *Parartemia*[12] are not found at salinities below ~10 g/l.

For *P. contracta*, the combined physiological adaptations to acid conditions and sodium excretion seem to have been the driving evolutionary forces.[21]

IV. HABITATS

All species of *Parartemia* are found naturally in shallow, ephemeral bodies of saline water. None occurs in fresh waters. Habitats dry more or less annually, but occasionally, as at Pink Lake, Victoria, which contains *P. zietziana*, longer periods may elapse between dryings. *Parartemia* does not occur in permanent salt lakes even though salinities are well within the range tolerated by the genus, e.g. 40 g/l in Lake Corangamite, Victoria, and 70 g/l in Lake Gnotuk, Victoria. Only *P. minuta* has so far been collected from salt lakes (e.g. Lake Eyre) which fill episodically and are dry for most of the time. However, *Parartemia*

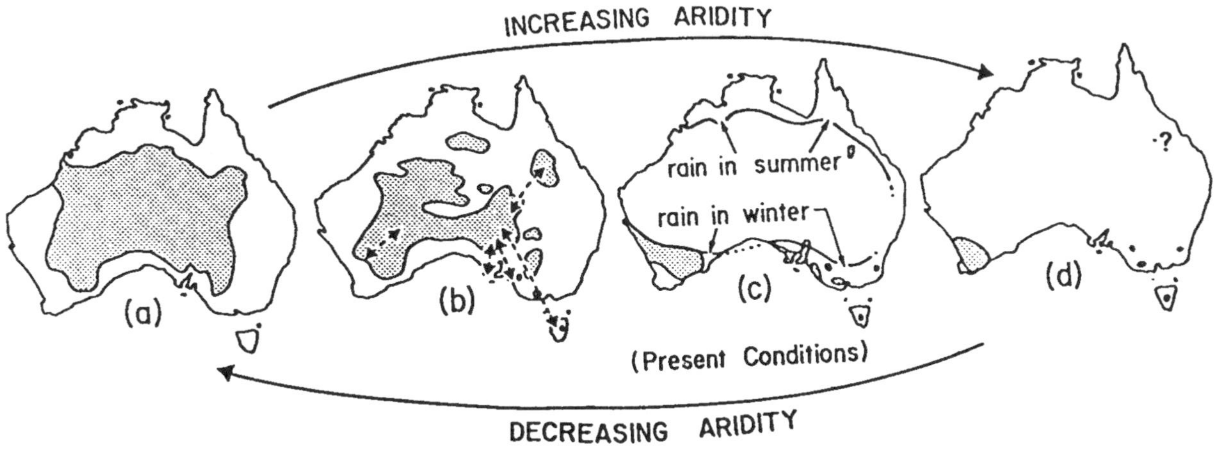

FIGURE 4. Presumed changes in the extent and position of regions containing predictably filled salt lakes (stippled areas) during different climates in Australia. (After Williams, W. D., *Verh. Int. Ver. Limnol.*, 22, 1208, 1984.)

does occur in permanent artificial waters, such as coastal solar ponds,[22] where *Artemia* might be found,[14] but, generally not in the same pond. Thus, over a four-year investigation of 15 ponds, it was found that *Artemia* occurred in ponds with salinities ~200 to 300 g/l, whereas *P. zietziana* was found in ponds of salinity <~200 g/l. Geddes[14] suggested that interaction between the genera, as well as physico-chemical factors, could be involved in the explanation of this differential distribution. Almost without exception, natural localities contain only one species of *Parartemia* on a continuous basis.

Typically, halophilic species of *Branchinella* also occur in ephemeral water-bodies. Unlike *Parartemia*, however, they may sometimes be found in permanent, moderately saline lakes as well. With one exception (northeastern South Australia), *Artemia* is restricted to more or less permanent artificial solar ponds located coastally. Figure 5 indicates the general nature of saline localities inhabited by Australian Anostraca. The localities illustrated have been selected to cover the type of habitats occupied and the taxa involved.

Because of the ephemeral nature of the natural habitats of *Parartemia*, their salinities fluctuate widely between seasons. They are frequently ~50 g/l or less during winter, but approach saturation, ~350 g/l, during summer. This is illustrated in Figure 6, which shows the pattern of seasonal fluctuation in salinity in Lake Eurack, Victoria, over a period of four years. *P. zietziana* lives in this lake, where, like other species of *Parartemia*, it occurs over wide ranges of the total salinity range of the habitat. This is indicated for several localities in Figure 7. Species of *Branchinella* are not subject to the same extent of variation in salinity, even at the upper end of the salinity range tolerated by *B. compacta*, for example, seasonal salinities in habitats occupied do not fluctuate widely.

Considerable information is available on field salinities tolerated by *P. zietziana*, but data for other species of *Parartemia* and most species of *Branchinella* are sparse. As a general summary of present knowledge,[12] most species of *Parartemia* occur between salinities of ~10 and 300 g/l. However, *P. cylindrifera* can occur at salinities as low as 3 g/l. *B. nichollsi* can tolerate a salinity of up to ~40 g/l, *B. simplex* up to ~21 g/l, and *B. compacta* has been collected at salinities between ~1 and 16 g/l. Most *Branchinella* species occur at salinities below ~2 g/l, although a few essentially freshwater species (e.g., *B. affinis* and *B. australiensis*) sometimes occur at slightly higher salinities.

No simple correlations can be drawn between the major ionic composition of Australian habitats and the species of *Parartemia* or *Branchinella* found in them, as has been done for *Artemia* populations (e.g., Bowen et al.[23,24]), for the simple reason that Australian salt lakes are remarkably homogeneous in major ion chemistry.[25] All are dominated by sodium and chloride. Some correlations, however, are possible between pH and species composition: *P. contracta* and an undescribed species of *Parartemia* are restricted to lakes with low pH values.[21]

V. PHYSIOLOGICAL FEATURES

The field range of salinities within which *Parartemia* occurs is reflected by the results of laboratory tolerance experiments. These have been confined to *P. zietziana*, but they confirm the wide salinity tolerance of at least this species in the field. They also show the close relationship between salinity tolerance and temperature, a relationship previously noted by many authors (e.g., Dorgelo[26]). The relationship is illustrated in Figure 8 based on Geddes.[27] In general, it is clear that wider ranges of salinity are tolerated at lower temperatures than at higher ones. For example, at 10°C the lower and upper LD_{50}s after 54 h of exposure are 5 and ~300 g/l, while at 30°C, these values are 13 and ~120.

Laboratory tolerance tests on *B. compacta*, though not as extensive as those on *P. zietziana*, also show reasonable concordance with field observations.[28] Lower and upper LD_{50}s with acclimation are <0.1 and 12 g/l; the field salinity range is 1.5 to 16.

FIGURE 5. Types of saline water-bodies inhabited by Anostraca in Australia. a, solar pond at Dry Creek, Adelaide (more or less permanent artificial locality inhabited by *Artemia* and *P. zietziana)*; b, Lake Eyre, South Australia (an episodically-filled salt lake inhabited by *P. minuta*); c, Lake Kariah, Victoria (a permanent, moderately saline lake inhabited by *B. compacta*); d, Lake Cundare, Victoria (an ephemeral, highly saline lake inhabited by *P. zietziana*).

FIGURE 5 (continued)

Parartemia can tolerate such wide ranges of salinity because it is a strong hypo-osmotic regulator, like *Artemia*.[29,30] We assume that the behavior of all *Parartemia* species in this respect is similar to that of *P. zietziana*. The mechanisms of hypo-osmotic regulation appear to be similar in *Parartemia* and *Artemia*, with water and NaCl being absorbed across the gut wall, and NaCl being excreted by the gills.[30] However, in *Parartemia*, the gut fluid,

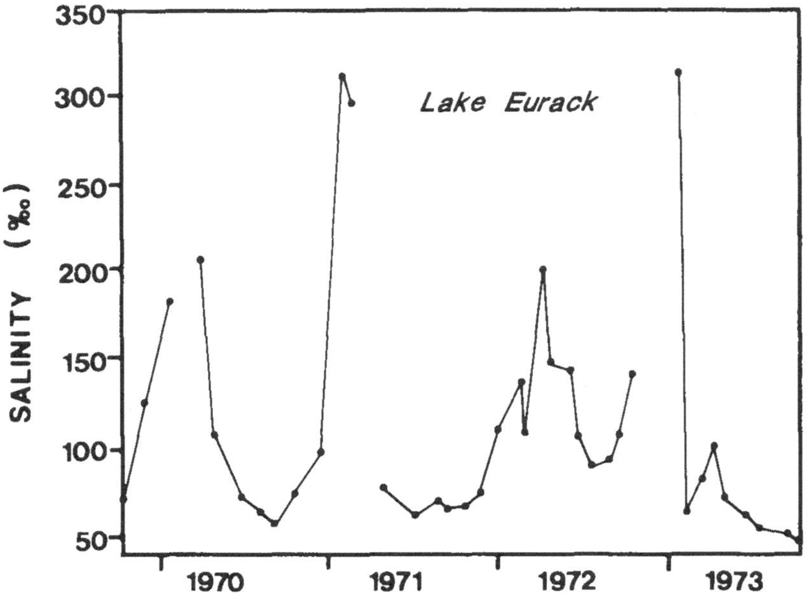

FIGURE 6. Patterns of fluctuation in salinity in Lake Eurack, Victoria. (After Williams, W. D. and Buckney, R. T., *Aust. J. Mar. Freshwater Res.*, 27, 367, 1976.)

although of different ionic composition than the hemolymph, is iso-osmotic with it; this seems not to be the case in *Artemia* (where it is hyper-osmotic). Geddes[30] suggested that this difference, if valid, provided evidence of the independent evolution of hypo-osmotic regulation in the two genera. Whatever the case, much awaits discovery about osmotic and ionic regulation in *Artemia*, as well as *Parartemia*. As Greenaway[31] recently noted for *P. zietziana*, the details of fluid transport into the hemolymph, ion fluxes, and the $Na^+ + K^+$-ATPase activity of the gut and gills have yet to be elucidated. Other species of *Parartemia* also await investigation. Osmoregulation in larval *Parartemia* is apparently achieved by the function of a larval salt gland similar to that described for *Artemia*.

B. compacta hyper-osmotically regulates at low salinities, but is an osmoconformer at higher ones;[28] the maximum salinity at which it can survive is determined by its tissue tolerance (or powers of cellular osmoregulation). Since it can survive at salinities well above the point of iso-osmoticity between its body fluids and the external medium (at which point most freshwater animals die), it is clear that while not able to hypo-osmoregulate it has nevertheless developed considerable physiological powers to combat osmotic stress. Even so, Geddes[28] suggested that it is a relatively recent colonizer of inland saline waters.

There are some fundamental differences in respiratory physiology between *Parartemia* and *Artemia*, and these are probably of critical significance at high salinities because of oxygen solubility/salinity relationships. To combat falling oxygen concentrations, *Artemia* synthesizes a variety of hemoglobins, with the amount synthesized related to ambient salinity. At least one hemoglobin has a high affinity for oxygen even in conditions of extreme hypoxia and at high temperatures,[32] so that respiration can be regulated at a more or less constant rate even at concentrations of oxygen less than 1 mg/l. At extremely low oxygen concentrations, however, respiration eventually ceases.

On the other hand, *P. zietziana* (and presumably all other *Parartemia* species since none ever develops a red coloration) cannot synthesize hemoglobin, at least in amounts sufficient for respiratory needs.[33] The small amount of hemoglobin manufactured by *P. zietziana* (~0.01% of hemolymph) may be more important for transport of heme than oxygen.[33] The

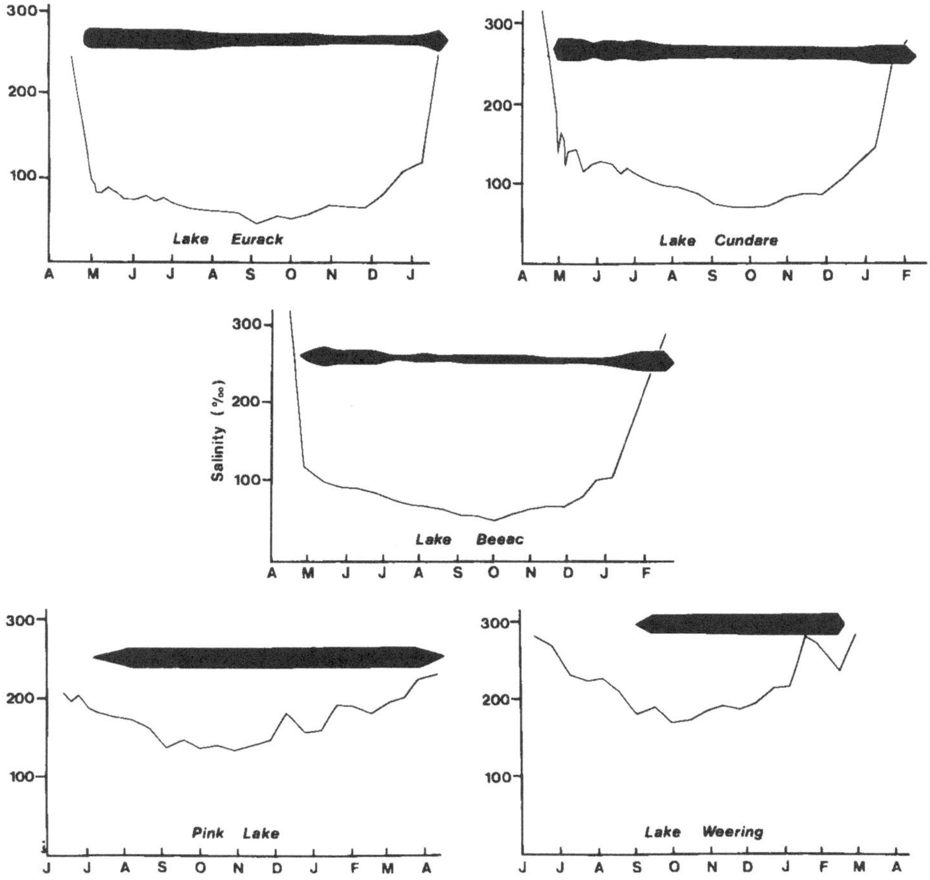

FIGURE 7. Seasonal occurrence of *P. zietziana* in relation to salinity in several localities. Width of line indicates relative abundance. (After Geddes, M. C., *Aust. J. Mar. Freshwater Res.*, 27, 1, 1976.)

inability of *P. zietziana* to manufacture significant amounts of hemoglobin at low oxygen concentrations means that, while it is a respiratory regulator like *Artemia*,[34] it cannot maintain respiration below concentrations of oxygen of about 1.8 mg/l. This difference between *P. zietziana* and *Artemia* may well explain the differential occurrence of the two taxa in the 15 solar ponds investigated by Mitchell and Geddes[22] and Geddes[14] referred to previously.

A matter of particular interest with regard to larval respiratory processes in *Parartemia* concerns these processes in *P. contracta*. As indicated, this species is found in highly acid salt lakes. Conte and Geddes[21] advanced evidence that the osmoregulatory systems of both larval *Artemia* and *P. zietziana* were similar and involved large amounts of ATP. In *Artemia*, the production of ATP is enhanced through a facultative pathway and involves aerobic glycolysis linked to a C-14 dicarboxylic shunt. The major source of CO_2 for the C-4 shunt must be bicarbonate and carbonate ions in the external medium. A similar aerobic glycolysis may allow *Parartemia* nauplii to meet their energy demands at high salinity. In acid salt lakes, bicarbonate ions are absent. Therefore, if larval *P. contracta* have a similar pathway, they must have evolved a way of producing ATP from endogenous CO_2 sources. Some form of proton pump to cope with high H^+ ion concentrations would also need to have evolved.

VI. ECOLOGICAL FEATURES

Unlike *Artemia*, which has both parthenogenetic and bisexual species (but not in the

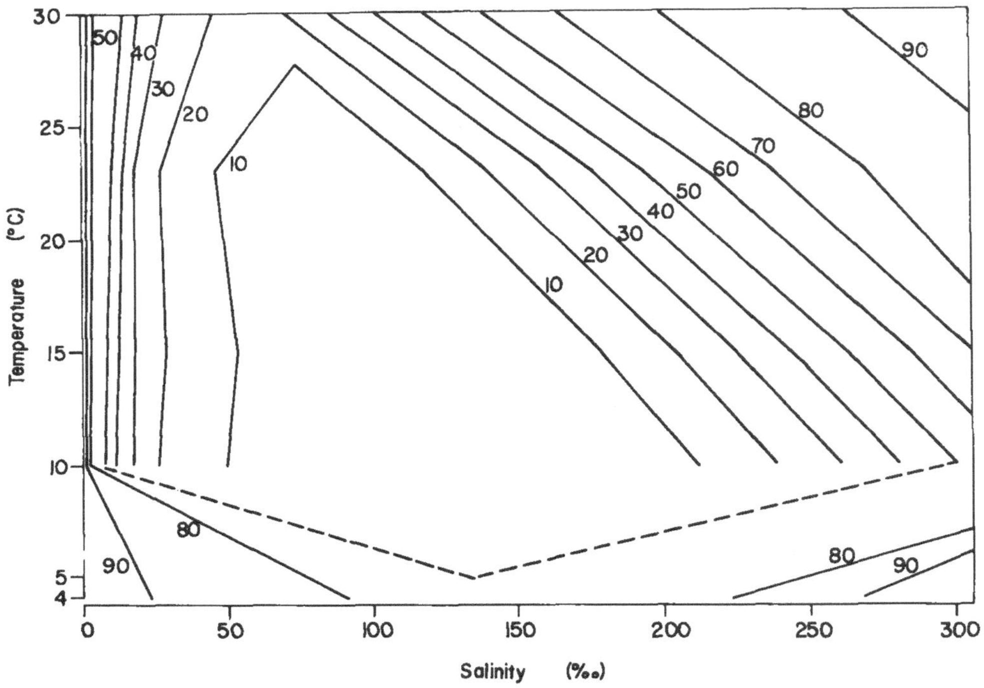

FIGURE 8. Relationship between tolerance and temperature in *P. zietziana*.

same population), all species of *Parartemia* and *Branchinella* appear to be bisexual. This may explain the species richness of these genera in Australia since the persistence of sexual reproduction would allow for a more rapid evolutionary response to climatically induced environmental change than would parthenogenesis.

However, like *Artemia*, but apparently unlike most Anostraca,[35] reproduction in *P. zietziana* and perhaps other *Parartemia* can be either oviparous or ovoviviparous. In *Artemia*, reproduction is either by ovoviviparity when free-swimming nauplii are released from the uterus, or by oviparity when the embryos are provided with a thick shell from the shell gland of the female and enter a state of dormancy before being deposited. The processes leading to oviparity and ovoviviparity in *Parartemia* are different, since it appears that embryos are always provided with a thick shell. Under some conditions they hatch from the shell within the ovisac, while under other conditions, development is arrested and cysts are laid. Little is known of reproductive events in halophilic species of *Branchinella*.

The life history of *P. zietziana* has been studied in some detail by Geddes[36] and Marchant and Williams[37] and the general pattern of events involved is (1) the production of cysts in spring at a time of increasing salinity before the habitat dries; (2) hatching from cysts when water reappears in autumn; and (3) subsequent production of generations by iteroparous females from ovoviviparous offspring. There may also, of course, be many interspecific differences. De Deckker and Geddes[20] found that *P. cylindrifera* in salt lakes near the Coorong, South Australia, produce only cysts and the life-cycle involves a single cohort.

While this is the general pattern, such details as the number of generations, the nature of the relationship between oviparity and ovoviviparity, fecundity, sex ratios, the sort of environmental cues needed to provoke the production of resting eggs, and salinities at which cysts hatch, seem to differ between populations of *P. zietziana*. For example, field observations indicate hatching of cysts apparently occurs at higher salinities in southeastern South Australian habitats (maximum recorded, 277 g/l) than in western Victoria (maximum, 202

TABLE 1
Amounts of Energy Assimilated by *P. zietziana* in Two Lakes in Western Victoria

Lake	Production	Respiration	Excretion	Assimilation
Pink	269 (16.5)	1,298 (79.5)	65 (4.0)	1,632
Cundare	24 (11.3)	179 (84.5)	9 (4.2)	212

Note: Excretion represents energy loss through nitrogenous wastes (assuming 27.3% of assimilable food is protein). Figures in parentheses are percent total assimilation. All other values as J × 10^3/m^2/yr.

After Marchant, R., Ph.D. thesis, University of Adelaide, 1976.

g/l).[20,36] Perhaps differences of this sort in *P. zietziana* populations have a genetic basis as they do in *Artemia* strains.[35]

On the basis of these field observations, it appears that there are differences between *Parartemia* and *Artemia* related to the maximum salinity at which cysts hatch. In *P. zietziana*, as noted, apparent maxima exceed 200 g/l; in *Artemia*, threshold values differ from strain to strain, but the maximum value for most strains[38] is ~85-90 g/l. The highest value recorded, 190 g/l, is for *Artemia* cysts from Mono Lake.[39] The *Parartemia* data relate only to field observations and represent measurements of open water salinity at the time of first collection of nauplii. No controlled laboratory studies on salinity and hatching have been undertaken.

For all Anostraca inhabiting salt lakes, an examination of intergeneric and interspecific differences in life-history details and how these relate to life-cycle strategies would be of much interest. This has not been done, but a valuable start has been made by Browne and co-workers[40-42] in laboratory studies of *Artemia*. No comparable studies have yet been undertaken using species of *Parartemia* or *Branchinella* from Australian salt lakes. A notable difference of particular interest in this respect between *Artemia* and *Parartemia* is that while most *Artemia* cysts float, those of *Parartemia* sink. This clearly will have profound consequences for dispersal and thus on life-cycle strategies and on speciation, as previously indicated.

So far as comparisons of secondary production in *Artemia* and *Parartemia* are concerned, so few data are available for populations of both genera that comparisons are meaningless. Somewhat surprisingly, given the intense commercial interest in the production of *Artemia* cysts, few estimates for production by this taxon are available.[35] One widely quoted value is that of Stephens and Gillespie[43] for production in the Great Salt Lake, Utah. They estimated production to be between 100 and 200 g dry weight/m^2/year. The only available estimates of production in *P. zietziana* are much lower than this estimate; Marchant and Williams[36] put forward values of only 11.3 and 1.0 g dry weight/m^2/year for the production of this species in two lakes in Victoria. Juvenile mortality accounted for most of the production in the two lakes studied. The complete energy budget for *P. zietziana* in the two lakes is indicated in Table 1.

The difference between the two values for production of *P. zietziana* derived by Marchant and Williams[37] was ascribed to the difference in the organic content of the sediments in the two lakes where populations were studied.[44,45] At both localities, *P. zietziana* appeared largely to ingest sediment as a source of food. This also seems to apply to a population of *P. cylindrifera* in a South Australian salt lake.[46] In feeding in both species, the phyllopods are used to disturb bottom sediments, and particles of high organic content are ingested. This method of feeding is perhaps a further significant ecological difference between *Parartemia* and *Artemia*, for while it has been suggested that certain *Artemia* strains may utilize sediment as a source of food, it seems for the most part that under natural conditions *Artemia* feeds upon algae and other particulate matter suspended in the water column. Comparative functional morphology studies are awaited with interest.

VII. CULTURE POTENTIAL OF *ARTEMIA* AND *PARARTEMIA*

Artemia are cultured throughout the world in extensive and intensive systems for the production of brine shrimp biomass and brine shrimp cysts. In Australia there is considerable potential for extensive culture of *Artemia* in solar saltponds; *Artemia* has been introduced into solar ponds to this end.[3] There has been little interest in investigating the potential of *Parartemia* as a culture species. The longer time to maturity due to the larger size of most *Parartemia* species, their limited respiratory capacity, and a lack of understanding of reproductive and feeding biology remain obstacles to intensive culture. As the cysts of *Parartemia* species are nonbuoyant there seems little possibility of cyst production from extensive pond culture. It is extremely difficult to collect even small numbers of cysts for laboratory study from the sediments of *Parartemia* habitats.

There remains the possibility of biomass production in natural salt lakes or solar ponds. As yet, we know little of the biomasses achieved in *Parartemia* habitats or of the value of *Parartemia* biomass as a food in the aquaculture or aquarium industries.

One further commercial consideration may relate to the genetic information contained in *Parartemia* species. Their independently evolved ability to tolerate extreme saline environments, and in some species extreme acidity, may relate to the possession of genetic material and enzyme systems that may be valuable in the new areas of molecular biology and biotechnology.

ACKNOWLEDGMENTS

We thank Miss Sandra Lawson, Department of Zoology, University of Adelaide, for typing assistance. This chapter was written while one of the authors (W.D.W.) was in receipt of a Scientific Merit Grant from the Australian Water Research Advisory Council. This Council is thanked for its support.

REFERENCES

1. **Hammer, U. T.,** *Saline Lake Ecosystems of the World,* Dr. W. Junk Publishers, Dordrecht, The Netherlands, 1986.
2. **Geddes, M. C.,** Occurrence of the brine shrimp *Artemia* (Anostraca) in Australia, *Crustaceana,* 36, 225, 1979.
3. **Anon.,** Intensive brine shrimp farms, *Austasia Aquaculture Mag.,* 1, 9, 1986.
4. **Hoverman, J. S.,** The application of the brine shrimp, *Parartemia zietziana,* to control turbidity in Victorian solar ponds, *M. Environmental Science thesis,* Monash University, Australia, 1987.
5. **Linder, F.** Contributions to the morphology and taxonomy of the Branchiopoda Anostraca, *Zool. Bidrag. Uppsala,* 20, 101, 1941.
6. **Geddes, M. C.,** Revision of Australian species of *Branchinella* (Crustacea: Anostraca), *Aust. J. Mar. Freshwater Res.* 32, 253, 1981.
7. **Sars, G. O.,** Descriptions of two new Phyllopoda from North Australia, *Arch. Math. Naturv. Christiania,* 18, 8, 1896.
8. **Sayce, O. A.,** The Phyllopoda of Australia, including descriptions of some genera and species, *Proc. R. Soc. Victoria,* 15, 224, 1903.
9. **Geddes, M. C.,** Studies on Australian anostracans (Crustacea: Branchiopoda), Ph.D. thesis, Monash University, Australia, 1973.
10. **Geddes, M. C.,** A new species of *Parartemia* (Anostraca) from Australia, *Crustaceana,* 25, 5, 1973.

11. **Geddes, M. C., De Deckker, P., Williams, W. D., Morton, D. W., and Topping, M.,** On the chemistry and biota of some saline lakes in Western Australia, *Hydrobiologia,* 82, 201, 1981.
12. **Geddes, M. C.,** Biogeography and ecology of Australian Anostraca (Crustacea: Branchiopoda), *Aust. Mus. Mem.,* 18, 155, 1983.
13. **Bowen, S. T., Burkin, J. P., Sterling, G., and Clark, L. S.,** *Artemia* hemoglobins: genetic variation in parthenogenetic and zygogenetic populations, *Biol. Bull.,* 155, 373, 1978.
14. **Geddes, M. C.,** The brine shrimps *Artemia* and *Parartemia* in Australia, in *The Brine Shrimp Artemia,* Vol. 3, *Ecology, Culturing, Use in Aquaculture,* Persoone, G., Sorgeloos, P., Roels, O., and Jaspers, E., Eds., Universa Press, Wetteren, Belgium 1980, 57.
15. **Clark, L. S. and Bowen, S. T.,** The genetics of *Artemia.* VII. Reproductive isolation, *J. Hered.,* 67, 385, 1976.
16. **Geddes, M. C. and Williams, W. D.,** Comments on *Artemia* introductions and the need for conservation, in *Artemia Research and Its Applications,* Vol. 3, *Ecology, Culturing, Use in Aquaculture,* Sorgeloos, P., Bengtson, D. A., Decleir, W., and Jaspers, E., Eds., Universa Press, Wetteren, Belgium, 1987, 19.
17. **Williams, W. D. and Kokkinn, M. J.,** The biogeographical affinities of the fauna in episodically filled salt lakes: a study of Lake Eyre, South Australia, *Hydrobiologia,* 158, 227, 1988.
18. **Timms, B. V.,** Limnology of Lake Buchanan, a tropical saline lake, and associated pools, of north Queensland, *Aust. J. Mar. Freshwater Res.,* 38, 877, 1987.
19. **Williams, W. D.,** Chemical and biological features of salt lakes on the Eyre Peninsula, South Australia, and an explanation of regional differences in the fauna of Australian salt lakes, *Verh. Int. Ver. Limnol.,* 22, 1208, 1984.
20. **De Deckker, P. and Geddes, M. C.,** Seasonal fauna of ephemeral saline lakes near the Coorong Lagoon, South Australia, *Aust. J. Mar. Freshwater Res.,* 31, 677, 1980.
21. **Conte, F. P. and Geddes, M. C.,** Acid brine shrimp: Metabolic strategies in osmotic and ionic adaptation, *Hydrobiologia,* 158, 191, 1988.
22. **Mitchell, B. D. and Geddes, M. C.,** Distribution of the brine shrimps *Parartemia zietziana* Sayce and *Artemia salina* (L.) along a salinity and oxygen gradient in a South Australian saltfield, *Freshwater Biol.,* 7, 461, 1977.
23. **Bowen, S. T., Fogarino, E. A., Hitchner, K. N., Dana, G. L., Chow, V. H. S., Buoncristiani, M. R., and Carl, J. J.,** Ecological isolation in *Artemia*: population differences in tolerance of anion concentrations, *J. Crust. Biol.,* 5, 106, 1985.
24. **Bowen, S. T., Buoncristiana, M. R., and Carl, J. R.,** *Artemia* habitats: Ion concentrations tolerated by one superspecies, *Hydrobiologia,* 158, 201, 1988.
25. **Hart, B. T. and McKelvie, I. D.,** Chemical limnology in Australia, in *Limnology in Australia,* De Deckker, P. and Williams, W. D., Eds., CSIRO, Melbourne, Australia, Dr. W. Junk Publishers, Dordrecht, The Netherlands, 1986, 3.
26. **Dorgelo, J.,** Salt tolerance in Crustacea and the influence of temperature upon it, *Biol. Rev.,* 51, 255, 1976.
27. **Geddes, M. C.,** Studies on an Australian brine shrimp, *Parartemia zietziana* Sayce (Crustacea: Anostraca). I. Salinity tolerance, *Comp. Biochem. Physiol.,* 51A, 553, 1975.
28. **Geddes, M. C.,** Salinity tolerance and osmotic and ionic regulation in *Branchinella australiensis* and *B. compacta* (Crustacea: Anostraca), *Comp. Biochem. Physiol.,* 45A, 559, 1973.
29. **Geddes, M. C.,** Studies on an Australian brine shrimp, *Parartemia zietziana* Sayce (Crustacea: Anostraca). II. Osmotic and ionic regulation, *Comp. Biochem. Physiol.,* 51A, 561, 1975.
30. **Geddes, M. C.,** Studies on an Australian brine shrimp, *Parartemia zietziana* Sayce (Crustacea: Anostraca). III. The mechanisms of osmotic and ionic regulation, *Comp. Biochem. Physiol.,* 51A, 573, 1975.
31. **Greenaway, P.,** Salt and water balance in the fauna of Australian athalassic waters, in *Limnology in Australia,* De Deckker, P. and Williams, W. D., Eds., CSIRO, Melbourne, Australia, Dr. W. Junk Publishers, Dordrecht, The Netherlands, 1986, 193.
32. **Decleir, W., Vos, J., Bernaerts, F., and Van den Branden, C.,** The respiratory physiology of *Artemia,* in *The Brine Shrimp Artemia,* Vol. 2, *Physiology, Biochemistry, Molecular Biology,* Persoone, G., Sorgeloos, P., Roels, O., and Jaspers, E., Eds., Universa Press, Wetteren, Belgium, 1980, 137.
33. **Manwell, C.,** Haemoglobin in the Australian anostracan *Parartemia zietziana:* evolutionary strategies of conformity vs regulation, *Comp. Biochem. Physiol.,* 59A, 37, 1978.
34. **Marchant, R. and Williams, W. D.,** Field estimates of oxygen consumption for the brine shrimp *Parartemia zietziana* Sayce (Crustacea: Anostraca) in two salt lakes in Victoria, Australia, *Freshwater Biol.,* 7, 535, 1977.
35. **Lenz, P. H.,** Ecological studies on *Artemia*: a review, in *Artemia Research and Its Applications,* Vol. 3, Ecology, Culturing, Use in Aquaculture, Sorgeloos, P., Bengtson, D. A., Decleir, W., and Jaspers, E., Eds., Universa Press, Wetteren, Belgium, 1987, 5.
36. **Geddes, M. C.,** Seasonal fauna of some ephemeral saline waters in western Victoria with special reference to *Parartemia zietziana* Sayce (Crustacea: Anostraca), *Aust. J. Mar. Freshwater Res.,* 27, 1, 1976.

37. **Marchant, R. and Williams, W. D.,** Population dynamics and production of a brine shrimp, *Parartemia zietziana* Sayce (Crustacea: Anostraca) in two salt lakes in western Victoria, Australia, *Aust. J. Mar. Freshwater Res.*, 28, 417, 1977.
38. **Lavens, P. and Sorgeloos, P.,** The cryptobiotic state of *Artemia* cysts, its diapause deactivation and hatching: a review, in *Artemia Research and Its Applications*, Vol. 3, Ecology, Culturing, Use in Aquaculture, Sorgeloos, P., Bengtson, D. A., Decleir, W., and Jaspers, E., Eds., Universa Press, Wetteren, 1987, 27.
39. **Thun, M. A. and Starrett, G. L.,** The effect of cold, hydrated dormancy and salinity on the hatching of *Artemia* cysts from Mono Lake, California, U.S.A., in *Artemia Research and Its Applications*, Vol. 3, Ecology, Culturing, Use in Aquaculture, Sorgeloos, P., Bengtson, D. A., Decleir, W., and Jaspers, E., Eds., Universa Press, Wetteren, 1987, 65.
40. **Browne, R. A.,** Reproductive pattern and mode in the brine shrimp, *Ecology*, 61, 466, 1980.
41. **Browne, R. A. and MacDonald, G. H.,** Biogeography of the brine shrimp, *Artemia*: Distribution of parthenogenetic and asexual populations, *J. Biogeogr.*, 9, 331, 1982.
42. **Browne, R. A., Sallee, S. E., Grosch, D. S., Segreti, W. O. and Purser, S. M.,** Partitioning genetic and environmental components of reproduction and life-span in *Artemia*, *Ecology*, 65, 949, 1984.
43. **Stephens, D. W. and Gillespie, D. M.,** Phytoplankton production in the Great Salt Lake, Utah, and a laboratory study of algal response to enrichment, *Limnol. Oceanogr.*, 21, 74, 1976.
44. **Marchant, R. and Williams, W. D.,** Field measurements of ingestion and assimilation for the Australian brine shrimp *Parartemia zietziana* Sayce (Crustacea: Anostraca), *Aust. J. Ecol.*, 2, 379, 1977.
45. **Marchant, R. and Williams, W. D.,** Organic content of some saline lake sediments in western Victoria, *Aust. J. Mar. Freshwater Res.*, 28, 269, 1977.
46. **Williams, W. D. and Kokkinn, M. J.,** Wetlands and aquatic saline environments, *Completion Report, AWRC Research Project* 84/160, Canberra, Australia, 1988.
47. **Marchant, R.,** The ecological energetics of *Parartemia zietziana* Sayce (Crustacea: Anostraca) in two saline lakes in western Victoria, Ph.D. thesis, University of Adelaide, Australia, 1976.
48. **Geddes, M. C.,** The brine shrimps *Artemia* and *Parartemia*. Comparative physiology and distribution in Australia. *Hydrobiologia*, 81, 169, 1981.
49. **Williams, W. D. and Buckney, R. T.,** Stability of ionic proportions in five salt lakes in Victoria, Australia, *Aust. J. Mar. Freshwater Res.*, 27, 367, 1976.

INDEX

A

B

C

H

I

J

K

L

M

U

V

W

X

Y